FLORA OF TROPICAL EAST AFRICA

COMMELINACEAE

ROBERT FADEN[1]

Annual or perennial herbs. Leaves basal and/or cauline, alternate, with closed sheaths; lamina simple, entire, often ± succulent. Inflorescences terminal, terminal and axillary, or all axillary, sometimes perforating the sheaths, sometimes becoming leaf-opposed, cymose, composed of helicoid cymes (cincinni), thyrsiform or variously reduced, sometimes enclosed in spathes. Flowers bisexual or bisexual and male, rarely bisexual, male and female, the plants then andromoenecious or polygamomonoecious. Sepals 3, free or connate, usually subequal and sepaloid, occasionally petaloid. Petals 3, free or connate, equal or unequal, petaloid, deliquescent. Stamens 6, all fertile or some staminodial or lacking (rarely all stamens lacking), filaments glabrous or bearded, anthers longitudinally dehiscent (rarely poricidally dehiscent). Ovary 2- or 3-locular, locules 1–many-ovulate, ovules uniseriate or biseriate (rarely up to 4-seriate); style simple, usually slender; stigma simple or rarely 3-lobed, enlarged or not. Fruits loculicidal capsules, rarely indehiscent or berries. Seeds 1–many per locule, hilum punctiform to linear, embryotega present and lid-like.

± 41 genera and 650 species. Cosmopolitan, mostly tropical; well-represented in Africa and Madagascar. East Africa has more species of Commelinaceae than any other regional flora or country flora worldwide.

The leaves of most Commelinaceae unfold in a characteristic manner, with both edges unrolling on the upper surface, like a scroll (involute vernation or ptyxis). This may be useful as a generic character, with all species of some genera (e.g. *Palisota*, *Pollia* and *Stanfieldiella*) always unfolding in this manner. In some genera (e.g. *Cyanotis*, *Murdannia* and *Anthericopsis*) the unfolding leaf always has one half of the lamina rolled around the other half (convolute or supervolute vernation). Both types of vernation may occur in some genera, such as *Commelina*, and then the differences may be useful at the species or species group level.

The flowers of Commelinaceae are very delicate and seldom are preserved in dried specimens. Our knowledge of their details is incomplete for many species. The flowers lack nectar and are open for only a few hours before they deliquesce. The opening and fading times are often characteristic of a species.

[1] Department of Botany, National Museum of Natural History, Smithsonian Institution, Washington. RF is grateful to the American Society for the Royal Botanic Gardens, Kew whose grant facilitated his research on this family for FTEA.
Bernard Verdcourt wrote up the cultivated taxa; Daniel J. Layton co-authored the *Coleotrype* account.

A number of tropical American species are grown as ornamentals:

Dichorisandra thyrsiflora Mikan with erect stems 1–2 m tall; leaves spirally arranged, 15–25 × 3–5 cm; inflorescence a large terminal thyrse of blue-violet flowers with white centres; stamens 6, two slightly longer than the others, anthers yellow; aril white.

This species is native to Brazil, and has been collected in Nyeri District: Mutamayu [Coverdale] Farm, 3 Jan. 2004, *Robertson* 7488; Faden has also observed this species cultivated in Nairobi on Ainsworth Hill.

Tradescantia *L.*

Perennial (rarely annual) herbs with fibrous or tuberous roots. Leaves spirally arranged or distichous. Inflorescences terminal or axillary, formed of a pair of sessile fused cymes enclosed in a pair of spathe-like or foliaceous bracts. Flowers bisexual, actinomorphic or almost so. Sepals usually equal, free, rarely not equal or joined, or accrescent and fleshy in fruit in one species. Petals blue, purple, pink or white, equal, usually free, sometimes unguiculate at base, rarely united in a tube. Stamens 6, all fertile, all similar and equal or subequal; filaments free or the antipetalous ones rarely epipetalous, glabrous or with moniliform hairs; anthers versatile. Ovary 3-locular, ovules 2 or rarely 1 per locule; stigma capitate. Seeds with a linear hilum or more rarely punctate with dorsal or lateral embryotega.

A fairly large genus of 70 American species. Several are widely cultivated throughout the World; several have been grown in East Africa, and some of these have become naturalized.

T. cerinthoides Kunth, Enum. Pl. 4: 83 (*Tradescantia blossfeldiana* Mildbr.)

Prostrate herb with ascending shoots; leaves dark purple beneath. Flowers 1.5 cm wide, pink with white centre.

KENYA. Nyeri District: Naro Moru, Hort. Aikman, 3 Jan. 2004, *Robertson & Aikman* 7486!

T. fluminenesis Vell., Fl. Flum. 140: iii, t. 152 (1825)

Forms dense mats; stem to 35 cm. Flowers with 3 prominent white ovate petals.

KENYA. Naturalized at Tigoni, hort. Luke (sight record)

T. pallida (Rose) D.R. Hunt in K.B. 30 (3): 452 (1975) (*Setcreasea pallida* Rose, 1911; *Setcreasia* [sic] *purpurea* Boom in Acta Bot. Neerl. 4: 167 (1055)). 'Purple heart'

Spreading herb with trailing stems and erect flowering shoots; stems and leaves succulent, dark purple with woolly sheath margins. Flowers mauve-pink.

KENYA. Nyeri District: Burguret, Mutamayo [Coverdale] Farm, Hort. Coverdale, 31 Dec. 2003, *Robertson* 7481!

T. spathacea Sw. (*Rhoeo spathacea* (Sw.) Stearn in Baileya 5: 198 (1957); *Rhoeo discolor* Hance, Walp. Ann. 3: 659 (1853))

Clump-forming perennial with erect stems. Leaves in rosettes, semi-erect, dark green above, deep purple beneath, oblong-lanceolate, 20–35 cm long, glabrous. Flowers in paired axilary cymes with persistent purple boat-shaped bracts; petals white.

From Central America.

T. zanonia (L.) Sw., Prod. Veg. Ind. Occ.: 57 (*Campelia zanonia* (L.) Kunth, Nov. Gen. 1: 264 (1816); *Commelina zanonia* L., Sp. Pl. 1: 41 (1753))

Robust clump-forming perennial herb to 1.2 m, branched, stem erect or decumbent. Leaves broadly elliptic to oblanceolate, glabrous or pubescent beneath. Flowers in long-pedunculate paired axillary cymes with 2 leafy bracts; petals white. Fruit berry-like, black.

From Mexico to Brazil and Bolivia, West Indies. The cultivar 'Mexican Flag' with leaves longitudinally streaked with bright yellow has been observed cultivated in Nairobi, Kenya.

T. zebrina Bosse, Vollst. Handb. Bl.-gartn., 2nd ed., 4: 655 (1849) (*Zebrina pendula* Schneizl. in Bot. Zeit. (Berlin) 7: 870 (1849); *Tradescantia pendula* (Schnizl.) D.R.Hunt in K.B. 36: 197 (1981))

Scrambling semi-succulent herb rooting at the nodes. Leaves glabrous, distichous, purplish beneath with two wide silvery stripes above. Flowers with 3 equal pink-purple petals fused basally.

Originally from Mexico and central America, naturalised and sometimes an invasive weed.

KENYA. Laikipia District: Nanyuki, Hort. J. Vernon, 30 Dec. 2003, *Robertson* 7478!

TANZANIA. Morogoro District: Kasanga Forest reserve, 1 Aug. 2000, *Mhoro* 199!

A cultivar with white striped leaves has been grown under the name of *T. albiflora* Kunth cv. *Albovittata*; *T. albiflora* is widespread in Australia.

Rooting at nodes. Leaves striped pale green and white.

KENYA. Nyeri District: Naro Moru, Hort. Aikman (sight record)

Jex-Blake, in Gardening E. Afr.: 89 (1957) mentions *T. virginiana* L. from eastern USA with blue flowers, but no specimens have been recorded.

Callisia *Loefl.*

Four species of this genus have been cultivated in East Africa and may be distinguished as follows:

1. Cymes fused in pairs or triads; flowers pedicellate with fleshy sepals; petals rose; robust perennial of bromeliad-like habit to 50 cm or more . . . . *C. warszewicziana*
   Cymes strictly fused in pairs 2
2. Petals with a conspicuous white lamina 4–7 mm wide . . . . . *C. gentlei*
   Petals with lamina translucent, inconspicuous 3
3. Leaves 1–4 × 1–2 cm . . . . *C. repens*
   Leaves to 30 × 7 cm . . . . *C. fragrans*

C. warszewicziana (Kunth & Bouché) D.R. Hunt

Robust rosette perennial to 50 cm or more. Leaves spiral and densely imbricate, frequently purple, fleshy, to 30 × 6.5 cm, sessile, acuminate, glabrous save for ciliate margin. Inflorescence terminal, up to 35 cm long, cymes fused at base into pairs or triads; bracteoles 7 mm long, pedicels 10–13 mm long. Sepals fleshy, 4–5 mm long, glabrous, persistent. Petals rose-purple, 6 × 6 mm. Stamens 5 mm long, with glabrous filaments. Ovary 3-locular; ovules 2 per locule. Capsule globose, 3.5 mm in diameter.

KENYA. Nanyuki town, 2 Jan. 2004, *Robertson* 7483!

C. gentlei Matuda var. *elegans* (H. Moore) D.R. Hunt

Decumbent herb 60–70 cm long, greenish or purple, minutely velutinous. Leaves subdistichous, greenish with pale longitudinal bands and often purple beneath, lanceolate or ovate, rounded to subcordate at base, margins ciliate, acute at apex. Inflorescences 6–15 cm long or more; bracteoles 5–7 × 1–2 mm, ciliate; pedicels to 1 mm long. Sepals ovate-cymbiform, 3.5–6.5 mm long. Petals white, obovate-spatulate, 7–9.5 × 4–7 mm. Stamens 6, white, subequal, filaments 3–7 mm long. Ovary 1–2 mm across, glabrous or minutely pilose at apex.

KENYA. Malindi, Robertson Plot, 13 Oct. 2002, *Robertson* 7443!

C. fragrans (Lindl.) Woodson

Creeping succulent perennial with main flowering stems up to 1.5 m, robust and bromeliad-like. Leaves of main stems usually bright green, narrowly elliptic-lanceolate, to 30 × 7 cm, subamplexicaul, acute, glabrous. Inflorescence an ample terminal panicle; bracts papery, to 2 cm; bracteoles 4–6 mm long; pedicels 3 mm long. Flowers fragrant. Sepals 3.5–5 × 1.5–2 mm. Petals whitish, lanceolate to ovate, 5–6 × 2.5–3.5 mm. Stamens 6, exserted, filaments 8–12 mm long; connectives 1.2–2.5 × 1.6–3 mm, white, more conspicuous than the petals. Ovary oblong-globose, 1.3 × 1 mm, minutely hairy at apex.

KENYA. Nairobi, Hort. Greensmith, 16 June 1977, *Gillett* 21359!

C. repens (Jacq.) L.

Spreading herb rooting at the nodes or with small erect stems with crowded leaves, forming a compact mat. Leaves quite variable, usually tinged purple beneath, thin or fleshy, narrowly to broadly ovate, 1–4 × 1–2 cm, base rounded to subcordate, apex acute, glabrous save for ciliate margins. Inflorescence spicate, usually ascending; bracteoles 5 mm long; pedicels 0.5–1.5 mm long. Sepals scarious, 2.5–3.5 mm long. Petals whitish, narrowly oblong, 3–5 × 1–1.5 mm, acute. Stamens 3–6, long-exserted, one or more frequently reduced to staminodes; connectives 1 × 1.5 mm. Ovary oblong-globose, 1 mm across, 2-locular, hairy in upper half; style 4–5 mm long.

KENYA. Nyeri District: Burguret R., Mutamayu farm, 31 Dec. 2003, *Robertson* 7480!; Laikipia District: Nanyuki, Hort. Vernon, 30 Dec. 2003, *Robertson* 7479!

KEY TO GENERA

1\. Inflorescences or partial inflorescences enclosed in, or closely subtended by, leafy or spathaceous bracts 2

Inflorescences or partial inflorescences neither enclosed in, nor closely subtended by, leafy or spathaceous bracts 5

2\. Spathes paired, the enclosed cincinni sessile and fused back to back; cultivated or naturalized . . . . *Tradescantia* (p. 2)

Spathes solitary or, if clustered or paired, then cincinni not as above; or bracts leafy, not spathaceous 3

3\. Bracts leafy; leaves succulent; flowers actinomorphic; petals equal, fused basally; stamens 6, equal, all fertile, with bearded filaments . . . . 2. **Cyanotis** (p. 12)

Bracts spathaceous; leaves herbaceous; flowers zygomorphic; one petal smaller than the other two, free; stamens (5–)6, 3 fertile and (2–)3 staminodial, with glabrous filaments 4

4\. Leaves always spirally arranged; spathes sessile, distichous on zigzag axes; flowers white; antherodes 2-lobed; forest understorey plants . . . 8. **Polyspatha** (p. 57)

Leaves often distichous; spathes usually distinctly stalked, solitary or variously clustered; flowers rarely white; antherodes usually 4–6-lobed; plants of various habitats, seldom in forest . . . . 11. **Commelina** (p. 127)

5\. Inflorescences all axillary and perforating the sheaths; flowers white (rarely purple) . . . . 3. **Coleotrype** (p. 31)

Inflorescences usually terminal or terminal and axillary, rarely all axillary and perforating the sheaths (in *Aneilema zebrinum*); flowers variously coloured 6

6\. Inflorescences large terminal thyrses; flowers blue with white centers; anthers dehiscing by terminal pores; seeds arillate . . . . *Dichorisandra* (p. 2)

Inflorescences and flowers various; anthers usually dehiscing longitudinally; seeds usually exarillate 7

7. Inflorescences composed of pairs of sessile cincinni fused back to back; cultivated .... *Callisia* (p. 3)
   Inflorescences various but not as above; native 8
8. Fruit a fleshy berry; stamens antepetalous, staminodes antesepalous; large and robust forest understorey herbs .... 1. **Palisota** (p. 6)
   Fruit capsular or, if berry-like, hard and crustaceous; stamens and staminodes variously arranged but not as above 9
9. Flowers and sometimes fruit present 10
   Only fruits present 17
10. Stamens 6, all fertile 11
    Stamens 6 (rarely 3 or 5) of which 3(–4) staminodial or rarely lacking 14
11. Lower petal reduced; upper 3 stamens differentiated from lower 3; ovary and capsule bilocular; locules 1-ovulate, 1-seeded .... 5. **Floscopa** (p. 40)
    Petals equal; stamens equal; ovary and capsule 3-locular; locules 2–several-ovulate and -seeded 12
12. Leaves succulent, sessile; petals fused basally, pink to purple or violet; filaments and often style bearded; plants of various habitats (rarely forest) 2. **Cyanotis** (p. 12)
    Leaves herbacous, petiolate; petals free, white; filaments and style glabrous; forest plants 13
13. Leaf sheath ciliate at apex (rarely not so); ovules and seeds uniseriate; fruits cylindric, brown, dehiscent .... 4. **Stanfieldiella** (p. 37)
    Leaf sheath ciliolate or eciliate at apex; ovules and seeds 2–4-seriate; fruits ellipsoid or spherical, metallic blue to blue-black, indehiscent; seeds 2–4-seriate .... 9. **Pollia** (p. 62)
14. Fertile stamens alternating with staminodes; leaves sessile, vernation convolute or folded 15
    Fertile stamens opposite staminodes; leaves often petiolate, vernation usually involute 16
15. Rosette plants; flowers ± 25 mm wide, white to pale pink; staminodes with antherodes unlobed; capsules 20–35 mm long .... 6. **Anthericopsis** (p. 49)
    Some or all leaves cauline; flowers ± 8–15 mm wide, blue to violet; staminodes with antherodes 3-lobed or staminodes lacking; capsules 1.5–9 mm long .. 7. **Murdannia** (p. 50)
16. Petals equal or unequal; fertile stamens equal with glabrous filaments; staminodes reduced and hidden within the flower; fruit berry-like, metallic blue or blue-black, indehiscent .... 9. **Pollia** (p. 62)
    Lower petal different in size and form from upper petals; medial stamen differentiated from lateral stamens; staminodes evident in flower, the medial sometimes reduced or lacking; fruit capsular, grey to brown, dehiscent (rarely indehiscent) .... 10. **Aneilema** (p. 66)
17. Fruit berry-like, metallic blue to blue-black, indehiscent; seeds 2–4-seriate .... 9. **Pollia** (p. 62)
    Fruit capsular, grey to brown, dehiscent (rarely indehiscent); seeds uniseriate (rarely 2-seriate) 18

18. Capsules bilocular or, if trilocular, the third locule reduced 19
    Capsules trilocular, locules equal 20
19. Bracteoles usually cup-shaped, perfioliate; capsules sometimes trilocular; seeds usually more than one per locule, rarely ribbed; plants terrestrial ..... 10. **Aneilema** (p. 66)
    Bracteoles neither cup-shaped nor perfoliate; capsules bilocular; seeds 1 per locule, usually ribbed; plants aquatic or of wet soil .... 5. **Floscopa** (p. 40)
20. Capsules subsessile; seeds with a terminal embryotega 2. **Cyanotis** (p. 12)
    Capsules distinctly pedicellate; seeds with dorsal to lateral embryotega 21
21. Leaves petiolate, vernation involute; forest plants ... 4. **Stanfieldiella** (p. 37)
    Leaves sessile, vernation convolute or folded; plants of various habitats, rarely in forest 22
22. Rosette plants; capsules 20–35 mm long .... 6. **Anthericopsis** (p. 49)
    Some or all leaves cauline; capsules 1.5–9 mm long 7. **Murdannia** (p. 50)

1. PALISOTA

Endl., Gen. Pl. 125 (1836); C.B. Clarke in DC., Monogr. Phan. 3: 130 (1881) & in F.T.A. 8: 27 (1901); Morton in J.L.S. 60: 204 (1967); Faden in Kubitzki, Fam. Gen. Vasc. Pl. 4: 119 (1998) & in K.B. 62: 133–138 (2007); *nom. cons.*

Perennial rhizomatous herbs, mostly rosette plants or suffrutices. Leaves spirally arranged, mostly in rosettes or pseudowhorls, rarely alternate or subopposite, generally petiolate, pubescent at least along the margins, base symmetrical. Inflorescences terminal or terminal and axillary, sometimes all axillary, pedunculate or subsessile thyrses, or rarely reduced to a single cincinnus. Flowers female and male or bisexual and male, rarely all bisexual, slightly or distinctly zygomorphic, pedicellate, small (mostly ± 1 cm wide). Sepals free, subequal, petaloid, glabrous or pubescent. Petals free, subequal or the anterior (outer and usually lower) one distinctly broader than the posterior (inner and usually upper) 2, usually similar to the sepals, not clawed, white or pink to maroon or violet. Staminodes 2–3, attached in front of the sepals, subequal or the posterior one reduced or lacking, each staminode consisting of a short, densely bearded filament and lacking an anther. Stamens attached in front of the petals, dimorphic, the 2 posterior ones with shorter filaments and often anthers of a different form from the anterior one, filaments usually glabrous. Ovary sessile, glabrous or pubescent, trilocular, locules equal, each with 2–6 uni- or biseriate ovules; style relatively short; stigma large, capitate to deltate. Fruits berries, orange to red or blue to black at maturity, locules 1–6-seeded. Seeds uniseriate or biseriate, triangular or polygonal in outline, individually surrounded by a form-fitting, thin, translucent layer that dries brown and papery, testa brown to grey, hilum ± punctiform, embryotega dorsal.

The pollen grains in the different anthers are dimorphic. Those from the posterior or upper stamens, which always have dehiscent pollensacs, are sterile, whereas those from the anterior or lower stamen are fertile. In some species the flowers that have a fully developed gynoecium do not release their fertile pollen and are thereby female. In other species a fully developed gynoecium and a release of fertile pollen may be present in the same flowers. Such flowers are bisexual or hermaphroditic. Virtually all species also have male flowers. The different sexed flowers are produced by all plants but not necessarily at the same time.

About 20 species of forest understorey and moist disturbed situations, in shade or partial shade, mainly in west and central Africa and the islands in the Gulf of Guinea. A number of species are grown as greenhouse ornamentals because of their bold foliage and attractive berries.

1. Rosette plants; leaves in larger plants up to 90(–150) cm long .... 2. *P. mannii*
   Suffrutescent plants; leaves to 40(–46) cm long 2
2. Inflorescences relatively lax to moderately dense; old flowers caducous, leaving scars on the inflorescence branches; fruiting pedicels erect, to 3 mm long; fruits ovoid, blue, eventually glaucous; seeds biseriate .... 1. *P. orientalis*
   Inflorescences very dense; old flowers persistent; fruiting pedicels curved or contorted, 4–8 mm long; fruits ± globose, yellow to red with darker spots; seeds uniseriate 3. *P. schweinfurthii*

1. **Palisota orientalis** *K. Schum.* in E.J. 36: 209 (1905). Types: Tanzania, Lushoto District: Eastern Usambaras, near Amani, *Engler* 592 (B!, syn.); near Ngwelo [Nguelo] in Handei Mountains, *Heinsen* 11 (B, syn.)

Suffrutescent herb to 1.5 m tall, shoots erect, sparsely branched; distal internodes to 14 cm long, sparsely pilose or glabrescent. Leaves in pseudowhorls, clustered mainly distally, petiolate to subsessile; sheaths to ± 2(–3.5) cm long, usually splitting, densely appressed white (rarely red) pubescent, long-ciliate with rusty hairs at the apex; petiole absent or up to ± 4.5 cm long, often broadly winged, long-ciliate; lamina narrowly elliptic to oblanceolate, 16–28 × (2–)3–8 cm, base narrowly cuneate, margins finely undulate, long-ciliate, apex acuminate; adaxial surface ± glabrous, except sometimes basally, abaxial surface patently pilose. Inflorescences solitary, terminal ± lax to ± dense thyrses; peduncle very short and concealed in the leaf sheaths or up to ± 7 cm long, sparsely appressed pubescent; thyrses cylindric to ellipsoid or narrowly ovoid, 3.5–12 × 1–4.5 cm, of numerous cymose branches (cincinni), but the inflorescence axis plainly visible; all axes with a scurfy pubescence, cymes to ± 2.5 cm long, bracts ovate to narrowly lanceolate, becoming reflexed, caducous or persistent, 2–6 mm long. Flowers bisexual, pedicels caducous after flowering, otherwise ± erect in fruit, 2–3 mm long in fruit. Sepals and petals marcescent, the sepals when young conspicuously yellowish farinose, in fruit 2–3 mm long, petals white, ± 3 mm long in fruit. Fruits ovoid berries, dark blue, becoming glaucous when ripe, 5–7 × 3–5 mm, conspicuously beaked, glabrous. Seeds biseriate, up to 5 per locule, ± polygonal, 2.5–3 × 2 × 1–1.5 mm, testa (with sac removed) dark brown or grey, radiately grooved and pitted.

TANZANIA. Lushoto District: Eastern Usambara Mts, Sangerawe, 11 Jan. 1947, *Greenway* 7914! & Kwamkoro Forest Reserve, 6 Jan. 1977, *Ruffo & Mmari* 2032!; Iringa District: Udzungwa Mountain National Park, Pt 243, Camp 244, 5 Oct. 2001, *Luke et al.* 8170!

DISTR. **T** 3, 6, 7; not known elsewhere

HAB. Rainforest understory, in shade; 900–1100 m

NOTE. The *Heinsen* specimen presently in the Berlin Herbarium does not appear to be the one that was cited in the type description although it has the same collection number. The cited specimen was sterile and was from 'Nguelo' whereas the specimen presently in the herbarium is in fruit and is from Derema. Moreover it lacks Schumann's handwriting on it.

This species closely resembles the west and central African *P. ambigua* (P. Beauv.) C.B. Clarke in its habit, moderately lax, solitary, terminal thyrse, ovoid blue fruits and biseriate seeds. According to Schumann *P. orientalis* differs by its longer, narrower leaves that are less distinctly ciliate on the margins, with longer bearded bases and the lack of a white underside on the leaf. Quoting notes of the collector Heinsen (unseen by me), Schumann states that the white flower colour differs from the blue flowers in *P. ambigua*. With a much greater range of specimens of both species available now, we can state that the tendencies in leaf size, shape and pubescence that Schumann noted do hold up to a large extent, although there are exceptions to every one. The leaves of *P. ambigua* tend to be smaller and more obovate, but there are some specimens that would be a reasonably good match for *P. orientalis*. Similarly, the appressed, often dense, whitish pubescence that is commonly present on the lower leaf surface in *P. ambigua* is never encountered in *P. orientalis*, but there are specimens of *P.*

ambigua with quite sparse pubescence. Thus leaf characters generally work to separate these species but they are not absolute. The white flower colour is also not absolute. While *P. ambigua* in west and central Africa has flowers in the blue to mauve, purple or violet range, plants of this species raised from seed from the Ituri Forest in eastern Congo–Kinshasa, the easternmost distribution of *P. ambigua* and the closest part of its range to that of *P. orientalis*, had white flowers, like *P. orientalis*. There are other differences that apparently are consistent. In the field it has been observed that the flowers of *P. ambigua* begin opening between 16:00 and 17:30. In cultivation, when grown in temperate greenhouses without supplemental light during the winter the flowers of *P. ambigua* do not open until after dark. In cultivation the flowers of *P. orientalis* have been found open as early as 12:15. The persistent fruiting sepals of *P. orientalis* are smaller than those of *P. ambigua* (2–3 mm vs. 3–4 mm long, as measured in herbarium specimens), although the length in fresh flowers would be a better comparison. Another difference is in the form of the inflorescence branches (cincinni). In *P. ambigua* they are distinctly thickened or clavate distally, which is probably related to their association with ants in the field. A similar thickening has also been observed in the cincinni of the Central African species *P. brachythyrsa* Mildbr. No such thickening is present in *P. orientalis* or in west and central African species such as *P. hirsuta* (Thunb.) K. Schum. and *P. thollonii* Hua.

In all likelihood *P. orientalis* evolved from *P. ambigua*, probably through long-distance dispersal and subsequent differentiation, or as a result of a vicariant event.

2. **Palisota mannii** *C.B. Clarke* in DC., Monogr. Phan. 3: 132 (1881) & in F.T.A. 8: 29 (1901); Morton in J.L.S. 60: 205 (1967); Brenan in F.W.T.A. 2nd ed., 3: 35 (1968); Malaisse, Fl. Rwanda 4: 142 (1988). Types: Equatorial Guinea, Bioko [Fernando Po], *Mann* 2340 (K!, syn.) (lectotype of Clarke in F.T.A. (1901)); Cameroon: Mt Cameroon, *Mann* 2139 (K!, syn.); Cameroon, Mopanga, *Kalbreyer* 163 (BM, K!, syn.)

Robust rosette perennial; stem short, covered by the leaf bases, unbranched; roots thick, not tuberous. Leaves with petiole narrowly winged, 12–40 cm long, margins long-reddish or tawny ciliate; lamina obovate to obovate-elliptic or oblanceolate, 26–110 × 8.5–22 cm, base narrowly cuneate, margin ciliate with white or tawny hairs, adaxial surface sparsely pilose to subglabrous, some hairs always present near the margin, apex abruptly acuminate; abaxial surface densely pubescent along the midrib, sparsely so elsewhere. Inflorescences scapose, pedunculate, 1-several per rosette, terminal and axillary; peduncle to 30 cm long, with one or more bract-like, reduced leaves at the base or higher, densely pilose at base, pubescent or sparsely pilose distally; thyrses cylindrical, very dense, 7–14 × 2–3.5 cm (in fruit to 23 × 5 cm), cincinnus bracts conspicuous in the young inflorescence, generally persistent, narrowly lanceolate to lanceolate, 10–14 × 2.5–5 mm, long-ciliate with rusty hairs; cincinni with old flowers persistent. Flowers functionally female and male, white or pale pink; pedicels 7–10 mm long in flower, becoming curved or spirally contorted in fruit. Posterior (upper) sepal oblong-elliptic to ovate-elliptic, 4–5 × 2–2.5 mm, anterior (lateral) sepals oblong-elliptic, 4–5 × 1.6–2 mm. Anterior (lower) petal lanceolate-ovate to oblong-elliptic, 4–4.5 × 2 mm; posterior (lateral) petals lanceolate to oblong-elliptic, 4.5–5 × 2–2.5 mm. Staminodes 3, subequal, filaments densely bearded; posterior (upper) stamens with filaments 1.5–1.8 mm long, anthers 1.2–1.4 × 0.8–1.2 mm; anterior (lower) stamen in male flower central in the flower, filament 3.5–4 mm long, anther ± 1 mm long and wide, pollensacs dehiscent, anterior (lower) stamen in female flower strongly decurved, filament 2.5–3 mm long, anther similar in size and shape to that in the male flower, but pollensacs indehiscent. Ovary ± 1 mm long, densely hirsute; style 2.5–3 mm long; stigma capitate, abortive gynoecium in the male flower ± 2.5 mm long. Berries globose or depressed globose, 6–11 × 6.5–13 mm, sparsely pilose, turning through bright yellow, orange, red, to purple-brown at maturity, locules at most 2-seeded, but usually most of the potential 6 seeds/fruit abortive or attacked by insects. Seeds ± polygonal, 3.5 × 3–3.5 × 2–2.5 mm, testa radiately, finely striate and pitted on all surfaces.

subsp. **megalophylla** (*Mildbr.*) *Faden* in K.B. 62: 137 (2007). Types: Cameroon, Lomie in Dja-bend, *Mildbraed* 5409 (B!, syn.) & on Dja R., 21 km N of Molundu, *Mildbraed* 3888 (B!, syn.)

UGANDA. Bunyoro District: Kitoba, May 1943, *Purseglove* P1562!; Mengo District: Kyewaga Forest, 14 Sep. 1949, *Dawkins* D370! & Kasa Forest, 17 Nov. 1949, *Dawkins* D448!
TANZANIA. Bukoba District: Minziro Forest Reserve, 5 Jul. 2000, *Bidgood et al.* 4855! & Minziro, Kagera, Minziro-Bulembe track, 28 Dec. 1994, *Congdon* 392; Mwanza District: Hodi River, Buhinda Forest Reserve, Nov. 1954, *Carmichael* 471!
DISTR. **U** 2, 4; **T** 1, 4?; Nigeria, Cameroon, Equatorial Guinea, Gabon, Angola (Cabinda), Central African Republic, Congo–Brazzaville, Congo–Kinshasa, Burundi, Sudan
HAB. Forest and swamp forest; 1150–1250 m
Flowering specimens have been seen from September through January and May.

SYN. *Palisota megalophylla* Mildbr., Z.A.E. 2: 52, 74 (1922), nomen, & in N.B.G.B. 9: 247 (1925), descr.

NOTE. As Morton (1967) has pointed out Clarke confused *Palisota mannii* and *P. schweinfurthii*. Among the three syntypes of *P. mannii*, *Mann* 2139 and *Kalbreyer* 163 are both *P. schweinfurthii*, although Clarke in F.T.A. treated them as a form of *P. mannii*. Fortunately, Clarke lectotypified *P. mannii* with *Mann* 2340 in F.T.A., and the specimen is a good one, so the application of the name is certain.

Purseglove P3063 from Uganda was identified as *P. barteri* Hook.f. by me many years ago because of its obovoid inflorescence that is 8 cm long and 5 cm wide. In F.W.T.A., ed. 2 (Brenan, 1968) such a plant would key out to *P. barteri*. However, in its leaf size (to 90 × 15 cm), pubescence and the dense, white-hirsute pubescence on the ovary it does not differ from Ugandan specimens of *P. mannii*. Another large-leaved plant, *Hall* JH007/94 from eastern Congo–Kinshasa, has similar, proportionally broad inflorescences (K sheet) or more typically cylindrical *P. mannii* inflorescences (US sheet). *Gereau et al.* 6270 & 6335 from Minziro Forest Reserve, Tanzania both have short, broad inflorescence but they are also small plants. I think that all of these specimens are just atypical *P. mannii* plants or that *P. mannii* has a different pattern of variation in eastern than in central Africa. I know of no records of *P. barteri* from eastern Congo–Kinshasa, and it is unlikely that this species would be disjunct from the western side of the Congo basin without having been collected in eastern Congo–Kinshasa. Until more convincing evidence in the form of additional collections, especially with mature fruits, is obtained, *Purseglove* P3063 is best treated as an atypical *P. mannii* specimen. Accepting it as such would raise the altitudinal limit for *P. mannii* in our region to 1370 m and would add August to flowering records for this species.

Kaj Vollesen (personal communication) says that he has seen sterile plants of a large rosette *Palisota* in Mpanda District, Tanzania. *Palisota mannii* is the only likely species to occur there, but I have seen no specimens from **T** 4.

This species is very distinctive in the field because of its very large leaves and rosette habit. Unfortunately, its size has discouraged many collectors from gathering normal-sized leaf material, and the small leaves often collected and the chopped off scapes have led to the confusion of this species with *P. schweinfurthii* in herbarium specimens.

A collection of fruits of this species preserved in spirit (*Dawkins* D448) shows that the fruits may be heavily attacked by boring insects of which I found both larvae and pupae inside. No other *Palisota* species has been observed to have such a high level of infected fruits, but I have seen relatively few spirit collections of field-collected fruits.

Our plants differ from *P. mannii* subsp. *mannii* in that they lack a dense white or brown, felty indumentum on the lower leaf surface, rusty hairs on the leaf margins, and a completely glabrous upper surface. That subspecies occurs throughout the range of subsp. *megalophylla* in central Africa, but it does not extend as far east as East Africa.

3. **Palisota schweinfurthii** *C.B. Clarke* in DC., Monogr. Phan. 3: 132 (1881) & in F.T.A. 8: 29 (1901); F.P.U.: 195 (1962); Morton in J.L.S. 60: 205 (1967); Brenan in F.W.T.A., 2nd ed.: 3: 35 (1968). Types: Sudan: Boddo River, *Schweinfurth* 3054 (K!, syn.) & Nabombisso R., *Schweinfurth* 3697 (K!, syn.); Congo–Kinshasa: Yuru River, *Schweinfurth* 3279 & 3281 (K!, syn.) [specimens = *P. mannii*]; on the Mbulu River, south of the Kibali River, *Schweinfurth* 3721 (K!, syn.) [specimen = *P. mannii*]; Angola: Golungu Alto, Quilombo–Quiacatubia, *Welwitsch* 6599 (BM!, syn.) [lectotype – selected here – the sheet with the full data label]; *Welwitsch* 6303b (BM!, syn.); São Tomé, Fazenda do Monte Caffè, *Welwitsch* 6602 (BM!, syn.) [specimen = *P. pedicellata* K. Schum.]

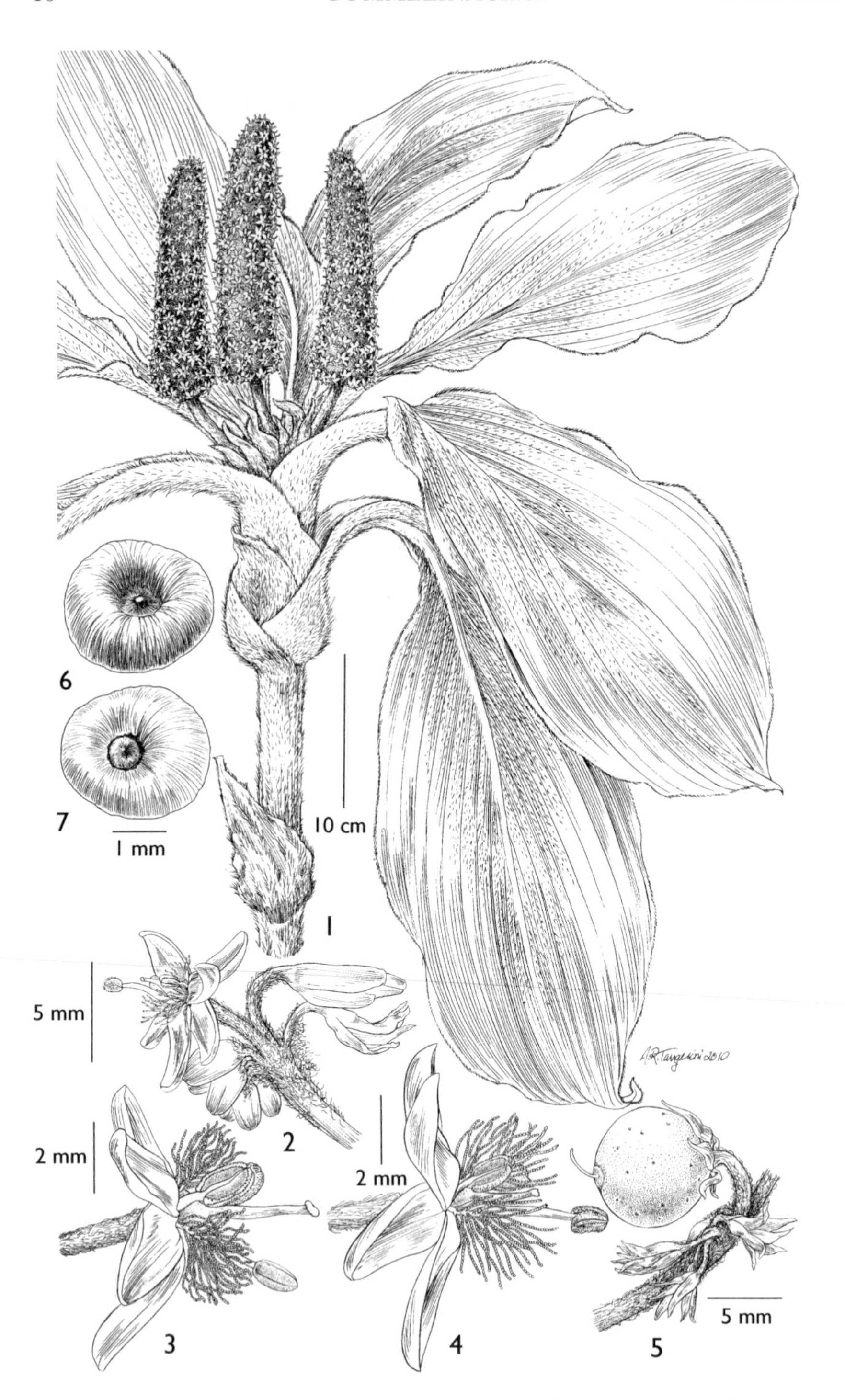

FIG. 1. *PALISOTA SCHWEINFURTHII* — **1**, flowering shoot; **2**, cincinnus with male flower and persistent old flowers; **3**, bisexual flower, lateral view; **4**, male flower, lateral view; **5**, fruit; **6**, seed, ventral view; **7**, seed, dorsal view. 1–4 from *Kahn* 93/26; 5–7 from *Harris* 3035. Drawn by Alice R. Tangerini.

Suffrutescent perennial 1–2 m tall; stems erect, sparsely branched, roots thick but not tuberous; distal internodes sparsely pubescent, sometimes also sparsely pilose. Leaves in pseudowhorls, petiolate; sheaths 1–2 cm long, often splitting to the base, densely pubescent, at least in patches; petiole usually narrowly winged, (2–)3–8 cm long, long-ciliate with red hairs, rarely ciliolate; lamina elliptic to oblong-elliptic, (13–)20–40 × (4–)7–15 cm, base cuneate, margins densely ciliate, apex acuminate; upper surface glabrous or sometimes sparsely pilose near margins and midrib, lower surface subglabrous except for midrib to sparsely pubescent, shortly hirsute or pilose throughout. Inflorescences 1–2 (or more) per terminal leaf whorl; peduncle up to 11 cm long, usually with one or more leafy bracts at or near the base, pubescent; thyrses cylindrical, very dense, 8–18 × 2–4 cm, cincinnus bracts conspicuous in the young inflorescence, generally persistent, 10–11 × 1–2 mm, ciliate, the distal part of each cincinnus densely covered at its base by long light brown hairs; cincinni with old flowers persistent. Flowers functionally female and male, ± 1 cm wide, opening in the afternoon, fragrant, white to pale pink; pedicels 3–5.5 mm long in flower, becoming curved or rarely spirally contorted and up to 6 mm long in fruit. Sepals sometimes strongly reflexed, subequal, oblong to ovate, 3.5–4.5 × 1.5–2 mm, hairy or subglabrous. Anterior (lower or medial) petal lanceolate-elliptic to obovate-spatulate, 3.5–5 × 2–2.5 mm, posterior (upper or paired) petals lanceolate-oblong to oblong-elliptic, 3.5–4.5 × 1.5–2 mm. Staminodes 3, the posterior (upper) one smaller than the others, filaments densely bearded; posterior (upper) stamens with filaments 1–1.8 mm long, anthers 1.5–1.7 × 0.5–0.8 mm, pollensacs dehiscent; anterior (lower) stamen in male flower held centrally in the flower, filament 3.2–3.7 mm long, glabrous, anther 1 × 0.8 mm, pollensacs dehiscent, anterior (lower) stamen in female flower deflexed against the anterior petal, filament 2–2.5 mm long, anther 1–1.2 × 0.8–1 mm, pollensacs indehiscent. Ovary 1–1.2 mm long; style 2.7–3 mm long; stigma large, capitate; abortive gynoecium in the male flower 2–2.2 mm long. Berries ± globose, yellow turning red or reddish orange, with small brown spots at maturity, 5.5–11 × 6–10 mm, sparsely pilose or glabrous, locules (1–)2-seeded. Seeds uniseriate, obtusely triangular to nearly round, 3.3–3.7 mm in diameter, testa (with sac removed), finely striate, dark grey. Fig. 1, p. 10.

UGANDA. Bunyoro District: Budongo Forest, Busingiro area, May 1933, *Eggeling* 1220(1327)!; Toro District: Sempaya, 26 Oct. 1953, *Dawkins* D813! & Sempaya, 29 Oct. 1951, *Osmaston* 1389!

TANZANIA. Mpanda District: Kungwe Mountain, Ntali [=Ntale] River, 17 July 1959, *Newbould & Harley* 4452!; Morogoro District: Nguru Mts, Liwale Valley, Manyangu Forest, 27 Mar. 1953, *Drummond & Hemsley* 1854! & Nguru South Forest Reserve at head of valley behind Mhonda Mission, 6 Feb. 1971, *Mabberley & Pócs* 696!

DISTR. **U** 2; **T** 4, 6; Nigeria, Cameroon, São Tomé, Equatorial Guinea (Rio Muni), Gabon, Chad, Central African Republic, Congo–Brazzaville, Congo–Kinshasa, Rwanda, Burundi, Sudan, Angola, Zambia

HAB. Rainforest, especially in wet sites such as along streams, gallery forest, forest and forest clearing edges; 700–1800 m

Flowering specimens have been seen from March to May, September and October.

NOTE. Many of the collections from our area originally determined as *P. schweinfurthii* are *P. mannii*. This is especially true of Tanzanian specimens from W of Lake Victoria. I have seen no specimens from that area that are unquestionably *P. schweinfurthii*.

Palisota schweinfurthii occurs in three distinct parts of our region: western Uganda, Mpanda District in western Tanzania, and the Nguru Mt in eastern Tanzania. To a large extent plants from these three areas have characteristic leaf pubescence. For example, all and only the three collections from Mpanda District have the leaves shortly hirsute or pilose beneath.

The only fruiting specimens examined from our area, *Mabberley* 696, and *Lovett & Congdon* 7073, both had glabrous fruits, although no flowers have been seen with glabrous ovaries.

2. **CYANOTIS**

D. Don, Prodr. Fl. Nepal: 45 (1825), *nom. conserv.*; Clarke in DC., Monogr. Phan. 3: 240–261 (1881); Clarke in F.T.A. 8: 78–84 (1901); Morton in J.L.S. 60: 191–199 (1967); Faden in Kubitzki, Fam. Gen. Vasc. Pl. 4: 120 (1998)

Perennial and annual herbs; shoots erect to decumbent or repent; underground storage organs, such as bulbs, corms, rhizomes and tuberous roots commonly present. Leaves spirally arranged or distichous, sessile, succulent. Inflorescences terminal or terminal and axillary, rarely all axillary, either elongate, with herbaceous usually sickle-shaped bracteoles arranged in two rows, or borne within the leaf sheath and not elongate, the elongate cymes often clustered, forming heads. Flowers bisexual, actinomorphic, sessile or subsessile; sepals generally united proximally, rarely ± free, usually keeled and pubescent; corolla tubular, with 3 shorter, spreading lobes; stamens 6, equal, fertile, greatly exceeding the corolla, generally free, rarely shortly connate at the base, densely bearded above the middle with long colorful moniliform hairs, usually with a apical swelling (rarely lacking), anthers basifixed, dehiscing by means of small basal slits; ovary sessile, trilocular, locules biovulate, style usually about the same length as the filaments, bearded like them or glabrous, with a similar apical swelling, stigma small to relatively broad. Capsules trilocular, trivalved, locules 2-seeded. Seeds uniseriate, with a basal punctatiform hilum and terminal embryotega, variously pitted.

A paleotropical genus of ± 50 species, with about 25 in Africa; mostly species of grassland and rocky places, but some occur in forest and the introduced *C. axillaris* is an aquatic.

Cyanotis repens subsp. *robusta* is sometimes cultivated in Kenya.

Cyanotis species are difficult to key out partly because of their variability, within and between species. Such characters as pubescence type (arachnoid versus non-arachnoid) and whether all of the individual cincinni are sessile are useful for defining some species, but may occur sporadically within many of the other species. Another problem with inflorescences in *Cyanotis* is to define exactly what an inflorescence is. If it is the individual cincinnus, then what to we call the more or less capitate clusters of several cincinni? They are not morphological equivalents. A further complication is that when stalked inflorescences are present, the pattern of branching soon becomes too complicated to describe or to use taxonomically.

Flowers, capsules and seeds are generally the most important organs for defining species of Commelinaceae. In *Cyanotis* they are only of limited use. In the flower the most useful character is whether the style is glabrous or bearded. The type of pubescence on the calyx and, to some degree, the shape of the calyx lobes may be useful, but they have not been explored fully. The shape and size of the anthers can be of use at times, as can the shape of the terminal swelling on the style and whether the stigma is broad or narrow. I have found no useful characters in the capsules, except occasionally the length. The length of the hairs on the capsule might be of use, but it was not explored. The only character that I have found in the seeds is the presence of a papery outer layer in *C. longifolia* seeds, which is lacking in the other species. Other species show variation – and some in seed size might be useful – but the testa patterns are so difficult to describe accurately that they have only been used in the species descriptions, not the keys.

The underground parts of *Cyanotis* species, which everyone hates to use in keys because they are frequently lacking in dried specimens, are often diagnostic. Even in these characters there is variation within species. For example, *Cyanotis paludosa* was described as being rhizomatous, but plants from southwestern Tanzania have bulbs. *Cyanotis foecunda* was described from a specimen that had a corm, but such organs seem to be lacking in virtually all collections from our area. In some instances their absence may be explained by the possibility that they were deeply buried and therefore left in the ground by the collectors, but in some parts of our area, particularly central and southern Kenya and northeastern Tanzania, I believe that corms are just not produced. Plants of *C. arachnoidea*, which normally have a thick base but no specialized underground structures, occasionally form a short rhizome, I would expect, in a rock crevice.

Cyanotis also presents a nomenclatural challenge in that two of our species were originally described from Asia, *C. barbata* from Nepal and *C. arachnoidea* from India. I am not completely convinced that African plants of *C. barbata* are conspecific with Asian plants, which lately have been called *C. vaga* (Lour.) Schult.f. in Asian floras, a name that Brenan (1968) pointed out seems to have misapplied. I am even less certain that African *C. arachnoidea* is the same species as Asian *C. arachnoidea*. That species is so variable in Asia that it is hardly worthwhile describing the African plants as a distinct species until the mess in Asia is sorted out. Wholly African problems include whether to recognize the non-cormose plant of *C. foecunda* at some taxonomic rank – I have chosen not to here – and what the status should be of the bulbous plants of *C. paludosa*. It is also necessary to decide at what level to recognize the bulbous type of *C. speciosa* that is present in our area from the rhizomatous type that occurs from Zimbabwe to South Africa.

1. Plants with underground storage organs (tuberous roots, rhizomes, bulbs or corms) 2
   Plants lacking underground storage organs 9
2. Plants with bulbs *and* tuberous roots; inflorescences all sessile; style glabrous . . . . 10. *C. speciosa* (p. 30)
   Plants with bulbs *or* tuberous roots or corms or rhizomes, not with more than one; inflorescences and style various 3
3. Roots tuberous; rhizomes, bulbs and corms lacking 4
   Roots thin, fibrous; rhizomes, bulbs or corms present 5
4. Bracts shorter than or equalling the cincinni; dwarf plants usually flowering before the central leaf rosette appears . . . . 2. *C. caespitosa* (p. 15)
   Bracts longer than the cincinni; usually robust (occasionally dwarf) plants flowering with a central leafy rosette . . . . 3. *C. longifolia* (p. 16)
5. Plants with rhizomes 6
   Plants with bulbs or corms 7
6. Robust erect plants growing in seasonally swampy places; pubescence usually not arachnoid . . . . . . 4. *C. paludosa* (p. 18)
   Small, spreading plants growing in rock crevices; pubescence almost always densely arachnoid . . . 7. *C. arachnoidea* (p. 24)
7. Underground parts small and hard, corms or corm-like, usually up to 1 cm in diameter 8
   Underground parts bulbs 2–3 cm in diameter . . . . 4. *C. paludosa* (p. 18)
8. Inflorescences all sessile and borne at up to 15 nodes on a zigzag axis, with the bracts decreasing in size distally; style usually glabrous; plants of low to medium elevations, 200–2150 m . . . . 8. *C. foecunda* (p. 26)
   Inflorescences sometimes stalked, borne at or from up to 4 nodes on the main shoots; bracts not as above; style bearded; plants of medium to high elevations, (1350–)1850–4150 m . . . . 5. *C. barbata* (p. 20)
9. Annuals 10
   Perennials 11

10. Aquatic plants, usually growing in water; all bracts single; style glabrous .... 1. *C. axillaris* (p. 14)
    Terrestrial plants of rocky and grassy places, sometimes growing near, but not in water; terminal bracts on the shoots paired; style bearded ...... 6. *C. lanata* (p. 22)
11. Inflorescences borne at up to 5(–8) nodes; stalked inflorescence often present; plants usually densely arachnoid-pubescent; style bearded .... 7. *C. arachnoidea* (p. 24)
    Inflorescences borne at up to 15 nodes, always sessile; plants very rarely arachnoid-pubescent; style usually glabrous 12
12. Inflorescences terminating the shoot growth; at least the distal bracts strongly distinct from the foliage leaves and decreasing in size distally on the flowering shoot .... 8. *C. foecunda* (p. 26)
    Inflorescences not terminating the shoot growth; bracts identical to the foliage leaves, neither recurved nor decreasing in size distally .... 9. *C. repens* (p. 27)

1. **Cyanotis axillaris** (*L.*) *Sweet*, Hort. Brit. 430 (1827); Clarke in DC., Monogr. Phan. 3: 244 (1881); Hook.f. in Fl. Brit. Ind. 6: 388 (1892) & in Trimen & Hook.f., Handb. Fl. Ceylon 4: 315 (1898); Vollesen in Opera Bot. 59: 97 (1980). Type: Tab. 174, fig. 3 in Plukenet, Phytographia sive stirpium icones, pars prior (1692!)

Annual starting from a short-lived rosette, the lateral shoots soon supplanting it; roots thin, fibrous; shoots trailing or scrambling, to 110 cm long. Leaves distichous, sheaths to ± 1 cm long, glabrous or pubescent, usually ciliate at the apex; lamina linear to linear-lanceolate, (1–)2–12(–14.5) × 0.3–1(–1.2) cm, base broadly cuneate to rounded, margins glabrous, scabrid or sparsely ciliate, apex acute to acuminate, surfaces glabrous or the lower surface sparsely pubescent proximally. Inflorescences enclosed within the leaf sheaths, bracteoles not visible without dissection, scarious, linear, 3–12 × 1–1.5 mm, glabrous. Sepals linear-oblanceolate, 6–9 × 1–1.5 mm, subglabrous; corolla pinkish purple, lilac or purple to pale blue or violet, 7–14 mm wide. Stamens exceeding the corolla by 6–7 mm, filaments white, densely bearded with lavender to blue-violet (distal tipped with white) hairs contrasting with the corolla, with a subapical white swelling, anthers yellow, dehiscing by a basal pore. Style equalling to slightly exceeding the stamens, white or concolorous with the corolla lobes, with an apical swelling, glabrous. Capsules yellowish or slightly darker, sometimes with darker brown flecks, stipitate, 3.5–6 mm long (including the beak but not the stipe), 1.5–2.3 mm wide, with a few minute hairs at the apex or the beak which is sometimes shortly bifid on each valve. Seeds ovate to elliptic in outline, 1.5–1.9(–2.4) × 1–1.5 mm, testa gray to gray-brown, foveolate to foveolate-reticulate on all surfaces, a few depressions sometimes longitudinally elongate; midventral ridge usually lacking or weakly developed.

KENYA. Kilifi District: Malindi–Garsen road, 2 km N of turnoff to Fundisa and Hadu, 25 July 1974, *Faden & Faden* 74/1211!

TANZANIA. Bagamoyo District: Bako swamp, 5 km S of Bagamoyo on Dar es Salaam road, 3 June 1973, *Wingfield* 2184!; Uzaramo District: Manyanamala, 2 km W of Commission for Science and Technology Building, Dar es Salaam, 7 June 1996, *Muasya* in *Faden et al.* 96/51!; Kilwa District: Selous Game Reserve, Kingupira, 18 May 1975, *Vollesen MRC* 2333!

DISTR. **K** 7; **T** 6, 8; Cameroon, Chad, Sudan, Malawi; India, Sri Lanka, Thailand, Malaysia, Vietnam, China, Indonesia, Philippines, Australia

HAB. Ricefields, marshes, edges of waterholes, wooded grassland, thicket edge, commonly growing in water; 5–300 m

Flowering [February] March–July.

SYN. *Commelina axillaris* L., Sp. Pl. 42 (1753)
Tradescantia axillaris (L.) L., Mant. 2: 321 (1771)
Amischophacelus axillaris (L.) R.S. Rao & Kammathy in J.L.S. 54: 305 (1966)

NOTE. This widespread Asian to Australian species has been reported from Africa only once in the literature. When and where it got introduced is uncertain, as is its means of introduction. Its collection as a weed in rice in Tanzania suggests that it has the potential to become a serious weed in wet habitats. Its recent discovery in Malawi shows that the plant is still spreading in Africa.

2. **Cyanotis caespitosa** *Kotschy & Peyr.*, Pl. Tinn. 48, t. 22A (1867); C.B. Clarke in F.T.A. 8: 82 (1901); Schnell in Bull. I.F.A.N. 19, sér. A: 748 (1957); Morton in J.L.S. 60: 192, fig. 7 (1967); Brenan in F.W.T.A. 2nd ed.: 3: 38 (1968); Faden in U.K.W.F. 2nd ed.: 306, t. 137 (1994); Ensermu & Faden in Fl. Eth. 6: 340, fig. 207.1 (1997). Type: Sudan, Bongo region, on the Djur River, Dec. 1863, *Tinne* 6 (W†, holo)

Geophytic perennials; roots with stipitate fusiform tubers to 7 × 1 cm; stem a short, thick, vertical caudex surrounded by the hairy sheaths of the old leaves. Leaves in a basal rosette, rarely persisting from the previous growing season, the new leaves not produced until after the inflorescences are fully developed; leaves lanceolate to lanceolate-oblong, to 14 × 2 cm (up to 21 × 3.5 cm elsewhere), margins densely ciliate, apex acute, adaxial surface glabrous or with sparse intramarginal hairs when young, abaxial surface densely pale-velutinous. Inflorescences scapose, usually several per plant, reddish purple or maroon, usually unbranched, borne in the axils of the old leaves (peripheral to the new rosette), erect to ascending or occasionally procumbent, not rooting, (4–)7–25(–37) cm long; each inflorescence bearing 1–7(–10) mostly stipitate simple laterally compressed cincinni, individual cincinni 0.5–0.8(–1) × 0.5–1.7 cm; bracts subtending cincinni (4.5–)6–15 mm long. Sepals fused basally, lanceolate-oblong to oblanceolate-oblong, (3–)4–6 × 1.5–2 mm, strongly keeled, densely pilose to sparsely puberulous, usually lacking an apical tuft; corolla bright blue, blue-purple, red-mauve or reddish purple, funnelform, 6.5–8 mm long, corolla tube 3.5–5 mm long, 3–4 mm wide at the apex, lobes 2.5–3 × 3–4 mm. Stamens exserted, filaments 7.5–9.5 mm long, densely blue- or mauve-bearded in the distal half; anthers 1–1.4 × 0.5–0.9 mm, yellow, dehiscing by basal slits. Ovary 1–1.3 mm long, densely pilose or puberulous, especially apically; style 7.5–9 mm long, bearded distally. Capsules yellowish, oblong-ellipsoid, 2.5–3 × 1.5–2 mm. Seeds ovoid to ovoid-ellipsoid, 1.5–2 × 1.2–1.5 mm, testa reddish brown, finely striate or with irregular longitudinal depressions.

UGANDA. Acholi District: Uganda Imatongs, Apr. 1938, *Eggeling* 3624! & N slopes of Lomwaga Mt, Aringa R. watershed, 8 Apr. 1945, *Greenway & Hummel* 7315!; Teso District: Serere, Apr. 1932, *Chandler* 575!

KENYA. Trans-Nzoia District: near Kipkarren, 28 Mar. 1952, *Cooke* 18! & near Kitale, Mar. 1936, *Thorold* in *A.D.* 3245! & 19.2 km from Kitale on Kapenguria road, Apr. 1967, *Tweedie* 3446!

TANZANIA. Buha District: just outside Kibondo on Biharomulo road, 16 July 1960, *Verdcourt* 2865!; Mpanda District: Mahali Mts, E side of Kabesi valley, 1 Sep. 1958, *Oxford University Tanganyika Expedition* 1998!; Iringa District: Sao Hill, Oct. 1959, *Watermeyer* 147!

DISTR. **U** 1–4; **K** 3, ?5; **T** 1, 4, 7; Sierre Leone, Ivory Coast, Ghana, Nigeria, Cameroon, Central African Republic, Congo–Kinshasa, Burundi, Sudan, Ethiopia, Angola, Zambia

HAB. Grassland, occasionally with scattered trees, and seasonally wet depressions; (900–)1100–2150(–2600) m

Flowering specimens have been seen from every month except January and November. Flowers have been recorded as opening very early in the morning and lasting up to midday.

NOTE. *Cyanotis caespitosa* flowers before the rains and then into the rainy season. It often flowers following burning of the grass, but it is not clear whether burning stimulates flowering or both are simply dry season phenomena. According to collectors, plants tend to occur as scattered individuals rather than in dense populations. Variation in flower color within populations seems to be common. Flowering is precocious, preceding the development of the leaves. Specimens with mature capsules are exceedingly rare – I have seen none from our area – as are ones with fully developed leaves. This pattern suggests that by the time leaves are full size, the plants have generally finished flowering and, although they may bear fruits, they have become inconspicuous to collectors. Flower colour is usually in the blue range. The reddish colored plants are generally mentioned as unusual individuals. Outside of our range, white-flowered plants have been noted.

The two collections from the Imatong Mountains in Uganda, *Eggeling* 3624 and *Greenway & Hummel* 7315, are from the highest elevations for this species. They have the largest cincinni and most arachnoid pubescence of any specimens from our area. *Thorold* 3245 has much of the lamina of one old leaf persisting on one plant, which shows that old leaves may occasionally persist until the next flowering season.

In Flora of Ethiopia this species is further recorded from Rwanda, Mozambique, Malawi, Zimbabwe, Botswana and Namibia, evidently in error. There are no specimens from any of these countries at Kew (K) or Harare (SRGH).

3. **Cyanotis longifolia** *Benth.*, Niger Fl.: 543 (1849); C.B. Clarke in DC., Monogr. Phan. 3: 259 (1881) & in F.T.A. 8: 81 (1901); Brenan in K.B. 7: 205 (1952); Morton in J.L.S. 60: 195 (1967); Brenan in F.W.T.A. 2nd ed.: 3: 37 (1968); Vollesen in Opera Bot. 59: 97 (1980); Cribb & Leedal, Mountain Fl. S Tanzania: 170, t. 47c (1982); Obermeyer and Faden, F.S.A. 4, 2: 55 (1985); Faden in U.K.W.F. 2nd ed.: 306, t. 137 (1994); Ensermu & Faden in Fl. Eth. 6: 340 (1997). Type: Congo–Kinshasa, *Curror* 1 (K!, holo.)

Perennial herb; roots clustered, with stipitate or ± sessile tubers to 8 mm thick; a central basal rosette present on a short thick stem, usually with long appressed white or tawny hairs at the base from the remains of old leaf sheaths. Rosette leaves with sheaths to 5 cm long, densely appressed pale-pubescent at the base, otherwise more sparsely villous; lamina complicate or occasionally planar, linear to linear-lanceolate, (5–)8–30(–37) × 0.4–1.4 cm, margin undulate, ciliate or glabrous, apex acute to acuminate, adaxial surface glabrous to villous, sometimes with an inframarginal band of hairs, abaxial surface white arachnoid-pubescent or densely villous; flowering shoots one to several per plant, peripheral to the current rosette but inside the scaly remains of old leaf sheaths, erect or prostrate and sometimes looping along the ground, not rooting, (4–)15–90 cm tall or long; cauline leaves apparently distichous, usually complicate, sheaths to 4.5 cm long, sparsely villous to densely white arachnoid-pubescent, ± ciliate at the apex; lamina linear to lanceolate-oblong, 2.5–20 × 0.3–1 cm, margin undulate, ciliate, apex acute to acuminate, adaxial surface glabrous to sparsely villous, usually with an inframarginal band of hairs, abaxial surface densely villous or white arachnoid-pubescent. Inflorescences borne at/from the distal 1–3 nodes on the flowering shoots, the terminal one or group sessile but appearing long-stipitate, those from the more proximal nodes sessile or stipitate; individual cincinni or groups 1–2.5 × 1–3 cm, the bracts subtending the inflorescences exceeding them, complicate, commonly reflexed, 1.5–6 cm long; individual cincinni strongly laterally compressed and circinnate, the groups of cincinni ± capitate in appearance; bracteoles biseriate, 5–8 in each series, lanceolate, falcate, 5–10(–14) × 1.5–3.5 mm, apex acute to ± acuminate, margins densely ciliate, surfaces usually arachnoid-pubescent, at least proximally. Calyx 5–6 mm long, fused at the base for 0.7–2 mm, or occasionally nearly free, lobes usually obovate, 3.5–4.3 × 1–2 mm, ± keeled, densely pubescent; corolla blue, blue-mauve to blue-violet, 7.5–9 mm long, tube 4.5–7 mm long, lobes broadly ovate, 2.5–4 × 3.5–5.5 mm. Stamens 6, equal, filaments free or rarely slightly connate at base, 9–11.5 mm long, exceeding corolla by 5–6 mm, with an apical swelling, densely bearded near the swelling, anthers usually dehiscing by small basal slits. Ovary oblong-elliptic, 1.5 mm long, ± trigonous, densely bearded; style 8–10 mm long, with a large terminal swelling, densely bearded with moniliform hairs below the swelling, stigma small. Capsules light brown, oblong-ellipsoid, 2–4 × 1.6–2.4 mm, appressed pubescent at least distally, with an apical tuft. Seeds ovoid to ellipsoid, (1.1–)1.2–1.5(–1.8) × (0.8–)1–1.5 mm, testa light orange-brown, sometimes with a silvery sheen, shallowly and irregularly pitted, not striate.

UGANDA. Toro District: Central Kibale, Oct. 1940, *Sangster* 693!; Busoga District: Lolui Island, 22 May 1964, *Jackson* 4144!; Masaka District: Bukakata, June 1953, *Makerere College* in *Lind* 168!

Kenya. Trans-Nzoia District: Milimani, Kitale, June 1969, *Tweedie* 3654!; West Pokot District: Kongelai [Kacheliba] Escarpment, 20 May 1969, *Napper & Tweedie* 2116!; Nairobi District: Kumbe Road, Langata, 29 June 1972, *Ng'weno in EA* 15137!

Tanzania. Mbulu District: Great North Road, Pienaars Heights or Dauar, beween Babati and Bereko, 200 km S of Arusha, 7 Jan. 1962, *Polhill & Paulo* 1082!; Musoma District: Serengeti, Seronera, 22 Mar. 1961, *Greenway* 9867!; Songea District: Matengo Hills, Miyau, 3 Mar. 1956, *Milne-Redhead & Taylor* 9008!

Distr. **U** 1–4; **K** ?1, 2–5; **T** 1–4, 6–8; Senegal, Gambia, Guinea Bissau, Guinea, Burkina Faso, Mali, Ivory Coast, Ghana, Benin, Togo, Nigeria, Cameroon, Gabon, Central African Republic, Congo–Brazzaville, Congo–Kinshasa, Rwanda, Burundi, Sudan, Ethiopia, Angola, Zambia, Malawi, Zimbabwe and Namibia

Hab. Grassland, sometimes with scattered trees or bamboo clumps, rocky places, *Brachystegia* and *Uapaca* woodland, bushland; (6–)500–2050(–2250) m

Flowering specimens have been seen from all months, but the peak in Tanzania is from December through March.

Syn. *Cyanotis longifolia* Benth. var. *bakeriana* C.B. Clarke in DC, Monogr. Phan. 3: 259 (1881). Type: Tanzania, Unyamwezi District: between Tura and Tabora, 1860, *Grant* s.n. (K, syn.); Angola, Huilla Prov., Apr. 1860, *Welwitsch* 6595 (K!, syn.); Pungo Andongo Prov., Jan. & Feb. 1857, *Welwitsch* 6651 (K!, syn.); Guisgungo, s. date, s. coll. (P, syn.)

C. djurensis C.B. Clarke in DC., Monogr. Phan. 3: 256 (1881) & in Durand & Schinz, Consp. Fl. Afr. 5: 433 (1895) & in F.T.A. 8: 82 (1901). Type: Sudan, Jur, *Schweinfurth* 1944 & Jur Ghatta, *Schweinfurth* III 217 (K!, syn.)

Note. The type consists of two small plants without roots, each with a basal rosette of leaves and a single subscapose flowering shoot with two inflorescences. In the left hand plant the inflorescences are subtended by bracts that barely exceed them, as in *C. caespitosa*, whereas in the right hand plant, the bracts are much longer than the inflorescences, as in typical *C. longifolia*. In the inflorescences of both plants there are some arachnoid hairs at the base of each cincinnus. The rest of the plant, including the rosettes, is densely covered by a tawny patent pubescence, without a trace of arachnoid pubescence. I have compared these plants with specimens of *C. caespitosa*, and they do not agree with them, especially the right hand plant, but the pubescence of the foliage and cincinni is very similar. Might this not be *C. longifolia* in the sense that everyone has been using the name, or is this part of the great range of variation that this species exhibits, especially in West Africa? One difference between these plants and *C. caespitosa* is that the rosettes on the type are clearly above the ground, whereas those of *C. caespitosa* seem to be more underground. But perhaps that is the wrong impression because the rosettes are not usually collected, since the plants flower precociously, whereas in the type of *C. longifolia* the rosettes are quite fresh, as are the flowering shoots.

This species is most distinctive because of its tuberous roots and usually abundant white arachnoid pubescence. This pubescence may be present on all aerial parts, or, in extreme cases, just be confined to the inflorescences. Small plants can be confused with *C. caespitosa* where both species occur, e.g. in W Kenya, but the bracts in *C. longifolia* always greatly exceed the inflorescences, whereas those in *C. caespitosa* equal to only slightly exceed the inflorescences. In *C. longifolia* a central basal rosette seems to be present during flowering, whereas in *C. caespitosa* flowering is normally precedes the growth of the rosette.

Although this species shows great variation in size and pubescence in our area there does not seem to be a case for formally naming any of the variants, unlike in F.W.T.A. ed. 2. When other variation is taken into account our plants should all be considered *C. longifolia* var. *longifolia*.

A *Wellby* collection at K from Lake Turkana [Rudolf] might have come from either side of the lake. If it were from the east side it would represent the only collection seen from **K** 1.

Rogers 216 from Nairobi District, where the species is decidedly uncommon, is especially dwarf, being only 4 cm tall.

Milne-Redhead & Taylor 8419 has the sepals nearly free at the base, which matches the desciption in Flora of Ethiopia and Eritrea (1997). However, in the other three spirit collections examined the sepals were fused basally.

Milne-Redhead & Taylor 9008 was the only collection examined in which the terminal swelling of the style was fusiform, not flask-shaped. The swelling was also smaller than in the other collections. This was also the only collection that showed longitudinal anther dehiscence (in one flower and in some anthers of a second flower) instead of the normal dehiscence by small basal slits (which may then be followed by longitudinal dehiscence). The specimen is unusually robust for plants from our area, with very large, broad bracts subtending the inflorescence clusters. The specimen also has fleshy roots, as noted by the collectors, but not distinct tubers, but perhaps the tubers were distal to the parts that were collected and were left in the ground.

The inflorescences always have some white arachnoid pubescence and often the internodes and leaves do as well. This pubescence varies from very dense to quite sparse and sometimes it is only present around the base of the bracteoles in the entire plant.

The seeds – checked in four populations – appear pale orange brown, sometimes with a silvery lustre. In every case it was found that the color is that of a thin papery layer that evenly covers the whole seed, except for the hilum. That layer can be easily rubbed off under which is a reddish brown layer that is more similar to the seeds of other species of *Cyanotis*.

4. **Cyanotis paludosa** *Brenan* in K.B. 15: 224 (1961); Faden in U.K.W.F. 2nd ed.: 306 (1994). Type: Kenya, Trans-Nzoia District: Kitale, *Wiltshire* 54 (K!, holo.)

Robust rhizomatous or bulbous perennial; roots thin, fibrous; shoots usually (always?) borne peripheral to a central, sterile rosette of leaves; shoots erect to ascending, sometimes decumbent and rooting at the lower nodes, sparsely or occasionally densely branched, (35–)50–120(–200) cm tall, internodes to 1 cm thick, usually inconspicuously pilose with hairs to 4 mm. Rosette leaves linear, to 43 × 2 cm, apex acuminate, surfaces glabrous or sparsely pilose, occasionally sparsely cobwebby-pubescent beneath; cauline leaves distichous on the flowering shoots, sheaths 1–5 cm long, occasionally splitting to the base, subglabrous to sparsely pilose, ciliate or eciliate at the apex; lamina linear to linear-lanceolate, 4–13(–30) × 0.5–1.2(–1.6) cm, base ± not narrowed into the sheath, margins glabrous or sparsely ciliate, apex acute to acuminate, surfaces similar in pubescence to the rosette leaves. Inflorescences 1–6 per (major branch of) flowering shoot, one terminal and with a peduncle to 10(–17) cm long, the others axillary from the distal 1–3 nodes, usually pedunculate but commonly one or more sessile; individual inflorescences capitate, 1.5–3 × 1.5–3.5 cm, bracts subtending the heads (1–)2, broad at the base, then abruptly constricted into a narrow blade, commonly recurved, 1–4 × 0.7–1.2 cm, margins ciliate or glabrous, surfaces like those of leaves; heads composed of several congested, sessile or very shortly pedunculate cincinni, the cincinni themselves bracteate at the base; bracteoles biseriate, ± 5–8 in each series, falcate, ovate or ovate-lanceolate, 7–12 × 1.5–5 mm, margins glabrous or ciliate, rarely white-cobwebby pubescent, apex acute, surface like that of leaves. Sepals oblong to narrowly elliptic, 5–6 × 2–3 mm, with a middorsal keel, glabrous to sparsely pilose, rarely densely long-hairy at the apex or white cobwebby pubescent; corolla pale blue to azure, mauve or violet, 10 mm long, with lobes 4.5 × 3 mm. Stamens equal, filaments ± 10 mm long, long-bearded apically with blue white-tipped hairs, with a subapical swelling; anthers ± 1 × 0.7 mm, yellow with a dark blue base, pollen orange, released by basal pores. Ovary oblong, trigonous, densely pubescent distally; style ± 11.5 mm long, bearded distally with hairs like those of the stamens, with a subapical swelling. Capsule yellowish or light brown, trilocular, trivalved, oblong-ellipsoid, 3–3.5 × 1.8–2.3 mm, sometimes with numerous ranks of tiny maroon flecks, pubescent at least distlally. Seeds 2 per locule, ovoid to ovoid-ellipsoid, 1.3–1.8 × 0.8–1.2 mm, testa light and dark brown, reddish brown or grey, finely striate with scattered small pits and sometimes grooves or longitudinally furrowed or pitted.

subsp. **paludosa**

Plant with thick, horizontal rhizomes to which may be attached bulbs; inflorescences lacking white arachnoid pubescence; rosette leaves, when present lacking white arachnoid pubescence beneath.

UGANDA. Masaka District: Lake Nabugabo, Aug. 1935, *Chandler* 1412! & Katera, 23 June 1935, *Thomas* 1287!; Mengo Distict: Kirerema and Kiruru, Sept.-Oct. 1914, *Dummer* 2691!
KENYA. Trans-Nzoia District: Aerodrome swamp, Kitale, Aug. 1967, *Tweedie* 3471! & 12.8 km W of Kitale on Elgon S road, June 1969, *Tweedie* 3655!; N Kavirondo District: Mumias, Lessos–Nandi Hills, 12 km from Nandi Hills, 13 Oct. 1981, *Gilbert & Mesfin* 6732!
TANZANIA. Bukoba District: Minziro Forest Reserve, W foot of Bulembe Hill, 15 Nov. 1999, *Gereau et al.* 6254!
DISTR. **U** 2, 4; **K** 3, 5; **T** 1; Congo–Kinshasa
HAB. Marshes, swamps, damp hollows, seasonally inundated grassland and open grassland, often at the edges of the wetter habitats; 1100–2000 m
Flowering June and August to November.

subsp. **bulbifera** *Faden* **subsp. nov.** a subspecie typica base bulbum ferentem rhizomate carenti differt. Typus: Tanzania, Ufipa District: Kigoma road, 17 km from Sumbawanga, 15 June 1996, *Faden, Phillips, Muasya & Macha* 96/226 (US!, holo., K!, iso.)

Main shoot with a bulbous base, a new bulb forming beneath the old one at the end of the growing season; rhizomes lacking; inflorescences and sometimes also the undersides of the rosette leaves with white, arachnoid pubescence; rosette leaves, when present, sometimes with white arachnoid pubescence beneath.

TANZANIA. Ufipa District: Mbizi Mts, 18 June 1996, *Faden et al.* 96/286! & Chapota Swamp, 6 Mar. 1957, *Richards* 8498!; Iringa District: Mufindi, edge of Lake Ngwazi, 7 May 1968, *Renvoize & Abdallah* 1987!
DISTR. **T** 4, 7; not known elsewhere
HAB. Swamps, marshes, swampy grassland, damp depressions, seepage areas and seasonally waterlogged sandy or sandy-peaty soils (sometimes in standing water) in grassland and woodland; 1450–2100 m; flowering January–June

NOTE (on the species as a whole). The nature of the underground parts in this species is not entirely clear. Plants collected in Tanzania (**T** 4 & 7) when they were dying back after the rainy season (*Faden et al.* 96/226 and 96/288) had bulbs and no rhizomes. Specimens from northern Tanzania (**T** 1), Uganda, Kenya and Congo–Kinshasa always have rhizomes, when underground parts have been collected. Although there are no specimens with bases collected from Uganda or Kenya at a season comparable to these two Tanzanian collections, label notes on three Kenyan collections suggest that bulbs might be produced in addition to rhizomes. *Tweedie* 3471, which lacks a base, states that "a root dug up had 2 or 3 round 'corms'... but they were lost in transit." *Gilbert & Mesfin* 6732, which also lacks a base, states that the flowering shoots are "bulbous based." Finally, *Faden & Evans* 69/716, which has a rhizome, states that the "leaf bases... [form] a bulb-like base at the end of the rhizome". The apparent difference in underground parts between the Tanzanian versus the Ugandan and Kenyan plants could be argued to be of sufficient magnitude to merit species distinction. However, the similarity in habitat among these plants, and difference from all other African *Cyanotis* species in this character, plus the lack of other strong differentiating characters, has led me to keep them together, at least provisionally.

The presence of a basal rosette of leaves may be general in this species, at least when plants are actively growing, but most collections do not show them. The two Tanzanian collections made at the end of the growing season (*Faden et al.* 96/228 and 96/288) lacked rosettes, which demonstrated either that the rosettes do not persist into the dry season or else that they were never present. *Gilbert & Mesfin* 6732 describes the rosette leaves as distichous.

Two points in the type description require comment. The capsule is described as 5.5 mm long, which is clearly a typo, because they are little more than half that long. The seeds are described as quadrate and 2–3 per locule in the type description. All the seeds that I have examined are ovoid to ovoid-ellipsoid, rounded at one end and truncate at the other. I have never seen more than two seeds per locule in any other *Cyanotis* species, and I have not found three in any capsule locule that I have examined in this species, so what Brenan saw would have been highly unusual, and the evidence appears to have disappeared.

Most heads are pedunculate. The type collection is unusual in having mostly sessile axillary inflorescences and very short bracts subtending the heads.

The two collections from **T** 7, *Renvoize & Abdallah* 1987 and *Watermeyer* 49, and *Richards* 15892 from **T** 4 have slightly cobwebby pubescence on the undersides of some of the leaves, particularly the rosette leaves. Because only one of the four additional collections examined from Tanzania has rosette leaves, it is unknown whether cobwebby pubescence is typically present on these leaves in subsp. *bulbifera*. In *Watermeyer* 49, a plant that had recently started flowering, the cobwebby pubescence in the inflorescence is much more prominent than in any other collection of subsp. *bulbifera*.

Tweedie 3655 mentions that in some plants the stems, leaves and bracts were "liberally tinged with maroon." How frequently that occurs in other populations is unknown, although *Richards* 8495 refers to the bracts as "brown and green".

The association of this species with marshes and other wet places seems to be consistent. It is more aquatic than any other *Cyanotis* except *C. axillaris*. Although it can grow in standing water for some time, it appears to occur only in habitats that dry up during the dry season, such as the edges of marshes and swamps.

5. **Cyanotis barbata** *D. Don*, Fl. Nepal.: 46 (1825); C.B. Clarke in DC., Monogr. Phan. 3: 2 (1881); Brenan in F.W.T.A. 2nd ed., 3: 38 (1968); Blundell, Wild Fl. E. Afr.: 414, fig. 817 (1987); Faden in U.K.W.F. 2nd ed.: 306, t. 137 (1994); Ensermu & Faden in Fl. Eth. 6: 341, fig. 207.2 (1997). Type: Nepal, 1821, *Wallich* s.n. (K!, K-W! 8993A)

Perennial herb arising from a buried corm ± 1 cm in diameter, the corms sometimes clustered; roots thin, fibrous; shoots solitary or several, (2–)6–45(–75) cm long, erect to ascending or decumbent at the base and rooting at the proximal nodes, basal leaf rosettes lacking; internodes usually with a line of matted hairs descending from the distal node, often also patently pilose to glabrescent, internodes, nodes and less often the leaves sometimes purple or reddish purple or tinged with purple or reddish purple. Leaves distichous; sheaths to 3 cm long, sparsely patently pilose, sometimes also or only with a line of twisted white hairs along the fused edge, ciliate at the apex; lamina usually complicate, lanceolate or narrowly lanceolate to linear, 2.5–19 × 0.3–1.1 cm, base narrowed or broadened into the sheath, margins ciliate, at least proximally, apex acute to acuminate, both surfaces pilose to subglabrous or less commonly glabrous. Inflorescences borne at or from the distal 1–3(–4) nodes, the shoot terminated by 1–4 sessile cincinni, the next node usually producing a sessile or shortly stipitate cincinnus or pair of cincinni, and the third node almost always producing a long-stipitate (rarely sessile) cincinnus or cincinnus pair; additional inflorescences also sometimes produced and the pattern too complex to describe simply; individual cincinni or groups of cincinni 1–2 × 1–3 cm; bracts ± complicate, often becoming reflexed, 1–8 cm long, similar in form and pubescence to the foliage leaves but sometimes with longer hairs proximally; bracteoles biseriate, 4–7 in each series, falcate, lanceolate-oblong to ovate, 5–12 × 1–4.5 mm, margins ciliate, apex acute to acuminate, surfaces like those of leaves. Flowers subsessile, 1–1.5 cm wide, open in the morning; calyx 4.5–6.5 mm long, fused basally for 1.5–3 mm, lobes subulate, keeled, 2.5–4 × 1–1.5 mm, apex acute, sparsely to ± densely pubescent (rarely densely arachnoid-pubescent? – see notes); corolla 9–11 mm long, tube 5–7 mm long, lobes blue to mauve, pink or magenta, rarely white or violet, broadly ovate, 3.5–4 × 3–5 mm. Stamens 6, equal, 8–12 mm long, densely bearded for ± 5 mm with long hairs (colour like corolla, or contrasting), with a glabrous apical swelling; anthers at least sometimes dark blue to brown or black at the base, the rest yellow, or entirely yellow to orange (according to collectors). Ovary oblong-ellipsoid, 1.1–1.6 mm long, densely pubescent distally; style 7.5–10 mm long, bearded below a large apical swelling; stigma small. Capsules light brown with darker flecks, ellipsoid, 2.5–3 × 2 mm, rounded at the apex, pubescent with an apical tuft of hairs and scattered hairs proximal to that. Seeds ovoid to ellipsoid, 1.1–1.6 × 0.9–1.1 mm, testa dark brown, with a few small longitudinal white streaks or patches, ± rugose-foveolate to scrobiculate, with irregular shallow depressions, longitudinally striate. Fig. 3/4–10, p. 34.

UGANDA. Karamoja District: Mt Moroto foot, 13 Sep. 1956, *Hardy & Bally* 10800!; Kigezi District: Mt Muhavura, Jan. 1947, *Purseglove* P2304!; Mbale District: Mt Elgon, western crater crest, Jan. 1918, *Dummer* 3383!

KENYA. Northern Frontier District: Mt Kulal, Narangani, 8 June 1960, *Oteke* 89! & Maralal, Lorok plateau 15–20 km N, 10 Nov. 1978, *Hepper & Jaeger* 6707!; Trans-Nzoia District: Kitale, by golf course, Aug. 1967, *Tweedie* 3469!

TANZANIA. Moshi District: Mt Kilimanjaro, Kibo, 22 Feb. 1933, *C.G. Rogers* 787!; Mbulu/Masai District: Oldeani Mt, W side, 16 Feb. 1961, *Newbould* 5701!; Mbeya District: Igoma–Katumba road, edge of Kitulo Plateau, 2 Mar. 1986, *Bidgood & Congdon* 113!

DISTR. **U** 1–3; **K** 1–6; **T** 2, 4, 7; Ghana, Cameroon, Bioko, Congo–Kinshasa, Rwanda, Burundi, Sudan, Ethiopia, Malawi, Mozambique, Zimbabwe; Yemen, India, Nepal, Bhutan, China

HAB. Grassland, scattered tree grassland, shallow soil in rocky places, forest edges and clearings, moorland, rarely in forest, thicket edges or savanna; (1350–)1850–3400(–4150 m)

Main flowering season in Tanzania January–March, in Kenya June–September, with a secondary peak in December; in Uganda, from which the number of collections is limited, it appears to be spread out from August to February of the following year.

SYN. *Cyanotis hirsuta* Fisch., C.A. Meyer & Avé-Lall., Ind. Sem. Hort. Petrop. 8: 57 (1841 [or 42]), *nomen subnudum*; C.A. Meyer, Observ. Bot.: 7 (1841) (fide Hasskarl (1867)); C.B. Clarke in DC., Monogr. Phan. 3: 254 (1881) & in F.T.A. 8: 78 (1901). Type: Ethiopia, *Schimper* 14 (B, holo.; K!, 2 sheets, iso.)

C. abyssinica A. Rich., Tent. Fl. Abyss. 2: 344 (1851); Hassk. in Commel. Ind., p.139 (1870). Type: Ethiopia, near Djeladjeranne, 1 Aug. 1841, *Schimper* 1556 (P holo., K! iso.)

C. abyssinica A. Rich. var. *glabrescens* A. Rich., Tent. Fl. Abyss. 2: 344, t. 98 (1851). Types: Ethiopia, Wajerat [Ouogerate], *Petit* s.n. (P, syn., K, iso.); Adua valley, *Quartin-Dillon* s.n. (P, syn., K!, isosyn.); Djeladjeranne, *Schimper* 1556 (P, syn., K!, 2-sheets, isosyn.)

Zygomanes parasitica Hassk. in Schweinfurth, Beitr. Fl. Aethiop.: 214 (1867). Type: Ethiopia, Semien Mts, Acallo Meda, *Schimper* 506 (five herbaria cited: G, P, W, also Buching, and Lenorm., holo. not designated; K! iso.)

Cyanotis parasitica (Hassk.) Hassk., Commel. Ind. 116 (1870); C.B. Clarke in DC., Monogr. Phan. 3: 256 (1881); Engl. Hochgeb. Fl. Trop. Afr.: 132 (1892); C.B. Clarke, F.T.A. 8: 79 (1901)

C. hirsuta Fisch., C.A.Meyer & Avé-Lall. var. *glabra* K. Schum. in P.O.A. C: 136 (1895). Type: Tanzania, Bumi ravine (?Kilimanjaro), *Volkens* 1915 (B†, holo.)

NOTE. The widespread African and Asian distribution of this plant has been recognized in many floras. In Africa the plant continues to be called *C. barbata*, e.g. in Flora of Ethiopia and Eritrea, whereas in Asia the older name *Cyanotis vaga* (Lour.) Schult.f. has been adopted for it in some floras, e.g. Flora of Bhutan. As Brenan (1968) pointed out, there is no type for Loureiro's plant, his description is clear enough to know that he had a species of *Cyanotis*, but hardly which one, and the plant is said to have come from around Canton, which is not a habitat for a high elevation species such as *C. barbata*. I don't think we will ever know for certain which species should be called *C. vaga*, so there is no compelling reason for adopting that name for African plants up to now called *C. barbata*.

The only specimens that I have found that might be considered types for *C. barbata* are *Wallich Herbarium* 8893A (K, K-W), for which the label reads, 'Nepal, 1821', when Wallich would have collected plants in Nepal (Stearn 1978, p. 8). However TL2 does not mention K or K-W as a depositories of types from *Prodromus Florae Nepalensis*, in which *C. barbata* is described. TL2 and Stearn (loc. cit.) record that Wallich collections formed part of the basis of Don's flora, so there is a strong possibility that the specimens at Kew are types.

The names *Cyanotis abyssinica* A. Rich. and var. *glabrescens* are problematic. Richard clearly was providing a new name in *Cyanotis* for the unpublished name *Commelina hirsuta* Hochst., which appeared on printed labels that were distributed with Schimper collections. However, the name *C. hirsuta* had already been published, so *C. abyssinica* may be a *nomen superfluum*, although it is based on a different type. It is possible that Richard was unaware of the other name in *Cyanotis* because he contrasted his new species with other species but not with *C. hirsuta*. Another problem with the name *C. abyssinica* is that Richard did not directly supply a description for the species when he described var. *glabrescens*. In that description, however, he contrasted the variety with the typical plant and thus indirectly described the species. A third problem is that although Richard typified the species with *Schimper* 1556, what is probably the same collection – a collection number is not cited in the latter case but the locality is the that of *Schimper* 1556 – is also listed under var. *glabrescens*. This could be taken care of by lectotypification, but the variety is not worth recognizing. The K sheet from Wajerat has the collector Quartin Dillon on the label, instead of Petit, but the name was added later, so it probably was just wrongly chosen.

Cyanotis pauciflora A. Rich. [*Zygomenes pauciflora* (A. Rich.) Hassk.] is cited as a synonym of *C. barbata* in Fl. Eth. 6: 341 (1997), with an isotype supposedly at Kew. When checked in Feb. 2007, the folder for *C. pauciflora* was lacking and I could not find the Quartin-Dillon collection cited in Fl. Eth. that matched the protocol of the type. From its description of having 'arachnoid' hairs, I would agree with Clarke (F.T.A. 8: 81, 1901) that this was likely a depauperate form of *C. lanata*, but I have been unable to confirm that.

The species is very variable in height, with a general trend toward reduction in plant height at higher elevations. However, the highest elevation collection by far, *Dummer* 3383, from more than 4000 m on Mt Elgon, is not especially dwarf, whereas the tallest collection, *Tweedie* 3469, while from a relatively low elevation for the species (1860 m), had to reach its height of 75 cm because it was growing among grasses of that stature, according to the collector's notes. *Tweedie* 783 states that the larger plant among those collected came from a wetter part of the same rock on which the population was growing, so the amount of moisture available might also affect the final size of the plant. The most dwarfed plants were described as *Cyanotis parasitica* from Ethiopia. They do have a distinctive appearance with their short internodes and crowded leaves. However, with abundant material available now, there is no sharp line between them and larger plants because there are many intermediates. Thus recognizing them at any taxonomic level is unjustified.

The presence of a dense arachnoid pubescence on the sepals of spirit material of *Tweedie* 783 is disturbing. I checked the herbarium specimen and it is correctly determined, so either this variation can occur in *C. barbata* or else the single flower was from a different species, presumably *C. longifolia*, although at 2045 m, this would be near the maximum elevation for *C. longfolia*. The sepals observed on the specimen do not show this kind of pubescence, but are typical for *C. barbata*.

In Fl. Eth. this species is separated from *C. foecunda* by it corms being produced just below the surface, whereas those in *C. foecunda*, which are rarely produced and/or almost never collected in our area, are deeply buried. Judging by the frequency with which corms have been collected in *C. barbata*, I would expect that they are mostly shallow in our region too.

6. **Cyanotis lanata** *Benth.*, Niger Fl.: 542 (1849); C.B. Clarke in DC., Monogr. Phan. 3: 258 (1881); K. Schum. in P.O.A. C: 137 (1895); C.B. Clarke in F.T.A. 8: 80 (1901); Schnell in Bull. I.F.A.N. 19, Sér. A: 739 (1957); Morton in J.L.S. 60: 194 (1967); Brenan in F.W.T.A. 2nd ed.: 3: 40 (1968); Faden in U.K.W.F. 2nd ed.: 306 (1994); Ensermu & Faden in Fl. Eth. 6: 342 (1997). Type: Nigeria, Quorra [= Niger River], *Vogel* 122 (K!, holo.)

Annual herbs; roots thin, fibrous, all or most from the base of the plant; stems and foliage often purple or pink; a central sterile rosette with white arachnoid-pubescent leaves to 10 cm long sometimes present but often apparently lacking or ephemeral; flowering shoots peripheral to the rosette (when present), tufted, erect to ascending or decumbent, occasionally rooting at the lower nodes, (2–)5–50 cm long or tall, internodes white arachnoid-pubescent or glabrous. Leaves distichous, sheaths to 2 cm long, white arachnoid-pubescent to subglabrous, ciliate distally; lamina linear to oblong-lanceolate, 1.5–9(–10.5) × 0.3–0.7(–1) cm, margins ciliate or glabrous, apex acute to acuminate, often mucronate, surfaces densely covered with white arachnoid hairs to subglabrous, the abaxial surface often more densely pubescent. Inflorescences terminal and axillary, borne at or from the 1–3(–5) most distal nodes on the flowering shoot, the shoot typically terminated by pair of cincinni subtended by a pair of bracts, each exceeding the cincinnus, axillary inflorescences of 1 or 2 cincinni, sessile (rarely shortly pedunculate) and partly enclosed in the subtending sheath; cincinni or cincinnus pairs 0.7–1 × 0.5–1.2 cm; bracts foliaceous, exceeding the cincinni, often reflexed, broad at the base and gradually narrowed into a linear-lanceolate to lanceolate-oblong lamina 0.5–3 cm long, pubescence like that of the leaves; bracteoles biseriate, 3–6 in each series, mostly falcate, 5–7 × 1–2.5 mm, margins ciliate, apex acute to acuminate, surfaces white arachnoid-pubescent, at least proximally. Flowers subsessile, very variable in color, mostly mauve to blue or pink, the corolla lobes often with a medial white stripe; calyx 4–5.5 mm long, fused proximally for 1.5–2.5(–3.5) mm, lobes 2–4 × (0.5–)1 mm, keeled (rarely not), ciliate, apex acute, surface pilose; corolla (6–)7.5–9 mm long, tube (4–)5–6.5 mm long, lobes broadly ovate to ovate-deltate, (2–)2.5 × (2–)2.5–3 mm, apex obtuse or rounded-obtuse to acute. Stamens 6, equal, filaments free (rarely adnate to the base of the corolla tube?), (6–)8.5–11 mm long, densely bearded near apex (very rarely entirely glabrous), with or without a slight terminal swelling; anthers (0.2–)0.5–0.9 mm long, dehiscing probably first by basal slits, sometimes later extending longitudinally. Ovary oblong-elliptic, 1–1.3 mm long, with an apical tuft of long hairs; style (5–)7–8 mm long, with a terminal swelling, bearded below the swelling, otherwise glabrous (very rarely entirely glabrous); stigma broad. Capsules light brown, oblong-ellipsoid, 2–3 × 1–1.5 mm, with an distal tuft of white hairs, the hairs extending proximally along the sutures, locules 2-seeded. Seeds ovoid (to ellipsoid), (1.1–)1.3–1.8 × (0.7–)0.8–1.3 mm, testa uniformly medium brown or grey-brown, striate and shallowly and irregularly pitted.

UGANDA. Teso District: Soroti–Lira road, km 19.2, *Verdcourt* 839!; Mengo District: near Kitamiro, 26 Sept. 1949, *Dawkins* 400! & Kinsinsi Point opposite Kaazi, 22 Feb. 1970, *Lye* 5079!

KENYA. Trans-Nzoia District: ENE slope of Mt Elgon, 24 Sept. 1962, *Lewis* 5976! & 35.2 km N of Kitale on Suam road, Aug. 1971, *Tweedie* 4105!; Kilifi District: Marafa, 19 Nov. 1961, *Polhill & Paulo* 787!

TANZANIA. Musoma District: Mara River Guard Post, 25 Feb. 1968, *Greenway & Turner* 13338!; Lushoto/Tanga District: Eastern Usambara Mts, lower Sigi Valley, 30 May 1950, *Verdcourt* 231!; Morogoro District: km 44 on Chalinze–Morogoro road, 7 June 1996, *Faden et al.* 96/61!

DISTR. **U** 1–4; **K** 1, 3–7; **T** 1, 3–8; Senegal, Gambia, Mali, Guinea Bissau, Guinea, Sierra Leone, Liberia, Ivory Coast, Burkina Faso, Ghana, Togo, Benin, Niger, Nigeria, Cameroon, Chad, Central African Republic, Congo–Kinshasa, Burundi, Rwanda, Sudan, Ethiopia, Angola, Zambia, Malawi, Mozambique, Zimbabwe, Botswana, Swaziland, South Africa; Yemen

HAB. Rock outcrops in open areas in grassland, bushland and woodland, often in rock crevices or in shallow soil, sometimes near but not in rock pools, shallow soil over laterite or hard pans; occasionally in grassland, wooded grassland or woodland (and apparently not associated with rocks); 50–2150 m

Flowering specimens have been seen from all months.

SYN. *Cyanotis schweinfurthii* Hassk., Commel. Ind.: 134 (1870). Type: Sudan, *Schweinfurth* s.n. (no coll. number or herbarium cited, but *Schweinfurth* 534, 535, 2334 & 2246 are all at K!)

C. lanata Benth. var. *schweinfurthii* (Hassk.) C.B. Clarke in DC., Monogr. Phan. 3: 258 (1881)

C. lanata Benth. var. *sublanata* C.B. Clarke in DC., Monogr. Phan. 3: 258 (1881). Syntypes: Nigeria, Abeokuta, *Irving* 1701 (K!, syn.); Seba on the Niger River [Kworra], *Barter* 1475 (K!, syn.) & Angola, without precise locality, *Welwitsch* 6648 (BM, syn.)

C. rubescens A. Chev. in Mém. Soc. Bot. Fr. 2, 8: 216 (1912). Type: Ivory Coast, Orodougou, between Touna and Siflé, 1 June 1909, *Chevalier* 21814 (P, not seen)

C. lanata Benth. var. *rubescens* (A. Chev.) Schnell in Bull. Inst. Franç. Afr. Noire, Sér. A, 19: 741 (1957)

NOTE. *Gillett* 14092, from Moyale, Kenya is unusual in having mostly single bracts below the inflorescences (or the second bract is very small and erect) and small inflorescences. It somewhat resembles *Cyanotis* sp. 7 in Fl. Eth. 6: 343 (1997), which in turn matches *Aké Assi* 6578 from Ivory Coast. The latter was mentioned in F.W.T.A. ed. 2, 3: 40 (1968) by Brenan in a note under *Cyanotis lanata*, who suggested that it might represent a distinct taxon. The *Gillett* collection differs from the others in having broader leaves, much denser arachnoid pubescence on all parts, and a single, larger, spreading bract versus two smaller, subequal, ascending bracts. It may be just an unusual form of *C. lanata*. *Faden et al.* 77/292 from Nairobi also has mostly single bracts, but otherwise it resembles typical *C. lanata*.

The flowers in the spirit material of *Milne-Redhead & Taylor* 9174 (K) from southern Tanzania are much smaller in all of their parts from the six other spirit collections studied from our area. The parenthetical smallest sizes for the floral parts in the description all come from this collection. Even more peculiar, the stamen filaments and style were completely glabrous. Although the label records that there were a 'few multicellular hairs occasionally spreading out from the filaments,' I could find none. This is contrary to the generic character of bearded stamen filaments in *Cyanotis*. Moreover, the style has been found in every other collection studied of this species to be bearded too. A further peculiarity was that the style was slightly longer than the stamens in this collection whereas in all others it is distinctly shorter than them. The great reduction or total loss of the floral parts that are normally associated with insect attraction and reward in *Cyanotis*, namely corolla lobes, stamen filament hairs and anthers, leads one to conclude that this population had all but abandoned out-crossing in favour of selfing. Even in 'normal' flowers of this annual species selfing is probably frequent because of the stigma's position below the anthers and the fact that the stamens and style become enclosed within the fading corolla. However, *Milne-Redhead & Taylor* 9174 would seem to have given up on even the occasional visitation by insects that can lead to out-crossing. In fairness, it should be pointed out that the spirit material of *Milne-Redhead & Taylor* 9174 consists mainly of tiny plants, the smallest only two cm high and unbranched, producing probably its only flower. What future would such a plant have if it didn't self? But there are also some shoot tips with inflorescences from larger plants, so the very tiny flowers are not restricted to the smallest individuals. Taxonomically, this population can only be treated as an aberrant one unless more such populations come to light. In their overall morphology the plants on the herbarium specimen in no manner look unusual for *C. lanata*. The unusual features of the flowers could only be determined from a liquid collection. I find no such collections at Kew from the countries directly south of our area (Mozambique, Malawi and Zambia) in which they should be looked for because of the provenance of this collection.

Collectors have frequently referred to reddish pigments in the vegetative parts of this species. They have been called pink, red, crimson, purple, reddish purple and reddish brown. The color may in fact vary but such coloration is more commonly present in *C. lanata* than in any other *Cyanotis* species.

7. **Cyanotis arachnoidea** *C.B. Clarke* in DC., Monogr. Phan. 3: 250 (1881); Morton in J.L.S. 60: 194, t. 2 fig. 1 (1967); Brenan in F.W.T.A. 2nd ed. 3: 38 (1968); Blundell, Wild Fl. E. Afr.: 414, fig. 863 (1987); Faden in U.K.W.F. 2nd ed.: 306 (1994). Type: India, Madras, Pulney Hills, Sep. 1836, *Wight* 2839 (lectotype K!, designated here)

Perennial herbs; roots fibrous, thin, produced from the base of the plant and occasionally from the proximal nodes of decumbent portions of flowering stems; central basal rosette present on a short thick stem or rarely terminal on a horizontal rhizome. Rosette leaves with sheaths to 3 cm long, white arachnoid-pubescent or occasionally nearly glabrous; lamina complicate or planar, narrowly lanceolate, 3–7.5 × 0.5–1.2 cm, margins usually densely ciliate, apex acute, sometimes ± mucronate, both surfaces densely white arachnoid-pubescent or occasionally sparsely pubescent to ± glabrous; flowering shoots several per plant, borne peripheral to the rosette, ascending or more commonly decumbent proximally, rarely rooting, with short thick internodes covered by overlapping leaf sheaths, and then the shoots ascending or occasionally prostrate, 6–25 cm tall or long, internodes densely white arachnoid-pubescent to pilose; cauline leaves distichous, sheaths to 1(–1.5) cm long, usually densely white arachnoid-pubescent, ciliate at the apex; lamina complicate or planar, lanceolate to ovate, 1–4 × 0.4–1.4 cm, base rounded, margins ciliate at least proximally, apex acute to acuminate, usually mucronate, adaxial surface usually sparsely white arachnoid-pubescent, occasionally densely so or subglabrous, abaxial surface usually densely white arachnoid-pubescent, rarely pilose to subglabrous. Inflorescences terminal and axillary, borne at or from the 1–5(–8) most distal nodes on the flowering shoot, often borne at the ends of short shoots produced from most nodes on prostrate flowering shoots, consisting of sessile or stipitate single cincinni or sometimes some paired, 0.5–1.2(–1.5) × 0.7–1.7 cm; bracts subtending the inflorescences equalling to slightly exceeding them, often strongly reflexed distally, pubescence like that of leaves; bracteoles biseriate, 3–5 in each series, ovate to lanceolate, falcate, 5–9(–11) × 1–3(–3.5) mm, margins ciliate, apex acuminate, surfaces sparsely to densely white arachnoid-pubescent. Flowers ± sessile; calyx 5–6.5 mm long, fused proximally for 1.8–3.5 mm, densely arachnoid-pubescent, lobes ovate to subulate, 2–4 × 1.5–1.8 mm, keeled, apex acuminate; corolla 7–9 mm long, tube 5–6 mm long, lobes blue, purple, mauve, blue-pink or violet-blue, broadly ovate, 2–3 × 3–3.5 mm, apex rounded. Stamens 6, equal, filaments free, 8–9 mm long, densely bearded for 3–3.5 mm, with an terminal swelling; anthers 0.8–1.1 mm long. Ovary oblong-ellipsoid, 1–1.4 mm long, densely appressed-pubescent; style 7–8 mm long, densely bearded distally below a glabrous terminal swelling; stigma small. Capsules light to medium brown, sometimes with darker flecks, oblong-ellipsoid, 2.5–3 × 1.6–2 mm, densely white pubescent distally, locules (0–)2-seeded. Seeds ovoid to ellipsoid, (1.1–)1.4–2(–2.2) × 0.8–1.3 mm, testa uniformly dark brown or the striations sometimes lighter brown and raised, striate and shallowly and irregularly pitted.

KENYA. Nairobi, Langata, Forest Edge Road, 6 Feb. 1977, *Faden, Faden & Ng'weno* 77/285!; Masai District: Ol'debesi–Lemoko, 28 Apr. 1961, *Glover, Gwynne & Samuel* 767!; Teita District: Taita Hills, Ngangao Forest, 11 Feb. 1977, *Faden & Faden* 77/316!

TANZANIA. Masai District: Longido Mountain, 16 Jan 1936, *Greenway* 4398!; Lushoto District: West Usambara Mts, near Magamba Secondary School, 2 June 1996, *Faden et al.* 96/17! & Mkuzi, 6.4 km NE of Lushoto, 8 Apr. 1953, *Drummond & Hemsley* 2060!

DISTR. **K** 3–7; **T** 1–3; Ivory Coast?, Ghana, Nigeria, Cameroon, Congo–Kinshasa, Angola, Zimbabwe?; India, Thailand, China

HAB. Rock outcrops, exposed rock faces, shallow soil over rocks; usually in full sun; 1500–2600 m Flowering specimens have been seen from all months except March and November.

SYN. *Cyanotis lanata* Benth. var. *lanuginosa* K. Schum. in E.J. 36: 209 (1905). Type: Tanzania, Lushoto District, West Usambaras, near Kwai, *Engler* 2249 (B!, holo.)

NOTE. There are at least 12 collections from India at Kew, and perhaps others elsewhere, that were annotated by Clarke as *Cyanotis arachnoidea* (Clarke, 1881). Clarke merely recorded "(Wight n. 2839); etc." *Wight* 2839, except for being somewhat larger than average, appears typical of the species. It also is a whole plant and has mature capsules and seeds in the packet. Therefore I have selected it as the lectotype of the species. The Indian specimens show several notable features. One is the presence of thick, fibrous, but not tuberous roots. A second is the very short bracts subtending the heads. Curiously, some of the specimens show very thick or bulbous bases. Clarke does not mention them except perhaps in the phrase "stems often thickened" when comparing *C. arachnoidea* with *C. barbata.*

I am not convinced that the African plants treated as *C. arachnoidea* by Morton (1967) and Brenan (1968) are truly conspecific with the Asian species. Morton (1967) first published this notion, which was based on discussions with J.P.M. Brenan at Kew, but neither Morton nor Brenan mentioned any characters, other than general appearance, for adapting the Indian name. I have examined the Indian and African material under *C. arachnoidea* at Kew, and while I find a resemblance, especially between some of the more robust forms from West Africa and India, my two field collections of *C. arachnoidea* from India and numerous collections of the closely related *C. thwaitesii* Hassk. from Sri Lanka, yielded plants that fell entirely outside the morphological range of African *C. arachnoidea.* However, Asian material of *C. arachnoidea* is very variable, and the technical characters generally used to distinguish species of Commelinaceae are usually lacking in herbarium specimens. Thus, until a more detailed study can be undertaken, I am provisionally accepting the Indian name for the African plant.

Drummond & Hemsley 4301, from the Taita Hills, Kenya is an especially robust specimen, with thick stems, large leaves, eight nodes with inflorescences, the largest inflorescences observed, as well as the longest and widest bracteoles. It also is the clearest specimen to show a rhizome. The leaf and bract pubescence are relatively sparse, and even the arachnoid pubescence in the inflorescences is less abundant than in most other specimens of the species. From its robustness the plant looks like it either was especially well fertilized – perhaps from rock hyrax droppings? – or that it might have been a polyploid. *Wood* 705 (from **K** 4) and Glover et al. (from **K** 6) are smaller plants that are even closer to being glabrous.

Many specimens appear to have an occasional thicker root among the thin ones, which would approach the Indian plants. However, on close examination, in every case these 'roots' turned out to be the remains of old flowering shoots. The lighter brown striations on the seeds were seen in two of the four populations in which the seeds were examined.

The seeds from 1-seeded capsule locules are especially long in this species (1.8–2.15 mm) as compared with seeds from 2-seeded locules ((1.1–)1.35–1.5 mm)

Cyanotis arachnoidea and *C. lanata* overlap in their altitudinal distribution by some 600 m. The main overlap occurs between 1520 and 1830 m, although *C. arachnoidea* is generally a higher elevation species than *C. lanata.* The two species grow together in Nairobi. *Cyanotis arachnoidea* is much more restricted to rocks than is *C. lanata.*

8. **Cyanotis foecunda** *Hassk.*, Commel. Ind.: 110 (1870); C.B. Clarke in DC., Monogr. Phan. 3: 255 (1881); K. Schum. in P.O.A. C: 137 (1895); C.B. Clarke in F.T.A. 8: 80 (1901); Obermeyer and Faden, F.S.A. 4, 2: 56 (1985); Blundell, Wild Fl. E. Afr.: 414, fig. 631 (1987); Faden in U.K.W.F. 2nd ed.: 306, t. 137 (1994); Faden in Fl. Som. 4: 81, Fig. 52 only, as *Cyanotis somaliensis* (1995); Ensermu & Faden in Fl. Eth. 6: 342 (1997). Type: Ethiopia, Sérraba in Uschan, *Schimper* 459 (G, holo.; K!, iso.)

Perennial herb with a distinct base (rarely lacking), sometimes arising from a small, deeply buried corm to ± 5 mm in diameter or cluster of corms; plants sometimes persisting through the dry season as a cluster of compact, short shoots or a prostrate mat; roots thin, fibrous, produced from the base of the plant and sometimes from decumbent or repent stems; shoots usually several, ascending to decumbent, rarely repent, occasionally ± scrambling in bushes, to 60 cm tall, unbranched to much branched from the lower nodes; internodes sometimes very short at the base of the plant and covered by overlapping sheaths, sparsely pilose to densely appressed white-pubescent. Leaves distichous; sheaths to 4 cm long basally, much reduced distally, sparsely pilose to densely appressed white-pubescent, ciliate at the apex; lamina planar to complicate, narrowly lanceolate to lanceolate-oblong, 2.5–12(–16.5) × 0.6–2 cm, base usually rounded, margins ciliate, apex acute (to acuminate), sometimes mucronate, adaxial surface glabrous except for submarginal bands of hairs or sparsely pilose to white-pubescent, abaxial surface whitish or greyish pilose. Inflorescences produced acropetally, the most basal one the oldest, all axillary or the distalmost one perhaps terminal, sessile or the proximal ones subsessile, 5–16 per shoot (excluding those on axillary branches), each inflorescence at first enclosed in the leaf sheath, becoming more exposed with age, ± spherical, to 1.5 cm wide, each inflorescence in the axil of a leafy bract; bracts strongly decrescent distally and becoming very strongly deflexed from its sheath, pilose to pubescent, rarely glabrous, the shoot axis often zigzag; bracteoles sometimes slightly falcate, linear-lanceolate to ovate-elliptic, 5.5–10.2 × 1.6–2.8(–4.2) mm, pilose, margins ciliate. Flowers with fruiting pedicels up to 3.5 mm long; sepals 5.5–8.4 mm long, fused at the base for 1–2 mm, strongly keeled, lobes obovate, 1–2 mm wide, ciliate, pilose along the keel, apex acute; corolla tube ± 4 mm long, lobes blue to violet, mauve or lilac, ovate-deltate, ± 3 × 4 mm, apex acute. Stamens with filaments 10–10.5 mm long, bearded distally with blue or violet hairs, with a large subapical swelling; anthers 0.8–1 mm. Ovary ± 1 mm long, densely bearded at apex; style 10 mm long, usually glabrous, rarely very sparsely bearded, with a distal swelling terminating in a fine point. Capsules oblong to oblong-ellipsoid, (2.3–)2.7–3.5(–3.9) × 1.3–1.9 mm, appressed pilose in the distal half (to $^2/_3$), with a white apical tuft of hairs. Seeds ovate to elliptic or oblong-elliptic in outline, (1–)1.2–1.8(–2.1) × 0.7–9(–1.1) mm, testa chestnut brown, strongly striate longitudinally, deeply transversely furrowed.

UGANDA. Karamoja District: Moroto, Kasunen Estate, June 1972, *Wilson* 2146b!; Ankole District: Ruizi River, 10 Feb. 1951, *Jarrett* 418! (mixture with *Cyanotis longifolia*); Teso District: Serere, June 1932, *Chandler* 788!

KENYA. Northern Frontier District: 10 km ESE of Baragoi, Gelai, 18 Nov. 1977, *Carter & Stannard* 507!; North Nyeri District: Nanyuki, 2 July 1956, *Verdcourt* 1520!; Masai District: Masai Lodge on Mbagathi River opposite Nairobi National Park, 9 Feb. 1977, *Faden & Faden* 77/300!

TANZANIA. Moshi District: Rongai Ranches, 20 Apr. 1957, *Greenway* 9189!; Lushoto District: Western Usambara Mts, Mombo–Lushoto road, 3 km, 1 June 1996, *Faden et al.* 96/10!; Dodoma District: 7.2 km E of Itigi Station, 7 Apr. 1964, *Greenway & Polhill* 11424!

DISTR. **U** 1, 2, 3; **K** 1, 3, 4, 6, 7; **T** 1–7; Cameroon, Congo–Kinshasa, Rwanda, Burundi, Sudan, Ethiopia, Angola?, Zambia, Malawi, Mozambique, Zimbabwe, Botswana, Namibia

HAB. Bare rocky ground, rock crevices (often near rivers), bushland, woodland, grassland, thickets, also on termite mounds, occasionally in swampy places; 200–2150 m

Flowering specimens have been seen from all months except October, but the main flowering period is March through July.

SYN. *Cyanotis montana* K. Schum. in Hochgebfl.: 156 (1892) & in P.O.A. C: 137 (1895). Types: Ethiopia, Berrechowa, *Schimper* 280 (B!, syn.) & plains of Keren, *Steudner* 1485 (B, syn.)
C. minima De Wild. in F.R. 12: 293 (1913). Type: Congo–Kinshasa, Shinsenda, *Ringoet* 402 (BR!, holo., photo K, US)

NOTE. The type has a very clear subterranean corm, but I have never found such organs on living plants within our area, even though I have looked for them repeatedly. Only one collection from the FTEA area shows such underground structures, *Greenway* 9189 from northern Tanzania, although *Tweedie* 3466 from western Kenya mentions that "the small corm [had] broken off the root" because "the connection is very brittle". The corms are illustrated in U.K.W.F. Many collections give the impression that corms might have been present and were not collected, but other collections, particularly from **K** 4, 6, 7, **T** 3 & 6, may fundamentally lack corms, the plants perennating by means of short shoots that remain above ground and survive the dry season. Such plants seem to have at the base shoots, with a series of short internodes covered in the dry season and early part of the rainy season by overlapping sheaths, that give rise to flowering shoots and to new short vegetative shoots. Whether these short shoots serve as perennating organs in every case or will yet give rise to inflorescences the same season is unknown. Certainly *Faden & Faden* 77/576 had persistent vegetative shoots and old infructescences when it was collected in very dry, thorny thicket clumps in **K** 7 at the height of the dry season. Corms were looked for but not found. Unfortunately, it is not possible to determine the habit of every plant collection of *C. foecunda* from dried specimens, so describing these plants as a distinct taxon would be premature. They should be more carefully studied from living material, especially under field conditions.

Cultivated material of *Faden & Faden* 77/576 was used for the illustration of *Cyanotis somaliensis* C.B. Clarke in Flora of Somalia (Vol. 4, Fig. 52, p. 81. (1995)) because of the mistaken assumption that that species might be synonymous with *C. foecunda*. It differs chiefly by the presence of a basal leaf rosette and it is endemic to northern Somalia.

Some specimens observed from our area, as well as specimens from Rwanda, Burundi and Congo–Kinshasa were initially thought to belong to *C. somaliensis* var. *uda* C. B. Clarke in De Wild., Ann. Mus. Congo. Bot., sér. 5, 1: 223 (1906) because of their mat-like habit. However, the type of that plant was restudied and determined to be either an aberrant collection of *C. repens* subsp. *robusta* or a closely related taxon that has not been collected again.

It is not evident when the flower color is listed whether this refers to the stamen filament hairs, the corolla lobes or both. Six collections from our area separately mention the color of the petals and filament hairs. In five, they were of different colors, in the sixth they were the same color.

Greenway and Polhill 11424 mentions that there were a few hairs on the style. I have also seen this in a collection from Kenya (*Faden* 69/1304). Other than these two records the style, insofar as it has been noted, is always glabrous.

9. **Cyanotis repens** *Faden & D.M. Cameron* in Novon 15: 110, fig. 1 (2005). Type: Kenya, Kilifi District, Mwarakaya [Nyara Kaya], *Brenan et al.* 14672 (K!, holo.; EA!, US!, iso.)

Perennial herb lacking a definite base, usually forming a ± prostrate mat; roots thin, fibrous; shoots branched proximally, unbranched distally, repent to decumbent and rooting at the nodes, internodes often red to maroon, densely pilose or sometimes slightly woolly to glabrous. Leaves distichous, not decreasing in size distally; sheaths 0.3–1.3 cm long, pilose or slightly woolly, ciliate at apex; lamina sessile, succulent, ovate to lanceolate-oblong, 1–6(–9.6) × 0.5–1.7 cm, base broadly cuneate to rounded or cordate, often amplexicaul, margins ciliate, apex acute, rarely acuminate, sometimes ± mucronate, surfaces subglabrous to densely pilose, the adaxial lustrous green, often dotted with maroon, the abaxial surface often maroon or suffused with maroon. Inflorescences normally all axillary, enclosed in the leaf sheaths, rarely a terminal inflorescence present; bracteoles scarious, subulate or linear-oblanceolate, 1.5–8.5 mm long, glabrous or pilose. Flowers 4–8 mm wide, sepals 5–8.2 mm long, fused proximally, lobes oblanceolate, elliptic to linear, densely pilose, apex acute; corolla tube 4–6 mm long, lobes bluish purple or blue-violet, rarely pink, ovate to obovate, 3–5 × 2.5–4.5 mm, apex acute to obtuse. Filaments 8–11.5 mm long, with a subapical swelling, densely bearded distally with blue or blue-purple hairs; anthers yellow, 0.5–0.6 mm long, dehiscing by basal pores. Ovary 1–1.7 mm long, densely pubescent at or near the apex, style 7.5–11 mm long, with a subapical swelling, glabrous. Capsules trilocular, trivalved, 2–3.5 × 1–1.5 mm, apex hirsute. Seeds (1–)2 per locule, oblong-ellipsoid to ovoid-ellipsoid, 1.1–2 × 0.7–1.1 mm, testa brown or reddish black, finely striate longitudinally, with transverse pits and grooves.

subsp. **repens**

Leaves 1–2.5(–4.3) × 0.5–1(–1.7) cm; sheaths 3–5 mm long, 1–4 mm wide (pressed); inflorescences 1–2-flowered; stamen filament hairs ± 1 mm long; capsules 2–2.6(–3.5) mm long; seeds 1.1–1.3(–1.5) mm long, testa sometimes reddish black. Fig. 2/1–12, p. 29.

UGANDA. Mbale District: Kapchorwa, 11 Sep. 1954, *Lind* 345!

KENYA. Teita District: 18 km on Voi–Taveta road from Nairobi–Mombasa road, 11 Feb. 1977, *Faden & Faden* 77/312!; Tana River District: Malindi–Garsen road, 0.8 km towards Garsen from turnoff to Kibusu, 22 July1974, *Faden & Faden* 74/1174!; Kilifi District: Cha Simba, 22 km SW of Kilifi, 14 Oct. 1974, *Adams* 98!

TANZANIA. Lushoto/Handeni Districts: Korogwe–Handeni road, 14 Jan. 1954, *Faulkner* 1323!; Handeni District: Bridge over Msangasi stream, ± 50 km S of Korogwe on road to Dar es Salaam, 28 Sep. 1972, *Flock* 463!; Tanga District: Perani Forest, SE Umba steppe, 12 Aug. 1953, *Drummond & Hemsley* 3714!

DISTR. **U** 3; **K** 7; **T** 3, 6; not known elsewhere

HAB. Dry or moist forest and forest edges, thickets, bushland, river banks, woodland, moist grassland, often associated with rocky outcrops, especially limestone, almost always in shade or partial shade; just above sea level to 2000 m; flowering plants have been seen from all months except March, with a slight peak from July through October

SYN. *Cyanotis* sp. A; Faden & Suda in J.L.S. 81: 301–325 (1980)
C. sp. nov.; Robertson & Luke, Kenya coastal forests 2: 88 (1993)

subsp. **robusta** *Faden & D.M. Cameron* in Novon 15: 113 (2005). Type: Kenya, Masai District, Ngerendei, near Mara River, *Glover, Gwynne & Samuel* 463 (K!, holo.; EA!, iso.)

Leaves (1.5–)2–6(–9.6) cm long (the longest generally at least 4 cm long), 0.8–1.5 cm wide; sheaths 5–13 mm long, 3.5–7 mm wide (pressed); inflorescences 5- or more flowered; stamen filament hairs ± 2.5 mm long; capsules 2.6–3.7 mm long; seeds 1.3–2 mm long; testa brown. Fig. 2/13, p. 29.

UGANDA. Mubende District: Wattuba, 13 Apr. 1970, *Katende* K95! & 10 km NW of Katera, 16 Mar. 1969, *Lye* 2329!; Mengo District: 13 km W of Lwampanga, 9 July 1956, *Langdale Brown* 2178!

KENYA. Kisumu–Londiani District: Kisumu, Nov. 1939, *Opiko* in *Bally* 664!; Masai District: Masai Mara Reserve, Telek River, 9 Sep. 1947, *Bally* 5269! & Masai Mara, near research station, 9 July 1979; *M.G. Gilbert* 5734b!

TANZANIA. Bukoba District: Ruasina [Rwasina] to Kiziba, no date, *Ford* 493!

DISTR. **U** 2, 4; **K** 5–6; **T** 1; ?Gabon, Congo–Kinshasa, Rwanda

HAB. Thickets, thicket edges, forest edges, stream banks, lake edges, cultivation, sometimes associated with rocks, almost always in shade or partial shade; 700–1850 m

Flowering plants have been seen from January, March, April, July, September and October.

SYN. *Cyanotis somaliensis* of hort. pro parte, not of C.B. Clarke (1895)
C. sp. 'B' in Faden & Suda, J.L.S. 81: 305, 311 (1980)
C. sp. 'A' in Faden, U.K.W.F. 2nd ed., 306 (1994)

NOTE. (on the species as a whole). This species has been long confused in the herbarium with *Cyanotis foecunda*, which differs in having a definite base, at least sometimes arising from a deeply buried corm; longer vegetative leaves, shoots rooting only at the base, if at all, flowering shoots always determinate, ending in an inflorescence, with leaves decreasing in size distally on the often zigzag flowering shoots.

Some plants of *Cyanotis repens* subsp. *robusta* has been in horticulture for many years under the name "Cyanotis somaliensis". That species however, which is confined to northern Somalia has a definite basal rosette and the flowering shoots have determinate growth.

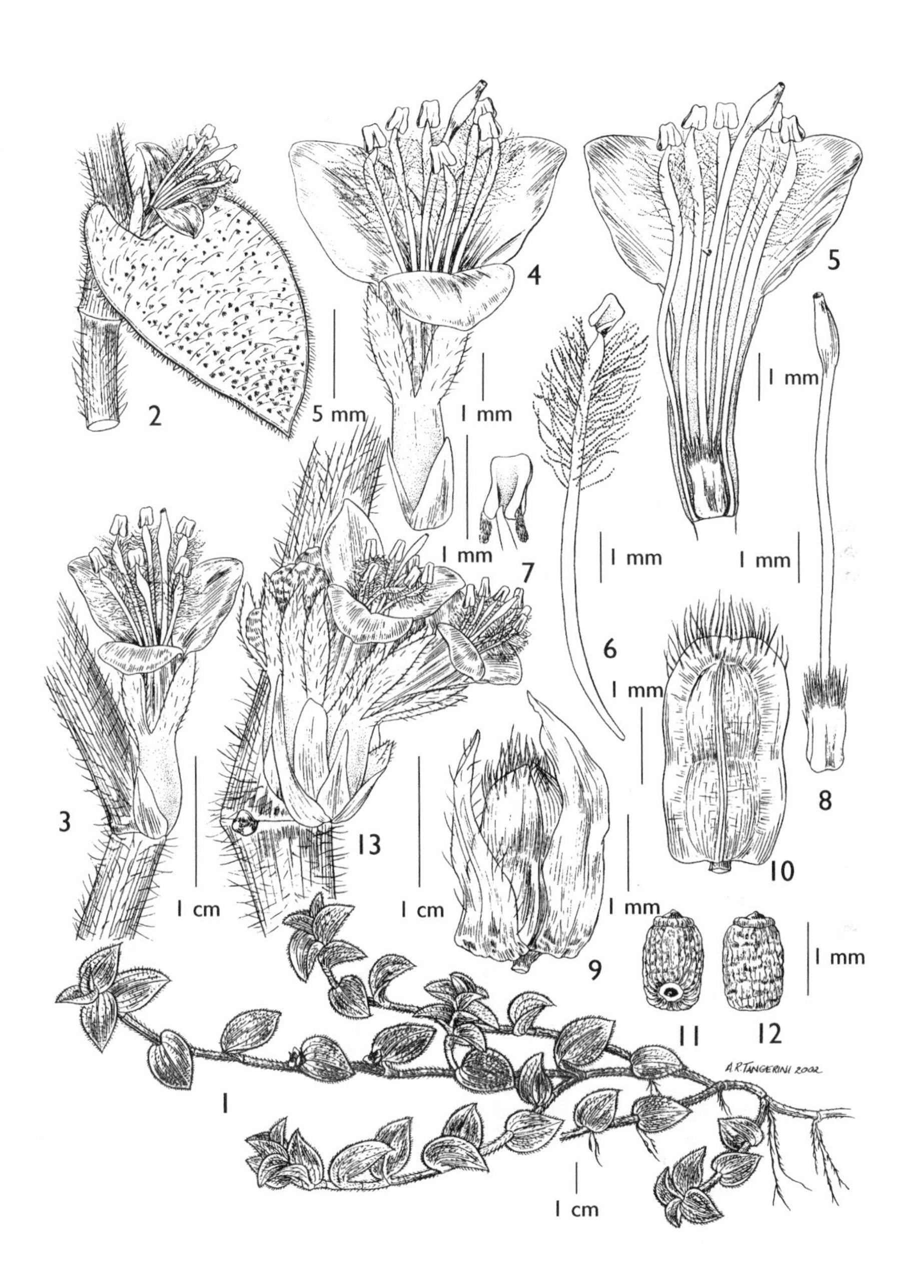

FIG. 2. *CYANOTIS REPENS* subsp. *REPENS* — **1**, habit; **2**, inflorescence; **3**, inflorescence, leaf removed; **4**, flower, lateral view; **5**, flower, front half removed; **6**, stamen; **7**, anther; **8**, gynoecium; **9**, capsule; **10**, capsule, sepals removed; **11**, seed, ventral view; **12**, seed, dorsal view. subsp. *ROBUSTA* — **13**, inflorescence. 1–12 from *Cameron* 02/02 ex *Faden & Faden* 74/1174, cultivated; 13 from *Cameron* 02/07 ex *Faden, Evans & Lye* 69/1067, cultivated. Drawn by Alice Tangerini, reproduced with permission from Novon 15: 111 (2005).

Cyanotis repens, *C. foecunda* and *C. somaliensis* form a natural group in Africa. They share strictly sessile inflorescences, a glabrous style and basic chromosome of $x = 13$. Subsp. *repens* is always a diploid whereas subsp. *robusta* is either tetraploid or hexaploid (Faden & Suda, 1980; Faden & Cameron, 2005), but there are only two published counts of the latter.

Archbold 2228 describes the stamens as being either glabrous or with a few hairs [on the filaments]. I have never seen this and would consider this an atypical population. Perhaps the glabrous 'filament' was actually the style!

Faden & Faden 74/1174 is unusually robust for subsp. *repens* with leaves to 1.7 cm wide (widest for the species as a whole) and leaf sheaths to 4 mm wide (widest for subsp. *repens*), but there is no other evidence that it does not belong to subsp. *repens*.

Lind 345 from Kapchorwa, Uganda is far out of the range of the other collections of subsp. *repens* but it is not separable from the plants of eastern Kenya and Tanzania. Moreover, it is quite distinct from the robust plants that it is geographically closer to, although it actually occurs outside the range of subsp. *robusta*. Curiously, this collection is from the highest elevation for the species as a whole (unless the vaguely recorded *Scott Elliot* 7813A "Ruwenzori, 6000–7000 ft" was collected at the upper part of this altitudinal range) and the habitat, damp grassland, also has not been mentioned by any other collector. However, these peculiarities shed no light why this populations occurs at this location.

The species is only known as a cultivated plant in **K** 4.

10. **Cyanotis speciosa** (*L.f.*) *Hassk.*, Commel. Ind.: 108 (1870); Obermeyer & Faden in F.S.A. 4, 2: 53, t. 14 (1985). Type: South Africa, Cape of Good Hope, *Thunberg* s.n. (LINN 406.8, holo.)

Perennial with a vertical or sigmoid rhizome or with a basal bulb surrounded by old papery long-hairy leaf-bases; roots scattered along the rhizome, thin or clustered at the base of the bulb, 2–5 mm thick and ± tuberous, rarely cord-like; shoots dimorphic: a central shoot consisting of a sterile basal rosette of leaves and 1–3 peripheral elongate flowering shoots; sterile basal rosette not elongating, rosette leaves with sheaths to 6.5 cm long; lamina linear, to 40 × 1–2.5 cm, apex acute to acuminate, mucronulate, adaxial surface sparsely pilose to glabrescent except for a dense inframarginal band of hairs, occasionally densely pilose, abaxial surface subglabrous to sparsely or densely pilose. Flowering shoots not rooting, erect to ascending, unbranched, 10–45 cm long, (3-)4–6(–7)-noded, with 1–2(–3) cauline leaves below the similar inflorescence bracts; internodes greatly reduced distally, the proximal ones to 17 cm long; cauline leaves with sheaths to 2.5 cm long; lamina planar to complicate, linear-lanceolate to lanceolate-elliptic or rarely ovate, (1.5–)3–11 cm × 0.5–1.5(–2.5) cm, base usually narrowed directly into the sheath, rarely rounded, margins densely ciliate, apex acute to acuminate, surfaces similar to rosette leaves. Inflorescences consisting of a terminal, ± opposite pair of sessile cincinni, and up to 3 sessile, axillary inflorescences developing basipetally from the distalmost leafy bracts just below the terminal inflorescence; shoot axis often strongly zigzag distally; inflorescence bracts similar to the cauline leaves, becoming reflexed with age; individual inflorescences capitate, 1–3 cm in diameter; bracteoles biseriate, narrowly lanceolate to lanceolate-oblong, (3.5-)5–6 × 1–1.5 mm, margins ciliate, surface glabrous to densely pilose. Sepals 5.2–9.5 mm long, fused proximally for ± 1–2 mm, strongly keeled, lobes obovate, 1.5–2 mm wide, margin ciliate, rarely eciliate, surface sparsely to densely pilose, apex acute; corolla tube ± 4.5 mm long, lobes pale to dark blue, broadly ovate, ± 3–3.5 × 4 mm. Filaments 11–12 mm long, densely bearded for the distal 4 mm, with a slight swelling just before the apex; anthers yellow to orange, 1–1.8 mm long. Ovary ± 1.5 mm long, densely bearded distally; style 9–10 mm long, with a subapical swelling, glabrous. Capsules dark brown, trilocular, trivalved, oblong-ellipsoid, 2.5–3 × 1.5–1.9 mm, with an apical tuft of white hairs, sometimes hairy along the capsule angles and sutures distally. Seeds 2 per locule, ovoid, 1.2–1.4 × 0.8–1.2 mm, testa grey to grey-brown, longitudinally striate, with scattered pits on both surfaces.

SYN. *Tradescantia speciosa* L.f., Suppl. Pl.: 192 (1782)
T. nodiflora Lam., Encycl. 2: 371 (1786). Type: South Africa, Cape of Good Hope, *Sonnerat* s.n. (P-LAM, holo.)
Commelina speciosa (L.f.) Thunb., Prodr. Pl. Cap.: 58 (1794)
Tonningia speciosa (L.f.) Raf., Fl. Tellur. 2: 16 (1837)
Cyanotis nodiflora (Lam.) Kunth, Enum. 4: 106 (1843); Hook. in Curtis's Bot. Mag. 29: t. 5471 (1864); Clarke in DC., Monogr. Phan. 3: 257 (1881) & in Fl. Cap. 7: 14 (1897) & in F.T.A. 8: 82 (1901); Brückner in E.& P. Pf. ed. 2, 15a: 167, fig. 62A (1930); Schreiber in Fl. SW Afr. 157: 10 (1969)
Tonningia nodiflora (Lam.) Kuntze, Rev. Pl. 2: 722 (1891)

subsp. **bulbosa** *Faden* **subsp. nov**. A subspecie typica bulbo basali vaginis papyraceis foliorum veterum plerumque pilis longis brunneolis appressis vestis obtecto, basi radicibus plerumque plus minusve tuberosis fasciculatis differt. Type: Zambia, Mpika District: Mpika–Kasama road, 16 km towards Mpika from crossing of Chambeshi River, *R.B. Faden & A.J. Faden* 74/120 (US!, holo.; K!, MO!, iso.)

Perennial with a basal bulb surrounded by old papery long-hairy leaf-bases, roots clustered, usually more or less tuberous.

TANZANIA. Ufipa District: Chapota, 4 Dec. 1949, *Richards* 2030!; Chunya District: North Lupa Forest Reserve, 27 Dec. 1962, *Boaler* 772!; Iringa District: Sao Hill, 15 Jan. 1965, *Chambers* 34!
DISTR. **T** 1, 4, 7; Burundi (?), Congo–Kinshasa, Angola(?), Zambia, Malawi, N Mozambique
HAB. Grassland, *Brachystegia* woodland, rocky hillsides, edge of termite mound, grassy roadsides; 1400–2250 m; flowering November to February but mainly in December

NOTE. The typical subspecies ranges from southern Mozambique, Zimbabwe and Botswana to the eastern Cape in South Africa and possibly also in Madagascar. It is characterized by a vertical or sigmoid rhizome with thin fibrous roots scattered along it (see F.S.A. 4, pt. 2, fig. 14, p. 54. 1985). The disjunct plants from Namibia are keyed out as having thin roots (Schreiber, 1967), so they probably belong to the typical subspecies. Specimens seen from Burundi and Angola lack bases but, from their distributions, they should belong to subsp. *bulbosa*.

I have never studied this species in the field or grown it in cultivation. Thus its phenology is uncertain to me, such as whether the basal rosette is permanent or is replaced annually and, if the latter, when that happens relative to the development of the flowering shoots.

The stamens appear to spread wider than the corolla, so it is difficult to know whether the flower color recorded by collectors is that of the stamen filament hairs, the corolla lobes or both, unless particularly stated by the collector. A few collectors in the Flora Zambesiaca area have noted a contrasting color between the filament hairs and the corolla lobes, but it is unknown whether this occurs in FTEA.

We have no information about flowering times in our area, but the label of *Pawek* 10539 from Malawi indicates that the flowers had not opened by noon but were open by 3 p.m, indicating an afternoon flowering period, which, if confirmed, would be highly unusual in *Cyanotis*.

3. COLEOTRYPE[2]

C.B. Clarke in DC., Monogr. Phan. 3: 120, 238, t. 8 (1881); G.P. 3: 851 (1883); Brenan in K.B. 7: 456 (1953); Perrrier, Not. Syst. Paris 5: 196 (1936); Morton in J. Linn. Soc. 60: 171, Fig. 1 (1967); Faden in Kubitzki, Fam. Gen. Vasc. Pl. 4: 120 (1998)

Perennial (in our species) or annual herbs. Leaves spirally arranged. Inflorescences axillary, perforating the leaf sheaths, sessile, capitate, consisting of 1-several congested and contracted cincinni and subtended by a series of bracts.

[2] by Robert B. Faden and Daniel J. Layton

Flowers sessile, actinomorphic (in our species) or zygomorphic. Sepals subequal, sepaloid, free (in our species) or connate. Petals equal (in our species) or unequal, united below into a narrow tube. Stamens 6, equal or unequal, filaments adnate to the corolla tube, then free (in our species) or connate above the attachment to the corolla tube, bearded or glabrous, anthers basifixed, dehiscing longitudinally or by apical pores. Capsules ± hidden among the bracts and sepals, 3-locular, 3-valved, dehiscent, locules 1–2-seeded. Seeds enclosed in a thin aril that becomes papery, hilum linear, embryotega lateral.

10 species; 6 in Madagascar and 4 in Africa.

Because the inflorescences are so contracted the bracts at their bases are difficult to interpret morphologically. The basal or outermost bract of the inflorescence is a prophyll, a bract that subtends most branches in monocots. The much smaller and narrower bract that precedes the first flower in a cincinnus is the cincinnus bract. How to term the bracts between the prophyll and cincinnus bract is unclear as is their constancy in number and form. Perrier (1936) points out that the bracts gradually diminish in size with the innermost very similar to the sepals.

We describe the papery layer that surrounds the seed as an aril for the first time because in liquid preserved material this layer completely surrounds the seed and is attached only to and presumably arises from the hilum. We believe that, when still moist, it may function to facilitate the extrusion of the pair of seeds from a locule by the basal contraction of the capsule when it dehisces at maturity. When the seeds are exposed the aril quickly dries and ceases to function, we believe. Thus, unlike typical arils, that in *Coleotrype* appears not to be further involved in seed dispersal. This dry aril/papery layer has been observed in the seeds of all four African species of *Coleotrype*. In *C. laurentii* it forms an appendage at one end of the seed in 2-seeded locules or at both ends of the seed in 1-seeded locules. Fruiting material of the Malagasy species has not been examined, so it is unknown whether such a layer occurs in their seeds.

1. Flowers pink-purple; leaves (2.4–)3–7 × 1.5–3.7 cm, variegated with 2 longitudinal, silvery bands above; **T** 7 ... 3. *C. udzungwaensis*
   Flowers white; leaves 8–22 × 2.5–6 cm, not variegated ... 2
2. Leaves 8–15 × 2.5–4.5 cm; inflorescences 1–1.5 cm long, to 1 cm wide; anthers creamy white, –3.5 mm long, dehiscing by apical pores; **K** 7, **T** 6, 7 ... 1. *C. brueckneriana*
   Leaves 15–22 × 3–6 cm; inflorescences 2–2.5 × 2–3.5 cm; anthers yellow, –1.5 mm long, dehiscing longitudinally; **U** 2, 4, **T** 1 ... 2. *C. laurentii*

1. **Coleotrype brueckneriana** *Mildbr.* in N.B.G.B. 13: 411 (1936); Luke in J. E.A. Nat. Hist. Soc. 94: 97 (2005). Type: Tanzania, Ulanga District: Mahenge area, Sali, *Schlieben* 1910 (B!, holo.; B!, BR! (photo K, US), LISC!, iso.)

Decumbent perennial; shoots arising to 30 cm, often rooting at lower nodes; roots thin, fibrous; internodes to 7 cm long, completely covered by the sheaths distally, glabrous. Leaves spirally arranged, well spread out along the stem or, less commonly, clustered towards the shoot apex; sheaths to 2 cm long, glabrous or sparsely pubescent, apex ciliate; lamina elliptic or narrowly elliptic to lanceolate-elliptic or rarely obovate, dull green, petiolate, 8–15 × 1.7–4.5 cm, base cuneate, margins undulate, scabrid distally, apex (acute to) acuminate; adaxial surface glabrous or sometimes pilose along the midrib, abaxial usually with an narrow inframarginal band of appressed minute hairs, occasionally pilose on the midrib proximally; petiole to 3 cm long, ciliate. Inflorescences 1–1.5 × 1 cm, of 1–2 cincinni; prophyll cup-shaped, perfoliate, pubescent marginally with uniseriate hairs, puberulous abaxially, outer bract ± reniform, ± 6 × 10 mm, ciliolate to ciliate, puberulous abaxially, other bracts broadly ovate to lanceolate-oblong, 7–9 × 4–9 mm, ciliate, puberulous abaxially; bracteoles lanceolate-oblong, up to 9.5 × 3.5 mm, ciliolate and with an apical tuft. Flowers bisexual. Sepals free, linear to linear-lanceolate, 7–10 × 0.5–1.5 mm, ciliolate, sometimes with an apical tuft of hairs, outermost one strongly hooded, carinate, ± 1 mm wide, apex acute, puberulous along keel, innermost one 1.5 mm wide, not hooded, apex rounded, boat-shaped but not carinate, puberulous abaxially. Corolla white, tube 6.5–10.5 mm long, lobes strongly recurved, oblong to oblong-elliptic, 4–5.5 × 2 mm, apex rounded, slightly hooded. Stamens 6, equal, filaments adnate up to the top of the corolla tube, free above, free portion of filaments 2–3.5 mm long, often curving outward, bearded with moniliform hairs, anthers creamy white, 2–3.5 × 0.7–0.8 mm, dehiscing by apical pores. Ovary sessile, trilocular, oblong-elliptic, 1.5–2.5 × ± 1 mm, puberulous distally, locules 2-ovulate; style exceeding stamens, 14–15 mm long, extending 5–7 mm above corolla tube; stigma capitate. Capsules 3–5-seeded, pale yellow, ellipsoid before dehiscence, 7–10 × 3–5 mm, rostrate, pubescent in distal half, valves spreading at maturity; locules 1–2-seeded. Seeds broadly ovoid or ovoid-deltoid to reniform, 2.7–4.7 × 1.9–2.6 mm, enclosed in a close-fitting papery aril that lacks an appendage; testa (beneath the papery layer) reticulate-alveolate to reticulate-foveate, lustrous brown-maroon, sometimes with white particles in the depressions.

Kenya. Kilifi District: Pangani "Kaya" Forest, 19 Nov. 1978, *Brenan et al.* 14587! & Pangani, crossing of Lwandani Stream on Chonyi-Ribe road, 17 Feb. 1977, *Faden, Faden, Gillett & Gachathi* 77/533!; Kwale District: Shimba Hills Nature Reserve, Mwele Forest, 25 Oct. 2001, *Luke & Luke* 8193!

Tanzania. Morogoro District: Kimboza Forest Reserve, 4 July 1970, *Faden, Evans & Pócs* 70/355! & Uluguru Mts, Kimboza Forest Reserve, 4 Apr. 1974, *Faden & Faden* 74/415!; Iringa District: Udzungwa Mountain National Park, Mizimu, 16 July 2003, *Luke, Luke & Arafat* 9509!

Distr. **K** 7; **T** 6, 7; not known elsewhere

Hab. Moist forests, often on limestone, sometimes in wet places such as river banks and swamps; 150–1100 m

Flowering specimens have been seen from March, April, July and possibly October.

Note. This rare species is known from only seven collections from five localities, two in Kenya and three in Tanzania. Its white flowers suggest a relationship with the only other white-flowered species, *C. laurentii*, but its narrow anthers with poricidal dehiscence might indicate otherwise. Some Madagascan species, e.g. *C. goudotii* C.B. Clarke, also have narrow anthers, but it is not clear from pressed flowers whether their dehiscence is poricidal or longitudinal. In either case, in *C. brueckneriana* the filaments are separately attached to the corolla tube as opposed to partially connate in the Madagascan species, which clearly aligns *C. brueckneriana* with the African, not the Madagascan taxa in the genus. Its anther dehiscence therefore places it in a unique position in the genus.

C. brueckneriana is easily separated from *C. laurentii* by the smaller size of the plants, smaller leaves and inflorescences and especially by the form, colour and dehiscence of the anthers.

Luke & Luke 8193 is unusual in having all of its leaves clustered at the shoot apex.

2. **Coleotrype laurentii** *K. Schum.* in E.J. 33: 377 (1903), as *Coleotripe*; De Wild., Pl. Bequaert. 5: 229 (1931); Brenan in K.B. 7: 456 (1953); Aké Assi, Contrib. 2: 222, t. 19 (1963); Morton in J.L.S. 60: 171, Fig. 1 (1967); Brenan in F.W.T.A., 2nd ed.: 3: 36 (1968). Type: Cameroon, Lobe River by Batanga, *Dinklage* 1223 (B!, syn.); Congo–Kinshasa: Kassai, Lulua and Sankurru [rivers], Nov. 1895, *Laurent* s.n. (B?, BR!, syn.)

Decumbent perennial; shoots arising to 1.2 m, but usually less, glabrous; roots thin, fibrous. Leaves spirally arranged, well spread out along the stem; sheaths to 3.5 cm long, surface glabrous, apex ciliolate; lamina lustrous, elliptic or narrowly elliptic, 15–22 × 3–6 cm, flat or slightly undulate, base attenuate into an almost winged petiole, margins ciliolate proximally, scabrid distally, apex subcaudate; adaxial surface with a narrow, inframarginal band of very short hairs, otherwise glabrous,

FIG. 3. *COLEOTRYPE LAURENTII* — **1**, habit, × $^1/_6$; **2**, inflorescence, × 1; **3**, infrutescence, × 1. *CYANOTIS BARBATA* — **4**, habit, × 1; **5**, flower, × 2; **6**, corolla, × 2; **7**, cincinnus, × 3; **8**, capsule, sepals removed, × 6; **9**, seeds, side view, × 6; **10**, seed, top view, × 6. Drawn by M. Grierson, reproduced with permission from J.L.S. 59: 353 (1966).

abaxial surface glabrous. Inflorescences 2–3 × 2–3.5 cm, of 3–8 cincinni; outer bracts ± reniform, 1–1.5 × 1.2–2 cm, broadly attached at base, sparsely ciliolate, otherwise glabrous, inner bracts and bracteoles carinate, at least distally, ovate to oblong, 0.5–2 × 0.4–0.8 cm wide, glabrous; cincinni mutiflowered. Flowers bisexual. Sepals free, linear-oblanceolate to oblanceolate-oblong, 12–15(–16 in fruit) × 1.5–3 mm, ciliolate distally, apex hooded, carinate, pilose along keel, outermost sepal most strongly hooded with large apical tuft, innermost sepal least hooded with apex sparsely pubescent. Corolla white, tube 10–12 mm long, lobes strongly recurved, oblong-elliptic, 6–9 × 2–3 mm, margins recurved, apex rounded. Stamens 6, subequal, filaments adnate up to the top of the corolla tube, free above, free portion of filaments 3–4 mm long, glabrous or sparsely bearded proximally, anthers yellow, 1–1.5 × ± 1.1 mm, dehiscence longitudinal, introrse. Ovary trilocular, 2–3 × 1–1.5 mm, densely pilose distally, locules 2-ovulate; style 13.5–16 mm long, extending 5.5–7.5 mm above corolla tube, ± equal in length or exceeding the stamens, curved to one side, stigma capitate. Capsules sessile, cylindrical-trigonous, ± 11 × 4.5 mm, apex and base rounded, pilose distally, style base persistent, locules 1–2-seeded. Seeds ellipsoid, 2.4–2.8 (not including aril appendage) × 1.6–1.9 mm, completely enclosed in a thin, papery, tan aril, which is drawn out into a ± linear appendage 0.8–1.4 mm long, apical or basal on the seeds (or at both ends in 1-seeded locule seeds), seeds of 2-seeded locules adherent by their arils and shed together, surface under papery aril reticulate-alveolate to reticulate-foveate, red-brown. Fig. 3/1–3, p. 34.

UGANDA. Bunyoro District: Budongo Forest Reserve, between the Nature Reserve and the Royal Mile, 29 Aug. 1995, *Poulsen, Nkuutu & Dumba* 880!; Kigezi District: Ishasha Gorge, 5 km SW of Kirima, 21 Sep. 1969, *Faden, Evans & Lye* 69/1191!; West Mengo District: Mpanga Forest Reserve and Research Station, 5 km E of Mpigi, 9 Sep. 1969, *Faden, Evans & Lye* 69/1008!.

TANZANIA. Bukoba District: Minziro Forest Reserve, Nyakabanga sub-village, W of Kagera River and Sinja sub-village, 15 Oct. 2000, *Festo & Bayona* 811!

DISTR. **U** 2, 4; **T** 1; Ivory Coast, Ghana, Nigeria, Cameroon, Gabon, Central African Republic, Congo–Brazzaville, Congo–Kinshasa

HAB. Forest and swamp forest; 700–1500 m

Flowering specimens have been seen from September. Flowers are said to be open from 11:00 to 15:00 (by Brenan, 1968).

NOTE. The almost caudate leaf apex, a unique feature in African Commelinaceae, makes this species easy to spot in the field, even when vegetative. At least that was our experience in Uganda in 1969, when we collected *C. laurentii* five times. The only prior collection from our area appears to have been *Chandler* 2170 from Uganda, made in 1938. This species is almost certainly much more common throughout its range than the available collections would indicate. It is easily overlooked because the small flowers are borne some distance below the shoot apex, so the plant may appear sterile from above. Old infructescences may be borne on the prostrate part of the stem.

The leaves can be considered either petiolate, with a long, winged petiole or else subsessile, with the lamina long-attenuate at the base.

The linear appendage(s) of the aril are very distinctive. It is hard to imagine what function they could serve. The papery aril appears to cling to the seeds quite tightly in this species, more so than in the other species, and does not seem to rub off very readily.

3. **Coleotrype udzungwaensis** *Faden & Layton* **sp. nov.** a *C. brueckneriana* et *C. laurentii* foliis parvioribus floribusque roseo-purpureis differt, a *C. natalensis* floribus multo parvioribus foliisque basi haud attenuatis differt, a speciebus omnibus tribus foliis in pagina superiore vittis duabus lateralis argenteis notatis recedit. Type: Tanzania, Udzungwa Mountain National Park, Pt. 224–pt. 226, 7°46'S, 36°50'E, *Luke, Mwangulango, Butynski, Ehardt, Kingdon & Kimaro* 7913 (EA!, holo; K!, NHT!, US!, iso.)

Decumbent perennial; shoots arising to 10 cm, sparsely branched, glabrous; roots thin, fibrous. Leaves spirally arranged, clustered toward shoot apex; sheaths 0.2–1 cm long, surface puberulous, striped vertically with 6–9 maroon stripes, apex ciliate; lamina petiolate, ovate to elliptic or obovate, (2.4–)3–7 × 1.5–3.7 cm, flat to slightly undulate, base rounded to cuneate, margins not to slightly undulate, ciliate proximally, apex acute to acuminate, mucronulate; adaxial surface light green marginally and medially with two longitudinal silvery bands in between, glabrous, abaxial surface purple-red except along margin and midrib where pale green (not visible on dried specimens), glabrous except for a narrow submarginal band of appressed minute hairs; petiole to 1 cm long, ciliate. Inflorescences 0.5–1 × ± 0.5 cm (wider with capsule), of 1–2 cincinni, each cincinnus apparently 2-flowered, prophyll puberulous abaxially, perfoliate; outermost bract broadly ovate to reniform, 3 × 4.5 mm, margins ciliate, surface glabrous; other bracts ovate to ovate-elliptic, 5.5–6.5 × 2.5–3 mm, puberulous abaxially, margins ciliate; bracteoles lanceolate to oblanceolate, 6.5–10 × 1–4 mm, boat-shaped, carinate, surface puberulous, keel with line of short hairs, apex acute, margins ciliate with long contorted hairs. Flowers bisexual. Sepals free, lanceolate to lanceolate-elliptic, 8–9 × 1.5–2 mm, subcucullate, carinate, margins ciliate with contorted hairs, surface puberulous, pubescent along keel, outer sepal proportionally broader and less hooded. Corolla pink-purple, tube ± 8 mm long, lobes ascending, sometimes recurved distally, ovate or ovate-elliptic to lanceolate, 3.5–5 × 1–2.5 mm, sometimes convexo-concave, hooded apically, apex rounded. Stamens 6, variable (see Notes), filaments fused to corolla tube either up to throat or slightly (up to 5 mm) down the tube (sometimes all attached below the throat), free portion of filaments (0–)1.2–4 mm long, sparsely bearded with concolourous, moniliform hairs, anthers 0.7–1.2 × 0.8–0.9 mm, dehiscence longitudinal, introrse to latrorse, pollen yellow. Ovary 1.75–3 × 1 mm, pubescent distally, puberulous elsewhere, locules 2-ovulate; style 8–10 mm long, equaling or exceeding stamens; stigma capitate. Capsule obovoid to broadly ellipsoid, 4–7 × 3–4 mm, rostrate, densely puberulous throughout, locules 1–2-seeded. Seeds ovate-deltoid to ± reniform, 2.9–4.3 × 1.9–2.2 mm, surrounded by a papery aril that is rounded at both ends and lacks appendages, the aril apparently rubbing off at maturity, testa (beneath the papery aril) reticulate-foveate, red-brown, pits and some ridges often with pale grey mealy material.

TANZANIA. Iringa District: Udzungwa Mountain National Park, Pt 224–pt 226, 7°46'S, 36°50'E, 28 Sep. 2001, *Luke et al.* 7913!

DISTR. **T** 7; known only from the type collection

HAB. Stream in montane forest; 1550 m

The specimen was in fruit in September; flowers were observed only in cultivation.

NOTE. With their two longitudinal, silvery stripes above and purple lower surface, the leaves of *Coleotrype udzungwaensis* bear a striking resemblance to those of the common variegated form of *Tradescantia zebrina* Bosse (*Zebrina pendula* Schnizl.). They are completely different in pattern and size from those of the other two *Coleotrype* species in our area. The pink-purple flowers, observed in cultivation, also readily separate this species from the white-flowered *C. brueckneriana* and *C. laurentii*. The only African species with similar-colored flowers to *C. udzungwaensis*, *C. natalensis* of Zimbabwe, Mozambique and South Africa, has much larger flowers (corolla tube ± 1.5 cm long; lobes broadly ovate to orbicular, 1–1.5 cm long and wide; stamen filaments free above the corolla tube for 5–9 mm, densely bearded their entire length, style 2–2.5 cm long), and proportionally narrower leaves with attenuate bases.

Perhaps the best account of the androecium in *C. udzungwaensis* is contained in a detailed description of the entire plant supplied by the collector, Quentin Luke, based on material of *Luke et al.* 7913 that he cultivated in Nairobi, Kenya. He describes the 6 stamens as "seemingly inserted at different levels in the corolla tube, possibly two at 4 mm, two at 4.5 mm and two at 5 mm; filaments 3.5 mm long, with long, purple hairs in a tangle for approximately 1 mm, some 0.5 mm from the base; anthers pale buff, 0.75 mm long, not exserted from corolla lobes, opening by longitudinal slits." In an image of a flower supplied by Luke the 3 antesepalous stamens appears to be slightly shorter than the antepetalous stamens and the stigma is borne at the level of the shorter stamen anthers. There is nothing in Luke's description or image that presages the irregularities that we observed in the flowers from our cultivated material of the same collection.

Our observations were based on five flowers of cultivated *C. udzungwaensis*. In these flowers the stamen filaments were found to be variable and irregular in length, curvature and point of attachment to the corolla tube. In a dried flower there were five long and one short stamens, the latter apparently fused to the style. No such adnation was observed in spirit collections, but in those flowers at least two filaments were slightly shorter than the others and were attached at the distal end of the corolla tube, whereas some of the longer filaments were attached slightly further down the tube. All of the flowers had one, or in one case two, of the shorter filaments completely fused to the corolla, with the distal end of the filament fused to the margin of one of the corolla lobes, with only the anther free. A further oddity, observed in one flower, was the complete connation of the filaments of two of the longer stamens above the attachment to the corolla.

It is impossible to be certain what the normal variation in androecial morphology is in this species because our sample size is so small. We believe that some if not all of the unusual morphologies that we observed in the androecium may have been teratological. All flowers in the spirit collections that we used had one to several very small mealybugs inside them. Perhaps their feeding activity at an earlier stage might have induced an abnormal development.

Even with the observed oddities in the androecium, *C. udzungwaensis* clearly belongs to the African section of *Coleotrype*. The corolla is actinomorphic, with equal petals, and the stamens, although possibly somewhat unequal, are never fused in the distinctive pattern of the Madagascan species.

The corolla was determined to be 'intense purple-violet' with the RHS Colour Chart (1995 edition).

4. STANFIELDIELLA

Brenan in K.B. 14: 283 (1960); Morton in J.L.S. 60: 207–209 (1967); Faden in Kubitzki, Fam. Gen. Vasc. Pl. 4: 124 (1998)

Perennial herbs; roots fibrous. Leaves spirally arranged, petiolate, vernation involute. Inflorescences terminal, terminal and axillary, or rarely all axillary, thrysiform or reduced to a single cincinnus, axes usually glandular-pubescent. Flowers bisexual, actinomorphic, pedicillate. Sepals free, subequal, sepaloid, usually glandular-pubescent. Petals free, equal, not clawed. Stamens 6, subequal, filaments glabrous, anthers small, connectives narrow, dehiscence longitudinal. Ovary trilocular, locules 2–10-ovulate; style slender. Capsules trilocular, trivalved. Seeds uniseriate, 2–10 per locule, hilum linear, embryotega semidorsal.

An African endemic genus of four species, centered in West and Central Africa, with one species extending to our area.

Stanfieldiella imperforata (*C.B. Clarke*) *Brenan* in K.B. 14: 284 (1960); F.P.U.: 196 (1962); Brenan in J.L.S. 59: 353, fig. 8 (1966); Morton in J.L.S. 60: 207 (1967); Brenan in F.W.T.A. 2nd ed.: 3: 23; Ensermu & Faden in Fl. Eth. 6: 346 (1997). Types: São Tomé and Principe: Principe [Princes Island], *Mann* s.n. (K!, syn.); Cameroon: Cameroon Mountain, *Mann* 1340 (K!, syn.); Angola, Cazengo, Mt Muxaulo, *Welwitsch* 6607 (BM, syn.)

Stoloniferous perennial herb with erect or ascending or occasionally decumbent shoots 15–60 cm tall; internodes with a fine pubescence continuous with that of the sheath above, otherwise glabrous; roots fibrous. Leaves spaced along the stem then clustered subterminally; sheaths 1–2 cm long, with a line of dense pubescence along the fused edge and a few to many long hairs at the summit (these occasionally lacking), otherwise glabrous; lamina elliptic to narrowly elliptic, lanceolate-elliptic or ovate-elliptic, (3.5–)5–13(–18) × (1–)1.5–4(–4.5) cm, base cuneate, margins scabrid apically, sometimes ciliate on the petiole, apex acuminate to acute or attenuate, surfaces commonly discolorous, with the abaxial purple or purple-tinged, usually glabrous, rarely pilose. Inflorescence a subsessile, compound terminal thyrse (occasionally 1–3 additional, separate axillary thyrses present below it), moderately dense, ovoid to broadly ovoid, 1–4 × 1–5 cm, usually of up to 10 cincinni (rarely many

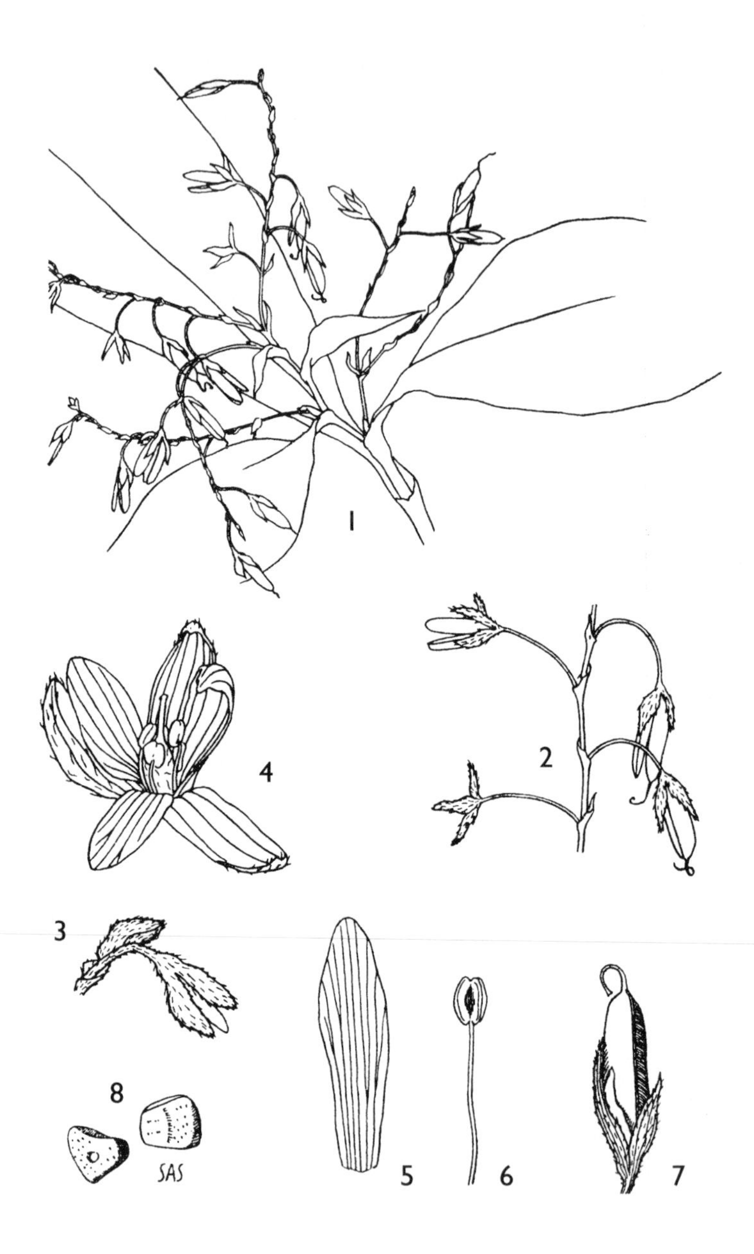

Fig. 4. *STANFIELDIELLA IMPERFORATA* — **1**, inflorescence, × 1; **2**, part of inflorescence, × 2; **3**, flower, × 3; **4**, opened flower, × 6; **5**, petal, × 6; **6**, stamen, × 7; **7**, capsule, × 3; **8**, seeds. × 10. 1–8 from *Brenan & Richards* 8832, Nigeria. Drawn by S.A.S. and reproduced with permission from K.B. 14: 282 (1960). "Modified from the original by Alice R. Tangerini."

more), peduncles of separate inflorescences up to 3.7 cm long; cincinni to 2.5 cm long and 10-flowered; lower cincinnus bracts foliaceous, the upper bract-like; bracteoles spaced 1–4 mm apart, ovate to ovate-lanceolate, amplexicaul, not perfoliate, 1.5–2.5 mm long, glabrous or glandular-pubescent, persistent. Flowers bisexual, pedicels 3–8 mm long, strongly declinate in fruit, glabrous or glandular-pubescent. Sepals lanceolate to oblong-lanceolate, 3.5–5 × 1–1.8 mm, glabrous or glandular-pubescent, sometimes reddish purple. Petals white (rarely pink), elliptic or oblanceolate-elliptic, 4–5 × 1.5–1.8 mm. Filaments ± 3.5 mm long, anthers ± 0.5 mm long. Style 2.5–3 mm long, recurved distally. Capsules lustrous dark brown, sometimes becoming dull grey-brown as they dehisce, terete or slightly trigonous, (4.5–)6–9(–10) × 2–2.5 mm, glabrous or glandular-pubescent. Seeds uniseriate, 5–10 per locule, dorsiventrally compressed, quadrate to rectangular or transversely elliptic in outline, 0.7–1.3 mm × 0.7–0.95 mm, testa pinkish white or pinkish grey, smooth, white-farinose or not, hilum linear-oblong. Fig. 4/1–8, p. 38.

var. **imperforata**; Brenan in K.B. 14: 284 (1960); Ensermu & Faden in Fl. Eth. 6: 346 (1997)

Sepals and pedicels glandular-puberulous.

UGANDA. Kigezi District: Ishasha Gorge, 21 Sep. 1969, *Faden* 69/1197!; Masaka District: Towa Forest, July, 1945, *Purseglove* 1755! & Lake Victoria, Bugala [Bugalla] Island, Sese [Sesse], 5 Oct. 1958, *Symes* 443!

TANZANIA. Lushoto District: between Amani and Monga, 25 May 1950, *Verdcourt* 212! & Amani–Kwamkoro road, 3.3 km SW of Amani, 25 July 1953, *Drummond & Hemsley* 3441!; Morogoro District: Nguru South Forest Reserve, head of valley behind Mhonda Mission, 6 Feb. 1971, *Mabberley* 698!

DISTR. **U** 2, 4; **T** 1, 3, 6; Sierra Leone, Liberia, Ivory Coast, Ghana, Nigeria, Cameroon, Bioko, São Tomé, Principe, Gabon, Central African Republic, Congo–Kinshasa, Ethiopia and Angola

HAB. Rainforest, often in open or disturbed situations such as glades and roadsides, often along streams, sometimes in rocky places, moderate to dense shade; 500–1450 m

SYN. *Buforrestia imperforata* C.B. Clarke in DC., Monogr. Phan. 3: 234, tab. 7 (1881) & in F.T.A. 8: 76 (1901)

B. minor K. Schum. in P.O.A. C: 136 (1895). Type: Tanzania, Lushoto District: E Usambara Mts, Nguelo, *Holst* 2280a (B, holo., of which K, photo!; K!, iso.)

NOTE. This is much less common than var. *glabrisepala* in our area (just the opposite of West Tropical Africa), except in the Eastern Usambara Mountains, where it has been collected frequently.

var. **glabrisepala** (De Wild.) Brenan in K.B. 14: 285 (1960). Syntypes: Congo–Kinshasa: Barumbu, *Bequaert* 1054 (BR, K!); Banalia, *Bequaert* 1401 (BR) & *Bequaert* 1440 (BR); Penghe, *Bequaert* 2508 (BR); Lesse, *Bequaert* 3203 (BR); Vemba Valley, near Kisali (Mayumbe), *Goossens* 1425 (BR); Musinga, *Claessens* 497 (BR, K!); environs of Bikoro (Lake Tumba), *Goossens* 2325 (BR) & *Goossens* 2344 (BR); Yembe (Lake Tumba), *Goossens* 2442 (BR); Bolobo (near Eala), *Goossens* 2463 [sic] (BR); environs of Busira, *Goossens* 2626 [sic] (BR); Lomela Valley, near Busanga, Nov. 1921, *Goossens* 2731 (BR), between Lisala and Likimi, Mar. 1914, *Goossens* 6302 (BR) + 32 more.

Sepals and pedicels glabrous.

UGANDA. Toro District: near Kirmia, W of Ntandi in Semliki Forest, 1 km N of Mantoroba Primary School, 23 Sep. 1969, *Faden* 69/1265!; Masaka District: Malabigambo forest, 8.4 km SSW of Katera, 3 Oct. 1953, *Drummond & Hemsley* 4602!; Mengo District: W of Kisubi on Kampala-Entebbe road, 7 Sep. 1969, *Faden* 69/954!

TANZANIA. Bukoba District: Kiamawa, Sep.-Oct. 1935, *Gillman* 462!; Minziro Forest, 19 Oct. 1955, *Willan* 266!

DISTR. **U** 2, 4; **T** 1; Ghana, Cameroon, Central African Republic, Congo–Kinshasa, Sudan, Ethiopia

HAB. Forest, forest edges and paths, sometimes in dense shade; 700–1400 m

SYN. *Buforrestia glabrisepala* De Wild., Pl. Bequaert. 5: 224 (1931)
B. imperforata C.B. Clarke var. *glabrisepala* (De Wild.) Brenan in Fl. Spermat. Parc Nat. Garamba 1: 160 (1956)

NOTE. *Lye* 4274, from the Semliki Forest is particularly robust, with leaves to 18 × 4 cm and very strongly developed axillary inflorescences. It also lacks long hairs at the summit of the sheaths.

NOTE (on the species as a whole). *Dawkins* 461 records the "rhizomes" as just below the surface of the soil. In my experience the stolons are always on the surface, but they might be subject to getting covered by debris. The same label records that the "inflorescences vary from pure translucent white throughout to pink due to that colour [being present] on [the] pedicels and sepals."

The only record of flowering times in our area is from *Faden* 69/1009 (var. *glabrisepala*) in which the flowers are noted open at 15:30. This does not agree with the record from F.W.T.A. or our experience in Cameroon in which at least var. *imperforata* is recorded as morning flowering.

In general the pubescence of the bracteoles, pedicels, sepals and ovary/fruit are all correlated. However, the presence of a few glandular hairs on the fruits is not uncommon in specimens of var. *glabrisepala*, so this is not a diagnostic character.

The leaves and sheaths have abundant glandular microhairs that are quite large for this hair type but are nevertheless much smaller than most macrohairs, such as the glandular hairs on the fruits. When viewed under at least 20× glandular microhairs may make these surfaces appear puberulous. However, these hairs are thin-walled, colorless and lie prostrate on the surface. They are normally so inconspicuous, except under high magnification, that they are not considered when pubescence is being described.

5. FLOSCOPA

Lour., Fl. Cochin. 192 (1790); C.B. Clarke in DC., Monogr. Phan. 3: 265–271 (1881); Clarke in F.T.A. 8: 84–88 (1901); Morton in J.L.S. 60: 199–202 (1967); Faden in Kubitzki, Fam. Gen. Vasc. Pl. 4: 124 (1998)

Perennial or annual herbs, shoots erect to decumbent; roots fibrous. Leaves alternate, spirally arranged, petiolate or sessile, base symmetric. Inflorescences terminal and axillary, thyrsiform or reduced to 1–2 cincinni, commonly the upper axillary inflorescences closely associated with the terminal inflorescence and forming a compound inflorescence; axes and bracteoles usually glandular-pubescent, rarely glabrous or eglandular-pubescent. Flowers bisexual, zygomorphic, pedicellate, very small (usually less than 5 mm wide). Sepals free, subequal, sepaloid, usually glandular-pubescent. Petals free, unequal, white, pink to purple or yellow to orange, not clawed, inner (upper) petals elliptic to obovate, outer (lower) petal oblong. Stamens 6, all fertile, upper 3 different in form from the lower 3; filaments glabrous, fused at the base, anthers small, dehiscence longitudinal. Ovary ± stipitate, glabrous, bilocular, locules 1-ovulate; style slender; stigma capitate. Fruits capsular, dehiscent, stipitate, bilocular, bivalved, broadly ellipsoid to discoid, locules 1-seeded. Seeds ellipsoid, usually ribbed, hilum linear, embryotega lateral.

A pantropical genus of about 20 species, with about 15 in Africa; mostly aquatic or semiaquatic herbs, either in forest or open habitats.

1. Inflorescence completely glabrous; seeds smooth . . . . 1. *F. leiothyrsa*
   Inflorescence pubescent; seeds usually ribbed or tuberculate, rarely smooth 2
2. Flowers yellow to orange 3
   Flowers pink to purple or violet, rarely white 5
3. Perennials; leaves mostly 7–26 cm long; sepals 3–4 mm long; paired petals 3.5–5 mm long . . . . 2. *F. tanneri*
   Annuals; leaves mostly 1–7 cm long; sepals 1.5–3 mm long; paired petals 1.6–3 mm long 4

4. Cymes 1–3 per inflorescence; leaves 2–7 mm wide; hairs on the sepals and inflorescence <0.2 mm long; capsule with persistent style base but not beaked. . . . . 3. *F. flavida*
   Cymes (2–)4-many per inflorescence; larger leaves 7–22 mm wide; longer hairs on the sepals >0.5 mm long; capsule with a prominent beak. . . . . 4. *F. tuberculata*
5. Leaves petiolate . . . . 5. *F. africana*
   Leaves narrowed or not basally, but not distinctly petiolate 6
6. Annuals rooted only at the base; seeds with 20–23 clear radial ribs . . . . 6. *F. polypleura*
   Perennials or annuals, often decumbent and rooted along the lower nodes; seeds with 11–18(–20) radial ribs, rarely ribs obscure to lacking 7
7. Lamina narrowed towards the sheath, not at all amplexicaul; seeds with 11–15(–16) faint ribs . . . . 7. *F. confusa*
   Lamina not at all narrowed at the base, base rounded and ± amplexicaul; seeds with 12–18(–20) usually well-marked ribs (rarely the ribs obscure to lacking) . . . . 8. *F. glomerata*

1. **Floscopa leiothyrsa** *Brenan* in K.B. 7: 206 (1952) & in F.W.T.A. 2nd ed. 3, 1: 27 (1968); Obermeyer & Faden in F.S.A. 4, 2: 58 (1985). Type: Tanzania: Dodoma District, Lake Chaya, *Burtt* 3802 (K!, holo.)

Annual herb 20–100 cm tall; roots thin, fibrous; shoots ± erect to ascending or decumbent and rooting at the lower nodes, simple or branched, internodes glabrous or nearly so. Leaves sparse; sheaths 1–1.5 cm long, glabrous; lamina narrowly lanceolate, 35–80 × 4–8 mm, the base passing directly into the sheath, margins sometimes undulate, scabrid distally, the apex acuminate-attenuate, surfaces glabrous. Inflorescences terminal on the main shoot and lateral shoots, a ± compound thyrse, lax, 4–15 × 2.5–6 cm, completely glabrous; lateral shoots with inflorescences ± 2–4 cm long, unbranched or 1–3-branched; cincinni curved, moderately densely flowered, naked at base, to 5 cm long and 8 mm wide; bracteoles brown, scarious, suborbicular, 1–1.5 mm in diameter. Flowers secund, bisexual, blue to pale violet or dark purple; pedicels 1–2 mm long, articulated at the base, glabrous. Sepals convexo-concave, elliptic, 2.2–4 mm long, apex obtuse, the outer (upper) sepal a little broader than the inner (lower) sepals, glabrous. Paired (upper) petals obovate-elliptic, ± 3 × 2.2 mm, apex rounded, outer (lower) petal linear-oblong, 3.5 × 0.4–0.5 mm. Stamens 6, 3 with anthersacs subparallel, with the anthers suborbicular, ± 0.6 mm in diameter, with the filaments ± 3 mm long, and 3 with the anthersacs divaricate, with the anthers ± 2 mm long and 1 mm wide. Ovary 1 × 0.8 mm, glabrous, with the base stipitate 0.5 mm long; style 3 mm long. Capsule pale olive or brownish olive, lustrous, subquadrate, inflated, 1.7–2.5(–3.5) × 1.5–2.1 mm, base stipitate with a stalk 0.4–1 mm long, apex truncate to obtuse. Seeds flattened ellipsoid, strongly compressed laterally, 0.9–1.7 × 0.5–0.8 mm, testa black (rarely dark reddish brown), alveolate, rarely smooth, not farinose.

TANZANIA. Dodoma District: Lake Chaya, 16 km W of Kazikazi, 2 July 1996, *Faden et al.* 96/512!; Mpanda District: 14 km on Inyonga–Tabora road, 18 May 2008, *Bidgood et al.* 6885!; Tabora District: 13 km from Koga Village (Ugalla River) towards Tabora, 19 May 2008, *Bidgood et al.* 6914!

DISTR. **T** 4, 5; Mali, Chad, Congo–Kinshasa, Zambia and Botswana

HAB. Lake margins, seasonally wet grassland and roadside ditches; full sun or shade (also known outside our area from marshes and shallow running water); 950–1300 m

This species likely flowers in the morning. Its flowers were found completely faded at 14:00.

NOTE. The seeds of this species, which were unknown to Brenan at the time he described the plant, have proven to be as distinctive as the loose glabrous inflorescence. Their lateral compression, usually black, alveolate testa and the absence of farinose granules are also distinctive. In addition, because of its lateral compression the ventral surface is so small and curved that the seed cannot be made to lie on this surface, which apparently is also unique to this species.

2. **Floscopa tanneri** *Brenan* in K.B. 15: 227 (1961). Type: Tanzania, Mwanza District, Mwanza, Bukumbi, Mwasonge, *Tanner* 1023 (K!, holo.)

Perennial herb, erect or decumbent only at the base, 12–45 cm tall; roots thickly fibrous, all arising from the caudex; stems single or occasionally two together, unbranched except for the inflorsecences, glabrous or subglabrous, but pubescent distally towards the inflorescence. Basal leaves the largest and ± in a rosette; sheaths 0.5–3.5 cm long, glabrous or pubescent, sometimes ciliate at apex; lamina linear-lanceolate, 7.5–26 × 1–2.1 cm, base gradually narrowed into the sheath, margins sometimes slightly undulate, scabrid apically, ciliate basally, apex acute to attenuate; surfaces glabrous or sparsely pubescent. Inflorescences a compound terminal thyrse, moderately dense, 2–8 × 2.5–10 cm, densely glandular-pubescent; lateral branches 1.5–5.5 cm long, simple or producing up to three branches; cincinni (the ultimate branches) densely floriferous up to nearly the base, sometimes shortly naked; bracts 0.4–3 × 0.2–1.2 cm; bracteoles scarious, obovate, 0.7–1.3 × 0.5 mm, margin usually erose or lacerate. Flowers secund, bisexual; pedicels 0.7–1.5 mm long, glandular-pubescent, articulated at the base. Sepals reddish purple to purple or brown, convexo-concave, elliptic, 3–4 × 2–2.5 mm, apex obtuse, densely and shortly glandular-pubescent, the outer sepal a little broader than the inner sepals. Petals yellow to orange, inner (upper) petals obovate, 3.5–5 × 2.5–4 mm, outer (lower) linear-oblong, 3.5 × 0.5 mm. Stamens 6, the upper 3 with filaments 2.5–4.5 mm long, the anthers 0.5 × 0.8 mm, the lower 3 stamens with filaments 3.5–7 mm long, the anthers 0.7–0.9 × 0.6–0.7 mm, anthersacs of all of the stamens subparallel. Ovary ± 1 mm in diameter, glabrous, with a stipe at the base ± 0.3 mm long; style ± 7 mm long; locules uniovulate. Capsules greyish brown or grey, lustrous, ± 3.5 × 2.5 mm, apex shortly acuminate, base narrowed into a stipe ± 0.5 mm long, the central seed-bearing part suborbicular, ± 2.5 mm in diameter. Seeds pale grey, reniform-semi-ellipsoid, 1.2–1.5 × 0.8 mm, with about 12–13 radiating ribs interrupted as in *F. tuberculata*.

TANZANIA. Chunya District: path to Muzibini village below Itigi Hill, 23 Mar. 1965, *Richards* 19829!; ?Mwanza District: Nyakaliro [=Nyakalilo?], 9 Apr. 1957, *Akiley* 72!; Ufipa District: 8 km on Namanyere–Chala road, 3 Mar. 1994, *Bidgood, Mbago & Vollesen* 2590!

DISTR. **T** 1, 4, 7; Congo–Kinshasa

HAB. Seasonally wet grassland, marshes, seasonal swamps near open water, and transition zone between grassland and *Brachystegia* woodland; 1100–1600 m

NOTE. Brenan, in the type description, called the flowers purple, following the collector's notes in the type collection. However, Brenan specifically mentioned that the plants were not in flower at the time they were pressed. Perhaps it was for this reason that the later collections of this species, by Mrs. Richards, with their clearly yellow to orange flowers were left undetermined to species by Brenan. Non-flowering specimens of *Floscopa* are typcially still colorful because of the colored sepals and/or the glandular hairs on them.

3. **Floscopa flavida** *C.B. Clarke* in DC., Monogr. Phan. 3: 269 (1881) & in F.T.A. 8: 87 (1901); Brenan in F.W.T.A. 2nd ed. 3, 1: 27, fig. 328 (1968); Obermeyer & Faden in F.S.A. 4, 2: 58, fig. 16 (1985). Types: Sudan, Djur, Agada, near Seriba Ghattas, *Schweinfurth* 2537 (K!, syn.) & Abu Jurun's Seriba, *Schweinfurth* 4286 (K!, syn.); Nigeria, Borgu, near the town of Fakun, *Barter* 760 (K!, syn.)

Annual 4–12(–18) cm tall; roots thin, fibrous; shoots usually densely tufted, erect to ascending, usually unbranched above the base except for the inflorescences, glandular-pubescent towards the inflorescence, at least in a vertical line. Leaves some or most basal, sometimes forming a ± distinct rosette; sheaths 3–5 mm long, glabrous or with a fine line of pubescence along the fused edge, eciliate at the apex; lamina linear-lanceolate to lanceolate-elliptic, 1–5(–7) × 0.2–0.7 cm, base scarcely narrowed into the sheath, margin scabridulous at apex, apex acute to acuminate; surfaces glabrous. Inflorescences terminal on the main shoot and on short axillary shoots, of a single cincinnus or of a simple thyrse composed of 2(–4) ascending to arcuate cincinni, to 2 cm long and 2.5 cm wide; cincinni usually floriferous to the base, to 2 cm long and 4 mm wide; bracteoles minute, scarious, persistent. Flowers secund, bisexual, 2.5–3 mm wide; pedicels 0.5–1.5 mm long, glandular-pubescent, maroon or reddish-purple. Sepals convexo-concave, elliptic to ovate- or obovate-elliptic, 1.5–2.3 × 1 mm, densely glandular-pubescent, maroon or reddish purple. Petals orange, yellow-orange or apricot; paired (inner) petals ovate-orbicular to obovate, 1.6–2 × 1.4–1.6 mm; lower (outer) oblong, 1.5 × 0.5 mm. Stamens yellow, 6, the filament bases fused, upper 3 stamens with filaments ± 1.3 mm long, anthers ± 0.3 × 0.4 mm, the lower 3 stamens with filaments 1.7–2 mm long, anther 0.3–0.4 mm long. Ovary subsessile, discoid, ± 0.5 × 0.6 mm; style ± 1.5 mm long, recurved apically. Capsules lustrous, usually violet, sometimes brown, 1.3–1.6 × 1.4–1.9 mm, apex acute, base with a stipe 0.2–0.3 mm long, the central portion much broader than long. Seeds reniform-semi-ellipsoid, 0.8–0.9 × 0.5–0.7 mm, testa white or grey, tuberculate, with white tubercles arranged in 13–19 radiate lines or not distinctly in lines, farinose with white or brown granules.

TANZANIA. Mbeya District: 14 km SW of Madibira on Madibira–Igawa track, 12 June 1996, *Faden et al.* 96/188!; Njombe District: Great North Road, Iyayi, 15 Apr. 1962, *Polhill & Paulo* 2008!; Songea District: ± 8 km W of Songea, 24 Apr. 1956, *Milne-Redhead & Taylor* 9698!

DISTR. **T** 4, 7, 8; Senegal, Guinea Bissau, Ghana, Nigeria, Chad, Congo–Kinshasa, Sudan, Zambia, Malawi, Angola

HAB. Damp seepage areas in woodland or dry scrub, moist sand pits and boggy grassland; sandy soil; 400–1650 m

SYN. *Floscopa pusilla* K. Schum. in Warb., Kunene–Sambesi Exped.: 185 (1903). Type: Angola, Kohi on the Kubango River, *Baum* 915 (B, holo.; K!, iso.)

NOTE. Buds on plants of *Faden et al.* 96/188 collected the previous day and kept in a plastic bag opened about 12:00 the next day, so this is likely to be the approximate opening time in the field.

Schlieben 2347 is unusually tall (to 18 cm long), with the stems solitary or only a few at the base, the leaves well distributed along the stems, and no obvious maroon colour on the pedicels, sepals and ovaries/immature capsules. The plants look etiolated, suggesting that they were growing in more shade than typical.

4. **Floscopa tuberculata** *C.B. Clarke* in F.T.A. 8: 87 (1901). Type: Malawi, Tanganyika Plateau, Fort Hill, no date, *Whyte* s.n. (K!, holo.)

Unbranched to sparsely branched annual (5–)7–20 cm tall; roots fibrous, yellow; shoots solitary or a few tufted, usually unbranched but sometimes producing lateral inflorescences, glandular-pubescent in upper internodes, becoming less so lower down. Leaves usually mostly cauline, occasionally mostly basal; sheaths 3–10 mm long, with a line of pubescence along the fused edge, usually ciliate at apex; lamina lanceolate-oblong to elliptic or ovate, (1.5–)2–7(–9) × (0.5–)0.7–2.2 cm, base rounded, margins scabrid at apex, sometimes ciliate at base, apex acute to acuminate; surfaces glabrous, except in the distalmost leaves which are sparsely pubescent. Inflorescences terminal on the main shoot and sometimes on short axillary shoots, consisting of a simple or compound thyrse of 2 to many cincinni, 1.5–4(–7) × 1.5–6(–8) cm, moderately dense; cincinni usually floriferous to the base, to 4(–7) cm long and 6 mm wide; bracteoles minute, scarious, persistent. Flowers secund, bisexual; pedicels 0.8–1.5 mm long, glandular-pubescent. Sepals convexo-concave, 2–3 mm long, glandular-pubescent, brownish-purple. Petals yellow to orange, upper (inner) 2 obovate to elliptic, 2.5–3 × 1.5 mm, lower (outer) one linear-oblanceolate, ± 3 × 0.5 mm. Stamens 6, filaments fused basally, lower 3 filaments 3–3.5 mm long, the upper 3 distinctly shorter, anthers <0.5 mm long, the upper three with reduced anther sacs. Style 2.5–3 mm long, recurved at apex. Capsules medium brown, lustrous, 2.1–2.3 mm long, 1.9–2 mm wide, apex acute, base cuneate, ± stipitate with a stipe up to 0.2 mm long. Seeds reniform-semi-ellipsoid, 1–1.2 × 0.7–0.8 mm, testa grey, with 11–14 radiate, transversely interrupted ridges or rows of tubercles, white or brown farinose granules present.

TANZANIA. Mpanda District: 19 km on Mpanda–Uvinza road, 14 May 1997, *Bidgood et al.* 3916! & Kasimba, Mpanda, no date, *Eggeling* 6169!; Ufipa District: 7 km on Namanyere–Kipili road, 3 May 1997, *Bidgood et al.* 3653!

DISTR. **T** 4; Zambia, Malawi

HAB. Seasonally wet grassland, marshes and seepage areas in woodland and at base of large outcrop; sandy soil; 900–1450 m

NOTE. This species appears intermediate between *F. tanneri* and *F. flavida*, the other two yellow- to orange-flowered species. It is in intermediate in plant size and leaf size and in sepal, petal, capsule and seed length.

This species is related to *F. schweinfurthii* C.B. Clarke, which is known only from the types from Sudan. That species, whose flower colour is unknown, differs from *F. tuberculata* in having its foliage and stems covered with glandular hairs, its pedicels longer, its capsules smaller and a dull, pale brown, and its seeds dark reddish brown (nearly black), proportionally broader, and with 14–17 fine ribs. It should be looked for in western Uganda in seasonally marshy places.

5. **Floscopa africana** (*P. Beauv.*) *C.B. Clarke* in DC., Monogr. Phan. 3: 267 (1881) & in F.T.A. 8: 85 (1901); Morton in J.L.S. 60: 199(1967); Brenan in F.W.T.A., 2nd ed.: 3, pt.1: 28 (1968). Type: Nigeria, Benin, *Palisot de Beauvois* s.n. (G, holo.)

Perennial herb 10–30 cm tall; roots thin, fibrous, arising from the lower nodes; internodes with a line of sparse to dense pubescence, at least distally, continuous with the pubescence of the sheath above. Leaves cauline; sheaths 5–10 mm long, with a line of pubescence along the fused edge, ciliate at the apex with hairs to 5 mm long, otherwise glabrous to sparsely pilose; lamina petiolate, narrowly lanceolate to elliptic or ovate, (2.5–)4–6(–7.5) × 1–2(–2.5) cm, base cuneate to rounded, margins finely undulate, scabrid apically, apex acute to acuminate-attenuate, surfaces glabrous or the adaxial with sparse long hairs. Inflorescence a terminal compound thyrse 2.5–5 × 2–4.5 cm, moderately dense, ± densely glandular- and eglandular-pubescent; smaller, separate inflorescences sometimes also present from the upper axils; lateral branches of the inflorescence to 3.5 cm long, several times branched, ultimate branches (cincinni) to 3 cm long and 5 mm wide, floriferous almost to the base; bracteoles scarious, mostly irregularly suborbicular, 0.2–0.5(–0.8) mm long, margins with an occasional hair, often erose. Flowers secund, bisexual, purplish; pedicels 0.7–1.5 mm long, glandular-pubescent, articulated at the base. Sepals oblong-elliptic to elliptic, 1.5–3.5 × 1–1.4 mm, glandular-pubescent (sometimes sparsely), purplish. Petals mauve-pink, the lower 1 much narrower than the other 2, filaments pinkish. Upper [?] three anthers yellow, lower 3 blue. Capsules brown, lustrous, 1.7–2.2(–2.8) × 1.9–2.2 mm, [the body of the capsule 1.3–1.5 × 1.9–2.2 mm], abruptly stipitate at the base, with the stipe 0.4–0.8 mm long. Seeds ellipsoid, 1.3–1.4 × 0.8–1 mm wide, testa brown, 16–17-ribbed.

SYN. *Aneilema africanum* P. Beauv., Fl. Oware 2: 57, t. 93 (1818) subsp. **petrophila** *J.K. Morton* in J.L.S. 60: 200 (1967); Brenan in F.W.T.A., 2nd ed.: 3, 1: 28 (1968). Type: Ghana, ± 6.7 km NNE of Kpene, Togo Plateau, 12 Nov. 1958, *Morton* A3444 (K!, holo.)

Shoots decumbent, the flowering branches ascending, unbranched or sparsely branched above the base. Leaves proportionally broad. Inflorescences small.

UGANDA. Bunyoro District: Budongo, Nov. 1935, *Eggeling* 2284!; Mengo District: Nakiza Forest, near Nansagazi, 27 Nov. 1950, *Dawkins* D.677!; Toro District: Buyayu–Bulanga road, 11 Dec. 1925, *Maitland* 1199!

DISTR. **U** 2, 4; Senegal, Liberia, Ivory Coast, Ghana, Cameroon, Congo–Kinshasa

HAB. Herbaceous undergrowth in forest, usually near streams; 950–1050 m

NOTE. Typical forms of subsp. *petrophila* with their low, creeping habit, short, proportionally broad leaves, and small inflorescences are quite distinct looking from typical subsp. *africana*, which is common in West and Central Africa. However, intermediates are frequent. I have used a somewhat broader concept of subsp. *petrophila* and would include all of our plants in it.

6. **Floscopa polypleura** *Brenan* in K.B. 7: 207 (1952); Morton in J.L.S. 60: 201 (1967); Brenan in F.W.T.A. 2nd ed., 3,1: 28 (1968). Type: Zambia, Nkonte Plain, near Mkupa, *Bullock* 2939 (K!, holo.)

Erect or suberect annual (7–)20–50 cm tall; roots thin, fibrous; shoots one or few from the base, unbranched other than for occasional, short lateral shoots bearing inflorescences; distal internodes laxly glandular-pubescent. Leaves few; sheaths 0.3–1.7 cm long, with or without a fine line of pubescence along the fused edge, apex ciliate or eciliate; lamina lanceolate to linear-lanceolate, 1.5–15 × 2–12 mm, base rounded or broadly cuneate, margins scabrid apically, sometimes ciliate basally, apex acuminate-attenuate; surfaces glabrous. Inflorescences terminal, compound thyrses 2–9 cm long, 1–6 cm wide, moderately dense, glandular-pubescent; smaller inflorescences sometimes also present from the lower axils; lateral branches of the terminal inflorescence 1–5 cm long, simple or with 1–4 branches, cincinni (the ultimate branches) erect or slightly curved, often densely floriferous nearly to the base, sometimes shortly naked; bracteoles scarious, obovate-elliptic or lanceolate, 0.5–1 mm long. Flowers bisexual, secund, ± 5 mm wide; pedicels 1–2.5 mm long, glandular-pubescent, articulated at the base. Sepals convexo-concave, elliptic, 3–3.5 mm long, 1.5–2 mm wide, apex obtuse, outer sepal a little broader than the 2 inner sepals, glandular-pubescent. Petals lilac to purple, mauve or violet, inner (paired) petals obovate-elliptic to ovate-orbicular, ± 3 × 2 mm, outer (lower) petal linear or oblong-linear, ± 3 × 0.5 mm. Stamens 6, filaments fused basally, upper 3 stamens with filaments ± 2.5 mm long, anthers with a broad connective from the base of which hang 2 anthers sacs ± 0.6 mm long and wide; lower 3 stamens with filaments ± 3 mm long, with suborbicular anthers with subparallel anther sacs, ± 0.6 × 0.7–0.8 mm. Ovary 0.6–0.7 × 0.8 mm, glabrous, with a basal stipe 0.5 mm long; style ± 3.5 mm long, curved at the apex; locules uniovulate. Capsules pale grey, lustrous, tranversely shortly oblong-ellipsoid, inflated, 1.3–1.6 × 2–2.3 mm, apex truncate or subtruncate, with a stipe 0.8–1 mm long. Seeds ellipsoid, ± 1 × 0.8 mm, testa grey, with 20–23 narrow radial ribs.

TANZANIA. Kigoma District: 35 km on Uvinza road from Kigoma–Kasulu road, 26 Apr. 1994, *Bidgood & Vollesen* 3200!; Mbeya District: Tunduma–Sumbawanga road, Ikana, 27 km towards Sumbawange from Ndalambo, 14 June 1996, *Faden et al.* 96/200!; Songea District: Kwamponjore valley, 20 June 1956, *Milne-Redhead & Taylor* 10852!

DISTR. **T** 4, 7, 8; Ghana, Zambia

HAB. Boggy grassland, sometimes near flowing water, often growing in standing water; 1000–1500 m

NOTE. *Faden et al.* 96/200 was found with flowers open at 13:30. They remained open during the entire period of observation (until 15:30). *Milne-Redhead & Taylor* 10852 records the absence of open flowers at 16:00.

7. **Floscopa confusa** *Brenan* in K.B. 15: 225 (1961) & in F.W.T.A. 2nd ed. 3, 1: 28 (1968). Type: Nigeria, Mambila District, Ngel Nyaki, *Hepper* 1719 (K!, holo.)

Diffusely spreading, perennial herb to 60(–150) cm tall; roots thin, fibrous, tufted, arising from the lower nodes; shoots decumbent, simple except for the inflorescences, glabrous or subglabrous towards the inflorescence, sometimes reddish pink. Leaves all cauline; lower sheaths to 3 cm long, glabrous on the surface or with a line of pubescence along the fused edge, ciliate at the apex; lamina linear-lanceolate, 5–15 × 0.5–2.5 cm, the upper decrescent and passing into foliaceous bracts, base narrowed into the sheath but scarcely petiolate, margins scabrid apically, apex longly attenuate-acuminate; surfaces inconspicuously puberulous (or subglabrous) especially along the midvein beneath. Inflorescences terminal, a compound thyrse, moderately dense and multiflowered, (5–)6–12 × (1.5–)3–6.5 cm wide, pubescent, not or inconspicuously glandular; lateral branches 1–7 cm long, simple or several times branched, ultimate branches (cincinni) to 6 cm long and 6 mm wide, floriferous almost up to the base; bracteoles scarious, irregularly suborbicular, ± 0.8 mm long and wide, margins ciliolate and a little erose. Flowers second, bisexual, bluish pink, lilac or pale purple; pedicels 2–3 mm long, glandular-pubescent, articulated at the base. Sepals ± convexo-concave, elliptic or oblong-elliptic, 2.7–3 × 1.2–1.7 mm, densely glandular. Upper (inner) 2 petals elliptic, 2.5–3 × 1.5 mm, apex acute, lower (outer) petal linear to linear-oblong, ± 2.5 × 0.4 mm, apex acute. Stamens 6, filaments fused basally, 2.5–3 mm long, anthers 0.3 × 0.5–0.8 mm, upper 3 anthers yellow, the lower 3 blue and yellow. Ovary ± 1 mm in diameter, glabrous, stipitate for 0.3 mm; style straight, ± 3 mm long, pinkish purple; locules uniovulate. Capsules greyish brown, lustrous, 2.5–2.7 × 2.2–2.5 mm, narrowed at the base into a stipe ± 0.5 mm, the central, seed-bearing part oblate, 2–2.3 × 2.2–2.5 mm. Seeds ellipsoid, 1.3–1.8 × 0.7–1.1 mm, testa grey-purple or reddish brown, marked with 11–15(–16) faint radial ribs that are not interrupted, densely white-farinose, particularly between the ribs.

UGANDA. Ankole District: Kalinzu Forest, 4 km NW of sawmill, W of Rubuzigye, 19 Sep. 1969, *Faden* 69/1131!; Bunyoro District: Bujawe, Nov. 1939, *Sangster* 574!; Masaka District: Lake Nabugabo, Aug. 1935, *Chandler* 1410!

TANZANIA. Biharamulo District: Kingombe Camp, *Rodgers in* MRC1769!

DISTR. **U** 2, 4; **T** 1; Nigeria, Cameroon, Central African Republic, Congo–Kinshasa

HAB. Lake edge, papyrus swamp, grassy swamp, swamp forest; in the open or under shade; 1100–1450 m

8. **Floscopa glomerata** (*Schult.f.*) *Hassk.*, Commel. Ind.: 166 (1870); C.B. Clarke in DC., Monogr. Phan. 3: 267 (1881) & in F.T.A. 8: 86 (1901); Morton in J.L.S. 60: 200 (1967); Brenan in F.W.T.A. 2nd ed. 3, 1: 28 (1968); Obermeyer & Faden in F.S.A. 4, 2: 56, fig. 15 (1985); Blundell, Wild Fl. E. Afr.: 414, fig. 632 (1987); Faden in U.K.W.F. 2nd ed.: 307 (1994); Ensermu and Faden in Fl. Eth. 6: 348, fig. 207.5 (1997). Type: Madagascar, without collector, Willdenow Herb. no. 6345 (B-W!, holo., microfiche)

subsp. **glomerata**

Annual or perennial herb; roots thin, fibrous, produced from the lower nodes; short rhizomes or underground stems sometimes present; shoots decumbent to erect, usually unbranched to sparsely branched, (10–)20–60(–100) cm tall, to 2 m long, soft and fleshy, sometimes spongy, often pink to reddish purple, glabrous or with a line of pubescence below the node to densely pilose. Leaves numerous,

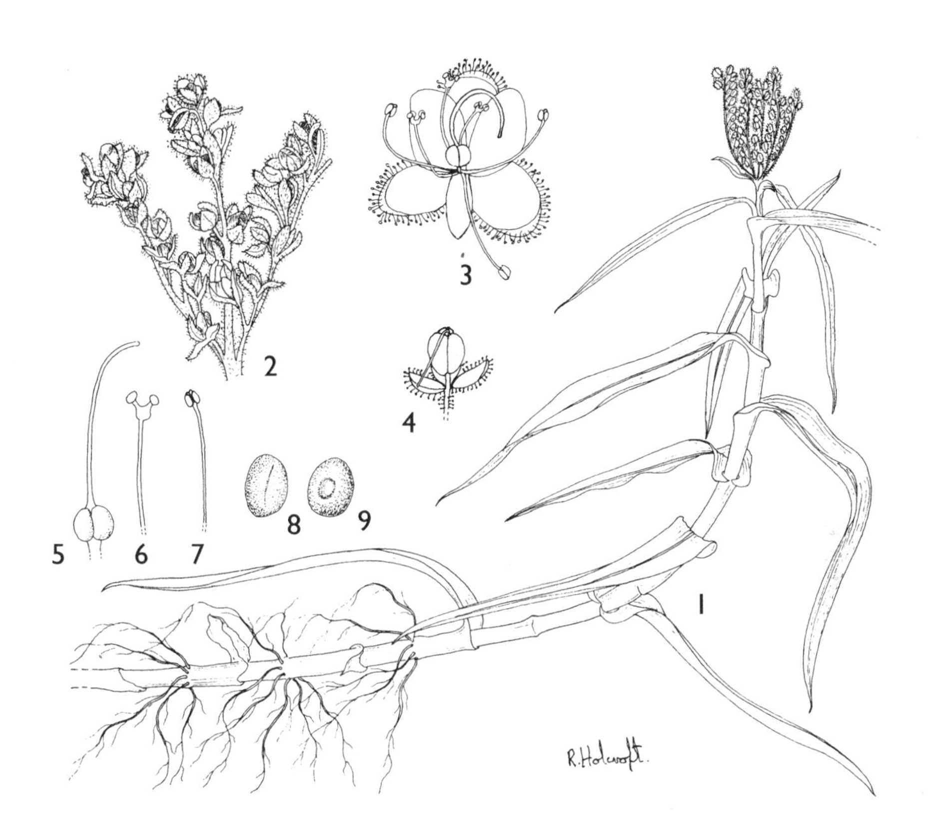

FIG. 5. *FLOSCOPA GLOMERATA* — **1**, flowering branch, × 0.6; **2**, part of infrutescence, × 2; **3**, flower, × 5; **4**, capsule with persistent sepals and style, × 3; **5**, gynoecium, × 6; **6**, staminode, × 6; **7**, stamen, × 6; **8**, seed with hilum, × 6; **9**, seed with embryotega, × 6. All from *Mauve* 5301. Drawn by R. Holcroft and reproduced with permission from F.S.A. 4, 2 fig. 15 (1985).

crowded distally below the inflorescence; sheaths 0.3–2 cm long, pubescent in a line along the fused edge, ciliate at the apex, elsewhere glabrous to pilose; lamina sessile, lanceolate-oblong to linear-lanceolate, commonly falcate and twisted, (2–)3–10(–15) × 0.6–2 cm, base cordate-amplexicaul, margins commonly strongly undulate, sometimes ciliate basally, elsewhere scabrid, sometimes purple, apex acute to acuminate; surfaces glabrous or upper surface scabrid to sparsely pilose and lower surface pilose or puberulous, sometimes scabrid on the midrib. Inflorescences terminal compound thyrses, ± dense, densely glandular-pubescent, (1–)1.5–5(–8) × (1–)1.5–5(–9) cm, smaller inflorescences sometimes also borne terminally on short lateral shoots, cincinni to 4.5 cm long, floriferous almost to the base, ascending or the lower ones becoming patent; bracteoles scarious, irregular, 0.5–1.3 mm long. Flowers secund, bisexual; pedicels 1.4–3(–3.8) mm long, glandular-pubescent. Sepals elliptic to ovate, 2–3 × 1.5–2 mm (to 4.7 mm long in fruit), glandular-pubescent, purple to violet, rarely white. Paired petals purple to mauve, lilac, lavender or violet, elliptic to ovate, 2–3.5 × 1.5–2 mm, lower petal oblong, 2.5–3 × 0.4–0.75 mm, concolorous. Filament bases fused, upper 3 (posterior) stamens with filaments 2.5–3.2 mm long, anthers 0.5–0.6 × 0.7–1 mm, lower 3 (anterior) stamens with filaments 3–3.5 mm long, anthers 0.5–0.6 × 0.5–1 mm. Ovary 0.7–0.8 mm long and wide, stipe to 0.6 mm long; style (2.5–)4.5 mm long, recurved at the apex; stigma slightly enlarged. Capsules bilocular, bivalved, tan, lustrous, transversely oblong-ellipsoid, 1.5–2.75 × 1.9–3 mm, occasionally apiculate, stipe to 0.8 mm long, rarely subsessile. Seeds ellipsoid to circular, 0.8–1.4(–1.6) × 0.7–1.1(–1.4) mm wide, testa dark brown to nearly black, rarely pale grey, with 12–18(–20) weak or strong radial ribs, rarely smooth, farinose with white to tan granules. Fig. 5, p. 47.

UGANDA. Acholi District: Patiko, Gulu, 12 Nov. 1941, *A.S. Thomas* 4013!; Ankole District: Igara, Feb. 1939, *Purseglove* 571!; Masaka District: Sango Bay, Sisal Estate, 17 Aug. 1951, *Norman* 37!

KENYA. Trans-Nzoia District: Kitale, 1 Apr. 1964, *Bogdan* 5674!; Kiambu District: Mtaro Estate along the Ndarugu River, ± 10 km W of Thika, 18 June 1972, *Ng'weno in* EA15128!; North Kavirondo District: Bungoma, near Uganda border, Nov. 1961, *Tweedie* 2251!

TANZANIA. Ufipa District: Chapota marsh, 6 Mar. 1957, *Richards* 8493!; Chunya District: Mbeya–Chunya road km 50, 1 km before Salangwe village, 30 June 1996, *Faden et al.* 96/476!; Iringa District: Mufindi, NE base of Irundi Hill, 21 Apr. 1989, *Lovett* 3255!

DISTR. **U** 1–4; **K** 3–6; **T** 1, 2, 4, 5, 7, 8; widespread in tropical Africa from Senegal to Ethiopia south to Namibia, Botswana and South Africa; Madagascar

HAB. Swamps, bogs and marshes (both seasonal and permanent), swampy grassland, stream and lake margins, dry river beds, damp hollows in grassland and woodland, rarely in seepage areas in bushland and in forest; usually in full sun, but occasionally in shade; (500–)900–2200 m

SYN. *Tradescantia glomerata* Schult.f., Syst. Veg. 7: 1175 (1830)

Aneilema rivulare A. Rich., Tent. Fl. Abyss. 2: 342 (1850), as *rivularis*. Type: Ethiopia, Chire, near Kouaieta, Mareb R., *Quartin-Dillon & Petit* s.n. (P, K!, syn.) & Addo-Hohot riverbank, Mareb river valley, *Quartin Dillon* s.n. (P, syn.)

Lamprodithyros rivularis (A. Rich.) Fenzl in Sitz. Ber. Akad. Wien 2: 139 (1865)

Floscopa rivularis (A. Rich.) C.B. Clarke in DC., Monogr. Phan. 3: 267 (1881) & in Durand & Schinz, Consp. Fl. Afr. 5: 436 (1895); K. Schum. in P.O.A. C: 137 (1895); C.B. Clarke in F.T.A. 8: 86 (1901); F.P.U.: 198 (1962)

F. rivularis (A. Rich.) C.B. Clarke var. *argosperma* C.B. Clarke in DC., Monogr. Phan. 3: 268 (1881). Type: Ethiopia, Fertit, *Schweinfurth* 212 & Tigre, *Schimper* 1226 (K!, syn.)

NOTE. This is far and away the most abundant species of *Floscopa* in our area. Large wet grassy depressions in Tanzania may appear pinkish purple at a distance due to this plant. A number of collectors have referred to its attractiveness, particularly en masse. The attractiveness is due almost entirely to the colour of the sepals, the petals adding relatively little because of their small size, relative sparseness (in comparison to the numbers of persistent calyces), and short anthesis.

Some plants are obviously annuals, but most are unclear. Rarely a short rhizome has been collected, e.g., *Bidgood et al.* 2738 and *Milne-Redhead & Taylor* 8062. However, underground parts of Commelinaceae are poorly collected. Thus for nearly all specimens there is no evidence of a perennating structure. Nevertheless, because plants sometimes grow in permanently wet situations, they might perennate without special organs.

The seeds of this species are very variable in number of ribs and how clearly the ribs are defined. This does not seem to be related to geography. The seeds of two collections, *Faden et al.* 96/163 and *Milne-Redhead & Taylor* 10901, are particularly unusual in being spherical and smooth

In 1996 we found two populations with flowers opening between 12:00 and 12:30 and a third population in which the flowers were fading about 14:30. In a fourth population the flowers were found already open at 10:50. *Milne-Redhead & Taylor* 10901 reports a single flower open at 11:00.

On the label of *Tanner* 4841 the flowers are described as "strongly aromatic," something that has not been noted by other collectors. On *Tanner* 731 the flowers are recorded as "not aromatic." *Carr* 846 refers to an unpleasant scent.

In addition to this subspecies other subspecies occur in West Africa, one of which is possibly the same as *F. glabrata* Hassk. from tropical America.

6. ANTHERICOPSIS

Engl. in P.O.A. C: 139 (1895); C.B. Clarke in F.T.A. 8: 75 (1901); Brenan in J.L.S. 59: 365 (1966); Faden & Inman in van der Maesen et al., *Biodiversity of African Plants*: 464–471 (1996); Faden in Kubitzki, Fam. Gen. Vasc. Pl. 4: 125 (1998)

Gillettia Rendle in J. Bot. 34: 56, t. 355, fig. b (1897)

Geophytes with tubers at the ends of thin, wiry roots. Leaves all or mainly in a basal rosette, spirally arranged, sessile. Inflorescences scapose or subscapose with scapes terminal and axillary, composed of 1–2 sessile, contracted, bracteolate cincinni at the summit of the scape, when 2 opposite, but not fused. Flowers pedicellate, bisexual, actinomorphic. Sepals free, equal, sepaloid. Petals free, equal, not clawed. Stamens 3, antesepalous, filaments glabrous, anthers basifixed; staminodes 3, antepetalous, filaments glabrous, antherodes small, unlobed. Ovary 3-locular. Capsules 3-locular, 3-valved, locules 8–12-seeded. Seeds uniseriate, hilum linear, embryotega semilateral.

A single species, restricted to eastern Africa. Brenan (1966) considered *Anthericopsis* sufficiently distinct morphologically to constitute its own informal taxonomic group within the family. However, the alternating stamen and staminode arrangement in *Anthericopsis* is identical to that in *Murdannia*, and Faden & Hunt (1991) placed both genera in tribe Commelineae. More recent anatomical (Faden & Inman in van der Maesen et al., The Biodiversity of African Plants: 464–471 (1996)) and molecular (Evans et al. in Systematic Botany 28: 270–292 (2003)) data have supported the close relationship between *Anthericopsis* and *Murdannia* within tribe Commelineae.

Anthericopsis sepalosa (*C.B. Clarke*) *Engl.* in E. & P. Pf. 1: 69 (1897); Blundell, Wild Fl. E. Afr.: 411, fig. 816 (1987); Faden in U.K.W.F. 2nd ed.: 308, t. 139 (1994) & in Fl. Somalia 4: 83, fig. 54 (1995); Ensermu & Faden in Fl. Eth. 6: 345, fig. 207.3 (1997). Syntypes: Kenya, Kitui, *Hildebrandt* 2640 (K!, syn.) & Malawi, mountains E of Lake Malawi [Nyasso], *Bishop Steene* s.n. (K!, syn.)

Completely glabrous perennial herb to 10(–30) cm; tubers enclosed in a hard, crustaceous layer, ovoid to ellipsoid, 0.7–2 × 0.4–1 cm. Leaves lanceolate to oblanceolate-elliptic, 4–31 × 1.2–3.5(–5.5) cm, base (in fully mature leaves) often cuneate, apex acute to acuminate. Inflorescence terminal and occasionally axillary from the rosette, sometimes 1–2 axillary from a leaf on the peduncle, sometimes sessile at onset of flowering, eventually with peduncles 6–20(–38) cm long at maturity; bracts and bracteoles herbaceous, narrowly lanceolate, 0.7–5 cm long. Flowers ± 2.5 cm wide; pedicels 1–5.5 cm long, ascending. Sepals linear-lanceolate to lanceolate-elliptic, (10–)15–25 × 2–4 mm. Petals white to pale pink or mauve or rarely bluish, elliptic, 15–25 × 13–18 mm. Stamens with filaments 3–6 mm long, anthers yellow, 3–5 mm long; staminodes with filaments ± 2 mm long, antherodes ± 1 mm long. Ovary 2.5–4 mm long; style 3–4.5 mm long. Capsules brown, oblong to linear-oblong, 20–30(–35) × 2–4 mm. Seeds brown to grey, transversely oblong, ± 1 × 1.5–3 mm wide, testa smooth, hilum raised within an oblong pit. Fig. 6, p. 50.

KENYA. Northern Frontier District: Dandu, 27 Mar. 1932, *Gillett* 12650!; Machakos District: Lukenya [Lukenia], 15 Dec. 1932, *Napier & Mainwaring* 2345!; Teita District: Voi–Taveta road, 3°23'S, 37°47'E, 6 Apr. 1974, *Faden & Faden* 74/417!

TANZANIA. Masai District: NW footslopes of Mt Longido, 30 Oct. 1961, *Greenway* 10299!; Lushoto District: 4.8 km NW of Mombo, 29 Apr. 1953, *Drummond & Hemsley* 2281!; Masisi District, just W of Bangala R., 17 Dec. 1955, *Milne-Redhead & Taylor* 7683!

DISTR. **K** 1, 4, 6, 7; **T** 2, 3, 5, 7, 8; Congo–Kinshasa, S Ethiopia, S Somalia, Zambia, Malawi, Mozambique

HAB. Wet and waterlogged grassland and thicket edges, also *Terminalia* woodland; (15–)400–1650 m

The flowers open in the morning and fade by about 12:30.

FIG. 6. *ANTHERICOPSIS SEPALOSA* — **1**, flowering branch, × 0.6; **2**, part of infrutescence, × 2. Drawn by M. Grierson and reproduced with permission from J.L.S. 59: 364 (1966), partly

SYN. *Aneilema sepalosum* C.B. Clarke in DC., Monogr. Phan. 3: 202 (1881); K. Schum. in P.O.A. C: 135 (1895)
Anthericopsis fischeri Engl. in P.O.A. C: 139 (1895); C.B. Clarke in F.T.A. 8: 75 (1901). Type: Tanzania, 'Masai Hochland', *Fischer* 258 (HBG, holo.)
Gillettia sepalosa (C.B. Clarke) Rendle in J. Bot. 34: 56, t. 355b (1897)
Anthericopsis tradescantioides Chiov. in Webbia 8: 40, fig. 13 (1951). Type: Ethiopia, Neghelli–Filtu road, *Corradi* 4693 (FT, holo.; K, photo!)

NOTE. The hard, crust-like layer around the tubers has not been observed in any other Commelinaceae. Plants often flower before the leaves are fully expanded. Such leaves may appear ovate, but they are only the tips of mature leaves. Brenan (1966) reported the presence of more than two cincinni in some inflorescences. No doubt this could occur, but two seems to be the usual number.

7. MURDANNIA

Royle, Illustr. Bot. Himal. 403, t. 95, fig. 3 (1840), *nom. cons.*; Brenan in K.B. 7: 179–208 (1952); Morton in J.L.S. 60: 202–204 (1967); Faden & Inman in van der Maesen et al., *Biodiversity of African Plants*: 464–471 (1996); Faden in Kubitzki, Fam. Gen. Vasc. Pl. 4: 124 (1998)

Aneilema R. Br., Prod.: 210 (1810) pro parte
Baoulia A. Chev. in Mém. Soc. Bot. Fr. 8: 217 (1912)

Perennials and annuals; rootstock fibrous or tuberous. Leaves spirally arranged or distichous, sessile. Inflorescences terminal and axillary thyrses of 2-many cincinni, sometimes very reduced and appearing as fascicles of 1-flowered cincinni, partially enclosed in the leaf sheaths; bracteoles persistent or caducous, sometimes lacking. Flowers pedicellate, actinomorphic to slightly zygomorphic, all bisexual or bisexual and male. Sepals free, subequal, sepaloid. Petals free, equal, not clawed. Fertile stamens 3, antesepalous, or sometimes one staminodial, filaments bearded or glabrous; staminodes 3(–4), antepetalous, rarely all lacking, sometimes a fourth antepetalous one present, filaments bearded or glabrous, antherodes 3-lobed or hastate. Ovary 3-locular. Capsules (2–)3-locular, (2–)3-valved, locules 1-many-seeded. Seeds uni- or biseriate, hilum punctiform to linear, embryotega lateral to dorsal.

Pantropical and warm temperate genus of ± 50 species, pantropical and warm-temperate, usually not in forests; 8–9 in Africa.

1. Cincinni several- to many-flowered, arranged in terminal and often axillary inflorescences 2
   Cincinni 1-flowered, appearing fasciculate in the axils of the upper leaves 4
2. Leaves flat, (1–)4–17 mm wide; bracteoles not perfoliate, caducous; fertile stamens 2, filaments bearded, free; plants of various habitats .... 1. *M. simplex*
   Leaves terete or semiterete, up to 2 mm wide in the middle; bracteoles perfoliate, persistent; fertile stamens 3, filaments glabrous, fused basally; plants of rock pools and seasonally waterlogged shallow soil on outcrops 3
3. Pedicels 4–13 mm long, erect or spreading laterally in fruit; sepals 4–4.5 mm long; petals 4–7 mm long; seeds 3–5 per locule .... 2. *M. clarkeana*
   Pedicels 2–6 mm long, strongly declinate in fruit; sepals 1.5 2.5(–3) mm long; petals 2–3.5 mm long; seeds 2–3 per locule .... 3. *M. semiteres*
4. Leaves 8–14 mm wide; pedicels (stalk above the articulation) 3–4 mm long; staminodes present; seeds biseriate, 16–24 per locule; coastal: **K** 7; **Z, P** .... 4. *M. axillaris*
   Leaves to 5 mm wide; pedicels 5.5–11 mm long; staminodes 0; seeds uniseriate, 2–3 per locule; inland above 950 m .... 5. *M. tenuissima*

1. **Murdannia simplex** (*Vahl*) *Brenan* in K.B. 7: 186 (1952) & F.P.N.A. 3: 326 (1955); F.P.U.: 195 (1962); Brenan in J.L.S. 59: 353, fig. 9 (1966) & in F.W.T.A. 2nd ed.: 3: 24, fig. 327 (1968); Hepper, W. Af. Herb. Isert & Thonning: 135 (1976); Vollesen in Opera Bot. 59: 97 (1980); Obermeyer & Faden in F.S.A. 4, 2: 47, fig. 11 (1985); Blundell, Wild Fl. E. Afr.: 415, fig. 633 (1987); Malaisse in Fl. Rwanda 4: 140, fig. 51.2 (1988); Faden in U.K.W.F. 2nd ed.: 309, Pl. 138 (1994); Faden in Fl. Som. 4: 83, fig. 53 (1995); Ensermu & Faden in Fl. Eth. 6: 346, fig. 207.4 (1997). Types: Guinea, *Isert* s.n. (C!, syn.) & *Thonning* s.n. (C!, syn., S!, isosyn.)

Caespitose to shortly rhizomatous perennial, with shoots tufted, unbranched or branched near the base, erect to ascending, rarely decumbent, 20–90 cm tall; shoot bases sometimes bulbous; roots uniformly thickened, not tuberous. Leaves basal and cauline, spirally arranged or distichous, widely spaced and strongly reduced distally on the flowering shoot, the distalmost bract-like; sheaths to 4 cm long, pilose to glabrous, ciliate at apex; lamina linear to lanceolate-oblong, often conduplicate, (2–)4.5–45 × (0.1–)0.4–1.7 cm, base cuneate to rounded, margins usually ciliate basally, apex acute to acuminate; surfaces glabrous to densely pilose or sometimes the adaxial pubescent only submarginally. Inflorescences usually compound, of 2 or more thyrses, terminal thyrse of (1–)2–5 elongate, ascending cincinni, other thyrses of 1–3 similar cincinni; cincinni to ± 3.5 cm long and 35-flowered; cincinnus bracts caducous; bracteoles approximate, scarious, elliptic, 3.5–6 mm long, caducous. Flowers secund, bisexual and male, to ± 16 mm wide; pedicels 4.5–8 mm long, erect

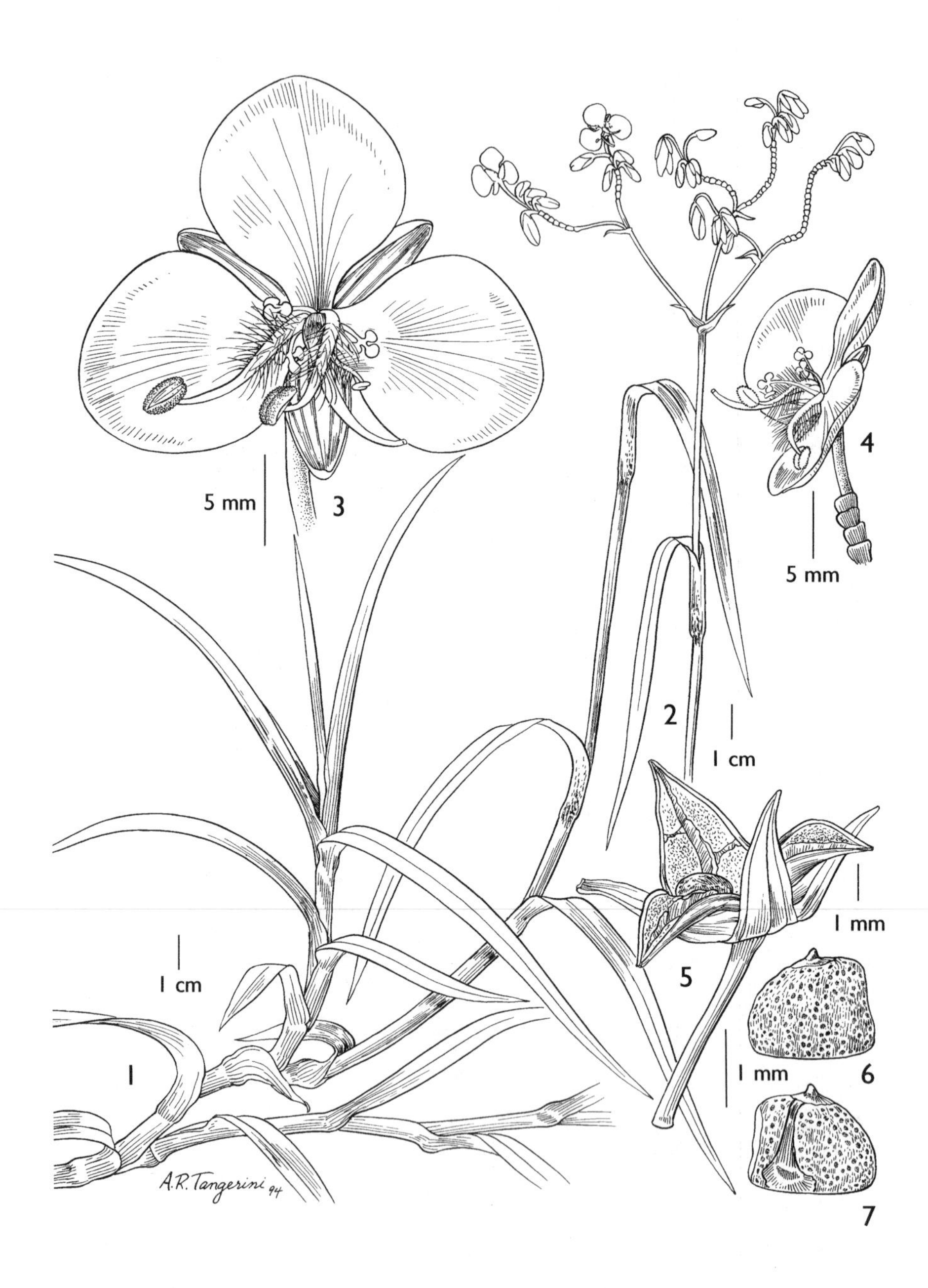

FIG. 7. *MURDANNIA SIMPLEX* — **1**, habit with base of flowering shoot; **2**, inflorescence; **3**, bisexual flower, side view; **4**, bisexual flower, front view; **5**, dehisced capsule; **6** & **7**, seed, dorsal and ventral view. Drawn by Alice Tangerini, reproduced with permission from Flora of Ethiopia, Vol. 6. Fig. 207.4.

in fruit, glabrous. Sepals lanceolate-elliptic to ovate-elliptic, 4.5–7 × 2–3.5 mm, glabrous. Petals blue or mauve to lavender, lilac or violet, ovate-orbicular to ovate- or obovate-elliptic, 5.5–11 × 4.5–8 mm. Fertile stamens 2, filaments both curved to one side (bisexual flowers) or divergent (male flowers), 5–7 mm long, bearded with long violet or purple hairs, anthers 1–1.3 mm long, blue; staminodes 4, 3 antepetalous, 1 antesepalous. Ovary 1.5–2.5 mm long; style curved to one side, 4–6 mm long. Capsules ovoid to broadly ellipsoid, apiculate, (3.5–)4–7.5 × 2.5–4 mm, glabrous, locules 2(–3)-seeded. Seeds uniseriate, transversely ellipsoid to ellipsoid, ovoid or rectangular, (1.1–)1.4–1.8 × 1.2–1.7 mm, testa completely covered by a brown, matted layer, smooth to faintly ribbed, alveolate, scrobiculate, or rarely foveolate, usually with whitish pustules radiately arranged, these on the higher areas between the depressions, when present. Fig. 7, p. 52.

UGANDA. Karamoja District: Kokumongole, 29 May 1939, *Thomas* 2876!; Mbale District: Elgon, Kapchorwa, Sep. 1956, *Lind* 450!; Mengo District: 25.7 km SW of Kampala, 6 Oct. 1962, *W.H. Lewis* 6016!

KENYA. Fort Hall District: Thika, N of Thika River, 7 May 1967, *Faden* 67/311!; Masai District: Melelo area 24 km from Olulunga, 7 June 1961, *Glover, Gwynne & Samuel* 1782!; Tana River District: 23 km S of Garsen hill on Garsen–Malindi road, 19 July 1972, *Gillett & Kibuwa* 20000!

TANZANIA. Musoma District: Campi ya Mpofu, Kleins Camp Track, 30 Mar. 1962, *Greenway et al.* 10561!; Lushoto District; West Usambaras, 0.8 km SW of Kwai, 12 June 1953, *Drummond & Hemsley* 2896!; Iringa District: Ngwazi, *Lovett* 1600!; Zanzibar: Massazine, 10 Sep. 1959, *Faulkner* 2362!

DISTR. **U** 1–4; **K** 1–7; **T** 1–4, 6–8; **Z**, **P**; widespread in Africa from Sierra Leone to Somalia and south to South Africa; Madagascar, tropical Asia (but see note)

HAB. Grassland, bushland, woodland, coastal forest, rocky outcrops, marshes, edges of bodies of water, pathsides, roadsides, rarely a weed in cultivation; clay to sand; full sun to shade; 30–2200 m

SYN. *Commelina simplex* Vahl, Enum. Pl. 2: 177 (1806); Thonning in Beskr. Guin. Pl.: 22 (1827)

Aneilema sinicum Ker-Gawl. in Bot. Reg. 8, t. 659 (1822), as *sinica*; C.B. Clarke in DC., Monogr. Phan. 3: 212 (1881) & in F.T.A. 8: 63 (1901); Hutch. & Dalziel, F.W.T.A. 2: 312, t. 286 (1936); U.O.P.Z.: 122 (1949). Lectotype: t. 659 in Bot. Reg. 8 (1822), based on a plant grown from seed from China, Canton, *Staunton* s.n. (BM, herb. R.Brown) [Name almost invariably attributed to Lindley, but he was not editor until later. A specimen at Kew was identified as the type by C.B. Clarke, but since it was not pressed until 1827, it could not have been studied by Ker-Gawler]

Commelina sinica (Ker-Gawl.) Roem. & Schult., Syst. Veg., Mant. 1, Addit. 1: 376 (1822)

Aneilema longifolium Hook., Exot. Fl. 3, t. 204 (1827), as *longifolia*. Type: Zanzibar, *Bojer* s.n. (K!, syn.) & *Helsinger* s.n. (ubi?) & "seeds of dried plants, communicated to Hooker", plants raised by R. Barclay at Bury Hill, Surrey (K!, syn.)

Commelina longifolia (Hook.) Spreng., Syst. 4, Cur. Post. 25 (1827)

C. hookeri Dietr., Sp. Pl. 2: 404 (1833), *nom illeg.*

Aneilema simplex (Vahl) Kunth, Enum. Pl. 4: 71 (1843)

A. sinicum Ker.-Gawl. var. *simplex* (Vahl) C.B. Clarke in DC., Monogr. Phan. 3: 212 (1881)

[*A. giganteum* C.B. Clarke in DC., Monogr. Phan. 3: 212 (1881), quoad syn. *Aneilema longifolium*, *Commelina longifolia*, *Commelina hookeri*, *non* (Vahl) R.Br.]

A. sinicum Ker-Gawl. var. *longifolium* (Hook.) C.B. Clarke in F.T.A. 8: 64 (1901)

Phaeneilema sinicum (Ker-Gawl.) G.Brückn. in E.J. 61, Beibl. 137: 69 (1926) & in N.B.G.B. 10: 56 (1927)

Murdannia sinica (Ker-Gawl.) G.Brückn. in E. & P. Pf., 2nd ed. 15a: 173 (1930)

M. stictosperma Brenan in K.B. 7: 187 (1952). Type: Zambia, S of Matonchi Farm, *Milne-Redhead* 3331 (K!, holo.)

NOTE. *Murdannia simplex* is exceedingly variable in plant size and diverse in its ecology. In African plants, basal shoots grow directly into flowering stems, unlike Asian plants in which they arise from the axils of a basal rosette. Thus the conspecificity of African and Asian plants has yet to be demonstated unquestionably.

Specimens with very narrow leaves and a bulbous base were described by Brenan (1952) as *M. stictosperma*, based on leaf width and seed size and surface features. With a much wider range of specimens now available, none of these characters has been found to hold up. The extremely narrow-leaved types, e.g., *Milne-Redhead & Taylor* 8537 from Songea

District (**T** 8), grade into typical, broader-leaved plants without any meaningful separation being possible. There are broad-leaved plants with somewhat bulbous bases. The seed size range in typical *M. simplex* encompasses that of *M. stictosperma*. And the supposedly unique testa features of Brenan's species all can be found within *M. simplex*. It is possible that the most extreme forms, such as the type of *M. stictosperma*, may represent an ecotype associated with laterite, but there seems to be no basis for recognizing this plant even at the varietal level.

Another extreme is found in some plants from coastal Kenya: Sokoke Forest, Jilori, 25 Nov. 1961, *Polhill & Paulo* 853. These are very gracile, with small leaves (to 11 × 0.5 cm) and long, decumbent flowering shoots. They have among the smallest capsules (3.5–4 × 2.5 mm) and seeds (1.2–1.3 × 1.1–1.2 mm) recorded in this species. Normal-sized plants also occur in this forest, so possibly the small ones come from nutrient poor areas of sand.

Several collections from the Kenya coast, e.g. *Gillespie* 207, have some capsules with 3-seeded locules. This feature has not been observed anywhere else for this species.

When flowering shoots become decumbent and root at the nodes they can give rise to new plants vegetatively. *Batty* 100 from Dar es Salaam is described as "layering itself and spreading rapidly."

The petals are finely bearded at the base, which has not been recorded before in any *Murdannia* species. The flowers open in mid-afternoon and close about 18:00.

2. **Murdannia clarkeana** *Brenan* in K.B. 7: 182 (1952); Blundell, Wild Fl. E. Afr.: 415, fig. 818 (1987); Faden in U.K.W.F., ed 2: 309 (1994) & in Krupnick & Kress, Plant Conservation: 108, Fig. 5.3B (2005) Type: Kenya, Nairobi, Nairobi R. valley near St. Austin's Mission Bridge, *Rayner* 305 (K!, holo.)

Slender, caespitose, rhizomatous, erect perennial 7–40 cm tall, completely glabrous; roots numerous, fibrous, thin or slightly thickened; shoots annual, unbranched except for distal lateral inflorescences. Leaves basal and cauline, spirally arranged, linear, succulent, terete or semiterete, basal leaves 3.5–16 cm long, with long, sheathing bases, cauline leaves with sheaths to 1 cm long, lamina to 8 cm long, all leaves 1–2 mm thick. Inflorescences terminal and axillary from the distal leaves, usually forming a compound inflorescence, moderately lax, individual inflorescence units composed of 1–3 elongate ascending cincinni, when more than 1 the cincinni in opposite pairs or in whorls of 3; cincinni to ± 3 cm long and 6-flowered; bracteoles asymmetrically cup-shaped, perfoliate, 1.5–3 mm long, apparently with a subterminal gland, usually purple- or violet-tinged. Flowers bisexual. Sepals oblong-elliptic, 4–4.5 mm long, blue- to violet-tipped. Petals blue to lavender, obtrullate(?) to obovate-elliptic, 4–7 × 3–4 mm, apex acute. Fertile stamens 3, filaments 3 mm long, fused basallly with those of two of the staminodes, glabrous, anthers ± 1.7 mm long, dark violet; staminodes pale yellow, filaments 2.5–3 mm long, glabrous, antherode ± 1 mm long, hastate. Ovary glabrous, ± 1.5 mm long; style ± 3 mm long. Capsules oblong, 3.5–4 × 1.5 mm. Seeds uniseriate, 3–5 per locule, irregularly quadrate, broadly ovate or oblong, 0.7–1.2 × 0.7–0.8 mm, testa slightly rugose, black finely variegated with pale brown.

KENYA. Laikipia District: 1.6 km S. of Lake Kelole [Kelele] Dam, 11 Mar. 1969, Magor 42!; South Nyeri District: 3 km N of Kiganjo, 12 Apr. 1977, *Hooper & Townsend* 1695!; Nairobi District: corner of Matumbato and Upper Hill roads, 6 Feb. 1977, *Faden, Faden & Ng'weno* 77/291!

DISTR. **K** 1, 3, 4; Central African Republic?, Chad?

HAB. Seasonally waterlogged, shallow soil on rock outcrops or seasonally wet murram and black cotton soils; 1500–1850 m

SYN. *Aneilema delicatulum* Jex-Blake, Some wild fl. Kenya: 113, t. 93 (1948), *nom. subnud.*

NOTE. An annual form of *M. clarkeana* was reported from Chad, e.g., *Audru* 1229, and Central African Republic by Lebrun et al. (Catalogue Pl. Vasc. Tchad Meridional, 187. 1972). I have examined the specimens and, at present, I cannot separate them from this species except by their habit. This is a very improbable distribution, however, so perhaps there are other important distinguishing features notable only in the living plant.

The type collection records the flowers open "from midday until 4 o'clock." My observations and, especially those of Mrs. F. Ng'weno over a long period of time, are that the flowers open at 11:00 and remain open for about two hours. The leaves turn maroon as the plants begin to desiccate. Non-rhizomatous plants are apparently first-year seedlings.

Blundell's record of the distribution of this species (and many other Commelinaceae) is erroneous.

3. **Murdannia semiteres** (*Dalz.*) *Santapau* in Poona Agric. Coll. Mag. 41: 284 (1951); Brenan in K.B. 7: 184 (1952); Faden in U.K.W.F.: 667, fig. 668 (1974); Blundell, Wild Fl. E. Afr.: 415 (1987); Malaisse in Fl. Rwanda 4: 140, fig. 51.1 (1988); Faden in U.K.W.F. 2nd ed.: 309, Pl. 138 (1994). Type: India, Concan, *Dalzell* s.n. (K!, lecto.)

Annual herbs 6–30 cm tall, caespitose, completely glabrous; rhizome lacking; roots fibrous, thin; shoots basally unbranched or few-branched, distally producing numerous long or short inflorescence-bearing axillary shoots. Leaves mainly cauline, spirally arranged, semiterete, linear to linear-lanceolate, succulent, basal leaves 3.5–18 cm long, with long, sheathing bases; cauline leaves similar, decrescent distally, sheaths to 2 cm long, often flushed with purple, lamina to 14 cm long; all leaves 1–2 mm thick in the middle. Inflorescences terminal and axillary from the middle and distal nodes, forming a compound inflorescence, ± lax to ± dense; individual inflorescence units composed of 1–3(–5) cincinni, when 2 or more, arranged in opposite pairs or whorls; cincinni to 4 cm long and 12-flowered; bracteoles scarious, asymmetrically cup-shaped, perfoliate, 1–2.5 mm long. Flowers bisexual; pedicels 2–6 mm long, strongly declinate in fruit. Sepals 1.5–2.5(–3) mm long, usually tipped with or occasionally entirely blue to violet. Petals dark blue-purple to azure blue, violet, pale lavender or white, obovate to obovate-elliptic, 2–3.5 × 1.1–1.5 mm; fertile stamens 3, filaments fused basally with those of 2 staminodes, ± 0.8–1.5 mm long, glabrous, anthers ± 0.6 mm long; staminodes 3, filaments ± 0.5–1 mm long, antherodes hastate, ± 0.6 mm long. Style ± 1 mm long. Capsules oblong-ellipsoid to obovoid-ellipsoid, (1.5–)2–3 × 1–2 mm, apex of valves retuse to emarginate, rarely truncate. Seeds uniseriate, 1–2(–3) per locule, broadly ovate to deltate, quadrate or ellipitc, occasionally some dorsoventrally compressed, 0.5–1.2 × 0.5–1 mm, testa slightly rugose, fundamentally black with varying amounts of a brown overlay, margins irregularly crenate-furrowed.

UGANDA. Teso District: Serere, Aug. 1932, *Chandler* 951! & Amuria, 14 Sep. 1946, *Thomas* 4533!; Mengo District; km 74 on Kampala–Masindi road, 1.6 km S of Kakinzi, 14 Sep. 1969, *Faden* 69/1053!

KENYA. Trans-Nzoia District, Caves of Elgon, 25 Jan. 1943, *Tweedie* 76! & Kipkarren, Aug. 1931, *Brodhurst Hill* 34!; Uasin Gishu District: ± 10 km Eldoret–Kitale, 8 Oct. 1981, *Gilbert & Mesfin* 6490!

TANZANIA. Bukoba District: Nshamba [Mshamba], Dec. 1931, *Haarer* 2371!; Dodoma District: Kazikazi, May, 1932, *Burtt* 3685!: Singida District: 6.4 km Kiomboi–Kasiriri road, 29 Apr. 1962, *Polhill & Paulo* 2239!

DISTR. **U** 3, 4; **K** 3, 5; **T** 1, 5; Congo–Kinshasa, Rwanda, Burundi, Zambia; India

HAB. Seasonally saturated, shallow soil on outcrops and in and about shallow, seasonal rock pools; 1050–2050 m

SYN. *Aneilema semiteres* Dalzell in Hook. J. Bot. Kew Gard. Misc. 3: 138 (1851)

NOTE. By their robustness some plants, e.g. *Burtt* 3685, appear perennial. However, populations found in Tanzania in 1996 in dried up rock pools, e.g. *Faden et al.* 96/501, were distinctly annual and, because perennating organs are lacking, it is probable that all African plants are annuals. The great reproductive effort shown by plants of *M. semiteres* (as compared to *M. clarkeana*) – specimens almost always have mature seeds – is often associated with an annual habit. The flowers open between 07:00 and 09:00 and are still open at 11:30.

4. **Murdannia axillaris** *Brenan* in Hook. Ic. Pl. ser. 5, 6: t. 3578 (1962). Type: Kenya, Lamu District: Utwani Forest Reserve, Mambosasa, *Greenway & Rawlins* 9374 (K!, holo.; EA!, iso.)

Annual herb to 30 cm tall, tufted, moderately robust; roots fibrous, thin; shoots simple or sparsely branched, ascending to decumbent, to 40 cm long, sometimes rooting at the nodes. Leaves cauline, distichous; sheaths 0.8–1.3 cm long, minutely pubescent along the fused edge, sometimes extended onto the internode below, ciliolate at apex; lamina oblong-lanceolate, 3–6 × 0.8–1.4 cm, base subamplexicaul, margins inconspicuously ciliolate basally, scabrid distally, apex acute, surfaces glabrous. Inflorescences terminal and axillary in the distal 1–4 axils, appearing as fascicles of up to 8 1-flowered cincinni basally enclosed within the leaf sheath, each cincinnus with a large, scarious bract ± 9.5 mm long at its base; bracteoles minute, caducous; peduncle plus pedicel (total stalk below the flower or fruit) erect, 8.5–17 mm long, the peduncle (part below the the articulation) 5–13 mm long, pedicel 3–4 mm long. Flowers bisexual. Sepals ovate-elliptic, 5–7 × 2.5–3 mm. Petals subequal, pink to mauve or white, obovate-elliptic, 5–7 × 3–4.5 mm. Fertile stamens 3, filaments 3–4 mm long, densely bearded, anthers 1.5–2 mm long, sometimes 1 reduced; staminodes 3, filaments 2–3 mm long, bearded, antherodes trilobed-hastate, ± 0.8 mm long. Ovary ± 2 mm long; style 2.5–3 mm long. Capsules pale brown, oblong, 6–9 × 2–3 mm, glabrous, style base persistent. Seeds biseriate, 16–24 per locule, irregularly quadrate to polygonal in outline, ± tetrahedral in shape, 0.4–0.5 × 0.8 mm, testa pale grey-brown or beige, with finely raised lines, ± muricate.

KENYA. Lamu District: Utwani Forest Reserve, Mambosasa, *Greenway & Rawlins* 9374! & Cogitani & Mkune swamps, Feb. 1957, *Rawlins* 341! & Witu swamps, Mar. 1957, *Rawlins* 380!
TANZANIA. Zanzibar I., Mtoni Swamp, 9 Aug. 1930, *Vaughan* 1438!; Pemba, Wesha road km 5, 30 Sep. 1929, *Vaughan* 687! & N of Ngezi Forest, Dec. 1983, *Rodgers, Mwasumbi & Hall* 2698!
DISTR. **K** 7; **Z**, **P**; not known elsewhere
HAB. In and at margins of fresh-water swamps, often among taller grasses; ± 10 m

NOTE. This is the only African species of *Murdannia* with biseriate seeds. It is very similar to the tropical Asian species *M. blumei* (Hassk.) Brenan, with which Brenan contrasted it. Of the two characters he used to separate them, the longer pedicels (actually pedicel plus peduncle, i.e., the entire stalk of the flower or fruit) in *M. axillaris* is consistent. The other character, filaments bearded in *M. axillaris* vs. glabrous in *M. blumei* is inconsistent, as the filaments in some specimens of *M. blumei*, e.g., *Faden & Faden* 77/172, from Sri Lanka, have bearded filaments. In addition to the "pedicel" difference, specimens of *M. axillaris* are, on average, more robust, less pubescent, and have more flowers per inflorescence than those of *M. blumei*. When both species become better known it is likely that more differences with be found.

The inflorescence in *M. axillaris* is actually a reduced thyrse. The single flowers are all attached to a short axis that is visible in a dissection of *Vaughan* 1438 but is otherwise hidden within the sheath. Each flower is attached to a one-flowered cincinnus and has an obvious articulation above the middle. Basal to the articulation is the peduncle of the cincinnus; distal to it is the true pedicel. The large scarious bracts are cincinnus bracts, not bracteoles.

In *Vaughan* 2292 the flowers are recorded as opening at noon and lasting for about two hours.

5. **Murdannia tenuissima** (*A. Chev.*) *Brenan* in K.B. 7: 189 (1952); Morton in J.L.S. 60: 202, fig. 8 (1967); Brenan in F.W.T.A. 2nd ed.: 3: 26 (1968); Ensermu & Faden in Fl. Eth. 6: 346 (1997). Type: Ivory Coast, between Amanikro [Manikro] and Kourakissikro [Tiegoriakro], *Chevalier* 22318 (P!, holo.; K!, iso.)

Perennial 30–90 cm tall, very slender; plant base apparently decumbent and rooting, giving rise to tufts of branched, ascending, flexuous shoots often supported by the surrounding vegetation; roots thin, fibrous. Leaves distichous or spirally arranged; sheaths 0.5–1.5 cm long, often partially split, prominently ribbed, puberulous along the fused edge, ciliate at apex; lamina linear-lanceolate,

2–6 × 0.2–0.5 cm, base cuneate to subrounded, margins usually ciliate basally, apex acuminate, surfaces glabrous or adaxial white-pilose. Inflorescences terminal and in the axils of the distal 1–3 leaves (sheaths of these leaves often split nearly to the base), appearing as fascicles of up to 7 1-flowered cincinni basally enclosed within the leaf sheath, each cincinnus with a large bract at its base; bracteoles caducous or persistent, or lacking, minute; cincinnus peduncle plus pedicel (total stalk below the flower or fruit) erect or sometimes spreading laterally in fruit, 13–26.5 mm long, the cincinnus peduncle (part below the the articulation) 7–17 mm long, pedicel 5.5–11 mm long. Flowers bisexual, 6–9 mm wide. Sepals lanceolate-elliptic to ovate-elliptic, 2.5–5 × 1–2 mm. Petals pale to dark lilac or mauve, ovate-orbicular to elliptic-suborbicular, 3–5.5 × 2.75–5 mm. Fertile stamens 3, filaments 1.5–2.3 mm long, bearded with white hairs, anthers 0.5–1.3 mm long, blue; staminodes lacking. Ovary 0.7–2 mm long; style tapering from the base, 0.5–1 mm long. Capsules pale yellow to light brown, oblong to oblong-ellipsoid, (3.5–)4.5–6 × 1.5–2 mm, style base persistent. Seeds uniseriate, (1–)2–3 per locule, quadrate to rectangular, slightly trapezoidal, broadly ovoid or ellipsoid, 1–1.3 × 0.8–1.1 mm wide, testa light brown, with numerous, often broken, raised lines radiating from the embryotega.

UGANDA. Bunyoro District: 85 km Masinde Port–Butiaba road, 3 Oct. 1935, *Eggeling* 2244!; Teso District, Lira Road swamp, 17 Sep. 1954, *Lind* 418!; Mengo District: Kings Lake, Kampala, 5 Dec. 1935, *Chandler & Hancock* 91!

TANZANIA. Ufipa District: Chapota Swamp, 6 Mar. 1957, *Richards* 8496! & 7 km Namanyere–Karange road, 4 Mar. 1994, *Bidgood et al.* 26208!; Songea District: 12 km W of Songea by Kimararampaka stream, 31 Dec. 1955, *Milne-Redhead & Taylor* 7969!

DISTR. **U** 2–4; **T** 4, 7; Guinea, Ivory Coast, Ghana, Nigeria, Cameroon, Central African Republic, Congo–Kinshasa, Ethiopia, Zambia and Angola

HAB. *Miscanthidium* swamps and swampy edges of streams and lakes, peat soils overlying laterite; 950–1500 m

SYN. *Baoulia tenuissima* A. Chev. in Mém.Soc. Bot. Fr. 2, 58, 8: 217 (1912)
Aneilema tenuissimum (A. Chev.) Hutch. in F.W.T.A. ed. 1, 2: 314 (1912); A. Chev. in Expl. Bot. Afr. Occ. Franç. 1: 668 (1920)

NOTE. The spotty distribution of this species is testimony to the almost certainty that it is the most difficult African Commelinaceae to spot in the field. Its grass-like foliage is inconspicuous among the taller grasses and sedges with which it grows, and its flowers are small and open only briefly. In West Africa they are recorded as opening about 10:30 and fading in the early afternoon.

The absence of staminodes led Chevalier to describe this species under the new generic name *Baoulia*. However, all later authors agree that it belongs in *Murdannia*.

8. **POLYSPATHA**

Benth., Niger Fl.: 543 (1849); C.B. Clarke in DC., Monogr. Phan. 3: 194–195 (1881) & in F.T.A. 8: 61–62 (1901); Morton in J.L.S. 60: 206 (1967); Faden in Kubitzki, Fam. Gen Vasc. Pl. 4: 125 (1998)

Stoloniferous perennial herbs with fibrous roots. Leaves spirally arranged, mainly clustered towards the shoot apex; lamina petiolate, base symmetric, ptyxis supervolute. Inflorescences terminal or terminal and axillary from the upper leaves, forming a terminal cluster, an axillary inflorescence sometimes perforating the subtending sheath, each inflorescence an apparent modified thyrse consisting of an elongate axis to which are attached several to many, sessile spathes arranged distichously, each spathe enclosing 1(–3) contracted, several-flowered cincinnus, spathe margins free; bracteoles persistent. Flowers bisexual, small (less than 1 cm wide), zygomorphic, shortly pedicellate. Sepals free, subequal, sepaloid, puberulous. Petals white (rarely pink?), unequal, upper 2 clawed, lower petal very reduced, not clawed. Filaments fused basally, glabrous; staminodes 3, borne on the upper side of the flower, antherodes 'V'-shaped, yellow, stamens 3, borne on the lower side of the flower, longer than the staminodes, the medial stamen shorter than the laterals, connective narrow, anther dehiscence longitudinal. Ovary sessile, glabrous, bilocular, locules l-ovulate; style slender; stigma capitate. Capsules dehiscent, bilocular, bivalved, glabrous; cells of the capsule wall ± isodiametric. Seeds elliptic, radiately ribbed, hilum linear, raised in a groove, embryotega semilateral to semidorsal.

African endemic genus of three species; plants of forest understory and disturbed situations, growing in shade.

1. Lamina lacking long, uniseriate hairs, often scabrous above; seeds ribbed, with (12–)14–18 smooth, uninterrupted ribs 1. *P. paniculata*
   Leaves with long, uniseriate hairs, at least on the adaxial surface, never scabrous above; seeds either shallowly ribbed-reticulate or deeply ribbed with (17–)18–23 prominent, knobby ribs that are transversely interrupted 2
2. Long, uniseriate hairs usually present on the both leaf surfaces, the internodes and sheath surfaces; spathes crowded, becoming deflexed against the inflorescence axis; cells of the spathe surface dull, neither brown nor bead-like under a 20× lens; seeds shallowly ribbed-reticulate, the ribs neither knobby nor transversely interrupted . . . . 2. *P. hirsuta*
   Long, uniseriate hairs present on the adaxial leaf surface and summits of the leaf sheaths, always lacking from the abaxial leaf surface and usually elsewhere; spathes well spaced, usually patent to slightly deflexed; cells of the spathe surface, at least near the midrib, but often all over, lustrous, brown, with bead-like cells under a 20× lens; seeds deeply ribbed, the ribs knobby and transversely interrupted . . . . . 3. *P. oligospatha*

1. **Polyspatha paniculata** *Benth.* in Niger Fl.: 543 (1849); C.B. Clarke in DC., Monogr. Phan. 3: 194 (1881) & in Durand & Schinz, Consp. Fl. Afr. 5: 429 (1895) & in F.T.A. 8: 61 (1901); F.P.U.: 196 (1962); Brenan in F.W.T.A. 2nd ed.: 3: 42 (1968). Type: Bioko [Fernando Po], *Vogel* 93 (K!, holo.)

Perennial with erect shoots 15–60 cm tall, stoloniferous; internodes puberulous with hook-hairs or a mixture of hook-hairs and short uniseriate hairs. Leaves subclustered terminally on the flowering shoot; sheaths (1–)1.5–2(–2.3) cm long, puberulous, ciliolate at apex; lamina petiolate, elliptic to ovate or obovate, (3.5–)7–17(–19) × (2.5–)3–5.5(–6.5) cm, base cuneate, margins planar, ± scabrid, apex acuminate (to acute or abruptly acute); adaxial surface glabrous or scabrid, abaxial sparsely puberulous. Inflorescence a terminal, compound thyrse 5–15 × 0.5–4 cm, composed of up to 7 erect to ascending,terminal and axillary (from the upper leaves) simple thyrses, one or more of the axillary thyrses sometimes perforating a leaf sheath, each consisting of a short peduncle and an elongate zigzag retrosely puberulous axis with 7–16 distichous spathes; spathes at first erect, then patent, soon becoming deflexed against the thyrse axis, usually slightly overlapping the one below (on the opposite side), brown or brownish, 8–11 × 3–5 mm (folded), apex acuminate to acute, puberulous, ciliolate; cymes up to 5-flowered. Pedicels 1.5–2 mm long, sparsely puberulous. Sepals 4–5 × 1–2 mm in flower, to 6.5 × 2.5 mm in fruit, glabrous or puberulous. Capsules tan, oblong-elliptic, 3.5–4 × 2–3 mm, apex rounded to truncate. Seeds elliptic in outline, 2.5–2.8 × 1.4–1.7 mm, testa (12–)14–18-ribbed, sometimes covered or partially covered by a matted, pale brownish material.

UGANDA. Bunyoro District: Massindi, Budongo Forest Reserve near Songo R., 31 Aug. 1995, *Poulsen et al.* 920!; Kigezi District: Ishasha Gorge, Aug. 1948, *Purseglove* 3065!; Mengo District: Mpanga Forest, 5 km E of Mpigi, Sep. 1969, *Faden et al.* 69/1006!

TANZANIA. Bukoba District: Minziro Forest Reserve, Nyakabanga, 16 Aug. 1999, *Sitoni, Festo & Bayona* 765!

DISTR. **U** 2, 4; **T** 1; Guinea, Sierre Leone, Liberia, Ivory Coast, Ghana, Nigeria, Cameroon, Equatorial Guinea, Central African Republic, Gabon, Congo–Brazzaville, Congo–Kinshasa, Cabinda, Sudan

HAB. Forest understory, especially in disturbed situations; 1150–1400 m

SYN. *P. paniculata* Benth. var. *glaucescens* C.B. Clarke in DC., Monogr. Phan. 3: 195 (1881). Type: Cameroon, Mt Cameroon, *Mann* 2138 (K, holo.)
P. glaucescens (C.B. Clarke) Hutch. in F.W.T.A. ed. 1, 2: 320 (1936) & in K.B. 1939: 244 (1939)

NOTE. Morton in J.L.S. 60: 206 (1967) states *P. glaucescens* (C.B. Clarke) Hutch. is merely a small form of *P. paniculata*, and not worthy of specific status. In fact C.B. Clarke himself effectively sank it in his remarks in F.T.A. 8: 52/62 (1901).

Flowers are reported in F.W.T.A., ed. 2 (1968) as being open 14:30–17:00. The flowers are normally white, but in *Sitoni, Festo & Bayona* 765 the petals are reported as pink.

2. **Polyspatha hirsuta** *Mildbr.* in N.B.G.B. 9: 256 (1925); Morton in J.L.S. 60: 206 (1967); Brenan in F.W.T.A. 2nd ed.: 3: 42 (1968). Type: Cameroon, Molundu, between Njombe and Nginda, 21 km N of Molundu, *Mildbraed* 4117 (B, holo.)

Perennial with erect shoots 15–60 cm tall, stoloniferous; internodes pilose or pilose-puberulous with ascending hairs. Leaves scattered along the shoots, with several clustered just below the inflorescence; sheaths 1–2 cm long, pilose-puberulous, long-ciliate at apex; lamina petiolate, elliptic to oblong-elliptic, (4–)6–12.5 × (1–)1.7–4.2 cm, base cuneate, margins ciliolate, scabrid, apex acuminate; both surfaces pilose-puberulous or occasionally the long hairs only present above. Inflorescence a terminal, compound thyrse (2.5–)3.5–5(–6) × 2–5 cm, composed of up to ± 12 erect to ascending terminal and axillary (from the upper leaves) simple thyrses, none perforating a leaf sheath, each consisting of a short peduncle and an elongate strongly zigzag retrosely puberulous axis, with up to 10 distichous spathes; spathes at first erect, then patent, soon becoming deflexed against the thyrse axis, overlapping the one below (on the opposite side) up to ± half its length, green, (5–)6–8.5(–11) × 4–6 mm (folded), apex rounded to obtuse or occasionally acute (in the lower spathes), pilose-puberulous, long-ciliate; spathes 2–5-flowered. Pedicels 1.5–2.5 mm long, with a few, scattered long hairs. Sepals 3.5–4.5 mm long, pilose. Capsules tan, oblong-elliptic to elliptic-orbicular, 2.5–3.5 × 2.3–3 mm, apex truncate to emarginate. Seeds ellipsoid, 1.8–2.3 × 1.4–1.6 mm, testa weakly ribbed, covered by a matted, pale brownish layer.

UGANDA. Bunyoro District: Budongo Forest Reserve, between the Royal Mile and Nature Reserve, Nyabisabo R., 4 Dec. 1996, *Poulsen et al.* 1246!; Toro District: near Kirmia, W of Ntandi, Semliki Forest, 23 Sept. 1969, *Faden* 69/1273! & Bwamba, Buyayu–Sempayo road, Oct. 1929, *Liebenberg* 924!

DISTR. **U** 2; Sierre Leone, Ivory Coast, Togo, Ghana, Nigeria, Cameroon, Congo–Kinshasa

HAB. Moist forest understorey; 700–1000 m

NOTE. In F.W.T.A., ed. 2 (1968) the flowers are reported as being open from about 14:30 until evening.

Some specimens from West Africa, particularly Ghana to Nigeria, are much less pubescent, sometimes lacking long hairs on the internodes and sheaths (commonly), abaxial leaf surface (commonly), adaxial leaf surface (rarely), and spathe surfaces (rarely). This tendency towards reduced pubescence has not been seen in our area.

Plants that I have interpreted as hybrids between *P. hirsuta* and *P. paniculata* have been seen from Ghana (e.g. *Thomas* D34) and Sudan. They have the elongate, laxer inflorescence of *P. paniculata* and broader, blunter and hairier spathes of *P. hirsuta*. The leaves are broadly elliptic, broader than the broadest *P. hirsuta*, and the adaxial surface may or may not have some long, uniseriate hairs. Perhaps most striking is the absence of capsules in such plants. Hybrids have not been seen from our area but they should be looked for. No hybrids have been recognized involving *P. oligospatha*.

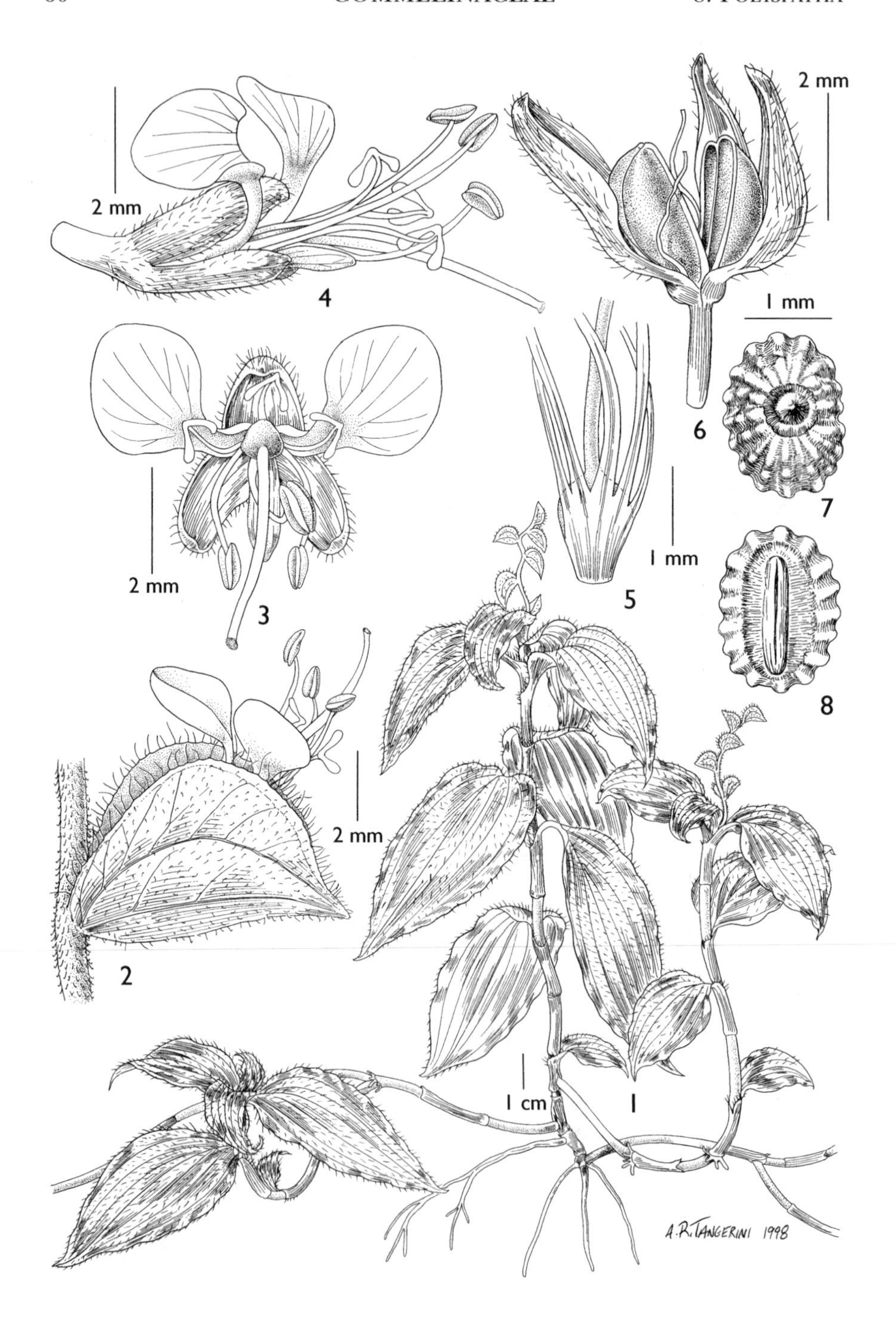

FIG. 8. *POLYSPATHA OLIGOSPATHA* — **1**, habit; **2**, spathe with open flower, side view; **3**, flower, front view; **4**, flower, side view; **5**, stamen and staminode filaments, showing basal fusion; **6**, dehisced capsule; **7** & **8**, seed, dorsal and ventral view. All from *Poulsen* 1275. Drawn by Alice R. Tangerini.

3. **Polyspatha oligospatha** *Faden* in Phytokeys 3: 10 (2011). Type: Uganda, Mengo District: Zintengeze [Zintengese], *Dummer* 5531 (US!, holo., K!, iso.)

Perennial with erect shoots to 5–20(–30) cm tall, stoloniferous; internodes puberulous with hook-hairs. Leaves usually subclustered terminally on the flowering shoot; sheaths 0.8–1.5 cm long, puberulous, long-ciliate at apex; lamina petiolate, elliptic to ovate or ovate-orbicular, rarely oblong-elliptic, 3–9(–11) × 2–4.5 cm, base cuneate to broadly cuneate, margins scabrid, apex acute to acuminate (to abruptly acute or mucronate); adaxial surface with scattered long hairs, abaxial puberulous. Inflorescence a terminal, simple or compound thyrse 2–7 × 7.5 cm, of up to 4 erect to patent or declinate terminal and axillary (from the upper leaves) simple thyrses, none perforating a leaf sheath, each consisting of a short peduncle and an elongate zigzag retrosely puberulous axis, with 4–8 distichous spathes; spathes at first erect, then patent, rarely becoming deflexed against the thyrse axis, usually not overlapping the one below, brown or brownish at least along the midrib, 6.5–10(–12) × (3–)4–7 mm wide (folded), apex acute to rounded, sometimes mucronate, sometimes the lower spathes acuminate, puberulous, ciliolate with hairs <1.5 mm long; cincinni ± 2–3-flowered; bracteoles ovate, with a few short hairs on the margins. Pedicels 1–2.5 mm long, glabrous or with an occasional short, uniseriate hair. Sepals ± 3–4.5 × 1–2.3 mm in flower, ± 4 mm long in fruit, puberulous. Corolla white, upper petals 5–7 × 2.5–3 mm, lower petal (2–)3–4.7 × (0.2–)0.8–1 mm. Staminodes with filaments 3–6 mm long, antherodes V-shaped; stamens with filaments 4.5–8.5 mm long, anthers 1–1.7 mm long. Style 5.5–8.5 mm long. Capsules tan, broadly ellipsoid, 2.5–3 × 2.5–3 mm, constricted between the seeds, apex emarginate. Seeds ellipsoid, 1.6–2.2(–2.5) × 1.3–1.5(–1.6) mm, testa with (17–)18–23 prominent ribs, the ribs ± knobby and transversely interrupted, surface tan, sometimes ± exposed except for some darker brown, matted material between the ribs, sometimes mainly covered by this material except for the rib tops. Fig. 8, p. 60.

UGANDA. Bunyoro District: Budongo Forest Reserve, 17 July 1969, *Stewart* in EA 14170!; Toro District: Bwamba, Buyayu–Sempayo road, oct. 1929, *Liebenberg* 922!; Kigezi District: Rwahumbura, Maramagambo Forest, 18 Sep. 1969, *Lye* 4117!

DISTR. **U** 2, 4; Ivory Coast, Cameroon, Congo–Brazzaville, Congo–Kinshasa, Sudan

HAB. Forest understorey; 750–1250 m

NOTE. *Poulsen* 1275 from Uganda, cultivated at the Smithsonian Botany Research greenhouse in 1997, had flowers open by 09:30 and fading at noon. Two collections from Cameroon, *Hall & Kahn* 75/93 and *Keating* 90–13 also record that this species flowers in the morning and fades about noon. Thus the flowering period for this species does not overlap with those of the other two species.

There are two sheets of *Liebenberg* 922 at Kew. Sheet 1 is a mixture of this species and *P. paniculata*, while sheet 2 is unmixed *P. oligospatha*. *Dummer* 5531 is unusual in having up to three cincinni per spathe instead of the normal one. *Daws* 683 is atypical is a number of characters: it is taller (to 30 cm tall) than any other specimen; its leaves are oblong-elliptic, as in *P. hirsuta*; they are less clustered subterminally on the shoots than in typical specimens of *P. oligospatha*; and they are the longest (to 10 cm) seen in any specimen. There is no compelling evidence that *Daws* 638 might be a hybrid with *P. hirsuta*, and, in view of the difference in flowering times, hybridization seems unlikely.

This species has been overlooked or misinterpreted because of the confusion caused by Clarke's form *P. paniculata* var. *glaucescens*, which was raised to a species in F.W.T.A, ed. 1, by Hutchinson, although it had already been abandoned by Clarke (1901) in F.T.A. From a study of the type of this variety, Brenan in F.W.T.A., ed. 2 (1968) correctly concluded that it was no more than a depauperate form of *P. paniculata*. This led him to annotate the only specimen at Kew of *P. oligospatha* from West Tropical Africa, *Aké Assi* 5707 from Ivory Coast, as "small form – I do not consider *P. glaucescens*... as distinct." On the other hand, it is clear that Brenan had begun to suspect that there was more variation in the genus than the two West African species by his note on the species folder for *P. paniculata* from Uganda, "Some of these Uganda sheets look distinct and should be investigated."

P. oligospatha is very distinctive because of the small size of the plants, the presence of long, uniseriate hairs on the upper side of the leaves, the inflorescences consisting of a few, well-spaced spathes that are usually patent to slightly deflexed, the cells of the spathe wall, at least near the midrib, but often all over, lustrous, brown and bead-like, the seeds deeply ribbed with the ribs knobby and transversely interrupted, and morning anthesis.

9. POLLIA

Thunb., Nov. Gen. Pl. 1: 11 (1781); C.B. Clarke in DC., Monogr. Phan. 3: 121–130 (1881) & in F.T.A. 8: 26–27 (1901); Faden in Kubitzki, Fam. Gen Vasc. Pl. 4: 125 (1998)

Perennial herbs, usually stoloniferous or rhizomatous. Leaves alternate, spirally arranged; lamina petiolate, base symmetric. Inflorescences very dense to lax terminal thyrses. Flowers bisexual and male, actinomorphic to slightly zygomorphic, small (usually 1 cm wide or less or less), pedicellate. Sepals free, subequal, sepaloid or petaloid. Petals free, equal or the lower one somewhat differentiated from the upper 2, usually white (sometimes blue or pink), upper 2 sometimes shortly clawed, lower one not clawed. Stamens 6, equal or irregularly unequal, all fertile, or upper 3 distinctly shorter than lower 3 and either fertile or staminodial; filaments free, glabrous; anther dehiscence longitudinal. Ovary sessile, glabrous, trilocular, locules equal or one somewhat reduced, ovules (1–)2(–4) seriate; style slender; stigma capitate. Fruits blue to blue-black or grey-blue, often lustrous and appearing metallic, berry-like but hard and crustaceous, indehiscent, trilocular, glabrous. Seeds (1–)2(–3) seriate, (1–)2–18 per locule, polygonal, grey or brown, smooth or shallowly pitted, hilum punctiform to oblong, embryotega dorsal.

A mainly paleotropical genus of ± 18 species, extending into temperate eastern Asia, with one record from the Neotropics (*P. americana* Faden from Panama); three species in Africa, and one in Madagascar.

1. Sepals glabrous; petals pinkish purple or white tinged with lavender; seeds 2–4 seriate, 7–19 per locule .... 2. *P. bracteata*
   Sepals puberulous; petals white; seeds 2-seriate, (2–)4–8 per locule 2
2. Inflorescence lax, panicle-like, composed of 4–10 cincinni, not covered by enlarged cincinnus bracts when young; leaves 4–13.5 × 1.5–3.3 cm; fruits ellipsoid, grey-blue when ripe; fertile stamens 6 .... 1. *P. mannii*
   Inflorescence very dense, head-like, composed of many cincinni, covered by enlarged cincinnus bracts when young; leaves 11–31 × 2.5–8 cm; fruits nearly spherical, lustrous metallic blue or blue-black; fertile stamens 3 .... 3. *P. condensata*

1. **Pollia mannii** *C.B. Clarke* in DC., Monogr. Phan. 3: 124 (1881) & in Durand & Schinz, Consp. Fl. Afr. 5: 422 (1895) & in F.T.A. 8: 26 (1901); Brenan in F.W.T.A. 2nd ed. 3: 32 (1968); Ensermu & Faden in Fl. Eth. 6: 348 (1997). Type: São Tomé, *Mann* 1098 (K!, lecto., selected here)

Perennial with a densely branched, repent shoot system giving rise to sheath-perforating, unbranched, ascending flowering shoots (10–)15–30(–50) cm tall. Leaves spirally arranged, not strongly clustered at the summit of the shoot; sheaths 1–3 cm long, puberulous, ciliolate at apex; lamina narrowly elliptic to lanceolate, 4–13.5 × 1.5–3.3 cm, base cuneate, apex acuminate-attenuate, surfaces puberulous to glabrous, the adaxial usually with midrib puberulous. Inflorescence a terminal lax ovoid thyrse 4–8.5 × 1.5–5 cm, of 4–10 alternate ascending cincinni; peduncle 2–3 cm

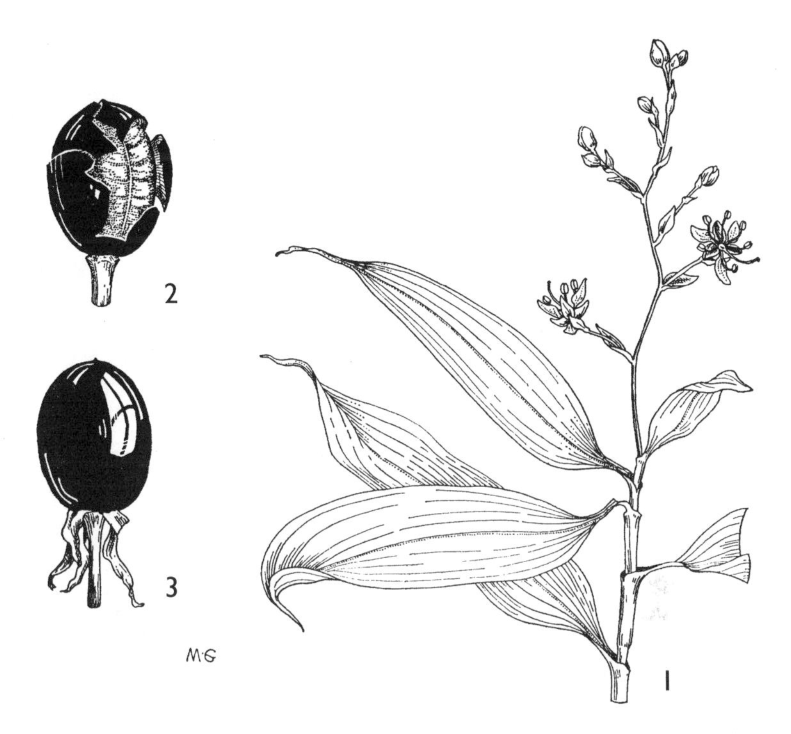

FIG. 9. *POLLIA MANNII* — **1**, habit, × 1; **2** & **3**, fruit, × 4. Drawn by M. Grierson and reproduced with permission, from J.L.S. 59: 359 (1966).

long, puberulous; cincinni to 2.5 cm long and 4-flowered; cincinnus bracts appressed to the cincinnus peduncles, persistent, lanceolate to elliptic, 4–10 mm long, puberulous; bracteoles persistent, asymmetically cup-shaped, perfoliate, 2.5–4 mm long, puberulous. Flowers bisexual, pedicels ± 4 mm long in flower, to 6.5 mm long in fruit, puberulous. Sepals oblong to oblong-oblanceolate or elliptic, 4–5 mm long, white, puberulous. Petals exceeding the sepals, white, obovate. Stamens 6, subequal, all fertile. Style ± 6.5–7.5 mm long. Fruits grey-blue when mature, ellipsoid, 5–6 × 4–4.5 mm. Seeds biseriate, 7–8 per locule, dorsiventrally compressed, polygonal in outline, ± 1.5 mm in diameter, testa dark brown or grey, ± smooth. Fig. 9.

UGANDA. Bunyoro District: Budongo Forest, 16 Sep. 1969, *Faden* 69/1078!; Toro District: near Kirmia, W of Ntandi in Semliki Forest, 23 Sep. 1969, *Faden* 69/1267; Mengo District: Mabira Forest, 30 Aug. 1950, *Dawkins* 628!

TANZANIA. Bukoba District: Minziro Forest, Sep. 1952, *Procter* 95! & Apr. 1958, *Procter* 883! & Minziro Forest, 26 Apr. 1994, *Congdon* 361!

DISTR. **U** 2–4; **T** 1; Ivory Coast to S Sudan and SW Ethiopia to Congo–Kinshasa and Angola

HAB. Forest understory and forest edge; 700–1250 m

NOTE. A part of *Welwitsch* 6604 was cited by Clarke (1881: 124) as belonging to this species, but that was in error, for both sheets are *P. condensata* (see Clarke, F.T.A. 8: 27, 1901)

The habit of this distinctive species is unlike that of any other of the genus, being neither stoloniferous nor rhizomatous. The diffusely spreading stems that give rise to ascending shoots more closely resemble those of some species of *Commelina* than of other *Pollia* species. Two collections from Uganda record the flowering time as 16:00. whereas in F.W.T.A. ed. 2 it is recorded as 10:00. Whether the flowers are open for a relatively long period, e.g. at least 10:00 to 16:00, or they open at different times in West and East Africa is unknown.

Faden 69/1078 notes that sometimes several of the stamens may have shorter filaments or be sterile. There are insufficient observations to know whether this might occur in other populations.

In the original description of this species C.B. Clarke also cites *Welwitsch* 6604 partim as well as *Mann* 1098, but in F.T.A. 8: 27 (1901) he states "part of *Welwitsch* 6604 was referred by me erroneously to *P. mannii* in the [protologue]"; so the inclusion of this specimen as a type in Fl. Eth. 6: 348 (1997) was an error.

2. **Pollia bracteata** *K. Schum.* in E.J. 33: 375 (1903). Type: Tanzania, Morogoro District: Uluguru Mts, Tegetero [Tegetiro], *Stuhlmann* 9031 (B!, holo.)

Perennial to 1.5 m tall, stoloniferous. Leaves clustered or not at the summit of the stems; sheaths 1.5–4 cm long, puberulous, ciliolate at the apex; lamina oblong-elliptic to narrowly elliptic, 15–34 × (2.2–)4–6.5 cm, base narrowly cuneate, apex caudate-acuminate; adaxial surface scabrid, the hairs sometimes ± confined to the midrib towards the base, abaxial surface puberulous, especially on the midrib towards the base. Inflorescence terminal, thyrsiform; peduncle bracteate, 9–12.5 cm long, puberulous; thyrse ovoid, ± dense to lax, 4.5–9.5 × 4–7 cm, of 13–28 mostly subverticillate ascending cincinni to 5 cm long; cincinnus bracts persistent, becoming reflexed, boat-shaped, lanceolate-elliptic, 7–12 mm long, puberulous; bracteoles perfoliate or not, 1.5–3.5 mm long, puberulous proximally. Flowers bisexual and male, 8–9 mm wide; pedicels 2–7 mm long, glabrous. Sepals white with pale lavender margins, ovate-orbicular to obovate, deeply cup-shaped, 3–4 × 2–3 mm, glabrous. Petals slightly unequal, pinkish purple or white tinged with lavender basally and on lower margin, paired petals sub-clawed, 3.5 × 2.5–3 mm, limb deeply cup-shaped, lower petal spatulate, 3 × 2–2.5 mm. Stamens 6, slightly exceeding the sepals, unequal, the upper 3 and sometimes the medial lower one with filaments 4–5 mm, the lower 3 all (or sometimes only the lateral 2) 6–8 mm long, anthers 0.8–1 mm long, upper 3 entirely yellow, lower 3 with blue-black pollen sacs. Ovary 1.5 mm long, ovules up to 4-seriate, ?6–19 per locule; style 5.5–6.7 mm long. Fruits dark blue, lustrous but sometimes not clearly metallic, subspherical, not trigonous, 5–7 mm in diameter. Seeds 2–4-seriate, 7–19 per locule, 1–1.5(–1.65) mm in diameter, 0.5–1 mm thick, outer surface polygonal, 4–6-sided, testa minutely papillose, light grey to brown.

TANZANIA. Lushoto District: Amani Zigi Forest Reserve, 6 Feb. 1999, *Sallu* 281! & Kwepima Hill between Ambangulu Forest Reserve and Kwemasimba public forest, 15 Apr. 1999, *Mwangoka* 437!; Morogoro District: N Uluguru Mts, N valley of Lumanga, S of Mgolole Mission, 8 Nov. 1972, *Pócs & Kornas* 6811/C!

DISTR. **T** 3, 6, 7; not known elsewhere

HAB. Forest understory; 850–1150(–1400?) m

NOTE. In fresh fruits and flowers of a collection from Udzungwa National Park (*Festo, Mwangoka & Luke* 2149) up 17 ovules in 4 rows were found in a locule of one ovary and up to 18 seeds in 3 rows were found in a single fruit locule. In *Jannerup & Mhoro* 0305, from Mkungwe Catchment Forest Reserve, the seeds were 4-seriate up to 19 per locule. In *Sallu* 281, from the Eastern Usambara Mountains, however, at least 21 seeds were counted in one nearly mature fruit, with a maximum of 10 seeds in one locule. With the small number of collections available and this great variance in seed number, it is impossible to know the typical number of seeds per locule or fruit in this species. The seeds in *Festo, Mwangoka & Luke* 2149 and *Jannerup & Mhoro* 0305 are tetrahedral, not at all dorsiventrally compressed, but those in *Sallu* 281 are more dorsiventrally compressed, a shape that is typical of other species of *Pollia*. The 4-seriate seeds observed in at least some locules in two populations of this species seems to be unique in *Pollia*.

This species approaches *P. gracilis* C.B. Clarke of Madagascar and the Comoro Islands in its moderately dense to lax inflorescences, numerous, mostly subverticillate cincinni, and six fertile stamens. That species differs from *P. bracteata* by its ellipsoid inflorescences with shorter peduncles, puberulous sepals, violet petals, and smaller, ellipsoid fruits that are whitish at maturity.

3. **Pollia condensata** *C.B. Clarke* in DC., Monogr. Phan. 3: 125 (1881) & in Durand & Schinz, Consp. Fl. Afr. 5: 421 (1895); K. Schum. in P.O.A. C: 134 (1895); C.B. Clarke in F.T.A. 8: 27 (1901); F.P.U.: 195 (1962); Brenan in F.W.T.A. 2nd ed.: 3: 33

(1968); Faden in U.K.W.F. 2nd ed.: 307 (1994); Ensermu & Faden in Fl. Eth. 6: 349 (1997). Types: Bioko [Fernando Poo], *Mann* 93; *Barter* 1518, 2020; & Angola, Cazengo, Muxaulo, *Welwitsch* 6604 (K!, syn.)

Perennial, robust, stoloniferous, forming colonies with erect flowering shoots 30–130(–200) cm tall, the shoots sometimes supported by prop roots from the lower nodes, the upper stem covered by overlapping sheaths. Leaves spirally arranged, mainly clustered towards the summit of the shoot; sheaths 2–6 cm long, puberulous, summit eciliate; lamina oblanceolate-oblong to obovate, 11–31 × 2.5–8 cm, base narrowly cuneate, apex acuminate; upper surface glabrous, lower sparsely puberulous to glabrous. Inflorescence a solitary terminal, very dense ovoid thyrse 1–5 × 1.5–3 cm, of numerous congested subsessile cincinni to 1(–1.5) cm long; peduncle 0.5–3.5 cm long, usually bearing 1–2 foliaceous bracts; cincinnus bracts large, ovate, covering the inflorescence in early flowering, then caducous, puberulous; bracteoles caducous. Flowers bisexual, pedicels 2.5–3.5 mm long in fruit. Sepals green, ovate, 4–5.5 mm long, with a distinctly thickened glandular apex, puberulous. Petals white, obovate, ± 6 × 4 mm, not clawed. Stamens 3, long-exserted, ± 7 mm long; staminodes 3, ± 1.5 mm long, included. Style exserted, ± 7 mm long. Fruits metallic blue or blue-black, sometimes tinged with violet or purple, spherical to subspherical, obtusely trigonous, (3–)4.5–5.5 × (3–)4–5 mm. Seeds biseriate, (2–)4–6 per locule, dorsiventrally compressed, polygonal or trapezoidal in outline, 1.5–2.5 mm in diameter, testa smooth, grey to dark grey-brown.

UGANDA. Kigezi District: Maramagambo Central Forest Reserve, 18 Sep. 1969, *Faden* 69/1115!; Toro District: Kibale Forest, 29 Aug. 1970, *Rwaburindore* 394!; Mengo District: Kasa Forest, 4 Oct. 1949, *Dawkins* 422!

KENYA. Kakamega District: Isiukhu R., S to SSW of Wibakale on Kambiri–Vihiga road, 27 Dec. 1969, *Faden & Faden* 69/2107! & Kakamega Forest, along Kubiranga stream, 16 Mar. 1977, *Faden & Faden* 77/844!

TANZANIA. Kigoma District: Gombe Stream National Park, 17 Dec. 1971, *Harris* 6029!; Lushoto District: Eastern Usambara Mts, Derema [Nderema], 24 Dec. 1956, *Verdcourt* 1722!; Morogoro District: Mkungwe Hill, N. Uluguru Mts, 5 July 1970, *Faden, Evans & Pócs* 70/364!

DISTR. **U** 2, 4; **K** 5; **T** 1, 3, 4, 6, 7; Sierra Leone to Cameroon, Bioko, São Tomé, Principe, Congo–Kinshasa, Sudan, Ethiopia, Angola, Mozambique

HAB. Lowland to submontane forest, sometimes growing near streams or lakes; 300–1700 m

SYN. *Pollia cyanocarpa* K. Schum. in E.J. 33: 375 (1903). Types: Tanzania, East Usambaras, Amani, Nderema, *Heinsen* 24 (B†, K!, syn.) & Uluguru Mts, Rubwe, *Stuhlmann* 9041 (B!, syn.)

Aclisia condensata (C.B. Clarke) G. Brückn. in N.B.G.B. 10: 56 (1927); F.P.N.A. 3: 330, t. 47 (1955)

NOTE. The citation of the Welwitsch syntype as "*Welwitsch* 6604 pro parte" by Clarke (1881) was erroneously based on Clarke having annotated one sheet of this collection as "*P. mannii*". That specimen, however, was also *P. condensata* and the collection was not mixed. Clarke corrected this error in F.T.A.

Pollia condensata is the only African species of the genus that has three fertile stamens and three staminodes. Its very dense inflorescences normally conceal the bracteoles and inflorescence axis. When young the inflorescences are nearly enclosed by the large cincinnus bracts and are often covered by a glistening secretion resembling the liquid in the spathes of *Commelina* species. The inflorescence is so dense that only the petal tips and ends of the stamens and style are exserted. The petals are ± erect, not spreading in the usual manner for the genus. The staminodes are so short that they are not visible without dissection. The staminode anthers are either not developed or else they are merely slight enlargements at the ends of the minute, slender filaments.

It is noteworthy that extreme forms of *P. condensata* occur in the Uluguru Mts of Tanzania which are known for their large numbers of endemics. *Pollia cyanocarpa* appears to be nothing more than *P. condensata* with a somewhat less dense inflorescence. *Harris & Pócs* 4520, also from the Uluguru Mts (Kinole), has a long zigzag peduncle and shows some bracteoles and part of the inflorescence axis. Nevertheless it has the broad cincinnus bracts of *P. condensata* and likely is only a variant of it.

10. ANEILEMA

R. Br., Prod.: 270 (1810); C.B. Clarke in DC., Monogr. Phan. 3: 195–231 (1881) pro parte, & in F.T.A. 8: 62–75 (1901) pro parte max.; Brückner in E.J. 61, Beibl. 137: 1–70 (1926) & in E. & P. Pf., ed., 2, 15a: 159–181; Morton in J.L.S. 59: 431–478 (1966); Faden in Smithsonian Contr. Bot. 1–166 (1991); Faden in Kubitzki, Fam. Gen. Vasc. Pl. 4: 126 (1998)

Lamprodithyros Hassk. in Flora 46: 388 (1863) & in Peters, Reise Mossamb., Bot. 1: 529 (1864)
Ballya Brenan in K.B. 19: 63 (1964)

Perennial or annual herbs; roots fibrous or tuberous. Leaves spirally arranged or distichous, lamina petiolate or sessile, ptyxis usually involute. Inflorescences usually thyrses, sometimes reduced to a single cincinnus, terminal, terminal and axillary, rarely all axillary. Flowers bisexual and male, rarely some female, strongly zygomorphic, pedicellate. Sepals free, unequal, sepaloid. Petals free, unequal, posterior 2 clawed, usually larger, anterior petal usually greatly reduced (sometimes large), not clawed. Staminodes 2–3 (medial sometimes lacking), posterior, shorter than the stamens, filaments glabrous, antherodes bilobed or reduced; stamens 3, anterior, the medial one usually shorter than the others, filaments free or basally fused, glabrous or those of the lateral stamens bearded, anthers with narrow connectives or rarely the medial one with an expanded connective, dehiscence longitudinal. Ovary (uni-), bi- or trilocular, when trilocular, locules unequal, locules 1–6-ovulate. Capsules (uni–), bi- or trilocular, bivalved, (rarely indehiscent), locules 1–6-seeded. Seeds uniseriate, hilum linear, embryotega lateral.

Sixty-four species mainly in tropical Africa, but 2 species in the Arabian Peninsula (of which one is African), 1 endemic to Madagascar, 5 (or 6) in Australia and surrounding islands, and 1 (or 2) in tropical America. DNA evidence supports the inclusion of the genus *Rhopalephora* (1 species in Madagascar, 3 in tropical Asia) in *Aneilema*, which would increase *Aneilema* to 68 species and greatly broaden its range. The genus was divided into seven sections in Faden (1991), six of which occur in our area.

The uppermost leaf or bract on the flowering shoot is sometimes functional reproductively, either surrounding and protecting the developing inflorescence or serving as a background for the open flowers. Because of these functions it was called the "inflorescence bract" in Faden (1991). For the sake of uniformity the term has been applied to the last bract or leaf that is borne proximal to the inflorescence in all species. The peduncle then is measured from the second node below the inflorescence to the base of the inflorescence. The nature of the inflorescence bract – whether it is foliaceous or bract-like – is sometimes useful taxonomically, as is its position on the peduncle.

In section *Lamprodithyros*, the cup-shaped lower (medial) petal often retains the lateral stamen filaments for some time after the flowers open (Faden in Bothalia 14: 997–1002. 1983). Differences in the development of the stamen retention mechanism are sometimes useful in distinguishing species of this section where they occur together.

1. Inflorescences all axillary; lamina sessile, with contrasting white veins above; plants mat-forming .... 22. *A. zebrinum* (p. 104)
   Some or most inflorescences terminal; lamina petiolate or sessile, veins not contrasting; plants rarely mat-forming 2

2. Inflorescence consisting of a single contracted cincinnus enclosed by two leafy bracts; lower petal cup-shaped; stamen filament bases fused . . . 3
   At least the larger inflorescences panicle-like thyrses, consisting of several to many cincinni, not enclosed by leafy bracts; lower petal shape various; stamen filaments free or fused . . . 5
3. Lamina sessile, base asymmetric, internodes commonly swollen . . . 21. *A. succulentum* (p. 103)
   Lamina usually petiolate, base symmetric, internodes not swollen . . . 4
4. Perennials; flowers mostly 15–20 mm wide; pedicels puberulous distally; medial anther yellow with yellow pollen . . . 19. *A. clarkei* (p. 101)
   Annuals; flowers mostly 10–15 mm wide; pedicels pilose-puberulous distally; medial anther white with white pollen . . . 20. *A. lamuense* (p. 102)
5. Flowers yellow to orange . . . 6
   Flowers white to pink, purple, blue or violet . . . 10
6. Inflorescences lax; lateral stamen filaments straight or undulate, bearded or glabrous . . . 7
   Inflorescences dense; lateral stamen filaments ± S-shaped, bearded . . . 27. *A. welwitschii* (p. 112)
7. Inflorescence axis and cincinnus peduncles and axes glabrous; stamen filaments glabrous; roots with distal tubers . . . 4. *A. johnstonii* (p. 76)
   Inflorescence axis and cincinnus peduncles and axes puberulous (at least near the nodes); stamen filaments glabrous or bearded; roots lacking tubers . . . 8
8. Perennials; leaves distichous; stamen filaments bearded . . . 9
   Annuals; most leaves spirally arranged; stamen filaments glabrous . . . 3. *A. ephemerum* (p. 74)
9. Sheaths with whitish hairs at the apex; flowers 19–29 mm wide; pollen orange; seeds 3 per ventral locule . . . 1. *A. aequinoctiale* (p. 70)
   Sheaths with orange or reddish orange hairs at the apex; flowers 8–15 mm wide; pollen white; seeds 1–2 per ventral locule . . . 2. *A. nyasense* (p. 72)
10. Inflorescences lax; flowers (15–)20–40 mm wide; lateral stamen filaments undulate . . . 11
    Inflorescences lax to dense; flowers 3.5–20 mm wide; lateral stamen filaments J- or S- shaped . . . 12
11. Bracteoles perfoliate; capsules puberulous, apex truncate or emarginate; stamen filaments glabrous . . . 5. *A. hockii* (p. 77)
    Bracteoles not perfoliate; capsules glabrous or subglabrous, apex rostrate; stamen filaments finely bearded in the proximal half . . . 6. *A. gillettii* (p. 82)
12. Roots tuberous, the tubers sometimes at the ends of thin roots; plants sometimes rhizomatous . . . 13
    Roots all fibrous; plants never rhizomatous . . . 16

13. Flowering shoots annual, dying to the ground in the dry season, the lower portions not covered by long, overlapping, papery sheaths 14
    Flowering shoots persistent at least basally, the lower portions covered by long, overlapping, papery sheaths .... 10. *A. brenanianum* (p. 86)
14. Root tubers on long stalks; inflorescences composed mainly of 20–40 cincinni 15
    Root tubers sessile or nearly so; inflorescences composed mainly of 6–10 cincinni .... 12. *A. pusillum* (p. 90)
15. Bracteoles not perfoliate; stamen filaments glabrous; capsules bi- to trilocular; ventral locules 2–3(–4)-seeded .... 11. *A. somaliense* (p. 89)
    Bracteoles perfoliate; stamen filaments finely bearded; capsules bilocular, locules 2-seeded .. 24. *A. lanceolatum* (p. 107)
16. All filaments glabrous 17
    Lateral stamen filaments bearded 26
17. Lower petal strongly reduced, neither cup- nor slipper-shaped, differently colored from the paired petals; medial anther broadly elliptic, spotted with maroon; capsules bilocular, rarely trilocular, dorsal locule, when present, usually empty 18
    Lower petal large, cup- or slipper-shaped, concolourous with the paired petals; medial anther saddle-shaped, yellow; capsules trilocular, the dorsal locule commonly 1-seeded 20
18. Shoots decumbent, rooting at the nodes; fruiting pedicels to 8 mm long; capsules 3–6 mm long 19
    Shoots mostly erect to ascending, not rooting; fruiting pedicels to 6.5 mm long; capsules 6–10 mm long .... 9. *A. rendlei* (p. 85)
19. Inflorescence composed of 1–13 cincinni; leaves to 10 cm long; capsules pale yellow, locules 2–3-seeded .... 7. *A. taylorii* (p. 83)
    Inflorescence composed of 14–20 cincinni; leaves to 16 cm long; capsules chocolate brown, locules 4-seeded .... 8. *A. usambarense* (p. 84)
20. Inflorescence lacking a central axis, composed of 1–8 clustered, usually contracted cincinni; medial staminode lacking or vestigial and lacking and antherode 21
    Inflorescence composed of several to many, usually elongate cincinni attached along a central axis; medial staminode usually well developed 22
21. Bracteoles with small marginal glands; medial sepal generally with marginal glands; lateral stamen pollen yellow .... 17. *A. tanaense* (p. 98)
    Bracteoles and medial sepal lacking marginal glands; lateral stamen pollen white .... 18. *A. calceolus* (p. 100)
22. Pedicels glabrous or with a few hairs at the apex, mostly erect in fruit; ventral capsule locules 2–3(–4)-seeded .... 16. *A. sebitense* (p. 97)
    Pedicels puberulous in the distal half or occasionally only at the apex, mostly recurved 180° or more in fruit; ventral capsule locules 2-seeded 23

23. Medial sepal with marginal (as well as a subterminal) glands; fruiting pedicel recurved 180°–270°(–360°); medial staminode lacking or vestigial and lacking an antherode .... 17. *A. tanaense* (p. 98)
Medial sepal with only a subterminal gland; fruiting pedicels recurved ± 180°; medial staminode with an antherode 24
24. Pedicels puberulous only at the apex; upper sepal glabrous; bracteoles ± symmetrically cup-shaped, glabrous; capsules mostly 3.5–5 mm long × 2–2.5 mm wide .... 15. *A. recurvatum* (p. 96)
Pedicels puberulous in the distal half, rarely only at the apex; upper sepal puberulous; bracteoles asymmetrically cup-shaped or not cup-shaped, commonly with 1-several long hairs at the base; capsules mostly 4.5–7.5 mm long × 2.3–5 mm wide 25
25. Bracteoles not cup-shaped; lateral stamen filaments crossing; capsules mostly 3.3–5 mm wide, dorsal capsule valve deciduous; dorsal and ventral locule seeds strongly dimorphic .... 13. *A. petersii* (p. 91)
Bracteoles asymmetically cup-shaped; lateral stamens filament ± parallel, not crossing; capsules mostly 2.3–3 mm wide, dorsal valve usually persistent; dorsal and ventral locule seeds scarcely dimorphic .... 14. *A. indehiscens* (p. 94)
26. Bracteoles cup-shaped, perfoliate; stamen filaments inconspicuously bearded on the lower surface; perennials 27
Bracteoles neither cup-shaped nor perfoliate; stamen filaments conspicuously bearded on the upper and lower surfaces; annuals (except 36. *A. leiocaule*) 30
27. Inflorescences lax or moderately lax, composed of 1–6 cincinni; flowers always white; fruiting pedicels declinate .... 26. *A. umbrosum* (p. 110)
Inflorescences dense, composed of many cincinni; flowers pink to lilac or lavender or white; fruiting pedicels erect to recurved 28
28. Locules 1-seeded; plants of montane forest, 1450–2600 m altitude .... 25. *A. dispermum* (p. 108)
Locules usually 2–3-seeded; plants of lowland or intermediate elevation forest or of savanna or bushland, below 1600 m altitude 29
29. Lamina petiolate or subpetiolate; capsules glabrous; forest plants .... 23. *A. beniniense* (p. 105)
Lamina usually sessile; capsules puberulous; savanna and bushland plants .... 24. *A. lanceolatum* (p. 107)
30. Cincinnus bracts and bracteoles lacking linear, gland-tipped apices (although sometimes apex glandular in *A. spekei*); inflorescences lacking long, uniseriate hairs 31
Cincinnus bracts and bracteoles with linear, gland-tipped apices; inflorescences commonly with long, uniseriate hairs 32

31. Leaves ovate to lanceolate-elliptic, mostly 0.7–3.5 cm wide; sheaths with long hairs at the apex; hilum in a groove . . . . 28. *A. spekei* (p. 113)
   Leaves linear to linear-lanceolate, 0.2–1 cm wide; leaf sheaths lacking long hairs at the apex; hilum in a pit . . . . 29. *A. richardsiae* (p. 115)
32. Leaves with lamina sessile [sometimes shortly petiolate in *A. hirtum*] 33
   Leaves with lamina petiolate. 34
33. Flowers 8–10(–12) mm wide; sepals usually lacking longer, uniseriate hairs (in addition to short hook-hairs), when present, such hairs only or mostly on the sepal base; stamen filaments bearded with maroon to purple hairs; capsule locules 1–2-seeded; seeds not 3-lobed . . . . 30. *A. hirtum* (p. 116)
   Flowers (9.5–)13–15 mm wide; sepals with uniseriate hairs scattered on the surface; stamen filaments bearded with yellow hairs; capsule locules 2-seeded; seeds ± 3-lobed . . . . 31. *A. chrysopogon* (p. 118)
34. Ventral locules usually 1-seeded; solitary cincinnus present at base of peduncle of terminal thyrse, widely separated from other cincinni; medial petal broader than either paired petal. . . . . . . 36. *A. minutiflorum* (p. 126)
   Ventral locules 2–3-seeded (1-seeded only through abortion); no cincinnus present at base of peduncle of terminal thyrse; medial petal narrower than a paired petal 35
35. Seeds 3 per ventral locule, 3-lobed . . . . 32. *A. nicholsonii* (p. 119)
   Seeds 2 per ventral locule, unlobed or 3-lobed. 36
36. Flowers 4–7.5 mm wide; medial staminode lacking; filaments bearded with reddish purple or maroon hairs; seeds 3-lobed . . . . 33. *A. termitarium* (p. 121)
   Flowers 9–13 mm wide; medial staminode present; filaments bearded with bluish purple hairs; seeds unlobed 37
37. Annuals; dorsal locule often well-developed, usually containing a seed; **T** 6, 7?, 8 . . . . 34. *A. pedunculosum* (p. 123)
   Decumbent perennials; dorsal locule not developed; **U** 2; **K** 1, 3, 4–7; **T** 2, 3, 6, 7 . . . . 35. *A. leiocaule* (p. 124)

Aneilema plagiocapsa K. Schum. has been collected on the Zambian side of the Kalambo River, e.g. *Richards* 24482, on the border between Tanzania and Zambia, but has never been collected in the FTEA area. In the key above it should easily key out to the choice between *A. hockii* and *A. gillettii*. It could occur with *A. hockii* from which it differs by its more numerous inflorescences on peduncles with several bracts, bearded lateral stamen filaments, and an indehiscent fruit enclosing the single dorsal locule seed, the fruit rounded at the apex (not truncate to emarginate) with cells of the capsule wall isodiametric (not transversely elongate). We failed to find it in 1996 near Kalambo Falls, but we were probably too late in the season for it.

1. **Aneilema aequinoctiale** (*P. Beauv.*) *G. Don* in Loudon, Hort. Brit., 15 (1830); Kunth, Enum. Pl. 4: 72 (1843); C.B. Clarke in DC., Monogr. Phan. 3: 222 (1881) pro parte excl. var. *minor* and var. *kirkii*; C.B. Clarke in F.T.A. 8: 65 (1901); U.O.P.Z.: 122

(1949); Brenan in K.B. 7: 191, fig. 15, 16 (1952); Morton in J.L.S. 59: 443, fig, 1, 4 (1966); Brenan in F.W.T.A., 2nd ed.: 3: 30 (1968); Faden in U.K.W.F.: 666 (1974); Obermeyer & Faden in F.S.A. 4, 2: 39 (1985); Blundell, Wild Fl. E. Afr.: 411, fig. 406 (1987); Faden in Smithsonian Contr. Bot. 76: 62, t. 1, i (1991) & in U.K.W.F. 2nd ed.: 307 (1994); Ensermu & Faden in Fl. Eth. 6: 351 (1997). Type: Nigeria, vicinity of Benin, *P. de Beauvois* s.n. (G!, holo.)

Perennial to 2(–3) m long; roots thin, fibrous; shoots scrambling or decumbent, usually with dense long hook-hairs. Leaves distichous; sheaths to 4.5 cm long, greatly reduced in distal leaves, densely hirsute, white-ciliate at apex; lamina sessile (distal leaves) or petiolate (proximal leaves), lanceolate to ovate or ovate-elliptic, (3–)5–16(–19) × (1–)2–6 cm, base rounded to cordate (distal leaves) or truncate to cuneate (proximal leaves), margins ciliolate-scabrid, apex (acute to) acuminate; surfaces puberulous and with longer hairs, adaxial surface sometimes scabrid near the margins. Thyrses terminal and sometimes also axillary, lax, ovoid to ellipsoid, 3.5–12 × 2.5–10 cm, with 2–12 cincinni, ascending or becoming patent to declinate; inflorescence axes puberulous; peduncles 3–18.5 cm long with a foliaceous bract ± halfway; cincinni to 8 cm long and 18(–24)-flowered, with broadly ovate, cup-shaped, perfoliate or not perfoliate, bracteoles (1.1–)1.8–3.5(–3.7) mm long, sometimes apparently glandular distally, puberulous at least proximally. Flowers bisexual and male, 19–29 mm wide; pedicels 3.5–5 mm long in flower, 7–12 mm long in fruit, erect in flower and fruit. Sepals entirely green or one or both lateral sepals with subapical-lateral red spots, slightly unequal, 7–8 × 2.5–3.7 mm, glandular subapically; paired petals bright yellow, ovate, 11–16 × 11–17 mm, base truncate to subcordate, claw with a maroon or red spot at base, apex rounded; medial petal reflexed, pale yellow or yellowish white, sometimes pale reddish at base, lanceolate to elliptic or oblong-elliptic, 7–9 × 3–3.5 mm; staminodes 3, yellow, antherodes bilobed; lateral stamens with filaments straight then recurved at apex, 8–18 mm long, bearded with short white hairs, anthers golden yellow to pale orange, ± 3 mm long, pollen orange; medial stamen with filament 4–8 mm long, anther yellow to orange, 1–1.2 mm long, pollen yellow. Ovary ± 3 × 2 mm, densely puberulous; style recurved at the apex and distinctly exceeding the stamens; stigma capitate, blue-violet. Capsules pale yellow to light brown, obovoid-oblong to oblong-ellipsoid in outline, (5–)7–10 × (3.5–)4–6 mm, puberulous; dorsal valve usually persistent; dorsal locule often developed, 1-seeded or empty, ventral locules 3-seeded (or fewer by abortion). Seeds ovoid to subquadrate or trapezoidal, 1.9–2.6 × 1.7–2.15 mm, testa brown to orange-tan, alveolate or faintly to shallowly foveolate-reticulate, not farinose. Fig. 10/3–4, p. 78.

UGANDA. Kigezi Diatrict: South Maramagambo Central Forest Reserve, just outside Queen Elizabeth [Rwenzoli] National Park, 8 June 1969; *Lock* 69/157!; Mengo District: Kasa Forest near Mityana, 12.8 km S of Mityana, N of Butayunja road, 17 Jan. 1950, *Dawkins* D491! & 1.5 km NE of Nansagazi, 11 Sep. 1969, *Faden, Evans & Lye* 69/1042!

KENYA. Northern Frontier District: Marsabit, 14 Feb. 1952, *Gillett* 15107! & Mathews Range, below Sitin, 10 June 1959, *Kerfoot* 1104!; Meru District: Lower Imenti Forest, 9.6 km from Meru along Meru–Mikinduri road, 1 Mar. 1970, *Faden & Evans* 70/118!

TANZANIA. Arusha District: Ngurdoto National Park, Ngurdoto Crater Gate, 22 Oct. 1965, *Greenway & Kanuri* 12189!; Morogoro District: Uluguru Mts below Morningside, 3–4 July 1970, *Faden, Evans & Kabuye* 70/317!; Songea District: 28 km N of Songea, by Luhimba River, 6 May 1956, *Milne-Redhead & Taylor* 10001!

DISTR. **U** 2, 4; **K** 1, 4, 7; **T** 1–8; Guinea, Liberia, Ivory Coast, Ghana, Nigeria, Cameroon, Central African Republic, Equatorial Guinea, Congo–Brazzaville, Congo–Kinshasa, Rwanda, Burundi, Sudan, Ethiopia, Angola, Malawi, Mozambique, Zimbabwe, Swaziland, South Africa

HAB. Moist or dry evergreen forest, mist and riparian forest, especially forest edges and glades, often in dense undergrowth, along rivers and streams and rarely lakes, *Brachystegia* woodland, thickets and thicket edges, rocky slopes, roadsides and roadside banks and other moist, disturbed places; 0–1950 m

Flowering plants have been seen from all months.

SYN. *Commelina aequinoctialis* P. Beauv., Fl. Oware 1: 65, t. 38 (1806)

Aneilema adhaerens Kunth, Enum. Pl. 4: 72 (1843). Type: South Africa, Cape Province, Port St. John's by the river, *Drège* 4466 (B†, holo.; FHO!, G!, lectotype!, L!, MO!, P!, S!, iso.)

Lamprodithyros aequinoctialis (P. Beauv.) Hassk. in Schweinfurth, Beitrag. Fl. Aethiop.: 211 (1867)

Aneilema aequinoctiale (P. Beauv.) G. Don var. *adhaerans* (Kunth) C.B. Clarke in DC., Monogr. Phan. 3: 222 (1881)

NOTE. The illustration accompanying the type description of *Commelina aequinoctialis* depicts the flowers as blue, instead of yellow. It is not clear what possessed the artist to choose that hue, but it is completely wrong. For further discussion see Brenan in K.B. 7: 194–195 (1952).

Two distinct types of *A. aequinoctiale* occur that are both morphologically distinct and geographically separate throughout much of their ranges. In West Africa, extending east to W Ethiopia, Uganda and at least to **T** 1 in Tanzania (e.g. *Festo & Bayona* 562), plants have bracteoles that are not perfoliate, flowers with short, straight, white styles ± equaling the stamens, and capsules often with very acute angled corners at the apex, which can be much wider than the rest of the fruit. In Kenya (all localities), at least **T** 2 in Tanzania (e.g. *Kayombo & Kindeketa* 2919), Mozambique and South Africa, plants of *A. aequinoctiale* have perfoliate bracteoles (sometimes splitting with age), long purple styles that exceed the stamens and are recurved at the apex, and capsules with less acute angles at the apex, which is not much (if at all) wider than the rest of the capsule. However, many if not most plants from Tanzania and Malawi appear to be intermediate between these two extremes, probably as a result of these once isolated taxa coming into contact in central to northeastern Tanzania and perhaps elsewhere. It has not been possible, working mainly from herbarium specimens, to sort adequately most of the Tanzanian specimens into those that exactly match one of the two extremes and those that are intermediates. I leave it for future workers, with access to a broad range of living plants, and perhaps molecular techniques, to characterize more fully the variation in this species throughout its range, and especially in Tanzania.

Aneilema aequinoctiale as a whole is very distinctive because of its scrambling habit; distichous leaves; 'sticky' internodes and leaf sheaths; absence of red hairs from the leaf sheaths; large, yellow flowers with bearded paired petal claws; shortly bearded stamen filaments; capsules up to 7-seeded, with a truncate apex and acute apical corners; and cells of the capsule wall isodiametric. The stickiness of the stems and sheaths, which is the caused by abundant velcro-like hook-hairs, enables *A. aequinoctiale* to climb through other plants. Only *A. nyasense* can be confused with *A. aequinoctiale*, differing by its smaller flowers which lacks hairs on the petal claws, smaller capsules with fewer seeds per locule, and by having red hairs on the leaf sheaths.

The lateral stamen filaments are dimorphic in bisexual and male flowers in this species (Faden 1991: 17). Those of the male flowers are distinctly longer than the comparable filaments of the bisexual flowers in the same populations.

Chimpanzees have been observed to swallow whole leaves of this species in the Kibale Forest of Uganda (*Wrangham* s.n.).

2. **Aneilema nyasense** *C.B. Clarke* in F.T.A. 8: 66 (1901); Brenan in K.B. 15: 213 (1961); Faden in Smithsonian Contrib. Bot. 76: 62, t. 1, j (1991). Type: Malawi, Kondowe to Karonga, *Whyte* 337 (K!, holo.)

Perennial up to 4 m long, decumbent or scrambling to semi-scandent; roots thin, fibrous, produced from rooted nodes; shoots hirsute with white or orange hook-hairs. Leaves distichous; sheaths (0.5–)1–2 cm long, hirsute, ciliate with long orange or red hairs near apex; lamina petiolate, ovate to lanceolate, 2–8.5(–10) × 1.5–3.5 cm, base asymmetric, rounded to truncate, margins long-ciliate, apex acute to acuminate; adaxial surface long-hirsute, abaxial hirsute with hook-hairs, sometimes with sparse long hairs intermixed. Thyrse terminal (rarely also axillary), lax, 3.5–8 × 3–8 cm, of (2–)3–9 slender cincinni to 8 cm long and 22-flowered; inflorescence axes puberulous with hook-hairs; peduncle 2.5–5 cm long with a leafy bract ± halfway; bracteoles 1.5–2.5 mm long, cup-shaped, perfoliate, glandular subapically and often along the margin, glabrous or with some minute hook-hairs at base. Flowers bisexual and male, 8–15 mm wide; pedicels 4–10 mm long in flower, erect or spreading to decurved and 6–14 mm long in fruit. Sepals green, upper sepal longer than the lateral, lanceolate to ovate-lanceolate, 4–6.5 (upper) or 3–5.5 mm (lateral) × 2.3–3 mm, glandular subapically, puberulous with hook-hairs at least medially; paired petals yellow, ± 7.5 × 6 mm of which the white claw ± 1.5 mm, glabrous; medial petal pale greenish yellow or cream, trough-shaped, lanceolate, 3–4 × 2 mm, apex strongly recurved; staminodes 3, the medial one occasionally vestigial, antherodes bilobed; lateral stamens with filaments straight, 4–6.5 mm long, finely bearded with short white hairs, anthers 1–1.4 mm long, pollen white; medial stamen filament 2.5–3 mm long, anther white, 0.2–0.3 × 0.5–0.6 mm. Ovary ± 1 mm long, glandular-pubescent; style white or pale lilac, 4–7.5 mm long; stigma capitate. Capsules pale yellow, bi- or trilocular, bi- or trivalved (the dorsal valve sometimes deciduous), obovoid, 3–5(–5.5) × 3–4(–4.5) mm, dorsal locule, when developed, 0–1-seeded, ventral locules 1(–2)-seeded. Seeds ovoid to ellipsoid, 1.8–2.8 × 1.6–1.9 mm, testa orange-tan to pinkish purple, alveolate-reticulate or foveolate-reticulate, white (or tan) farinose granules present only around the hilum; embryotega brown, contrasting with the testa; hilum in a broad, shallow groove or the whole ventral surface shallowly concave, dark brown or black, straight.

var. **nyasense**

Sepals in fruit (5–)5.5–6.5 mm long, longer than the fruit; capsules 4–5(–5.5) mm long; fruiting pedicels 10–14 mm long.

UGANDA. Bunyoro District: 35 km S of Hoima on Fort Portal road & 9.6 km N of Kabwoya, 17 Sep. 1969, *Faden, Evans & Lye* 69/1092! ; Toro District: Rwenzoli [Queen Elizabeth] National Park, Nyamugasani River, NW of Katwe, 17 Apr. 1969, *Lock* 69/83!; Mengo District: Buvuma, near Kitamiro, at S extremity of Kitamiro Landing clearing, 27 Sep. 1949, *Dawkins* 414!

KENYA. Kilifi District: Kaya Ribe, Mleji River, 16 Sep. 1997, *Luke et al.* 4735!; Kwale District: Shimba Hills, Mwele, 13 Nov. 1992, *Luke* 3379!

TANZANIA. Bukoba District: Minziro Forest Reserve, 5 July 2000, *Bidgood et al.* 4848!; Iringa District: Udzungwa Mountain National Park, Camp 244 (7°43'S, 36°54'E), 9 June 2002, *Luke & Luke* 8782!; Njombe District: Lupembe region, upper Ruhudje River, Ditima [Nditima], 2 Oct. 1931, *Schlieben* 1266!

DISTR. **U** 2, 4; **K** 7; **T** 1, 4, 6, 7; Congo–Kinshasa, Burundi, Malawi, Mozambique

HAB. Forest (lowland and submontane), including glades and edges, swamp forest, riverine forest, edges of lakes, streams and rivers, often in dense vegetation, thickets and swampy bushland, wet roadside banks, sometimes growing in tall grass; (50–)340–1200(–2000)m

Flowering specimens have been seen from all months except January and February.

SYN. *Lamprodithyros tacazzeanus* Hassk. in Flora 46: 390 (1863) & in Peters, Reise Mossamb., Bot. 1: 531 (1864); Brenan in K.B. 15: 215 (1961) (non *Aneilema tacazzeanum* A. Rich.) [*Peters* s.n. only; see notes]

Aneilema vankerckhovenii De Wild. in B.J.B.B. 5: 85, 160 (1915) & Pl. Bequaert. 5: 222 (1931); Troupin, Fl. Sperm. Parc Nat. Garamba 1: 159 (1956). Type: Congo–Kinshasa, Kasai: Saint-Trudon, 5 Apr. 1913, *Vankerckhoven* s.n. (BR!, holo.)

var. **brevisepala** *Brenan* in K.B. 15: 214 (1961). Type: Tanzania, Pangani District: Msubugwe [Musubugwe] Forest, *Milne-Redhead & Taylor* 7312 (K!, holo.; BR!, EA!, P!, iso.)

Sepals in fruit 3–4(–5) mm long, usually somewhat shorter than or subequaling the fruit; capsules 3–4 mm long; fruiting pedicels 6–11 mm long.

TANZANIA. Morogoro District: NE part of Northern Uluguru Mts, Mkungwe Hill, near summit, 5 July 1970, *Faden, Evans & Pócs* 70/369! & Uluguru Mts, NE side, Mkungwe Mountain, 23 May 1933, *Schlieben* 3973!; Lindi District: Tanda-ngongoro, 13 May 1903, *Busse* 2483!

DISTR. **T** 3, 6, 8; not known elsewhere

HAB. Dry evergreen forest at edge of semi-permanent water, woodland/evergreen forest transition, among grass under trees, swampy places; 150–1100 m

Flowering specimens have been seen from March to May, September and October.

NOTE. (on the species as a whole). Hasskarl's name *Lamprodithyros tacazzeanus* was used for a Peters collection from Mozambique, but the name was specifically cited by Hasskarl as a new combination based on *Aneilema tacazzeanum* A. Rich. Therefore *Lamprodithyros tacazzeanus* is nomenclaturally a synonym of *Aneilema tacazzeanum* [= *Aneilema forskalii* Kunth], and *L. tacazzeanus* cannot by typified as a distinct taxon, even though the specimen upon which it was based was a different species. Although the original Peters collection was apparently destroyed in Berlin a 'kleptotype' from Hasskarl's original herbarium, seen at L, confirmed that the Peters collection was *A. nyasense.*

This species bears a strong resemblance to *A. aequinoctiale* in its yellow flowers with bearded stamen filaments, scrambling habit assisted by long, hook-hairs on the sheaths and often the internodes, distichous leaves, and capsules with a truncate to retuse apex with usually two acute corners. Both species are also partial to moist habitats. *A. nyasense* differs by its much smaller flowers, with glabrous petals and white pollen; smaller, fewer seeded capsules; and especially by the presence of red or orange hairs at the apex and often on the surface of the sheaths and sometimes the internodes.

The capsules are unusual in this species. They are most commonly bivalved, 2-seeded, with one seed in each ventral locule and the dorsal locule not developed. The dorsal valve, although lacking a dorsal locule seed, is nevertheless very loosely attached and sometimes deciduous. When all three locules produce a seed, the capsule may have three acute angles at the apex and may split into three equal valves, all of which are persistent. In other species, when the dorsal valve is deciduous it almost always contains an indehiscent dorsal locule seed, the valve thus acting as a disseminule. While that may possibly occur in this species, it has not yet been observed. No other African *Aneilema* is known to have a fully trivalved capsule. Fruits with more than one seed per ventral locule are uncommon in *A. nyasense,* and only a single 5-seeded capsule has been observed from our area, *Rwaburindore* 1734 from Uganda. The ovaries of 25 flowers of cultivated material of *Faden, Evans & Pócs* 70/369 (var. *brevisepala*) were dissected over the course of a month and the ovules counted. Eighteen flowers had two ventral ovules – one in each locule – and no dorsal ovules; four flowers were like the preceding but also had a dorsal ovule; two flowers had three ventral and no dorsal ovules; and one flower had four ventral ovules but lacked a dorsal ovule. Thus even within a single plant there can be a large variation.

Schlieben 6233 (var. *brevisepala*) records the flowers as white, which is the only such record seen for this species. Either the flower colour is erroneous or else this is apparently a unique occurrence in the species.

I have several reservations about var. *brevisepala.* First, based on its distribution in eastern Tanzania, one would have predicted that any collections to be found along the Kenya coast would prove to be of this variety, but instead, the two recent Luke collections from there belong to var. *nyasense.* Second, *Faden, Evans & Pócs* 70/369, from the same hill, but a different elevation, as *Schlieben* 3973, which is typical var. *brevisepala,* has some fruiting sepals to 5 mm long, or longer than those recorded by Brenan (3–4 mm) for var. *brevisepala.* Third, both varieties occur in **T** 6, Ulanga District at about the same elevation. Fourth, the size difference for the sepals, which is the main distinguishing character, applies only to fruiting material, and even with that I have had to broaden the size ranges for the two varieties in order to accept them, so that now they are not as different as Brenan described them. Although var. *brevisepala* is maintained here provisionally, I would not be surprised if future collections even further blurred the weak morphological distinctions and supposed geographical separation between the two varieties.

Whole leaves are recorded as being eaten by chimps in the Budongo Forest, Uganda (*Tinka* s.n.).

3. **Aneilema ephemerum** *Faden* **sp. nov**., herbae annuae; folia sessiles; inflorescentiae laxae cincinnis (2–)4–6 elongatis compositae; flores armeniaci vel flavi plerumque trichomatibus minutis basi petalorum instructis; filamenta staminum glabra; capsulae apice retusae loculis 2–3 seminalibus; semina arenulas similia. Type: Northern Frontier District: Dadaab–Wajir Road, 6 km N of Sabue Airstrip, *Brenan, Gillett, Chomba & Kanuri* 14819 (US!, holo.; EA!, K!, iso.)

Annual to 50 cm tall; roots thin, fibrous; shoots erect to ascending, sometimes shortly decumbent and rooting at the lower nodes, hirsute-puberulous with long and short hook-hairs. Leaves spirally arranged or partly distichous on the lateral shoots; sheaths 0.8–1.7 cm long, hirsute-puberulous, long-ciliate at apex; lamina sessile, lanceolate to lanceolate-elliptic, 4–10.5 × 1–3.7 cm, base narrowly cuneate (proximal) to rounded (distal), margins ciliolate proximally, scabrid distally, apex acute to acuminate; both surfaces hirsute-puberulous with a mixture of long and short hook-hairs, the abaxial more densely so. Thyrses terminal and sometimes also axillary, lax, ovoid, 4–7.5 × 2.5–5 cm, with (2–)4–7 ascending cincinni; inflorescence axes puberulous with hook-hairs; cincinni to 4 cm long and 10-flowered; peduncles 4.5–11.5 cm long with a foliaceous bract; bracteoles green sometimes tinged with violet, cup-shaped, perfoliate, 1–1.4 mm long, not glandular subapically, but usually with small marginal glands or the whole margin glandular. Flowers bisexual and male, 7–18 mm wide; pedicels horizontal to ascending and 7.5–10.5 mm long in flower, ascending to erect and to 13 mm in fruit. Sepals green proximally, violet-maroon distally, the upper larger and more hooded than the others, 2.7–5 × 1–2.1 mm, glandular distally; paired petals orange-yellow or yellow, 4–11 × 4–9 mm of which the white claw (0.5–)1–2 mm; medial petal white or greenish white tinged with orange-yellow or yellow, slightly convexo-concave to planar, elliptic to oblong, 3–7 × 1.5–2.3 mm; staminodes yellow, antherodes bilobed; lateral stamens with filaments gently undulate, 4–10 mm long, glabrous, anthers 1.3–1.8 × 0.6–0.9 mm, pollen orange; medial stamen with filament 3.5–6 mm, anther shield-shaped, pollen yellow. Ovary ± 1.6 × 1.2 mm, covered with appressed glandular hairs; style undulate, 5–6 mm long; stigma pinkish purple, capitate. Capsules pale yellow with darker flecks or greenish tan without darker markings, bilocular, bivalved, obovoid-ellipsoid to oblong, 3–6 × 2.5–3 mm, sparsely puberulous, dorsal locule abortive, sometimes represented by a keel, ventral locules 2–3-seeded. Seeds ovoid to ± 3-lobed, 1.2–2.2(–2.5) × 1.3–1.8 mm, testa tan or orange-tan, with irregular darker spots and streaks, embryotega elliptic, chocolate brown, with a small, white, central apicule, hilum flush with the ventral surface, dark reddish brown, ± straight, farinose granules very sparse, confined to the hilum and around the embryotega.

KENYA. Northern Frontier District: Wajir, 25 Dec. 1943, *Bally* 3740! & 63 km on Mado Gashi–Garissa road, 13 May 1978, *Gilbert & Thulin* 1698! & Tana River District: Garissa–Thika road, 7.5 km towards Thika from the turnoff to Galole, 10 June 1974, *Faden & Faden* 74/794!

DISTR. **K** 1, 4?, 7; not known elsewhere

HAB. Open or dense mixed bushland in sandy soil or in shallow soil on rock; 170–450 m

Flowering specimens have been seen from May, June, November and December; flowers from plants collected in the evening opened the following morning by 08:00 and faded 11:30.

SYN. *Aneilema ephemerum* Faden in Smithsonian Contr. Bot. 76: 62, t. 5, d (1991), *nom. nud.*

NOTE. Six collections have been seen of this species, all from the very dry country of eastern and northeastern Kenya. *Mungai & Rucina* 228/84, from the Kora Game Reserve, was either from **K** 4 or **K** 7. I have a note indicating that this species occurs in Kitui District (**K** 4), but I do not cite a specimen, so the record is unconfirmed.

Aneilema ephemerum is very distinctive because of its annual habit and yellow or orange-yellow flowers. Among the other yellow- to orange-flowered species with lax inflorescences, *A. aequinoctiale*, *A. nyasense* and *A. johnstonii*, all perennials, it is likely more closely related to the first two species (and differs from *A. johnstonii*) because of the long, several-celled hook-hairs on the internodes and sheaths that make them feel 'sticky' and its capsule with cells of the capsule wall ± isodiametric. It agrees with *A. johnstonii* but not the other two species only in having glabrous stamen filaments, a variable character in section *Amelina*, to which all four species belong. Overall, *A. ephemerum* shares more characters with *A. aequinoctiale* and *A. nyasense*, decumbent to scrambling perennials of moist to mesic, shady habitats, than with the tuberous-rooted *A. johnstonii*, a species that occurs in mesic to dry, often open habitats. From all three species it differs by its annual habit, by the presence of minute hairs at the base of the lower petal, and by its ecology and distribution.

4. **Aneilema johnstonii** *K. Schum.* in P.O.A. C: 135 (1895); C.B. Clarke in F.T.A. 8: 67 (1901); Obermeyer & Faden in F.S.A. 4, 2: 40, fig. 7–3 (1985); Faden in Smithsonian Contrib. Bot. 76: 62, t. 1d, 7f (1991); Faden in U.K.W.F. 2nd ed.: 307 (1994); Ensermu & Faden in Fl. Eth. 6: 351, fig. 207.6 (1997). Type: Tanzania, Kilimanjaro [Kilimanscharo], below Marangu, *Volkens* 2146 (B, lecto!, chosen by Faden in Bothalia 15: 100 (1984), BM!, G, K!, iso.)

Perennial geophyte; roots thin, with distal fusiform tubers; shoots tufted, 25–100 cm tall, usually erect to ascending, rarely straggling but not rooting at the nodes, puberulous. Leaves all cauline, spirally arranged; sheaths 1–2.5 cm long, puberulous or hirsute-puberulous, white-ciliate at apex; lamina sessile to subpetiolate, linear-lanceolate to lanceolate-elliptic, 5–12(–17) × (0.8–)1.2–3 cm, base rounded (distal leaves) to cuneate, margins ciliolate proximally, scabrid distally, apex acuminate; both surfaces puberulous with hook-hairs and/or prickle-hairs, often scabrid, sparse longer hairs often also present. Thyrses terminal, lax, ovoid to narrowly ovoid, 4–12 × 2.5–7 cm, of 6–25 ascending cincinni arranged in (1–)2–6 whorls; inflorescence axes glabrous except for the puberulous peduncle; peduncle 4–9.5 cm long, with a basal foliaceous bract; cincinni to 4.5 cm long and 9-flowered; bracteoles cup-shaped, perfoliate, glandular, glabrous. Flowers bisexual and male, 12–18(–25) mm wide, pedicels erect to ascending and 4–7 mm long in flower, 6–11.5 mm long in fruit. Sepals green and maroon, brown or deep violet glabrous, upper sepal ovate, paired sepals lanceolate-elliptic to oblong, 4–4.5 mm long, with a green, subapical gland; paired petals orange or orange-yellow, 7–9.5 × 6.5–8(–12) mm, rounded and slightly hooded at the apex, glabrous, of which the purple or maroon-based claw ± 3.5 mm long; medial petal greenish orange or greenish white to white tinged with orange, with a reddish spot at base, lanceolate to elliptic or slightly obovate, 4.5–5.5 mm; staminodes 3, antherodes bilobed; lateral stamens with filaments undulate to nearly straight, 9–12 mm long, glabrous, anthers 2–2.2 mm long, pollen yellow; medial stamen with filament ± 7–8.5 mm long, anther yellow usually with a transverse maroon or purple band, 1.8–1.9 mm long, pollen yellow. Ovary ± 1.7 × 1.4 mm, glabrous; style J-shaped, 10–13.5 mm long; stigma capitate. Capsules bi- or trilocular, bivalved (the valves persistent), ellipsoid to obovoid, (3.5–)4.5–7 × 3–4.5 mm, glabrous; dorsal valve persistent, dorsal locule (0–)1-seeded; ventral locules (1–)2–3(–4)-seeded; ventral locule seeds ovoid to deltoid, 1.6–2.3 × 1.7–2.6(–3.2) mm, testa smooth to faintly alveolate-reticulate, beige, tan or buff-orange, white-farinose around the hilum. Fig. 10/5–7, p. 78.

KENYA. Northern Frontier District: Subata Microwave Relay Station, NE part of Ol Lolokwe (Ol Donyo Sabachi), 2 Nov. 1978, *Gilbert, Gachathi & Gatheri* 5308! & Dandu, 4 Apr. 1952, *Gillett* 12698!; Teita District: Voi–Taveta road at Voi 30/Taveta 42 mile post, 16 Mar. 1969, *Faden, Evans & Siggins* 69/323!

TANZANIA. Pare District: Kiruru, May 1928, *Haarer* 1444!; Mbeya District: Chimala, 3 Jan. 1962, *Boaler* 395!; Songea District: ± 7 km W of Songea, 1 Jan. 1956, *Milne-Redhead & Taylor* 8023!

DISTR. **K** 1, 4, 6, 7; **T** 2–8; Ethiopia, Mozambique, Zambia, Malawi, Zimbabwe, Botswana

HAB. *Brachystegia* (miombo) woodland, wooded grassland, grassland, *Acacia-Commiphora* woodland, scrub or thorn bush, inselbergs and other rocky habitats, termite hills, edge of mbugas and flooded depressions, riverine thicket; sandy or clayey soils; light shade to full sun; 200–2100(–3200) m

Flowering specimens have been seen from November to June. The flowers begin opening 0900–0930 hr in our area.

NOTE. Three collections were cited by Schumann (1895) when he published *Aneilema johnstonii*. Among those collections *Von Höhnel* 159 from the Pare Mountains is *A. hockii* and the Berlin collection of *Johnston* s.n. was apparently destroyed, so *Volkens* 2146 (B) was chosen as the lectotype. Unnumbered *Johnston* collections from Kilimanjaro, 2000–3000 ft at K and BM are probably isotypes.

The flowers have always appeared orange or orange-yellow to me, but they are often recorded as yellow by collectors. The occasional record of flowers of blue or similar hues probably reflects that the species was growing at the same site as *A. hockii* and might well have been part of a mixed collection.

This species is very distinctive because of its orange to orange-yellow flowers with glabrous filaments and a completely glabrous inflorescence. The dark sepals and often inflorescence axis serve as a good background for the petals. The short, subequal staminodes, whose antherodes commonly resemble upside down horseshoes, are unique in the genus. The inflorescence bract and uppermost foliage leaf are often similar in size and form a pair of leaves at the base of the peduncle.

5. **Aneilema hockii** *De Wild.* in F.R. 12: 290 (1913); Brenan in K.B. 7: 190, fig. 1–14 (1952); Schreiber et al. in F.S.W.A. 157: 2 (1967); Obermeyer & Faden in F.S.A. 4, 2: 37, fig. 7-1 (1985); Blundell, Wild Fl. E. Afr.: 411, fig. 628 (1987); Faden in Smithsonian Contrib. Bot. 76: 62, t. 1, g, h, l, t. 5, e, j (1991) & in U.K.W.F. 2nd ed.: 307, t. 138 (1994); Ensermu & Faden in Fl. Eth. 6: 352, fig. 207.7 (1997). Type: Congo–Kinshasa. Haut-Katanga, Lubumbashi [Élisabethville], 1911, *Hock* s.n. (BR!, holo.)

Perennial; roots fibrous, cord-like or tuberous; shoots decumbent or erect to ascending, ± 20–120(–150) cm tall, lower portions sometimes covered by long, overlapping sheaths, hirsute or puberulous with hook-hairs. Leaves usually spirally arranged, sometimes distichous; sheaths to 6 cm long, hirsute with hook-hairs making them feel sticky, or puberulous with minute hook-hairs, white-ciliate or eciliate at the apex; lamina sessile or the proximal ones sometimes petiolate, linear-lanceolate to oblong-elliptic, (4–)9.5–22 × (0.15–)1–5 cm, base cuneate to rounded, margins planar to undulate, scabrid, sometimes ciliate near the base, apex acuminate (to acute); surfaces puberulous and often scabrid. Thyrses terminal and sometimes also axillary, lax, ovoid to cylindric, 4.5–16 × 2–11 cm, with 3–30 cincinni; inflorescence axes puberulous with hook-hairs; peduncles 7–25 cm long, with a bract above or near the middle; cincinni to 6 cm long and 16-flowered; bracteoles cup-shaped, perfoliate, 2–3.5 mm long, eglandular or with small glands along the rim and sometimes subapically, sparsely puberulous or glabrous. Flowers female, bisexual and male, (15–)20–40 mm wide, fragrant; pedicels 5–13 mm long. Sepals green, striped with maroon or bluish purple, lanceolate to ovate, 5.5–10(–14 in fruit) × 1.9–5 mm, hooded at the apex, glandular subapically; paired petals mauve to blue, lavender or bluish purple, rarely white, often with a white triangle at the base of the limb, 12.5–20 × 9.5–17 mm, of which the claw 3.5–6 mm long; medial petal pinkish white to pinkish purple, trough-shaped, lanceolate, 7–12 × 2.5–3 mm; staminodes 3, yellow, antherodes bilobed; lateral stamens with filaments fused basally, undulate, sharply recurved distally, (10–)16.5–30 mm long, glabrous, anthers 3–4 mm long, pollen dirty orange; medial stamen filament 10–12 mm long, anther 1.2–2.5 × 1.4 mm, pollen yellow. Ovary 2–3.5 × 1.3–1.7 mm, glandular-pubescent; style straight, strongly recurved distally, (9–)16–26 mm long; stigma capitate. Capsules pale yellow, bilocular, bivalved (valves persistent), oblong to oblong-ellipsoid, (5–)7–15 × 3–4.2(–5) mm, sparsely puberulous to glabrous, locules 3–6-seeded. Seeds rectangular to ovoid or trapezoid, (1.3–)1.5–2.9 × (1.4–)1.7–2.9 mm; testa very variable, smooth to alveolate or rugose, light grey to tan to orange, with or without dark brown granules, sometimes also with white to tan granules around the hilum. Fig. 10/1–2, p. 78.

1. Inflorescence axis usually very short, the distal cincinnus peduncles usually exceeding it; bracts at the bases of the cincinni usually patent to ± reflexed; seeds 2.2–2.9 × (2.3–)2.5–2.9 mm; roots tuberous; **T** 6, 8 . . . . a. subsp. *hockii*

Inflorescence axis always elongate, usually longer than the cincinnus peduncles; bracts at the bases of the cincinni usually closely appressed to the cincinni, not reflexed; seeds (1.3–)1.5–2.1(–2.4) × (1.4–)1.7–2.1(–2.5) mm; roots variable but rarely tuberous (subsp. *longiaxis*) 2

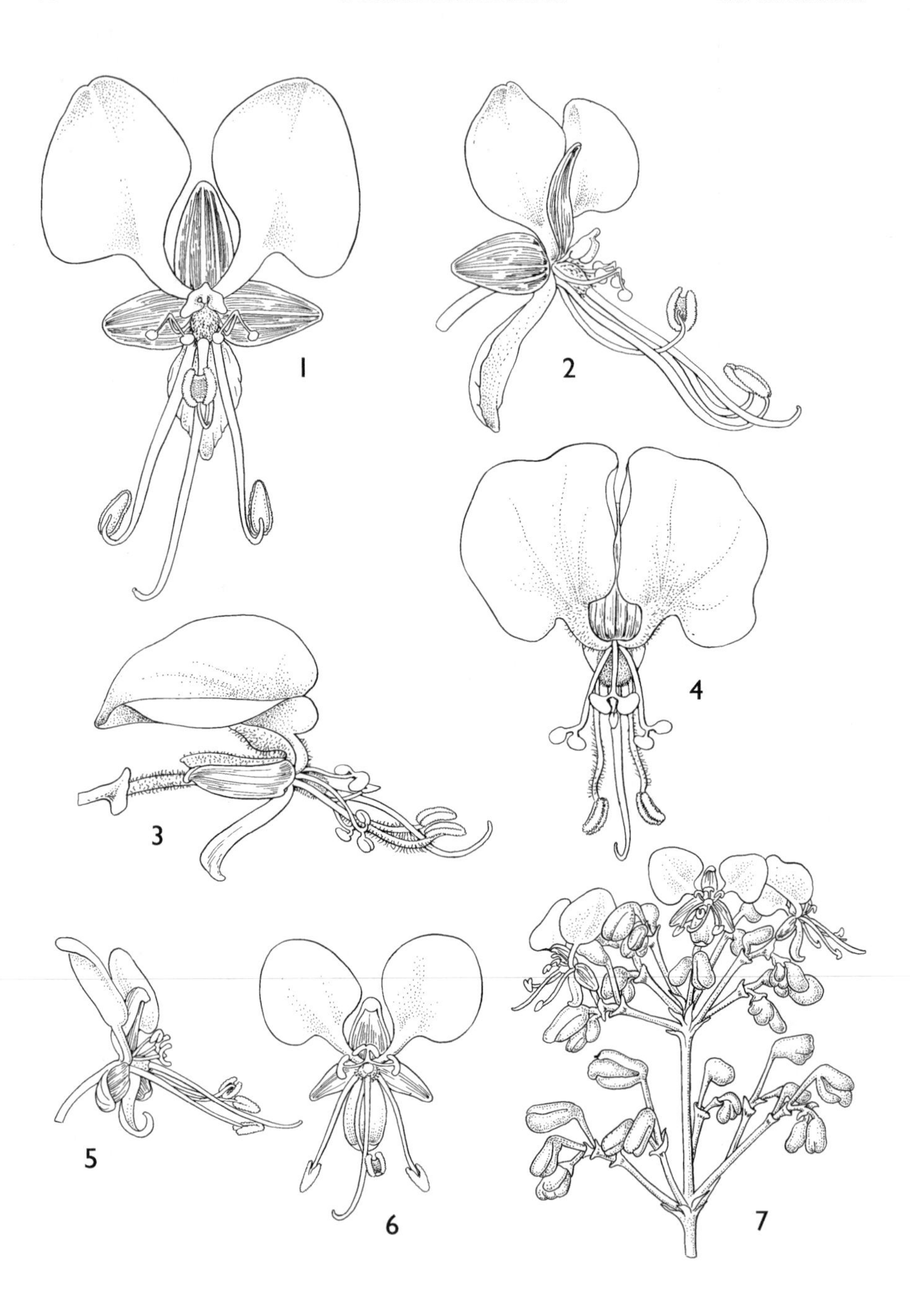

FIG. 10. *ANEILEMA* flowers and inflorescences. *ANEILEMA HOCKII* — **1**, bisexual flower, front view; **2**, bisexual flower, side view. *ANEILEMA AEQUINOCTIALE* — **3**, bisexual flower, side view; 4, bisexual flower, front view. *ANEILEMA JOHNSTONII.* — **5**, inflorescence; **6**, bisexual flower, side view; **7**, bisexual flower, front view. 1–2 from *Faden & Faden* 74/190; 3–4 from *Faden & Faden* 74/199; 5–7 from *Pawek* 12327. Drawn by Alice R. Tangerini, and modified by her from F.S.A. 4, 2 fig. 7.

2. Plants with elongate rhizomes; shoots usually erect, to ± 30 cm tall; leaves densely crowded at the base of the shoots, usually up to 7(–10) cm long, with strongly undulate margins; inflorescences composed of 3–8 mostly alternate (2 lowermost sometimes opposite) cincinni; plants of seasonally waterlogged soil; **K** 3, 4, 6; **T** 2 .... b1. var. *rhizomatosum*

Plants usually not rhizomatous or sometimes a few, adjacent shoot bases forming a very short rhizome; shoots mostly ascending to decumbent, to ± 100(–150) cm tall; leaves usually scattered along the shoots, the longest usually more than 10 cm long, margins not strongly undulate; inflorescences composed of 4–30 mostly opposite or whorled cincinni; plants of various habitats but rarely of seasonally waterlogged soil; widespread .... b2. var. *longiaxis*

a. subsp. **hockii**

Plants with a definite base; roots tuberous; leaves narrowly lanceolate, the proximal ones sometimes petiolate; inflorescence axis usually very short, the distal cincinnus peduncles usually exceeding it; distal cincinnus peduncles usually much longer than the proximal ones; cincinnus bracts usually patent to ± reflexed, often violet or partly violet; seeds 2.2–2.9 × (2.3–)2.5–2.9 mm, testa alveolate to rugose, orange-tan to tan or slightly darker, with dark brown particles in the depressions.

TANZANIA. Masasi District: Chidya, 15 Dec. 1968, *Leonhardt* 210!; Songea District: ± 11 km W of Songea, 2 Feb. 1956, *Milne-Redhead & Taylor* 8966!; Lindi District: Ruponda, 12 Dec. 1948, *Anderson* 246!

DISTR. **T** 6, 8; Congo–Kinshasa; Angola, Zambia, Malawi, Mozambique, Zimbabwe, Namibia, Botswana, South Africa

HAB. Woodland, sometimes by streams or on rocky ridges; on red loam, red sandy soil or in shallow soil; 15–650 m

Flowering plants have been seen from December to February.

SYN. *Aneilema aequinoctiale* (P. Beauv.) G. Don var. *kirkii* C.B. Clarke in DC., Monogr. Phan. 3: 222 (1881). Type: Mozambique, Chupanga [Shupanga], 10 Jan. 1863, *Kirk* s.n. (K!, lecto., chosen by Brenan, 1952); same locality, 1 Jan. 1859, *Kirk* s.n. (K!, syn.); Sena [Senna], 3 Jan. 1860, *Kirk* s.n. (K!, syn.)

A. aequinoctiale sensu C.B. Clarke in F.T.A. 8: 65 (1901) pro parte

A. aequinoctiale auctt., e.g. Norlindh & Weimark in Bot. Not. 48: 24 (1948) quoad *Norlindh et al.* 4054 and E.P.A.: 1516 (1971), *non* (P. Beauv.) G. Don

A. wildii Merxm. in Trans. Rhod. Sci. Assoc. 43: 152 (1951). Type: Zimbabwe, Marandellas, *Dehn* 251 (M!, holo.)

NOTE. Although typical plants of subsp. *hockii* are very distinctive, others are sometimes difficult to separate from certain plants of subsp. *longiaxis*. Plants are usually collected without bases, thus the difference in the roots can rarely be used, and there are at least some specimens of subsp. *longiaxis* with tuberous roots, e.g. *Schlieben* 3327. Some specimens of subsp. *hockii*, e.g. *Milne-Redhead & Taylor* 8966, can have a somewhat elongated inflorescence axis, with one long internode, suggesting subsp. *longiaxis*, which is normally multi-noded. Furthermore, some plants of subsp. *longiaxis* may have small, few-noded inflorescences that might suggest subsp. *hockii*, except for their smaller, appressed cincinnus bracts, elongate inflorescence axis and the geographic location of the collection. Chromosome counts have been obtained for eleven populations of subsp. *longiaxis* (nine from Kenya and two from Tanzania) and four of subsp. *hockii* (two from Malawi and one each from Zambia and South Africa). All populations of subsp. *longiaxis* were diploid and all four populations of subsp. *hockii* were tetraploid. Thus however tricky some specimens may be to recognize from herbarium specimens, these do seem to be distinct taxa.

b. subsp. **longiaxis** *Faden* **subsp. nov.**, a subspecie typica axe inflorescentiae semper elongato plerumque pedunculis cincinnorum longiore bracteis cincinnorum juxta pedunculo cincinnorum arte appressis seminibus parvioribus et radicibus plerumque non tuberosis differt. Typus: Tanzania, Lushoto District: Western Usambara Mts, Mombo–Lushoto road, 3 km, *Faden, Phillips, Muasya & Macha* 96/8 (US, sheet 1!, holo.; EA!, K!, MO!, NHT!,US!, sheet 2!, iso.)

Plants with or without a definite base, sometimes rhizomatous; roots variable but rarely tuberous; leaves linear-lanceolate to lanceolate, lanceolate-elliptic, narrowly elliptic or oblong-elliptic; inflorescence axis always elongate, usually longer than the cincinnus peduncles; cincinnus peduncles usually ± uniform within the inflorescence (occasionally the distal ones distinctly longer than proximal ones); cincinnus bracts usually closely appressed to the cincinni, not reflexed.

b1. var. **rhizomatosum** *Faden* **var. nov.**, a varietate typica rhizomatibus elongatis surculis florentibus brevibus plerumque erectis foliis basi surculorum florentium plerumque dense aggregatis marginibus valde undulatis inflorescentiis cincinnis 3–8 pro parte maxima alternis compositis differt. Typus: Kenya, Machakos District: Athi Plains, *Polhill* 369 (K!, holo.; BR!, EA!, iso.)

Plants producing thick, white, elongate rhizomes; shoots usually erect, to ± 20–30(–60) cm tall; leaves all sessile, usually densely crowded at the base of the shoots, usually up to 7(–10) cm long, with strongly undulate margins, often with long hairs on abaxial surface; inflorescences composed of 3–8 mostly alternate (2 lowermost sometimes opposite) cincinni.

KENYA. Nairobi/Machakos Districts: near Athi River, 14 May 1961, *G.R. Williams in EA* 12355!; Masai District: S shoulder of Ngong Hill on Nairobi–Magadi road, 25 May 1958, *Napper* 730! & 32 km N of Kajiado on Athi Plains, 11 Dec. 1995, *Verdcourt* 2511!

TANZANIA. Moshi District: Engare Nairobi, Block L 6, 5 June 1944, *Fuggles Couchman* 126! & same locality, 19 June 1944, *Greenway* 6863!

DISTR. **K** 3, 4, 6; **T** 2; not known elsewhere

HAB. Grassland, sometimes with scattered *Acacia drepanolobium*, on seasonally waterlogged black cotton soils; also recorded in vlei and in an old wheat field on "black cracking clay"; 1250–2150 m

Flowering specimens have been seen from February to June, August and December.

NOTE. This is a very distinctive plant because of its rhizomatous habit, short, usually erect shoots with leaves crowded towards the base and small inflorescences usually with mostly alternate cincinni. It always associated with seasonally waterlogged soils, particularly volcanic derived, black cotton soil. A few sterile plants were found in this habitat on the Laikipia Plateau on Mpala Farm, ± 2 km WNW of Kurunga Dam, and collected as living plants under *Faden* 95/19. A specimen was never pressed. Because this represents the extreme northern record of this taxon and the only record from **K** 3, it should be sought and documented from this location.

Faden & Faden 74/734 from Kangonde (Kangondi), in Machakos District, Kenya, is atypical in having leaves more spread out along the stems and cincinni more commonly opposite or even occasionally whorled. However, plants have the distinctive, thick, white rhizomes of var. *rhizomatosum*.

Var. *rhizomatosum* is common is suitable habitats around Nairobi. It is often associated with *Commelina latifolia* var. *undulatifolia* and *C. eckloniana* subsp. *nairobiensis*, and where one is found, the others should be looked for.

Var. *rhizomatosum* is perhaps just an ecotype of subsp. *longiaxis*. Plants of var. *longiaxis* are not infrequently found in the area in different habitats. However, the rhizomatous habit of var. *rhizomatosum* is maintained in cultivation, regardless of the potting mix used. Plants propagated from cuttings will soon produce typical subterranean rhizomes.

b2. var. **longiaxis**

Plants with or without a definite base, usually not rhizomatous or sometimes a few, adjacent shoot bases forming a very short rhizome; shoots mostly ascending to decumbent, to ± 100(–150) cm tall; leaves usually scattered along the shoots, the longest usually more than 10 cm long, margins not strongly undulate; inflorescences composed of 4–30 mostly opposite or whorled cincinni; plants of various habitats but rarely of seasonally waterlogged soil.

UGANDA. Karamoja District: Kidepo Valley National Park, 27 May 1969, *Harrington* 519! & E Matheniko County, Turkana Escarpment, Mar. 1959, *Wilson* 706!
KENYA. Northern Frontier District: Moyale, 10 July 1952, *Gillett* 13557!; Kilifi District: 4 km S of Dakabuko, 2 Dec. 1990, *Luke & Robertson* 2565!; Tana River District: Garsen, 13–14 July 1974, *Faden & Faden* 74/1065!
TANZANIA. Masai District: Clairadad, S side of Ngorogoro Crater Rim, 2 Feb. 1962, *Newbould* 5954!; Chunya District: Lupa N Forest Reserve, 152 km N of Mbeya on Itigi road, 17 Nov. 1962, *Boaler* 712; Iringa District: Msembi–Causeway Track, 5.9 km, 23 Mar. 1970, *Greenway & Kanuri* 14172
DISTR. **U** 1; **K** 1, 3, 4, 6, 7; **T** 1–7; Ethiopia, Zambia, Malawi, Zimbabwe, South Africa
HAB. Bushland and thickets, often with *Commiphora* and succulents, rocky hills and hillsides, woodland, including miombo and other types, grassland, edge of mbugas and other wet places; in sandy soil, sandy loam, red soil, and black cotton soil mixed with sand; 5–2150 m
Flowering specimens have been seen from all months except August.

NOTE. Although few collections from our area of var. *longiaxis* have been seen with definite bases and thick to tuberous roots, all seven such collections were from Tanzania. In my experience such plants do not occur in Kenya. Quite possibly they are typical of this taxon in Tanzania and further south, but that needs additional field work and better collections and observations.

Although this variety most commonly occurs in well-drained soils it sometimes may be found in black cotton soil in various parts of its range within Kenya and Tanzania. In 1974 we found a series of populations in Tana River District, Kenya (*Faden & Faden* 74/1065, 74/1177, 74/1180 & 74/1181), growing at very low elevations (5–20 m) where the species had evidently not been collected previously. The plants were growing in black cotton soil mixed with sand, in bushland and thickets of various species composition. The plants of these populations all had a very distinctive look to them, with decumbent stems, the proximal parts covered or partially covered by very long, sticky sheaths, glabrous or subglabrous inflorescence axis, glabrous cincinnus peduncles, very pale lavender flowers and small, smooth seeds (see Faden 1991, t. 1j). Despite the substrate, these plants never formed rhizomes. These populations were so distinctive and so removed from other populations that we thought they might constitute a distinct taxon. Perhaps further collecting will discover other collections that link these to the more common forms of var. *longiaxis* or provide more evidence to give these plants a formal taxonomic status.

The single specimen seen from **T** 4 (*Edward & Rashid* in MRC 2171) is from Rungwa Game Reserve, which straddles the intersection of **T** 4, 5 & 7.

NOTE (on the species as a whole). This is one of the most widespread and most variable of the *Aneilema* species in our area. Its confusion with *Aneilema aequinoctiale* over a long period has been well documented by Brenan (1952). This mix-up began with the colored illustration for *Commelina aequinoctialis* P. Beauv. in *Flore d'Oware et de Benin* (Vol. 1, t. 38. 1806) in which the flowers are shown as blue. Presumably this led Clarke and later workers to believe that the flower colour in *A. aequinoctiale* was variable. However, we now know that that is erroneous, and that *A. hockii* is totally distinct from *A. aequinoctiale* not only in flower colour but also in characters such as (in the former) fused stamen filament bases, glabrous stamen filaments, glabrous paired petals, and transversely elongate outer capsule wall cells. *A. hockii* occurs in much drier habitats than *A. aequinoctiale* and the two species only occur near one another when their habitats are in close proximity.

Plants are very variable in height, ranging from very dwarf plants of var. *rhizomatosum*, which can be less than 20 cm tall, up to 1.5 m in height. The tallest plants are typically supported by other vegetation. Plants with decumbent stems that lack a distinct base have no underground storage. They persist during the dry season by slowly dying back from the shoot apex, the lower, thicker internodes often covered by long, overlapping, persistent, papery sheaths that help prevent desiccation. Some plants with tufted shoots and distinct bases, which may have underground storage in thick roots, also may have long sheaths proximally. Plants of var. *rhizomatosum*, which typically have overlapping sheaths at the bases of the shoots, usually die back to their subterranean rhizomes.

Aneilema hockii has a number of unusual reproductive characters. It is one of the few *Aneilema* species to have fragrant flowers, although the scent is not very strong. The pollen produced by the medial stamen, although abundant, is sterile (see Faden 1991: 32). This is also one of the few species to have female flowers (Faden loc. cit., t. 1l) – in addition to bisexual and male flowers – which are the first flowers to open in an inflorescence, and are never produced again within the inflorescence. In the female flowers, the medial stamen is fully developed, but the laterals, which produce the fertile pollen, are abortive. Thus the female flower provides approximately the same visual attractants to a pollinator as the other flower types, plus a fragrance, plus pollen as a reward. *Aneilema hockii* is one of the few species *not* in sect. *Lamprodithyros* to have the stamen filaments fused at the base. The lateral stamen filaments are strikingly longer in male than bisexual flowers (Faden in Smithsonian Contrib. Bot. 76: 62, t. 1g). The female flowers have especially short lateral stamens and styles.

Aneilema hockii can be confused only with *A. gillettii* in northern Kenya and with *A. plagiocapsa*, which occurs just outside our area in Zambia. For distinctions from *A. gillettii* see the main key (above) and to distinguish it from *A. plagiocapsa* see the discussion after that key. *A. hockii* may grow with *A. johnstonii* – one of the collections cited in the protologue for *A. johnstonii* was instead *A. hockii* – but the two are easily distinguished by the orange or yellow-orange flowers and completely glabrous inflorescence and flowers in *A. johnstonii.*

6. **Aneilema gillettii** *Brenan* in K.B. 15: 212 (1961); Faden in Smithsonian Contr. Bot. 76: 62, t. 1, e, f (1991); Ensermu & Faden in Fl. Eth. 6: 352 (1997). Type: Kenya, Northern Frontier District: Moyale, *Gillett* 14033 (K!, holo.; EA!, G!, iso.)

Robust perennial to ± 1 m tall; roots tufted, cylindrical, tuberous; shoots annual, erect to asending. Leaves spirally arranged; sheaths 1–3 cm long, pilose to puberulous, pilose along the line of fusion, white-ciliate at apex; lamina sessile, obovate-elliptic to ovate-elliptic, 7–11 × 3–6.4 cm, decrescent distally, base cuneate to subpetiolate (lower and middle leaves) to cordate (upper), margins undulate, ciliate at least basally, scabrid distally, apex acute to ± acuminate; surfaces pilose-puberulous. Thyrses terminal and often axillary, lax, narrowly ovoid to cylindric, (8–)15–23.5 × 4–7(–9) cm, with (5–)12–26 ascending (sometimes declinate at the apex) cincinni; inflorescence axes puberulous; peduncles 11–17 cm long; cincinni to 3.5 cm long and 7-flowered; bracteoles scarious, amplexicaul but not perfoliate, 2–3.5 mm long, puberulous, not distinctly glandular. Flowers bisexual and male; pedicels 5–10 mm long (to 17 mm in fruit), ascending to erect in fruit. Sepals ovate-elliptic (upper) to lanceolate-elliptic (paired), 6.5–9 mm long. Petals subequal, purple to blue or violet, paired petals obovate-elliptic, medial somewhat boat-shaped, ± 12 × 7–9 mm, the paired ones narrowed at the base into a claw ± 2–3 mm long; staminodes 3, unequal, antherodes bilobed; lateral stamens with filaments undulate, ± 17–24 mm long, minutely bearded below the middle, anthers ± 3 mm long; medial stamen with filament ± 13 mm long, anther ± 3 mm long. Ovary ± 3.5 mm long, subglabous; style ± 17 mm long. Capsules bilocular, bi- or partially trivalved (valves persistent), oblong-ellipsoid, 9–11 × 4–5 mm, glabrous; dorsal locule obsolete, ventral locules (2–)3–4-seeded. Seeds mostly ovoid to trapezoidal, 1.7–2.5 × 1.8–2.2 mm, testa yellowish brown to orange-brown, scrobiculate to nearly smooth, tan-farinose in the depressions and around the hilum; hilum brown, ± flush with the surface.

KENYA. Northern Frontier District: Moyale, 14 Oct. 1952, *Gillett* 14033! & same locality, 4 Nov. 1952, *Gillett* 14132!

DISTR. **K** 1; S Ethiopia

HAB. Montane scrub and degraded montane scrub; 1100 m

NOTE. As Brenan noted in his account, this species is indeed very distinctive and possibly confused only with *A. hockii* in its range. It differs from that species by its generally larger inflorescences, non-perfoliate bracteoles, and flowers with a different colour, with a much larger lower petal that is concolorous with the other petals, bearded stamen filaments, and pointed capsules. In the range of *A. gillettii*, *A. hockii* lacks tuberous roots.

Brenan's account of the capsules and seeds of this species are bracketed in the type description, perhaps because he did not think that his material was adequate. He recorded the seeds as 8–12 per capsule. I can find no more than four seeds per locule, and often fewer, so the maximum per capsule is eight. The dorsal valve often partially splits at the apex, along the middorsal suture. However, this never goes as far as the middle of the capsule, so the capsules never truly become trivalved.

In Ethiopia this species occurs from 1100–1700 m in *Acacia* woodland and in scrubland with *Sterculia*, *Sesamothamnus* and *Ochna*.

7. **Aneilema taylorii** *C.B. Clarke* in F.T.A. 8: 79 (1901), as *taylori*; Faden in Smithsonian Contr. Bot. 76: 64, fig. 45, t. 1m, n, t. 5a (1991). Type: Kenya, Kilifi District: Rabai Hills, Mombasa, Fimbine? [probably = Fumbini on Kilifi Creek], Jul.-Nov. 1885, *W.E. Taylor* s.n. (BM, holo.)

Perennial or annual herbs; roots fibrous; shoots decumbent, much branched, flowering shoots ascending, 15–30(–60) cm tall, internodes puberulous, occasionally sparsely pilose as well. Leaves spirally arranged; sheaths 0.5–1.3 cm long, puberulous and frequently sparsely pilose, ciliate at apex; laminae petiolate, lanceolate to ovate, 2–10 × 0.7–3(–4.3) cm, base cuneate, margin sometimes undulate, scabrid, apex acuminate to acute; adaxial surface hirsute-puberulous, abaxial surface pilose-puberulous; petioles to 1.5(–2) cm long, ciliate. Thyrses terminal and occasionally axillary, ± lax, ovoid, 1.5–5.5(–7) × 1.5–6(–12) cm, with (1–)2–10(–13) patent to ascending cincinni; inflorescence axes glabrous apart from the basally puberulous peduncle (1.5–)2–5.5(–7.5) cm long, with a bract halfway; cincinni to 5(–6.5) cm long and 17(–18)-flowered; bracteoles eccentrically cup-shaped, perfoliate, 1–1.5 mm long, apex with a prominent gland, glabrous. Flowers bisexual and male, (7–)9–15 mm wide; pedicels 3.5–5.5(–6.5) mm long in flower, to 8 mm long in fruit, persistent, glabrous. Sepals medial sepal elliptic to lanceolate-ovate, 2.5–3.7 × 1.8–2.4 mm wide, lateral sepals slightly smaller, subapical gland unlobed or bilobed; paired petals lilac to lavender, (5.2–)5.5–9.5 mm long, (3.7–)4–7 mm wide, with a claw 1.5–3.5 mm long, apex rounded, glabrous; medial petal reflexed, greenish-white, sometimes tinged with lilac, margins colorless, 3–5 × 2.3–3.7 mm, staminodes 3, yellow with reddish filaments, the laterals longer than the medial, antherode bilobed; lateral stamens ± dimorphic, filaments J- or S-shaped, 4–10 mm long, glabrous, anthers creamy yellow or grey-green, 0.9–1.4 × 0.5–0.8 mm, pollen yellow or orange-yellow; medial stamen filament 3–6 mm long, anther 1–1.7 × 0.7–0.9 mm, pollen yellow and/or white. Ovary 1–1.8 × 0.8–1.1 mm, glabrous except for a few glandular microhairs; style 3.5–8.5 mm long, J-shaped, stigma small. Capsules pale yellow, lustrous, bilocular, bivalved (the valves persistent), ellipsoid to oblong-ellipsoid, (2.8–)3–4.5(–6) × 2–3.2 mm, glabrous, locules 2–3-seeded. Seeds pale greyish brown, trapezoidal, 1.2–2 × 1.3–1.6 mm, testa scrobiculate, white-farinose in all depressions.

KENYA. Kilifi District: Kaloleni district [i.e. area; erroneously recorded as Kwale District], 1 Sep. 1959, *Verdcourt* 2410!; Kwale Distrct: Muhaka Forest, 2 Mar. 1977, *Faden & Faden* 77/611!; Marenji [Marenje] Forest Reserve, 6 Sep. 1957, *Verdcourt* 1921!

TANZANIA. Lushoto District: Eastern Usambara Mts, Muheza–Amani road, crossing of Kwamkuyu River, 30–31 Mar. 1974, *Faden & Faden* 74/371! & Amani, 22 Aug. 1929, *Greenway* 1668! & Mkomazi Game Reserve/Kwizu Forest Reserve, 10 km on Kisiwani–Same road, 3 May 1995, *Abdallah & Vollesen* 95/105!

DISTR. **K** 4, 7, **T** 3; not known elsewhere

HAB. Lowland, submontane and riverine forest, sometimes associated with limestone outcrops, frequently on boulders along streams and at the bases of waterfalls in the Eastern Usambara Mountains, described as, "the first invader in a cleared area in evergreen rain forest," on *Greenway* 1668; partial or dense shade; 30–900(–1300) m

Flowering specimens have been seen from March to June and August to October; flowers open around dawn and fade about noon.

SYN. *Aneilema rendlei* sensu Brenan in K.B. 7: 196 (1952) pro parte, *non* C.B. Clarke (1901)
Aneilema sp. nov. aff. *A. umbrosum* of Morton in J.L.S.: 59: 464. 1966, in adnot.

NOTE. This species has sometimes been confused with others. Brenan (1952) cited two specimens of *A. taylorii*, *Greenway* 1668 and *Verdcourt* 161 (cited as 151), under *Aneilema rendlei*, noting that they had smaller capsules with fewer seeds than other specimens of *A. rendlei*. Morton (1966) identified another collection of *A. taylorii*, *Verdcourt* 2410, which had been determined in the Kew Herbarium as *A. umbrosum* var. *ovato-oblongum*, as an undescribed species closely related to *A. umbrosum*. Although pressed specimens of *A. taylorii* bear a striking resemblance to some West African plants of *A. umbrosum*, this similarity is superficial.

Aneilema taylorii exhibits a great deal of morphological variation, particularly in size of floral parts. I previously noted that all of the Kenya collections, as well as the lowest elevation collection from Tanzania, *Organ* in *EA* 15079, were annuals, in contrast to plants from the Usambaras and elsewhere in Tanzania (Faden, 1991). This difference could support the recognition of two subspecies in *A. taylorii*, but hard and fast characters usable in herbarium specimens have remained elusive.

All Tanzanian collections of *A. taylorii* from Lushoto District are from a small area of the Eastern Usambara Mountains centered around Amani, where the species is apparently fairly common between 500–900 m altitude. It is decidedly uncommon throughout the rest of its range. A new collection from Tanzania, *Abdallah & Vollesen* 95/105 (cited above), has extended the range of this species further inland, to a higher elevation (1300 m), a different habitat (riverine forest) and much closer proximity to the related *A. rendlei*. Another recent record from even further inland, *Luke & Luke* 7396 from Ngaia Forest Reserve, Meru District, Kenya, places *A. taylorii* well within the range of *A. rendlei*. The two species are ecologically isolated but have been found fully interfertile when crossed in cultivation.

8. **Aneilema usambarense** *Faden* in Smithsonian Contr. Bot. 76: 69, fig. 46 (1991). Type: Tanzania, Eastern Usambara Mts, W slope of Mt Mlinga, *Greenway* 6058 (K!, holo.; EA!, PRE, iso.)

Perennial herbs; roots not seen, probably thin, fibrous; shoots decumbent, the flowering shoots ascending, to 90 cm tall, puberulous to glabrescent. Leaves spirally arranged; sheaths 0.7–2 cm long, puberulous, ciliate at apex; laminae petiolate, lanceolate-elliptic, 6.5–16 × 2–4.5 cm, base cuneate, margin scabrid, apex acuminate; adaxial surface hirsute-puberulous, abaxial surface pilose-puberulous; petioles to 3.5 cm long, ciliate. Thyrses terminal (rarely also axillary from the inflorescence bract), ± dense, ovoid, 4–7 × 4.5–6 cm, with 14–20 ascending cincinni; inflorescence axis glabrous, but for the puberulous peduncle 1.5–2.8 cm long, with a reduced bract halfway; cincinni to 4 cm long and 9-flowered; bracteoles eccentrically cup-shaped, perfoliate, 1.2–1.5 × 0.4–0.7 mm, glabrous, with a prominent subapical gland. Flowers bisexual and male; pedicels 6–8 mm long, aligned with the cincinnus axis in flower, erect in fruit, glabrous; sepals: medial sepal ± 3 mm long, lateral sepals oblong-elliptic, ± 3 mm × 1.5 mm, glandular near the apex, glabrous; paired petals mauve, ± 7 mm long, medial petal ovate, 4–5 mm long; staminodes 3, antherode bilobed; lateral stamens with filaments declinate, 7–8 mm long, glabrous, anther 1.5 × 0.7 mm, medial stamen with filament straight, then gently recurved, ± 5 mm long, anther 1.8 mm long. Ovary ± 2 mm long, apparently glabrous; style J-shaped, 8.5 mm long; stigma small. Capsule sessile, chocolate broken, lustrous, oblong-ellipsoid, bivalved, bilocular, 5–6 × 2.5–3 mm, glabrous, valves persistent; dorsal locule obsolete, ventral locules 4-seeded. Seeds trapezoidal, 1.1–1.7 × 1.5–1.6 mm, testa pale greyish brown, scrobiculate on all surfaces, white-farinose in all depressions.

TANZANIA. Tanga District: Eastern Usambara Mts, W slope of Mt Mlinga, 4 Dec. 1940, *Greenway* 6058!

DISTR. **T** 3; known only from the type

HAB. Shady path sides in evergreen forest with *Parinari excelsa* etc., growing with *Sacciolepis curvata*; ± 620 m

SYN. *Aneilema rendlei* sensu Brenan in K.B. 7: 196 (1952) pro parte

NOTE. *Aneilema usambarense* is most closely related to *A. taylorii*, but differs in its taller flowering shoots, longer leaves with longer petioles, more numerous cincinni in the inflorescence, and darker brown capsules with more numerous seeds per locule. It also has somewhat longer cincinnus peduncles and more widely spaced bracteoles than *A. taylorii*. Possibly other distinguishing characters are present in the flowers, but they could not be determined from the dried specimens.

On casual inspection *Aneilema usambarense* appears intermediate between *A. taylorii* and *A. rendlei*. It is somewhat intermediate only in the number of cincinni per inflorescence, capsule size and number of seeds per locule. In most other characters, including habit, leaf shape, cincinnus peduncle and pedicel length, seed width, distance between bracteoles and degree of lateral stamen dimorphism, *A. usambarense* is much more similar to *A. taylorii* than to *A. rendlei*. Its distribution and ecology also suggest a closer relationship with *A. taylorii*; a hybrid origin is unlikely. Artifical hybrids made between cultivated plants of *A. taylorii* and *A. rendlei* differed from *A. usambarense* in having broader, pale yellow capsules (3.2–3.9 mm wide) and broader seeds (1.6–1.8 mm wide). They were indeed truly intermediate between the two species.

The final status of this species can only be determined from living material. At attempt by the author to find in 1974 was unsuccessful. Unfortunately, the habitat was being destroyed rapidly. Conservation Status: Possibly extinct.

9. **Aneilema rendlei** *C.B. Clarke* in F.T.A. 8: 68 (1901), excl. *Donaldson Smith* 346; Chiovenda in Webbia 8: 38 (1951) pro parte majore; Brenan in K.B. 7: 196 (1952) pro parte; E.P.A.: 1518 (1971), *Donaldson Smith* excl. 346; Faden in Smithsonian Contr. Bot. 76: 71, fig. 45, t. 1o, p, t. 2d; t. 5f (1991) & in U.K.W.F. 2nd ed.: 308 (1994); Ensermu & Faden in Fl. Eth. 6: 353 (1997). Type: Ethiopia [Somaliland on label], 23 Apr 1895, *Donaldson Smith* s.n. (BM!, holo.)

Perennial herbs; roots fibrous, generally produced only at the lower nodes; shoots erect to ascending or occasionally decumbent, generally 25–100 cm high, much branched at the base, unbranched or sparsely branched above, shoot bases swollen; internodes (2.5–)4–6(–9) cm long, glabrescent to puberulous. Leaves spirally arranged; sheaths 1–2.5(–3) cm long, puberulous or pilose, ciliate at apex; lamina petiolate, narrowly lanceolate to elliptic, 2.5–11 × 1–3.5 cm, base cuneate, margin scabrid and sometimes ciliate, apex acuminate or occasionally acute; both surfaces dull, the adaxial hirsute-puberulous, abaxial pilose-puberulous; petioles to 1.5(–1.8) mm long, ciliate. Thyrses terminal and occasionally axillary, ± dense, ovoid, (2–)3–5(–5.5) × (1.5–)2–3(–6) cm, with (9–)13–20(–32) ascending cincinni; inflorescence axes glabrous, except for the basally puberulous peduncle (1.5–)4–6(–10.5) cm long, with a bract halfway to subapical; cincinni to 3.5 cm long and 13-flowered; bracteoles eccentrically cup-shaped, perfoliate (rarely not perfoliate), 1.2–2.2 × 0.4–1.2 mm, glabrous, with a prominent subapical gland. Flowers bisexual and male, slightly fragrant, 9.5–17 mm wide; pedicels 2.5–4(–5.5) mm long, to 6.5 mm long in fruit, persistent, glabrous; sepals: medial lanceolate-ovate to ovate-elliptic, 3–4.5 × 2–2.6 mm, laterals slightly narrower, glabrous, with a subapical bilobed gland; paired petals lilac with darker veins shading to greenish-yellow at base, 7.5–11 × 4.5–8 mm, of which the claw 2.5–3 mm, apex rounded, glabrous; medial petal not strongly reflexed, greenish-white, sometimes flushed with lilac ovate to ovate-elliptic, 4.5–5.5 × 3–3.5 mm; staminodes 3, yellow, antherode bilobed; lateral stamens strongly dimorphic, filaments glabrous, in bisexual flowers 5.5–8 mm long, broadly U-shaped, in male flowers 9–13 mm long, J-shaped; anthers 1.5–1.7 × 0.6–0.8 mm, pollen orange-yellow; medial stamen with filament 3.5–5 mm long, anther 1.7–2 × 1–1.4 mm, pollen yellow. Ovary 2–2.5 × 0.9–1.1 mm, glabrous except for some glandular microhairs; style 7.5–10 mm long, J-shaped; stigma small. Capsules pale yellow, lustrous, 2(–3)-locular, bivalved (valves persistent), oblong to oblong-ellipsoid, (5–)6–10 × (3.1–)3.2–4.2 mm, glabrous; dorsal locule obsolete or, if present, empty (rarely 1-seeded), ventral locules each 3–5-seeded, rarely less by abortion. Seeds trapezoidal, 1.5–2.2 × 1.5–2 mm, testa scrobiculate on ventral and dorsal surfaces, grey-brown to flesh-pink, white-farinose in all depressions.

KENYA. Northern Frontier District: Mathews Range, Ngeng, 14 Dec. 1958, *Newbould* 3184!; Laikipia District: Colcheccio Ranch, NE corner, 8–9 Apr. 1985, *A. Faden* 27/85! Teita District: Taveta–Voi road, at mile post Taveta 36/Voi 36, 16 Mar. 1969, *Faden et al.* 69/322!

TANZANIA. Lushoto District: Western Usambara Mts, Lushoto–Mombo road, 2.5 km SW of Gare turnoff, 16 June 1953, *Drummond & Hemsley* 2928!; Pare District: 3 km Hedaru–Same road, 28 June 1970, *Kabuye & Evans* 163!; Bagamoyo District: Pongwe, 8 Apr. 1970, *Harris et al.* BJH4377!

DISTR. **K** 1, 3, 4, 7; **T** 2, 3, 6; Ethiopia

HAB. Dry deciduous bushland and bushland thicket, often on rocky slopes, occasionally along seasonal streams; usually in sandy soil; generally in partial shade; 300–1200 m

Flowering specimens have been seen from all months except August, September and November; in the field the flowers were observed to open ± 08:00.

SYN. *Aneilema octospermum* C.B. Clarke, in sched., as *octosperma*; Brenan in K.B. 7: 196 (1952), *nom. nud.* pro syn.

NOTE. On the same day that he collected the type of *A. rendlei*, Donaldson Smith also collected his number 346 which is cited by Clarke (1901) under *A. rendlei* but noted to differ from the type in several characters. *Donaldson Smith* 346 is *A. sebitense*.

Brenan (1952) supplied a detailed description of *A. rendlei* but unfortunately based it in part on several specimens that do not belong to this species. *Greenway* 1668 and *Verdcourt* 161 (cited as 151) are *A. taylorii* and *Greenway* 6058 is *A. usambarense*.

Collections with atypical features: *Verdcourt* 3901 and *Newbould* 3184 have unusually narrow leaves. *Faden & Evans* 70/788 and *Faden & Faden* 72/240, both from the same population, have bracteoles that are regularly not perfoliate, a feature otherwise unknown in this species. In all other characters these four collections are typical of *A. rendlei*.

Capsules are almost always confined to the lower cincinni in the inflorescence. This indicates that the upper cincinni rarely produce bisexual flowers. Although plants cultivated in Nairobi did occasionally produce some perfect flowers on the upper cincinni, further field observations are required to determine whether this occurs under natural conditions.

Small, erect, unbranched plants 1.5–5.5 cm tall have been found associated with larger plants of this species. These small plants have the characteristic swollen shoot bases of *A. rendlei* and are considered to be first year seedlings. Such seedlings are able to survive long, dry periods because they store water in the stem and have reduced transpiration in the dry season due to the die off of the laminae and to the presence of overlapping, papery, persistent sheaths covering the stem. A number of such plants, apparently in good condition, were found in the Tsavo area of Kenya in March, 1974, at the end of a longer than usual dry season. Plants of this species may take several growing seasons to mature.

Aneilema rendlei is most closely related to *A. brenanianum*, *A. taylorii* and *A. usambarense*. *Aneilema brenanianum* differs from *A. rendlei* in its rhizomatous habit, tuberous roots, more erect shoots, sheaths becoming papyraceous and completely covering the older parts of the shoots (occurring in *A. rendlei* only in seedlings), sessile laminae usually lacking long uniseriate hairs, white to very pale bluish purple paired petals, longer pedicels, broader capsules and larger seeds. *Aneilema taylorii* diverges from *A. rendlei* in its decumbent habit, non-swollen shoot bases, less dense thyrses with fewer cincinni, less dimorphic lateral stamens, often decurved fruiting pedicels, smaller capsules with fewer, smaller seeds. For differences between *A. usambarense* and *A. rendlei* see the key.

10. **Aneilema brenanianum** *Faden* in Smithsonian Contr. Bot. 76: 73, fig. 47, t. 2, a-c, t. 3, b (1991); Blundell, Wild Fl. E. Afr.: 411, fig. 23 (1987), *nomen nudum*; Faden in U.K.W.F. 2nd ed.: 308 (1994). Type: Kenya, Masai District, km 56–57 on Nairobi–Magadi road, 14 Dec 1969, *Faden, Evans & Lye* 69/2069 (US!, holo.; EA!, K!, MO!, iso.)

Perennial herbs; roots tuberous, to 7 mm thick at base, slightly tapering or ± uniform in thickness, to 20 cm long; rhizomes subterranean, moniliform, each segment producing 1-several roots and usually one aerial shoot; aerial shoots erect to ascending, rarely straggling, 15–60(–100) cm tall, the ends dying back to a stiff perennial, sometimes semi-woody, basal portion; internodes in lower perennial part of shoots completely covered by overlapping persistent leaf sheaths, those near the inflorescences exposed, puberulous. Leaves spirally arranged; sheaths 1–8.5 cm long, in lower part of shoot overlapping, becoming grey and papyraceous, puberulous or glabrescent, sparsely ciliate at apex in young leaves; lamina sessile, narrowly lanceolate-elliptic, (3–)6–15 × (0.4–)0.6–2.1 cm, base narrowly cuneate, margin scabrid, apex acute to acuminate; both surfaces puberulous, abaxial rarely also with a few longer hairs. Thyrses terminal, ± dense, ovoid to ovoid-ellipsoid, 2–5(–7.5) × 1.5–4(–5) cm, with (5–)7–20(–24) ascending cincinni; inflorescence axes

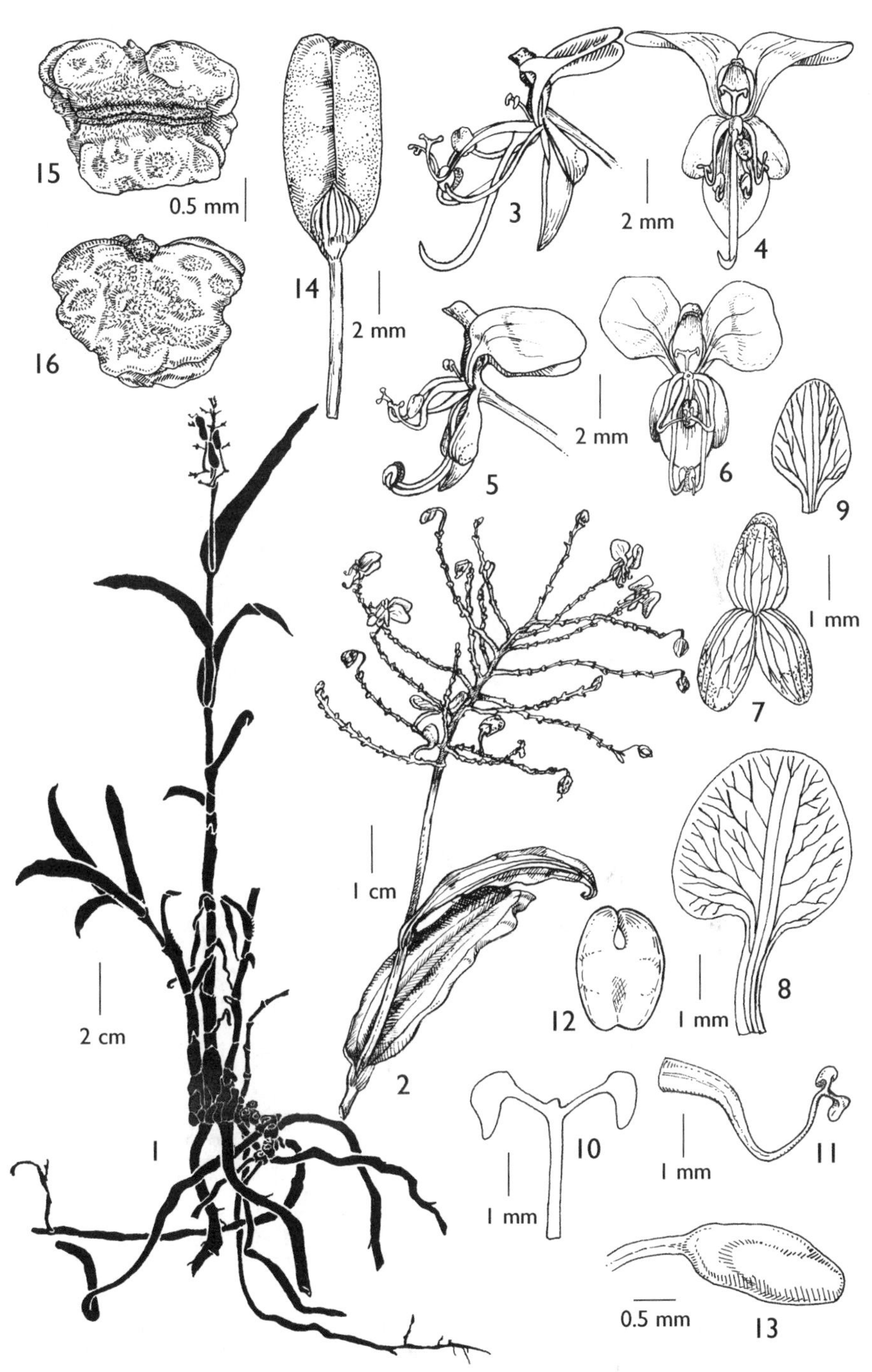

FIG. 11. *ANEILEMA BRENANIANUM* — **1**, habit; **2**, inflorescence; **3** & **4**, bisexual flower, side and lateral view; **5** & **6**, staminate flower, side and front view; **7**, calyx, front view; **8** & **9**, lateral and medial petal; **10** & **11**, medial and lateral staminode; **12**, medial stamen anther; **13**, ovary and base of style; **14**, capsule; **15** & **16**, seed, ventral and dorsal view. 1, 3–6, 10, 13–16 from *Faden & Faden* 72/166; 2, 7–9, 11–12 from *Faden et al.* 69/2069. Drawn by Tana Acton, reproduced with permission from Smithsonian Contr. Bot. 76, Fig. 47.

puberulous; peduncle 2.5–8(–10.5) cm long, with bract ± halfway; cincinni to 3 cm long and 9-flowered; bracteoles eccentrically cup-shaped, perfoliate, 1.8–2.3 × 0.8–1 mm, with a prominent gland near the apex and usually smaller glands along the margins. Flowers bisexual and male, slightly fragrant, 12–14.5 mm wide; pedicels 5–7.5 mm long in flower, to 12 mm long in fruit, persistent, glabrous or with a few minute hairs at apex. Sepals ovate to oblong-elliptic, 3–3.7 × 2–3 mm wide, the laterals slightly narrower than the medial, ± glabrous with a prominent subapical gland; paired petals white or whitish (very pale bluish-purple when fading), 7–10.5 × 4–8.5 mm, of which claw 2–4 mm long, glabrous; medial petal reflexed, greenish-white, ovate to ovate-elliptic, 3–5 × 2–4 mm; staminodes 3, yellow, antherode bilobed; lateral stamens dimorphic, 5–7 mm long, J-shaped, in bisexual flowers, similarly curved but strongly deflexed near the base and with a larger hook near the apex in the male flowers, glabrous, anthers 1.3–1.9 × 0.7–1 mm, pollen creamy white; medial stamen with filament 3–4 mm long, anther 1.4–1.9 × 1–1.4 mm wide, pollen creamy yellow. Ovary 2–2.3 × 1–1.3 mm, glabrous except for some glandular microhairs; style 6–7 mm long, J-shaped; stigma small. Capsules pale yellow, lustrous, bilocular, bivalved (valves persistent), oblong-ellipsoid, 6–11 × 4–5 mm, glabrous; dorsal locule obsolete, ventral locules each 3(–4)-seeded. Seeds trapezoidal, 2.1–2.5(–2.9) × 1.8–2.4 mm, testa scrobiculate on all surfaces, tan, white-farinose in all depressions. Fig. 11, p. 87.

KENYA. Masai District: Nairobi–Magadi road 55.4 km, 27 June 1971, *Faden & Evans* 71/513! & 28 km from Kiserian towards Magadi on Nairobi–Magadi road, 1 Mar. 1969, *Faden & Napper* 69/251! & 61.4 km from Nairobi on Magadi road, 12 Apr. 1960, *Verdcourt et al.* 2667!

TANZANIA. Masai District: Longido, just below water tank of Longido (town), 20 June 1971, *Faden & Evans* 71/498! & bottom slopes of Mt Longido near new water tanks, 12 Dec. 1959, *Verdcourt* 2531!

DISTR. **K** 6; **T** 2; not known elsewhere

HAB. Rocky hillsides with *Acacia-Grewia* or *Acacia-Commiphora* bushland, or occasionally dry plains; sandy soil; partial shade; (600?–)1150–1500 m

Flowering specimens have been seen from December-April and June; flowers have been observed to open at early as 11:45 in the field on a hot, sunny day or as late as 17:00 on a cool, overcast day.

SYN. *Aneilema* sp. A, Faden in U.K.W.F.: 664, 666 (1974)

NOTE. *Aneilema brenanianum* is most distinctive because of its persistent shoot bases covered by overlapping, papery sheaths. These shoot bases can become very stiff and semi-woody, with the consistency of dead grass culms. Overlapping sheaths, but without stiff shoots, have been observed in our area only in occasional plants of *A. hockii* and seedlings of *A. rendlei*.

The EA and K sheets of *Verdcourt* 2531 do not clearly exhibit the papery, overlapping sheaths that characterize *A. brenanianum*. The flowering shoots have grown directly from the rhizome, and the lower leaves still have a lamina. Apparently the plants were collected fairly early in the growing season because only immature capsules are present. The sheaths have not yet turned grey or papyraceous. The PRE sheet of this collection is more typical of the species: its flowering shoots have arisen from an older shoot base covered by grey, papery sheaths.

Plants of *A. brenanianum* nearly always grow with their roots in crevices between boulders, making them almost impossible to extract whole. This adaptation allows them to accumulate additional moisture in the form of runoff and probably tends to reduce water loss in the dry season.

The Kenyan collections, with the possible exception of *Heriz-Smith* s.n., are all from an 8 km section of the Nairobi–Magadi road. The Heriz-Smith specimen might also have come from this area, as the name "Magadi" is often used in a general sense for the whole region of Lake Magadi and the dry, low altitude area to the north of it. If the Heriz-Smith collection came instead from the vicinity of the town of Magadi, then the 600 m altitude, recorded with uncertainty above, would be correct.

11. **Aneilema somaliense** *C.B. Clarke* in F.T.A. 8: 69 (1901); Chiovenda, Result. Sci. Missi. Stef.-Paoli, Coll. Bot.: 167 (1916) & in Webbia 8: 38 (1951); E.P.A.: 1518 (1971); Blundell, Wild Fl. E. Afr.: 412, fig. 857 (1987); Faden in Smithsonian Contr. Bot. 76: 78, t. 2e–g, t. 5g (1991) & in Fl. Somalia 4: 85 (1995); Ensermu & Faden in Fl. Eth. 6: 353 (1997). Type: Ethiopia [Somaliland on label], Harradigit, Mar. [1885], *James & Thrupp* s.n. (K!, holo.)

Perennial herbs; roots tuberous, to 60 cm long, with distal fusiform to cylindrical tubers to 24 cm long; perennial shoot bases subterranean, short, erect, sometimes a few connected, forming a short rhizome; flowering shoots annual, erect, lower parts sometimes prostrate or looping on ground but not rooting at nodes, shoots disarticulating at base and nodes at end of growing season, 7.5–40 cm tall, puberulous and rarely sparsely pilose. Leaves spirally arranged; sheaths 0.3–1(–1.5) cm long, puberulous and rarely pilose along the fused edge, ciliate at apex; lamina sessile, ovate to elliptic, (1.5–)2.5–7.5(–13) × (1–)2–3.5(–5.5) cm, base cuneate to rounded, rarely subcordate, margin scabrid and frequently ciliate, especially towards the base, apex acuminate to acute, rarely obtuse or rounded and then mucronate; both surfaces lustrous grey-green, puberulous. Thyrses terminal, ± dense, ovoid to cylindrical, (2–)3–9.5 × (1.5–)2–4.5 cm, with (8–)17–40(–50) patent to ascending cincinni, occasionally a solitary cincinnus from the axil of the inflorescence bract; inflorescence axes puberulous; peduncle 3–9 cm long, puberulous, with a bract halfway; cincinni up to 2.5 cm long and 8-flowered; bracteoles cup-shaped, amplexicaul but not perfoliate, 0.9–1.4(–1.7) mm long, with a prominent subapical gland. Flowers female, bisexual and male, slightly fragrant, (7.5–)10–12.5(–17) mm wide; pedicels 2.8–4.5(–5.5) mm long in flower, to 7 mm long in fruit, persistent, ± glabrous. Sepals ovate to lanceolate, (2–)2.8–3.5 × 1.6–3.1 mm, the laterals slightly narrower than the medial, puberulous, with a prominent subapical gland; paired petals white to very pale lilac or pale blue, ovate, 4.8–9 × 3.2–7 mm wide, of which claw 2–3 mm, glabrous; medial petal reflexed, green with whitish margins, sometimes tinged with red apically, 2.8–6 × 1.6–4.5 mm; staminodes 3, yellow, antherode bilobed; lateral stamens with filaments 5–6.5 mm long, S-shaped, glabrous, anthers 0.6–1.3 × 0.5–1 mm, pollen yellowish white; medial stamen filament 3.5–5 mm long, anther 0.9–1.1 × 0.9–1 mm, pollen orange-yellow. Ovary 0.9–2.5 × 1.1–1.3 mm, with dense glandular hairs; style 5.5–6.2 mm long; stigma small. Capsules pale yellow to brown, lustrous, oblong-ellipsoid, bi- or rarely somewhat trivalved, bi- to trilocular, (5–)6–9 × 3.2–4(–4.8) mm, sparsely puberulous; dorsal locule empty or 1-seeded, ventral locules 2–3(–4)-seeded. Seeds mostly trapezoidal, 1.6–2.7 × (1.2–)1.3–1.6(–2) mm, testa pinkish brown to tan, furrowed, whitish-farinose in all the depressions.

KENYA. Northern Frontier District: Matakweni Hill, 3 km WSW of Wamba, 2–3 July 1974, *Faden & Faden* 74/939! & Dandu, 12 Mar. 1952, *Gillett* 12651! & Moyale, 13 Oct. 1952, *Gillett* 14032!

DISTR. **K** 1, 4; Ethiopia, Somalia

HAB. *Acacia* or *Acacia-Commiphora* bushland or woodland, dwarf shrubland or degraded montane scrub or bushland, sometimes on rocky slopes or on limestone; sandy, silty, loamy or clayey (black cotton) soils; generally growing in partial shade; 750–1400 m

Flowering specimens have been seen from March–May and October–December; in cultivation the flowers are open in the morning and fade in the early afternoon, but exact opening and fading times have not been determined.

SYN. *Aneilema smithii* C.B. Clarke in F.T.A. 8: 70 (1901); E.P.A.: 1518 (1971). Syntypes: Ethiopia, Lake Rudolf to Gondokoro, Jara, 23 Oct. 1899, *Donaldson Smith* s.n. (BM, lectotype); Ethiopia, Lake Rudolf to Gondokoro, Anole, 21 Oct. 1899, *Donaldson Smith* s.n. (BM, holo.)

Aneilema tacazzeanum sensu Chiovenda, Miss. Biol. Borana, Racc. Bot. 4: 305 (1939), pro *Cufodontis* 319, *non* A. Richard (1850)

NOTE. Within a small geographic area *A. somaliense* can occupy quite diverse ecological sites. *Faden & Faden* 74/939 was growing on a steep rocky slope beneath a massive granitic hill. It was found under shrubs of *Acalypha fruticosa* and an undescribed, spiny species of *Barleria* and was associated with the fern *Actiniopteris radiata*. Only one kilometer away, *Faden & Faden* 74/923 was found in level *Acacia tortilis* subsp. *spirocarpa-A. brevispica* bushland growing in sandy soil with much bare ground between the trees and thicket clumps. *Faden & Faden* 74/951, from the same area as the two previous collections, was growing in *Commiphora* bushland, in which the shrubs *Euphorbia cuneata* and *Boscia coriacea*, both absent from the other two localities, were common. In all three of these habitats, plants of *A. somaliense* occurred in protected spots under trees and shrubs.

12. **Aneilema pusillum** *Chiovenda*, Result. Sci. Miss. Stef.-Paoli, Coll. Bot.: 167 (1916); E.P.A.: 1518 (1971); Faden in Smithsonian Contr. Bot. 76: 87, t. 2i–k, t. 5h (1991) & in Fl. Somalia 4: 85 (1995); Ensermu & Faden in Fl. Eth. 6: 353 (1997). Type: Somali Republic, Dafet Mission between Uanle Uein and Ildùc Uein, *Paoli* 1277 (FT, holo.)

Perennial herbs; roots tuberous, fusiform or uniformly thickened, to 5 × 0.7 cm; rhizome absent or to 2 cm long, old basal sheaths persistent or not; flowering shoots annual, erect, 5–15(–18) cm tall, glabrous or sparsely puberulous. Leaves spirally arranged or distichous, all cauline or mostly basal; sheaths often split to the base, to ± 0.8(–2) cm long, glabrous or sparsely puberulous, glabrous or sparsely ciliolate at apex; lamina succulent when fresh, sessile, linear or ovate-elliptic, 2–9 × 0.1–2 cm, base broadly cuneate to rounded, margin scabrid or not, sometimes ciliate basally, apex acute to acuminate; adaxial surface ± glabrous, abaxial glabrous to puberulous with hook-hairs. Thyrses terminal on main shoot and sometimes also on lateral shoots, ± lax to ± dense, narrowly ovoid to pyramidal, 1.5–6.5 × (0.7–)1–3.5 cm, with (2–)6–10(–13) cincinni; inflorescence axis often zigzag, puberulous with hook-hairs; peduncles 2–7(–8.5) cm long, with a bract halfway; cincinni to 3 cm long and 8-flowered; bracteoles eccentrically cup-shaped, perfoliate, 1.1–3 × 0.3–1.3 mm (very rarely split to base), glandular near apex, puberulous. Flowers bisexual and male, 6–9 mm wide; pedicels (1.5–)3–5 mm long in flower, to 6 mm long in fruit. Sepals reflexed or not, green, sometimes tinged with purple or maroon, subequal, lanceolate to ovate, 2–4 × 1.3–2 mm, puberulous, glandular or eglandular subapically; paired petals pinkish red pale lilac or blue, ovate or to suborbicular (or obovate), 3–5.4 × 2.5–4.5 mm of which the yellow-green or whitish claw (0.3–)0.7–1 mm; medial petal greenish tinged with pinkish red and pinkish red at apex, or pink or white, lanceolate to ovate, 2.5–4.5 × 1.5–2 mm; staminodes 3, yellow, antherode bilobed; lateral stamens with filaments 2.8–6.5 mm long, S-shaped, glabrous, anthers 0.8–1.5 mm long, pollen yellow; medial stamen filament 2–5.2 mm long, anther 0.5–1.4 mm long. Ovary 0.8–1.3 × 0.6–1 mm, puberulous; style 3.5–6.5 mm long, arcuate-decurved for most of its length; stigma small or slightly enlarged. Capsules pale yellow (sometimes streaked with brown) or brown, lustrous, ellipsoid to oblong-ellipsoid, dehiscent, bivalved, trilocular, 2.4–5 × 1.5–2.5 mm, puberulous; dorsal locule prominent, 1-seeded or empty, ventral locules each 2-seeded (or less by abortion). Seeds transversely ellipsoid (dorsal) or subdeltoid to ovoid (ventral), 1–1.9 × 0.7–1.3 mm wide, testa tan or orange-tan.

subsp. **variabile** *Faden* in Smithsonian Contr. Bot. 76: 88 (1991) & in Fl. Somalia 4: 86 (1995). Type: Kenya, Northern Frontier District, 6 km S of El Wak, *Gilbert & Thulin* 1251 (US!, holo.; F!, iso.)

Roots fusiform; leaves 0.3–0.8 cm wide, all or mostly cauline; flowers pale lilac.

KENYA. Northern Frontier District: 6 km S of El Wak, 30 Apr. 1978, *Gilbert & Thulin* 4659! & Beila [Bela] camp on the Daua R., *Ruspoli & Riva* 457 (844) (438)!
DISTR. **K** 1; Ethiopia, Somalia
HAB. Shallow soil over limestone rocks in *Acacia-Commiphora* woodland; 420 m

Flowering specimens have been seen from April in our area and from May and June further north.

NOTE. Field work, mainly in the 1980s, especially by Thulin and Kuchar, has increased the total number of collections of *A. pusillum* from the four seen for Faden (Biosystematic Study of Aneilema, Ph.D. Thesis, 1975) to 23. I recognize four subspecies in Somalia, of which only one, subsp. *variabile*, extends beyond its borders (Faden, 1991). The taxa are still somewhat ill-defined due to incomplete data, and thus this classification is by no means final.

Despite all of the variation, *Aneilema pusillum* is a very distinctive species because of the plants' small size, very succulent leaves, few-branched inflorescences, small, late-opening flowers, deciduous dorsal capsule valves, seeds with a short hilum completely surrounded by a crenate-margined ridge, and a chromosome number ($2n = 28$) unique for the genus.

13. **Aneilema petersii** (*Hassk.*) *C.B. Clarke* in DC., Monogr. Phan. 3: 225 (1881) pro parte & in F.T.A. 8: 70 (1901) pro parte; Durand & Schinz, Consp. Fl. Afr. 5: 431 (1895) pro parte; Schumann in P.O.A. C: 136 (1895) pro parte; Brenan in K.B. 7: 195 (1952); Faden in U.K.W.F.: 666 (1974); Blundell, Wild Fl. E. Afr.: 412, fig. 629 (1987); Faden in Smithsonian Contr. Botany 76: 93 (1991) & in U.K.W.F. 2nd ed.: 308 (1994); Ensermu & Faden in Fl. Eth. 6: 354 (1997). Type: Mozambique, in moist sites, [undated], *Peters s.n.* (B, holo., L!, fragment)

Perennial or annual herbs; roots fibrous; shoots erect or ascending to decumbent, sparsely to densely branched, sometimes thickened at the base, 20–60(–100) cm tall. Internodes (0.7–)2–15.5(–19.5) cm long, puberulous at least below the nodes (rarely completely glabrous). Leaves spirally arranged (rarely distichous); sheaths 0.5–2.5 cm long, puberulous or pilose-puberulous, ciliate at apex; lamina petiolate (the upper sessile), narrowly lanceolate to ovate, (2.5–)3.5–11(–13.5) × (0.6–)1–3.5(–4.5) cm, base cuneate to rounded, sometimes to cordate-amplexicaul in upper, margin scabrid and ciliate, apex acute to acuminate (rarely obtuse); both surfaces usually lustrous, adaxial pilose-puberulous, abaxial puberulous or pilose-puberulous. Thyrses terminal and axillary, ± lax, ovoid, to 7(–9.5) × 6.5(–9.5) cm, with 1–16(–19) patent to ascending cincinni; inflorescence axes puberulous; peduncles (1.7–)3–18.5(–21.5) cm long, with a bract halfway; cincinni to 4(–7) cm long and 16(–33)-flowered; bracteoles ovate to ovate-lanceolate, sometimes slightly perfoliate at base, (1.1–)1.3–1.6 × 0.5(–0.7) mm, glandular near the apex or eglandular, puberulous (rarely subglabrous), and long-ciliate near base. Flowers bisexual and male, (6.5–)11–19 mm wide; pedicels (2.2–)2.5–5(–5.8) mm long in flower, to 7 mm long and recurved in fruit, puberulous at least apically. Sepals subequal, lanceolate to ovate-elliptic or ovate-orbicular, 2–5.2 × (1.7–)2–3 mm, subapical gland prominent in medial, puberulous; paired petals white to lilac or pinkish-purple, ovate to ovate-deltate, 4.5–9.5 × (3.5–)4.5–11 mm, of which the claw (1–)1.5–3(–4) mm; medial petal same colour, cup-shaped, ovate to obovate, 3.5–7.5 × 3–8 mm wide, 1.5–3.7 mm deep; staminodes 3, yellow, antherode bilobed; lateral stamens with filaments (3.8–)4.5–10 mm long, S-shaped, usually crossing, anthers 0.5–1.3 × 0.4–1.1 mm wide, pollen yellow (rarely whitish); medial stamen with filament 3–7 mm long, anther 0.9–2.7 × 0.6–1.3 mm, pollen same colour as laterals. Ovary 1.2–2.1 × 1–1.5 mm, with patent glandular hairs; style 2.5–9.8 mm long, arcuate-decurved ; stigma capitate. Capsules stipitate, pale grey or brown with dark spots, obovoid to subquadrate (occasionally obovoid-oblong), dehiscent, bivalved, trilocular, (3.4–)4.5–7.5 × (2.4–)3.3–5 mm, carinate middorsally, puberulous; dorsal locule prominent, 1-seeded or empty, ventral locules each 2-(or, by abortion, 1-)seeded. Seed of dorsal locule ± hemispherical, 2.3–3.4 × 2–2.6(–3.1) × 1.6–2.1 mm, testa orange-buff; ventral locule seeds ovoid to oblong or rectangular, (1.3–)1.5–2.6(–2.9) × 1.6–2.3(–2.6) × 1–1.8 mm, testa orange-buff.

subsp. **petersii**; Faden in Smithsonian Contr. Bot. 76: t. 2m, n and Fig. 35a, b (1991)

Perennials; short, perforating, inflorescence-bearing shoots absent from the lower nodes of the main shoots; flowers 15–19 mm wide. Sepals (3.3–)3.5–5.2 mm long; paired petals lilac to pinkish-purple, 7–9.5 mm long, with claws 2–3(–4) mm long; styles (6.5–)7.5–9.8 mm long.

KENYA. Northern Frontier: Moyale, 25 Oct. 1952, *Gillett* 14091!; Kwale District: Kwale, Aug. 1929, *R.M. Graham* 1903!; Lamu District: Mambosasa Game Forest Post along Ziwa la Mkuni, 19 July 1974, *Faden & Faden* 74/1142!

TANZANIA. Morogoro District: 4.8 km N of Tunungo, 48 km SE of Morogoro,15 July 1962, *Boaler* 626!; Tanga District: 8 km SE of Ngomeni, 2 Aug. 1953, *Drummond & Hemsley* 3610! & 9 km Tanga–Gombero road from Tanga–Mombasa road junction, 27 Mar. 1974, *Faden et al.* 74/326!

DISTR. **K** 1, 7; **T** 3, 6, 8, **Z**; Ethiopia, Mozambique

HAB. Grassland with scattered shrubs, bushland, thicket margins, wooded grassland, woodland, lowland forest margins, roadsides, cultivation, occasionally in damp situations, rarely in rocky places; sandy soil; full sun to dense shade; 0–500(–1100) m

Flowering specimens have been seen from all months except December; in the field the flowers open from 06:00–06:30 and fade between 10:30–11:30.

SYN. *Lamprodithyros petersii* Hassk. in Flora 46: 389 (1863) & in Peters, Reise Mossamb., Bot. 1: 529 (1864)

Aneilema tetraspermum K. Schum. in P.O.A. C: 136 (1895). Syntypes: Tanzania, Tanga, *Volkens* 175 (B†; K, lecto!; BM!, G!, isotypes); *Stuhlmann* 6062 and 6458 (both B†, syn.)

A. leptospermum K. Schum., as *leptasperma*, based on *Volkens* 175!, *nomen nudum pro syn.*

A. sacleuxii Hua in Bull. Mus. Hist. Nat. Paris 1: 121 (1895); C.B. Clarke in F.T.A. 8: 74 (1901). Type: Zanzibar, seaside, *Sacleux* 1142 (P!, holo.; photo K) [cited as "*Sacleux* 1192" by Brenan in K.B. 7: 195 (1952)]

A. chrysanthum K. Schum. in E.J. 33: 376 (1903). Type: Tanzania, by Mengwa, *Stuhlmann* 8611 (B†, holo.)

A. tacazzeanum sensu Chiovenda, Miss. Biol. Borana, Racc. Bot. 4: 305. 1939, pro *Cufodontis* 705, *non* A. Richard (1850)

subsp. **pallidiflorum** *Faden* in Smithsonian Contr. Bot. 76: 97, t. 2o, p, t. 6a (1991) & in U.K.W.F. 2nd ed.: 308 (1994). Type: Kenya, Teita District, Irima Rock above Irima Water Hole, *Faden & Faden* 74/237 (US!, holo.; BR!, EA!, K!, MO!, PRE!, WAG!, iso.)

Annuals; short, perforating, inflorescence-bearing shoots usually produced from the middle and lower nodes of the main shoots; flowers (6.5–)11–15 mm wide. Sepals 2–3.4(–3.8) mm long; paired petals white to very pale lilac, 4.5–7(–7.8) mm long, with claws (1–)1.5–2 mm long; styles 2.5–6.5 mm long.

UGANDA. Bunyoro District: 37 km N of Butiaba turnoff on Butiaba–Murchison Falls National Park road, 16 Sep. 1969, *Faden et al.* 69/1068!

KENYA. Northern Frontier District: Dandu, 9 May 1952, *Gillett* 13132!; Kitui District: Thika–Garissa road, 6 km Nguni–Enziu, 7 June 1974, *Faden & Faden* 74/751!; Taita District: Tsavo National Park E, Irima Rock above Irima Water Hole, 29 May 1972, *Faden & Faden* 72/284!

TANZANIA. Mbulu District: Lake Manyara National Park, Bagayo River, 18 June 1965, *Greenway & Kanuri* 11871!; Shinyanga District: Shinyanga, 1932–33, *Bax* 368!; Kilwa District: Selous Game Reserve, Kingupira, 10 Apr. 1975, *Vollesen* 1964!

DISTR. **U** 2; **K** 1, 3, 4, 7; **T** 1, 2, 8; S Sudan, Zambia

HAB. Deciduous *Commiphora* and *Acacia* bushland and bushland thickets, frequently on rocky hills and outcrops, also recorded from grassland, woodland, riverine thickets, roadsides and in swampy situations; on sandy soil; partial shade to full sun; (125–)500–1000 m

Flowering specimens have been seen from February–June and August (Kenya and Tanzania) and September (Uganda); flowers open from 05:45–06:30 and fade 10:30–11:00.

NOTE. The great variation in vegetative morphology in *A. petersii* is due to the occurrence of both annuals and perennials in the species. Subsp. *petersii* is always perennial; it typically has a definite base and produces ascending shoots; occasionally it may also form longer, decumbent shoots that give rise to new plants. The main shoots are thickened at the base, at least during the dry season. Subspecies *pallidiflorum* is consistently annual despite some collectors' notes (including mine on *Faden & Faden* 72/284) to the contrary. Although all plants in some populations are very reduced and clearly annual, some from other populations may have decumbent shoots with their habits impossible to determine by casual inspection. Repeated field observations indicate that even these decumbent plants live less than a year and flower only once before dying. Their phenology is discussed below.

Distichously arranged leaves have been noted in *A. petersii* only in two populations, both of subsp. *pallidiflorum* (*Faden et al.* 69/617 and *Faden et al.* 69/1068). All the lateral shoots of the latter have this arrangement, at least in plants raised from seeds.

The leaves of subsp. *petersii* are typically narrower (to 3.2 cm wide) than those of subsp. *pallidiflorum* (to 4.5 cm wide). They are also proportionally narrower, never being ovate. Subcordate- to cordate-amplexicaul leaves are present only in subsp. *pallidiflorum*.

The presence of numerous short lateral inflorescence-bearing shoots in subsp. *pallidiflorum* but not in subsp. *petersii* is concordant with the former's annual habit, which requires greater emphasis on seed production. This is accomplished in subsp. *pallidiflorum* partly through an increased number of inflorescences and flowers and partly through self-pollination. The short shoots usually emerge from the mouths of the sheaths of the upper leaves but perforate those of the lower leaves. The inflorescence bract of the terminal inflorescence of the main shoot is generally longer in subsp. *pallidiflorum* than in the typical subspecies. In populations of the former in the Voi area of Kenya the inflorescence bracts are borne close to the inflorescences and are held erect. They thus form a dark green background for the white flowers, increasing the flowers' visibility.

The size ranges of floral parts are greater in *A. petersii* than in any other species of section *Lamprodithyros*. This is due in large measure to the reduced flower size in subsp. *pallidiflorum*, an attribute commonly associated with self-pollination in *Aneilema*. The last flowers produced in plants of this subspecies just prior to their death are particularly small. Flowers of the two subspecies differ consistently in size and petal color. The most reliable distinctions are included in the key.

Differences in the capsules and seeds between the two subspecies are largely qualitative. Their reliability is uncertain due to insufficient material from some critical areas such as Mozambique (subsp. *petersii*), northern Tanzania (subsp. *pallidiflorum*) and the Kenya-Ethiopia border region (both subspecies). On the basis of available material, the capsules of subsp. *pallidiflorum* are more obovate (versus subquadrate), more deeply emarginate apically and browner (versus greyer) than those of subsp. *petersii*. The ventral locule seeds of subsp. *pallidiflorum* are regularly blackish-farinose, while those of subsp. *petersii* often completely lack farinose granules. The apical ventral seeds are usually more pointed in the former.

The difference in flowering phenology between the two subspecies is striking and is largely due to the rainfall distribution patterns in the regions in which they occur (Jackson, 1961; National Atlas of Kenya, 1970). Throughout the range of subsp. *petersii* there is no consistently dry month. Hence it may be expected in flower in every month in some part of its range. The lack of flowering specimens from December is almost certainly an artifact. *Faden et al.* 72/38, collected on January 10, has mature fruits and must have been flowering in December.

For subsp. *pallidiflorum*, the flowering phenology is clear only in E Kenya. There are two distinct rainy seasons, March to May and October to December. Observations in the Tsavo area indicate that the plants come up with the October to December rains but do not flower then. Many of them survive the following dry season and flower during the March to May rains. All of them die during the subsequent dry season. In the Ugandan locality the rainy season spans the coincident dry months of the Kenyan and Tanzanian sites, so it is not surprising that the single collection from Uganda was flowering at a very different time of year from the other collections. In 1977, the rains in Kenya, which should have ended in December, 1976, continued into January. Consequently, a few plants of subsp. *pallidiflorum* in Tsavo National Park were found in flower in February, 1977 in what would normally have been the height of the dry season.

The two Mafia Island collections (*Greenway* 5236; *Wallace* 810) have unusually short bracteoles and narrow capsules. They are considered to represent a minor geographic variant not worthy of taxonomic recognition.

The subspecific determinations of the four collections from the Kenya-Ethiopia border region are all somewhat doubtful. *Bally & Smith* 14873, *Cufodontis* 705 and *Gillett* 14091 are morphologically similar to one another. Because preserved flowers of *Gillett* 14091 are in the size range of subsp. *petersii* (sepals 3.9–4.9 mm long; lateral stamen filaments 9 mm long; style 7.7 mm long) but are too large for subsp. *pallidiflorum*, all three of these collections have been treated as subsp. *petersii*. The most striking features of the fourth collection, *Gillett* 13132, that immediately separate it from the other three collections are its very large leaves (to 13.5 × 4.5 cm) and capsules (to 9 × 5.5 mm), the maximum dimensions for both of these organs in *A. petersii*. While capsule size shows considerable variation within both subspecies, the leaf width of *Gillett* 13132 is approached only by specimens of subsp. *pallidiflorum*. The Gillett collection also agrees with this subspecies in flowering sepal length (3–3.2 mm). Only the lack of perforating inflorescence-shoots would distinguish this collection from typical subsp. *pallidiflorum*, but perhaps this is merely a function of age, as has been noted in other collections of this subspecies, e.g. *Ament & Glover* 263. Overall, *Gillett* 13132 much better conforms with subsp. *pallidiflorum* than with subsp. *petersii*. The occurrence of both subspecies of *A. petersii* in the Kenya-Ethiopia border region at first seems unlikely in view of their apparently different precipitation requirements and nearly total geographic isolation in other areas. However, closer study reveals that *Gillett* 13132, treated as subsp. *pallidiflorum*, is geographically separate from the other three collections that were considered subsp. *petersii*, and that its locality, Dandu, is situated in a zone of lower mean annual rainfall (< 255 mm) than Moyale (687 mm), where *Gillett* 14091 was collected (National Atlas of Kenya, 1970). Thus the pattern of distribution for these populations of the two subspecies conforms to that found in other parts of their ranges, at least with regard to rainfall.

Aneilema petersii is a highly distinctive species. It can be separated from all other taxa of section *Lamprodithyros* by its mostly non-perfoliate bracteoles, some or all with long, uniseriate hairs basally, crossed lateral stamen filaments not retained by the medial petal, longly stipitate, dehiscent, broad capsules with a humpbacked, deciduous, dorsal valve, and strongly dimorphic seeds, the dorsal locule seed particularly large (largest seed in the section). Both subspecies have all of these attributes and are easily separated in the field using the key characters, many of which, unfortunately, do not preserve well in herbarium specimens.

14. **Aneilema indehiscens** *Faden* in Bothalia 15: 96 (1984); Obermeyer & Faden in F.S.A. 4(2): 40, fig. 9–2a, b (1985); Faden in Smithsonian Contr. Bot. 76: 100, fig. 48, t. 3a, b, t. 6b (1991) & in U.K.W.F. 2nd ed.: 308 (1994); Ensermu & Faden in Fl. Eth. 6: 354 (1997). Type: Kenya, Tana River District, Garsen–Malindi road, 1.5 km towards Malindi from turnoff to Oda, *Faden & Faden* 74/1184 (US!, holo.; BR!, EA!, FT!, K!, MO!, PRE!, WAG!, iso.)

Perennial herbs; roots fibrous; vegetative shoots trailing and often looping along the ground, occasionally rooting at the nodes, sometimes straggling through shrubs, to 3 m long, flowering shoots produced irregularly, erect to ascending, to ± 60 cm tall, to 5 mm thick, glabrous or very sparsely puberulous. Leaves spirally arranged; sheaths 0.5–2 cm long, puberulous and sometimes sparsely pilose, ciliate at apex; lamina shortly petiolate, narrowly lanceolate to ovate-elliptic, rarely ovate, (2.5–)3–10(–13) × (0.7–)1–2.5(–3.5) cm, margin scabrid, ciliate at least basally, apex acuminate or acute; both surfaces lustrous, puberulous. Thyrses terminal and frequently axillary, lax to ± dense, ovoid to broadly ovoid, (2–)2.5–5(–8) × (1.5–)2-(–7) cm, with (1–)3–9 ascending cincinni; inflorescence axis puberulous; peduncle 2–5(–8) cm long, with bract ± halfway; cincinni to 7.5 cm long and 27-flowered; bracteoles eccentrically cup-shaped, usually perfoliate, 1.3–2.6 × 1 mm, with a prominent subapical gland, puberulous and often long-ciliate on the fused edge. Flowers bisexual and male, (9–)13–17.5 mm wide; pedicels 3.8–6(–8) mm long in flower, to 10 mm and recurved in fruit, puberulous. Sepals ovate to lanceolate, 2.4–4.3(–4.9) × 1.8–3.2 mm, the laterals wider than the medial, glandular near apex, puberulous; paired petals white to very pale lilac, broadly ovate to ovate-deltate, 7.3–9.5 × 6–8.5 mmof which the white or whitish claw 2–3.5 mm long, glabrous; medial petal cup-shaped, 6–8 × 4–6(–7.5) mm, 3–5 mm deep, same colour; staminodes 3, yellow, antherode bilobed; lateral stamens with filaments7.7–8.5 mm long, gently S-shaped, glabrous, anthers 0.6–1.3 × 0.6–1 mm, pollen yellow to orange or dirty white; medial stamen with filament 5–7 mm long, anther 1.5–2.4 × 1.2–1.7 mm, pollen yellow to orange-yellow. Ovary 1.5–2.3 × 1.2–1.6 mm wide, with patent glandular hairs; style 8–9.3 mm long, straight or gently arcuate-decurved; stigma capitate. Capsules subsessile to stipitate, chestnut brown or mottled dark and light brown or grey-brown, lustrous, obovoid-ellipsoid to oblong, dehiscent and bivalved (occasionally partially trivalved) or indehiscent, trilocular, (4–)4.5–6(–6.8) × (1.9–)2.3–3(–3.4) mm, puberulous; dorsal valve 1-seeded or, by abortion empty (very rarely 2-seeded), ventral locules each 2- or, by abortion, 1-seeded (very rarely 3-seeded). Seeds ellipsoid, 2–2.9 × 1.3–1.7(–1.9) mm (dorsal) or ovoid to trapezoidal and 1.5–2.2(–2.5) × 1.3–1.8 mm (ventral), 0.65–1 mm thick, testa usually orange-buff.

subsp. **indehiscens**

Petals white; capsule usually chestnut brown.

KENYA. Tana River District: Garsen, 13–14 July 1974, *Faden & Faden* 74/1066!; Teita District: 3 km E of Bura Railway Station, 17 Jan. 1972, *Gillett* 19562! & Voi, 8 May 1931, *Napier* 973!
TANZANIA. Bagamoyo District: 4.5 km towards Mbwewe from crossing of Milgoji River on Korogwe–Dar es Salaam road, 1 Apr. 1974, *Faden & Faden* 74/380! Lushoto District: Mazinde, *Drummond & Hemsley* 2337!; Zanzibar: Chumbuni, 1931, *Vaughan* 1851!
DISTR. **K** 7; **T** 3, 6; **Z**; not known elsewhere
HAB. Bushland and thickets; on sandy or clayey soils; usually in partial shade; 10–1050 m
Flowering specimens have been seen from January, March–May, July and October; the flowers open 06:00–06:30(–06:45) and fade 11:00–12:30.

subsp. **keniense** *Faden* in Smithsonian Contr. Bot. 76: 104 (1991); Ensermu & Faden in Fl. Eth. 6: 354 (1997). Type: Kenya, West Suk District, Kerio Valley, Tot–Sigor road, 4.8 km before crossing of Weiwei River, collected sterile, cultivated at the University of Chicago, *Faden & Faden* 77/788 (US!, holo.; EA!, F, K!, iso.)

Petals pale lilac; capsules mottled light and dark brown.

KENYA. Baringo District: Samatian Island, Lake Baringo, 31 Dec. 1978, *Lavranos* 17063!; Elgeyo District: Tambach–Tot road, 10 km N of junction with Kabarnet–Tambach road, 23 Nov. 1970, *Faden, Evans & Gibson* 70/894A!; South Kavirondo District: Lambwe Valley, Riamkanga, 16 Oct. 1969, *Faden* 69/1300A!
DISTR. **K** 2, 3, 5; S Ethiopia, Rwanda
HAB. Thickets and thicket edges; 1000–1250 m
Flowering specimens have been seen from May and October to November.

NOTE (on the species as a whole). The third subspecies, *A. indehiscens* subsp. *lilacinum* Faden, occurs in southern Zimbabwe, southern Mozambique, Swaziland and South Africa.

Subsp. *keniense* differs from subsp. *indehiscens* by its pale lilac flowers and paler, mottled capsules that are more distinctly stipitate. In these characters it agrees with subsp. *lilacinum*, from which it differs by its capsules with the dorsal capsule valve usually terminating in a ridge, a scarcity of uniseriate hairs on the bracteoles, the relatively short pedicels (4.5–6.5 mm long), and concolorous pollen in all three anthers.

The long-trailing or straggling habit of *A. indehiscens* is characteristic and apparently constant. I have noted it in all 13 populations of subsp. *indehiscens*, both populations of subsp. *lilacinum* and one population of subsp. *keniense* observed in the field. It is also maintained in cultivation. Plants persist during the dry season as thick, leafless shoots. A habit similar to that of *A. indehiscens* has been noted elsewhere in the genus only in *A. recurvatum* and in a single sterile plant believed to belong to *A. hockii*.

The bracteoles are usually perfoliate. Occasionally, however, one or more may be split to the base by the pedicels. *Faulkner* 1160 and *Vaughan* 1851 are unusual in having all of the bracteoles regularly non-perfoliate and thus similar to *A. petersii* in this character.

The medial petal usually does not retain the lateral stamens when the flower opens. In one population (*Faden & Faden* 74/489), however, the stamens were held briefly but not in the usual manner. Only the filaments were contained in the petal; the anthers were shortly exserted. Furthermore, when the margins of the petal unfurled, the stamens rose slowly, indicating that they had not been under tension.

In one population of subsp. *indehiscens* (*Faden & Faden* 74/1184) the lateral staminodes often have small anther sacs that never contain any pollen. Because this small population may well consist of a single genotype, the presence of such an atavism in it may not be significant. On the other hand, the tendency for the medial antherode to be lacking, noted in the same population, is more important, because it has been found in all three Tana River District populations of subsp. *indehiscens* encountered, but not in any other populations of the species.

The lateral stamen filaments usually appear parallel. Commonly they are slightly bowed outward in the middle and converge again towards the apex. Very rarely they may cross near the apex. The lateral anthers are generally held very close together.

The presence of two ovules in the dorsal locule of the ovary has been noted in two populations, both of subsp. *indehiscens*. Of the five examples that have been found, four had had only the basal seed develop while the fifth had had both ovules produce seeds. In the latter case these seeds were very similar to those of the ventral locules in size and shape. Three seeds were found in one ventral locule of a single capsule of *Faden & Faden* 74/208.

Hook-hairs have been noted on the ovary of a single flower of subsp. *indehiscens* (*Faden & Faden* 74/380) and on two of 13 capsules (one hook-hair on each) from cultivated plants of the same collection. Ovarian hook-hairs are otherwise unknown in this species, although they do occur in other species of section *Lamprodithyros*.

Capsules of subsp. *indehiscens* are frequently indehiscent or only partly dehiscent along the lateral sutures. Occasionally the dorsal locule also splits for a short distance along the middorsal suture near the apex. A specimen of subsp. *keniense* (*Faden et al.* 70/894A) appears to have regularly dehiscent capsules, while cultivated plants of another collection (*Faden & Faden* 77/788) have indehiscent capsules. The extent of dehiscence is difficult to determine in pressed specimens because their preparation might produce or increase the split.

Aneilema indehiscens is most closely related to *A. petersii* and *A. recurvatum*. Indeed it may prove to have arisen as an allopolyploid hybrid between the two. All three species have been found in the Kerio Valley of western Kenya and, at two of the four known localities for *A. indehiscens* subsp. *keniense* in Kenya, it was found growing with *A. recurvatum*. In eastern Kenya subsp. *indehiscens* occurs not far from both subspecies of *A. petersii* but is ecologically distinct from each. Among the 13 populations of *A. indehiscens* subsp. *indehiscens* and 22 of *A. petersii* (both subspecies) observed in the field, only one instance has been found of the two species growing together. That was in a disturbed ecotonal situation in Tana River District (see Faden, 1983b). *Aneilema indehiscens* may be distinguished from *A. petersii* by the former's trailing habit, often falcate antherode lobes, usually slightly elongate connectives, lateral stamen filaments usually not crossing, narrower, often indehiscent capsules, and only slight dimorphism between dorsal and ventral locule seeds. From *A. recurvatum A. indehiscens* differs by the latter's less symmetric, puberulous bracteoles that lack marginal glands, larger flowers with puberulous sepals, more rounded paired petals, less divergent and less strongly curved lateral stamen filaments that are usually not retained by the medial petal, and generally larger capsules that are often indehiscent.

15. **Aneilema recurvatum** *Faden* in Smithsonian Contr. Bot. 76: 106, fig. 49; t. 3, c, d, t. 6, c (1991) & in U.K.W.F. 2nd ed.: 308 (1994); Ensermu & Faden in Fl. Eth. 6: 355 (1997). Type: Uganda, Bunyoro District, 37 km N of Butiaba turnoff on Butiaba–Murchison Falls Nat. Park road, 10 km N of Sambiye River crossing, near Kisansya, *Faden, Evans & Lye* 69/1066 (EA!, sheet 1, holo.; BR!, EA! sheet 2, K!, MHU!, iso.)

Perennial herbs; roots fibrous; vegetative shoots decumbent to subscandent, profusely branched; flowering shoots ascending, sparsely branched, to 30 cm tall or more; internodes glabrous or rarely sparsely puberulous. Leaves spirally arranged; sheaths 0.5–1.5(–2) cm long, puberulous, ciliate at apex; laminae shortly petiolate, narrowly lanceolate-elliptic to ovate-elliptic, 2.5–7(–8) × 0.5–2.2(–2.9) cm, margin scabrid, occasionally ciliate, apex acuminate to acute; both surfaces lustrous and sparsely puberulous. Thyrses terminal and frequently axillary, ± lax, ovoid to broadly ovoid, 2–4.5 × 2–6.5 cm wide, with (1–)2–7(–9) ascending cincinni; inflorescence axes puberulous; peduncles (1.5–)2.5–5(–6.8) cm long, with bract above the middle; cincinni to 4.2 cm long and 15-flowered; bracteoles ± symmetrically cup-shaped, perfoliate, 1–2 × 0.5–1.3 mm, prominently glandular subapically and with smaller inconspicuous glands along the margin, usually glabrous. Flowers bisexual and male, 10–14 mm wide; pedicels (4.4–)5–8(–9) mm long in flower, to 10 mm long and recurved in fruit. Sepals subequal, the laterals slightly larger, elliptic to ovate, 2.4–3.5(–3.8) × 1.6–2.4 mm, the medial usually glabrous, the lateral puberulous, prominently glandular near apex; paired petals lilac, 6.5–9(–11) × 4–6.5(–7.5) mm of which the whitish claw 1.5–3(–4) mm; medial petal same colour, whitish medially, slipper-shaped, 5–7.5(–8.5) × 3–4(–4.8) mm, 3.2–4.3(–5) mm deep. Staminodes 3, yellow, antherode bilobed; lateral stamens with filaments 6.2–8.6 mm long, S-shaped, glabrous, anthers 0.8–1.2 × 0.5–0.9 mm wide, pollen yellow; medial stamen filament 4.2–6 mm long, anther 1.5–2.2 × 0.9–1.4 mm wide, pollen yellow. Ovary 1.3–1.8 × 0.9–1.45 mm, with patent glandular hairs; style 7–8.5 mm long, straight to arcuate-decurved; stigma capitate. Capsules substipitate or subsessile, pale grey to greyish tan, lustrous, oblong-ellipsoid to obovoid-oblong, dehiscent, bivalved, trilocular, (3–)3.5–5(–5.5) × 2–2.5 mm, puberulous; dorsal locule prominent, 1-seeded or, by abortion, empty, ventral locules each 2-(or, by abortion, 1–) seeded. Seed of dorsal locule ellipsoid, 1.9–2.6 × 1.2–1.5 × 0.7–0.9 mm, testa brown; ventral locule seeds ovoid to trapezoid, 1.4–2.3 × 1.2–1.6 × 0.7–0.9 mm, testa brown.

UGANDA. Bunyoro District: 38 km N of Butiaba turn off on Butiaba–Murchison Falls Nat. Park road, near Kisansya, 16 Sep. 1969, *Faden et al.* 69/1066! & near Kisansya on Butiaba Flats, 16 Sep. 1969, *Lye et al.* 3981!; Teso District: 3 km W of Wera, 29 Sep. 1962, *Lewis* 5999!

KENYA. Elgeyo District: Kerio Valley, Kabarnet–Tambach road, 2.8 km before junction with Tot road, 14 Mar. 1977, *Faden & Faden* 77/785!; Kisumu–Londiani District: Kisumu, 28 Aug. 1969, *Hindorf* 806!; South Kavirondo District: Lambwe Valley, Riamkanga, 16 Oct. 1969, *Faden* 69/1300!

TANZANIA. Musoma District: Ushashi Rest House, Ushashi, 16 Mar. 1959, *Tanner* 4011!

DISTR. **U** 2–4; **K** 3, 5; **T** 1; southern Ethiopia

HAB. Grassland, bushland and bushland thickets, sometimes near streams; sandy or clayey soils; partial shade; 700–1300 m

Flowering specimens have been seen from March, April and July–November; flowers are open from about 09:00 to 13:45.

SYN. *Aneilema rendlei* sensu Chiovenda in Webbia 8: 38 (1951) pro *Corradi* 2159 pro parte, *non* Clarke (1901)

A. tacazzeanum sensu Lewis in Sida 1: 279 (1964), *non* A. Rich. (1850) [= *A. forskalii* Kunth]

A. sp. *C*, Faden in U.K.W.F.: 666, 667 (1974)

NOTE. This species was not correctly typified when it was published (Faden, 1991) because specimens that were pressed from the cultivated material of the type collection at a later date were erroneously identified as isotypes. These collections are not types. Only specimens of the original field collection are types. So as to not confuse this matter further, and to conform with the Code, those later collections will henceforth be given a different collection number, *Faden* 69/1066bis.

Aneilema recurvatum approaches *A. indehiscens* in habit. However, the long, trailing shoots of the former tend to be much more branched than those of the latter. The differences are maintained in cultivation. In the Lambwe and Kerio valleys in Kenya *A. recurvatum* and *A. indehiscens* subsp. *keniense* grow together, and I twice made mixed collections before I learned to recognize them. When inflorescences are present, the species are easily distinguishable by the the more symmetic, less pubescent bracteoles with glandular margins and less pubescent pedicels in *A. recurvatum*.

Aneilema recurvatum is a distinctive species because of its nearly symmetric, glabrous bracteoles, narrow, pointed paired petals, and small capsules and seeds. The stamen retention mechanism (see Bothalia 14: 1001 (1983)) is well developed in this species. In cultivated plants the stamens are held by the medial petal for about one to two hours.

Aneilema recurvatum is very common in the Lambwe Valley of southwestern Kenya, where it occurs at the edges of moist thickets composed of a variety of trees and shrubs, including *Mystroxylon aethiopicum, Euphorbia candelabrum, Scutia myrtina, Erythroxylum fischeri* and species of *Cadaba, Euclea, Grewia, Rhus* and *Scolopia*. In the Kerio Valley of western Kenya it occurs in drier thickets with *Acacia mellifera, A. brevispica, Teclea pilosa, Grewia* sp., *Sansevieria ehrenbergii* and *Cissus rotundifolia*.

16. **Aneilema sebitense** *Faden* in Smithsonian Contr. Bot. 76: 110, fig. 50, t. 3e (1991). Type: Kenya, West Suk District, near Sebit on Parua road, along Sebit River, *Faden & Faden* 77/803 (US!, holo.; EA!, F, K!, iso.)

Perennial herbs; roots fibrous; shoots ascending, sometimes rooting at the lower nodes, to ± 1 m tall, puberulous. Leaves spirally arranged; sheaths 1–2.5 cm long, pilose-puberulous, ciliate at the apex; lamina petiolate (except in uppermost), lanceolate-elliptic to ovate, 3–13.5 × (0.7–)1.5–4 cm, margin scabrid and ciliate, apex acuminate to acute; both surfaces ± dull, densely puberulous, sometimes also with scattered uniseriate hairs. Thyrses terminal and sometimes axillary, ± lax, ovoid to broadly ovoid, 4–10 × 3.5–10 cm, with (2–)5–9(–12) ascending cincinni; inflorescence axes glabrous to sparsely puberulous (at least basally); peduncles

(3–)5–13 cm long, with halfway bract; cincinni to 5 cm long and 13-flowered (to 6 cm long in cultivation); bracteoles nearly symmetrically cup-shaped, perfoliate, 1.8–2.7 × 1–1.6 mm, prominently glandular subapically and with numerous smaller glands along the margin, glabrous. Flowers bisexual and male, 10–20 mm wide; pedicels 6.5–10.5 mm long in flower, to 13 mm long and erect to slightly recurved in fruit. Sepals subequal, the laterals slightly wider, ovate-elliptic to ovate-lanceolate, 3.5–4.5(–6) × 1.9–3.5 mm, glandular near the apex, glabrous; paired petals lilac, ovate, 9–10.5 × 6–10 mm of which the white claw 2–3 mm; medial petal same colour or slightly paler, cup-shaped, 7.5–13 × 5.5–8 mm, 5–7 mm deep; staminodes 3, yellow, antherode bilobed; lateral stamens with filaments 8.7–11.5 mm long, S-shaped, glabrous, anthers 1.1–1.6 × 0.5–1 mm, pollen creamy yellow to dirty white; medial stamen filament 6–7.5 mm long, anther 1.7–2.5 × 0.9–1.5 mm, pollen golden yellow. Ovary ± 2 × 1.3–1.5 mm, with patent, glandular hairs; style ± 10.5 mm long, arcuate-decurved; stigma capitate. Capsules stipitate, tan or greyish tan, sometimes marked with blue apically or flecked with darker spots, lustrous, obovoid-oblong to oblong-ellipsoid, bivalved, trilocular, (4–)4.5–7.5 × (2.3–)2.8–3.5(–3.7) mm, puberulous to subglabrous; dorsal locule 1-seeded or, by abortion, empty, ventral locules each 2–3(–4)-seeded. Seed of dorsal locule ellipsoid, 2.1–2.9 × 1.5–1.7 mm, 1–1.2 mm thick, testa buff or buff-orange; ventral locule seeds ovoid to trapezoid, (1.3–)1.5–2(–2.9) × 1.5–1.9(–2) mm, 1–1.3 mm thick, testa tan or yellowish brown.

KENYA. West Suk: Near Sebit on Sebit–Parua road, along Sebit River, 15 Mar. 1977, *Faden & Faden* 77/803! & same locality and date, *Faden & Faden* 77/803A! & Cherangani Hills, R. Sebit, 1°1'N [sic: changed to 1°24'N on EA sheet], 35°19'E, 31 July 1969, *Mabberley & McCall* 88! and 88a! (part of same collection according to Mabberley *in litt.*)

DISTR. **K** 2; Ethiopia

HAB. *Acacia-Commiphora-Combretum* woodland and dry hillsides above river; rocky areas; 1500–1550 m

The single flowering collection was made in July.

SYN. *Aneilema rendlei* C.B. Clarke, F.T.A. 8: 69 (1901), pro *Donaldson Smith* 346

NOTE. This species was not properly typified when it was published in Faden (1991) because the original field collection as well as specimens pressed from cultivated material under the same collection number at later dates were all cited as types. To correct this, only the four field collected sheets are here designated as types. The specimens pressed on 17 August 1977 will be renumbered *Faden & Faden* 77/803 bis and those pressed on 27 September 1977 as *Faden & Faden* 77/803 ter. Thus many of the institutions cited in Faden (1991) as holding isotypes of *Aneilema sebitense* will not in fact have types. *Faden & Faden* 77/803A is a separate collection made on the same date as *Faden & Faden* 77/803 and was not cited as part of the type collection.

Donaldson Smith collected his number 346, which is *A. sebitense*, in Ethiopia on 23 April 1895, the same day he collected the type of *Aneilema rendlei*. Clarke (1901) treated this specimen as *A. rendlei* but noted that it could hardly be distinguished from the Australian *A. acuminatum* R.Br. The resemblance to the latter is purely superficial.

Aneilema sebitense most closely resembles *A. forskalii* Kunth of Ethiopia, Eritrea, Djibouti, Sudan, Yemen and Oman. It differs from that species by its perennial habit, ciliate leaf margins, generally less widely spaced bracteoles, narrower, more shortly stipitate capsules with up to four seeds per ventral locule, and smaller seeds. *Aneilema sebitense* also resembles *A. recurvatum* but can be distinguished by its glabrous pedicels that are generally less reflexed in fruit, less divergent lateral stamen filaments, and larger capsules with larger seeds that are more dimorphic.

17. **Aneilema tanaense** *Faden* in Bothalia 15: 98 (1984) & in Smithsonian Contr. Bot. 76: 125, Fig. 27a, fig. 53; t. 3i–l; t. 6g (1991). Type: Kenya, Tana River District, Garissa–Malindi road, 16 km N of junction for Garsen, *Gillett* 19528 (US!, holo.; B, BR!, EA!, FT!, K!, MO!, PRE, iso.)

Annual (rarely perennial) herbs; roots fibrous; primary shoot erect or ascending, much branched at the base, 15–35 cm tall, lateral shoots decumbent, or prostrate initially and then ascending, puberulous. Leaves spirally arranged on main shoot, distichous, at least initially, on lateral shoots, sometimes becoming spirally arranged again towards the inflorescences; sheaths (1–)3–7 mm long, puberulous or pilose-puberulous, apex ciliate; laminae sessile or shortly petiolate, lanceolate to ovate, 1–6.5 × 0.8–2.5(–3) cm, margin scabrid and sparsely ciliate basally, apex acute to obtuse or rounded (in smaller leaves), both surfaces lustrous, puberulous, sometimes mottled with maroon. Inflorescences terminal on main and lateral shoots, ultimately produced from nearly all nodes; thyrses ± dense, broadly ovoid, 1–2(–3) × 1.5–3(–5) cm, with up to 8 ascending cincinni; smaller inflorescences consisting of 1-several clustered cincinni, lacking a distinct axis and not clearly thyrses; inflorescence axes puberulous; peduncles of larger inflorescences (1–)2–5(–8.5) cm long, with bract in upper part; cincinni to 2.2 cm long and 10-flowered; bracteoles symmetrically or eccentrically cup-shaped, perfoliate, 1.4–1.8 × 0.8–1 mm, glandular near the apex and with smaller glands along the margin, puberulous near base. Flowers bisexual and male, (9–)10–14.5 mm wide; pedicels (4–)5.5–10(–11) mm long, recurved and often spirally twisted in fruit. Sepals subequal, elliptic to ovate, 2.5–3 × 2–2.5 mm, prominently glandular near the apex, puberulous except for margins; paired petals pink or pale lilac, (4.2–)6.5–8 × 4.8–7 mm of which the white claw 1.5–2 mm; medial petal same colour, slipper-shaped, 4.7–6 × 3–4.7 mm, 3–3.6 mm deep; staminodes 3, yellow, antherode bilobed; lateral stamens with filaments 5.5–6.5 mm long, S-shaped, glabrous, anthers 0.7–1.2 × 0.5–0.9 mm, pollen yellow or orange-yellow; medial stamen filament 3.5–4 mm long, anther 1–1.5 × 1.1–1.3 mm, pollen as in laterals. Ovary 1–1.2 × 0.8–1 mm, with patent, glandular hairs; style 5.5–6.5 mm long, arcuate-decurved; stigma capitate. Capsules substipitate to shortly stipitate, grey to tan, sometimes with a dark grey band around the base of the dorsal locule, frequently with irregular brown spots and stripes, lustrous, obovoid (to ovoid), bivalved, trilocular, (2.4–)2.7–3(–3.4) × (1.1–)1.5–2.1 mm, puberulous; dorsal locule very prominent, often with a seed, ventral locules each 2-(or, by abortion, 1-)seeded. Seed of dorsal locule hemispherical, 1.1–1.6 × 1–1.4 × 0.8–1 mm, testa tan; ventral locule seeds subtriangular, 1.2–1.4(–1.7) × 1.2–1.3(–1.5) × 0.8–0.95 mm, testa grey or greyish tan.

KENYA. Kwale District: Mombasa–Nairobi road, 2.5 km towards Mombasa from turnoff of Maji ya Chumvi Railway Station, 22 Feb. 1977, *Faden & Faden* 77/582! & Lungalunga–Ramisi road, ± 6.5 km from Vanga, 8 Mar. 1977, *Faden & Faden* 77/738!; Tana River District: Garsen–Malindi road, 1.5 km towards Malindi from turnoff to Oda, 22–24 July 1974, *Faden & Faden* 74/1185!

DISTR. **K** 7; only known from the lower reaches of the Tana River and Kwale District

HAB. Deciduous or semi-evergreen bushland and thickets; 10–250 m

Flowering specimens have been seen from January, February, July and August; the size of the largest inflorescences in *Gillett* 19528 indicates that the species must have been flowering in December. In the field the flowers open 08:30–09:00 and fade 13:00–13:30.

SYN. *Aneilema clarkei* Rendle in J.L.S.: 30, t. 34, Fig. 8 tantum, figs. 7 & 9–12 et descr. excl. (1895)
A. calceolus sensu Brenan in K.B. 15: 223 (1961), pro *Gregory* s.n.

NOTE. The first inflorescences produced by plants of *A. tanaense* are invariably distinct thyrses with a short but clear inflorescence axis. Later inflorescences consist of fewer cincinni (sometimes only one) and may lack an axis, giving them a very strong resemblance to inflorescences of *A. calceolus*. They differ from the latter chiefly in having more elongate cincinni with more widely spaced bracteoles that have glandular margins.

In Tana River District, *A. tanaense* is restricted to a peculiar type of alluvium found along the lower reaches of the Tana River. This soil type is referred to as "black cotton soil with sand" by local agricultural officers. The trees and shrubs most commonly associated with *A. tanaense* are *Acacia bussei*, other *Acacia* spp., *Commiphora campestris*, *Dobera* sp., *Combretum hereroense*, *Grewia tenax* and *Sansevieria powellii*. In Kwale District, *A. tanaense* occurs in two habitats: 1) termite mounds and thicket edges, the thickets containing trees and shrubs such as *Euclea* sp., *Diospyros consolatae*, *Manilkara* sp., *Carissa bispinosa*, *Haplocoelum inoploeum* and *Erythroxylum emarginatum* as well as succulents; and 2) bushland and thickets dominated by *Acacia* spp. and succulents. The only two perennial populations are from the last habitat near Maji ya Chumvi.

Aneilema tanaense is transitional between the species of section *Lamprodithyros* that regularly have thyrsiform inflorescences and those that never have them. It is most closely related to two coastal species, *A. calceolus* from Kenya and Tanzania and *A. benadirense* Chiov. from Somalia. Preliminary crosses between *A. tanaense* and *A. calceolus* suggested a high degree of interfertility.

18. **Aneilema calceolus** *Brenan* in K.B. 15: 223 (1961), excl. *Gregory* s.n.; Faden in Smithsonian Contr. Bot. 76: 128, fig. 35, c, d; t. 3, m-o, t. 6, f (1991). Type: Tanzania, Tanga District: near Kange Limestone Gorge, *Milne-Redhead & Taylor* 7285 (K!, holo.; BR!, EA!, iso.)

Annual (rarely perennial) herbs; roots fibrous; vegetative shoots repent, much branched, forming mats to 1 m or more in diameter, flowering shoots ascending, to 15(–20) cm tall, puberulous. Leaves distichous; sheaths 2–6 mm long, puberulous and with a few long hairs towards the base, ciliate at apex; lamina sessile or shortly petiolate, ovate to ovate-elliptic, 1–3.5(–5) × 0.5–1.5(–2) cm, margin scabrid, sparsely ciliate basally, apex acute or occasionally obtuse, often mucronulate; both surfaces lustrous, puberulous and pilose. Inflorescences terminal on main and lateral shoots, reduced thyrses (see discussion below), to ± 2 × 2 cm consisting of 1–4(–6) cincinni that appear fasciculate in the axils of bracts or reduced leaves; inflorescence axis ± not developed; peduncles to 7.5 cm long, puberulous; cincinni to 7 mm long and 14-flowered ; bracteoles eccentrically cup-shaped, perfoliate, 1.1–1.9 × 0.4–1 mm high, inconspicuously glandular near the apex, ± glabrous. Flowers bisexual and male, (10–)12–15 mm wide; pedicels 7–11 mm long in flower, to 13(–16?) mm and recurved and frequently somewhat spirally twisted in fruit. Sepals subequal, lanceolate to ovate-elliptic, 3–4 × 2–2.5 mm, glandular near the apex, puberulous; paired petals lilac, ovate, 5–9 × 3.5–7.5 mm of which the whitish claw 1–3 mm; medial petal same colour, slipper-shaped, 5–7.5 × 2.5–4 mm wide, 2.5–3.7 mm deep; staminodes 2, the medial usually missing, yellow, antherode bilobed; lateral stamens with filaments 4–8 mm long, S-shaped, glabrous, anthers 0.7–1.3 × 0.5–0.8 mm wide, pollen white; medial stamen with filament 3–5 mm long, anther 1.4–2 × 1.2–1.5 mm, pollen yellow. Ovary 0.9–1 × 0.9–1.1 mm wide, with patent, glandular hairs; style 5.5–7 mm long, very gently arcuate-decurved; stigma slightly capitate. Capsules substipitate, light brown or pale greyish brown, spotted and striped with dark brown, obovoid, (2.3–)2.5–3.6 × (1.4–)1.9–2.5(–2.8) mm, bivalved, trilocular, puberulous; dorsal locule very prominent, usually with a seed, ventral locules each 2-(or, by abortion, 1-)seeded. Dorsal locule seed hemispherical, 1.2–1.6 × 1–1.4 × 0.8–1.1 mm, testa tan, smooth; ventral locule seeds subtriangular, 1.2–1.6(–2) × 1.1–1.5 × 0.7–1 mm, testa tan, pinkish tan or greyish tan, shallowly scrobiculate.

KENYA. Kwale District: Lungalunga–Ramisi road, 1 km before turnoff to Kinango, 14 Feb. 1977, *Faden & Faden* 77/378!; Tana River District: Garsen–Witu road, near Nyangoro Bridge, 14 July 1974, *Faden & Faden* 74/1069! & 20 Feb. 1977, *Faden & Faden* 77/565!

TANZANIA. Tanga District: 8 km SE of Ngomeni, 31 July 1953, *Drummond & Hemsley* 3561! & Kange Gorge, ± 5 km E of Tanga, 29 Mar. 1974, *Faden & Faden* 74/333 & near Kange Limestone Gorge, 13 Nov. 1956, *Milne-Redhead & Taylor* 7285!

DISTR. **K** 7; **T** 3; not known elsewhere

HAB. Dry or moist, lowland evergreen forest or thicket; growing on Kambe limestone, often in shallow soil, in the Tanga area of Tanzania; partial or dense shade; 10–450 m

Flowering specimens have been seen from June, July and September–November; in the field the flowers open 09:00–10:00 and fade 13:00–13:30.

NOTE. *Gregory* s.n., cited by Brenan (1961) as this species, differs from the Tanzanian specimens in being more erect and in having shorter pedicels, more elongate cincinnus axes that are puberulous where exposed, and especially in its bracteoles with small glands along the margin and generally sparsely puberulous. It also has small marginal glands on the medial sepal. This specimen is treated above as *A. tanaense*.

Aneilema calceolus is generally annual. Only completely or nearly completely dead plants of *Faden & Faden* 74/324, 74/333 and 77/378 were found in the field, and 74/324 and 77/378, when grown from seed, behaved as annuals, i.e. after flowering and fruiting they ceased vegetative growth and were unable to root from cuttings. However, the northernmost, isolated population in the Tana River District of Kenya (*Faden & Faden* 74/1069 & 77/565) is definitely perennial and has been cultivated for more than 30 years.

In *A. calceolus* the stamens are retained by the medial petal longer than in any other species of section *Lamprodithyros*. Field observations of *Faden & Faden* 74/1069 indicate that the stamens are held for an average of two and one-third hours or approximately half of the flowering period.

The most closely related species to *A. calceolus* is *A. tanaense*. They agree in flower and capsule structure, pedicel curvature and pubescence and seed dimorphism. Reduced inflorescences of *A. tanaense* are very similar to inflorescences of *A. calceolus*. The two species differ in habitat, habit and in the characters given in the key.

19. **Aneilema clarkei** *Rendle* in J.L.S.: 30: 430, t. 34, fig. 7 & 9–12, *non* fig. 8 (1895); Schumann in P.O.A. C: 136 (1895); C.B. Clarke in F.T.A. 8: 73 (1901); Brenan in K.B. 15: 224 (1961), in adnot.; Faden in Smithsonian Contr. Bot. 76: 130; Fig. 35, e; t. 3p, t. 4d, t. 6h (1991). Type: Kenya, Tana River District, Lake Dumi, 13 Feb. 1893, *Gregory* s.n. (BM!, holo.)

Perennial herbs; roots fibrous; vegetative shoots procumbent to repent, much branched; flowering shoots erect to ascending, generally unbranched, 15–30(–40) cm tall, glabrous to puberulous. Leaves distichous; sheaths 2.5–10 mm long, puberulous and sparsely pilose, apex ciliate; lamina petiolate, lanceolate to elliptic or ovate, 1.5–5(–7.8) × (0.4–)0.8–1.5(–2.5) cm, margin scabrid, sparsely ciliate basally, apex acuminate to acute; both surfaces lustrous, pilose-puberulous. Inflorescences terminal on the flowering shoots and on short shoots produced from the uppermost 1–3 nodes, each inflorescence consisting of a short solitary cincinnus subtended by a pair of bracts; peduncles 0–2.5 cm long, puberulous, with a terminal bract; cincinni to ± 5 mm long and 10-flowered; bracteoles eccentrically cup-shaped, perfoliate, ± 3–5 × 2 mm, apparently eglandular, the outer sparsely puberulous. Flowers bisexual and male, (12.5–)15–20 mm wide; pedicels 7–11.5 mm long, recurved near apex in fruit. Sepals subequal, lanceolate-ovate to ovate-deltoid or ovate-orbicular, 3.2–5.5 × 2.5–4.5 mm, glandular near the apex, glands unlobed; paired petals lilac to lavender, ovate to ovate-subreniform, (7.5–)9–12 × (5.5–)7–11 mm of which the white to lavender claw 2.5–4 mm; medial petal same colour, cup-shaped, 6–9 × 3–7 mm, 2.5–5 mm deep; staminodes (2–)3, yellow or white, antherode bilobed; lateral stamens with filaments (5.5–)6.5–8.5 mm long, S-shaped, glabrous, anthers 1–1.6 × 0.5–1.1 mm, pollen white; medial stamen filament 3.5–5.5 mm long, anther 1.4–2.5 × 1–1.6 mm, pollen yellow. Ovary (1–)1.5–1.9 × (1–)1.1–1.3 mm, with patent, glandular hairs; style 7.5–11 mm long, nearly straight or gently arcuate-decurved; stigma capitate. Capsules substipitate, tan or greyish tan, with small, dark brown spots, obovoid to oblong-ellipsoid, bivalved, trilocular, 3.6–4.5 × 2.3–3 mm, puberulous; dorsal locule very prominent, 1-seeded (or occasionally empty), ventral locules each 2-(or, by abortion, 1-)seeded. Seeds broadly ellipsoid, 1.8–2 × 1.4–1.5 mm (dorsal locule seed) or ovoid to subtriangular, 1.5–1.7(–2.1) × 1.5–1.6 mm (ventral), 0.9–1.1 mm thick, testa light brown.

KENYA. Kilifi District: Sokoke Forest, road to Jilore Forest Station, 3.2 km from turnoff on Kilifi–Malindi road, 28 July 1971, *Faden & Evans* 71/714!; Kwale District: Mwena River S of Marenje Forest, N side of river, 8 Mar. 1977, *Faden & Faden* 77/749!; Lamu District: Mambasasa, Utwani Forest Reserve, 16 Oct. 1957, *Greenway & Rawlins* 9344!

TANZANIA. Uzaramo District: near Magogoni, July 1936, *Vaughan* 2395!; Pemba: Ngezi Forest, 30 Aug 1929, *Vaughan* 621! & 22 Jan. 1933, *Vaughan* 2064!

DISTR. **K** 7; **T** 6; **P**; not known elsewhere

HAB. Lowland evergreen forest, sometimes on Kambe limestone, *Brachystegia* forest or woodland, river slopes, *Hyphaene-Albizia* scrub, and moist thickets; dense or partial shade; probably always on calcareous soils; 0–250 m

Flowering specimens have been seen from January–March, May and July–November; in the field the flowers open 08:45–09:30 and fade 12:00–13:00.

NOTE. Plants of *A. clarkei* tend to form mats where locally common. Vegetative shoots can attain at least one meter in length.

The lateral stamen filaments are usually retained by the middle petal when the flowers of *A. clarkei* open. Field notes on *Faden & Faden* 71/809 indicate that in a rare flower the stamens are not held. Rarely one stamen is not released.

The presence of dorsal capsule valves in mature fruiting specimens of *A. clarkei* shows that this valve is not regularly deciduous, as it is in some other species of sect. *Lamprodithyros*.

Similar inflorescence structure and fruiting pedicel curvature to those of *A. clarkei* are found only in *A. succulentum* and *A. lamuense*, which must be considered its closest relatives. *Aneilema clarkei* differs from *A. succulentum* by the former's petiolate leaves with symmetric bases and larger capsules and seeds, from *A. lamuense* by its perennial habit, lack of uniseriate hairs on the pedicels and sepals, and yellow medial anther with yellow pollen, and from both in its larger, bluer flowers, better developed stamen retention mechanism, more divergent lateral stamen filaments and less regularly deciduous dorsal capsule valves.

20. **Aneilema lamuense** *Faden* in Smithsonian Contr. Bot. 76: 133, fig. 54, t. 4a-c, t. 6i (1991) & in Fl. Somalia 4: 162, fig. 55 (1995). Type: Kenya, Lamu District, Kitwa Pembe Hill and vicinity, *Faden & Faden* 74/1083 (US!, holo.; BR!, C, EA!, FT!, K!, MO!, PRE, UPS, iso.)

Annual herbs; roots fibrous; shoots erect to ascending or sometimes decumbent, usually much branched, to 20(–40) cm tall, all shoots eventually terminating in inflorescences, puberulous (sometimes only just below the nodes). Leaves spirally arranged on main shoot, distichous on lateral shoots; sheaths 3–9 mm long, puberulous to densely pilose-puberulous (uppermost leaves), apex ciliate; lamina shortly petiolate or sessile, ovate-elliptic to lanceolate, (1–)2–4.5(–5.8) × 1–2(–2.5) cm wide, margin scabrid, sparsely near base, apex acute to acuminate; adaxial surface dull, abaxial lustrous, both surfaces pilose-puberulous. Inflorescences terminal and on short shoots from the uppermost 1–3 nodes on the major shoots, each inflorescence of a short solitary cincinnus partially enclosed in a subopposite pair of bracts; peduncles to ± 1 cm long, puberulous, with a terminal bract; cincinni to ± 5 mm long and 10-flowered; bracteoles lanceolate-ovate to ovate, often eccentrically cup-shaped, perfoliate or not, ± 4–6 × ± 2 mm, eglandular but the apex thickened, outer ones pilose-puberulous. Flowers bisexual and male, 10.5–14(–15.5) mm wide; pedicels 7–10 mm long, apically recurved in fruit. Sepals with the laterals longer and wider than the medial, lanceolate to ovate-elliptic, 2.8–4.3 × 1.6–2.6(–2.9) mm, the medial generally glabrous, the lateral pilose-puberulous near base, not clearly glandular; paired petals lilac, ovate, 6–9 × 5–7(–8) mm of which the basally whitish claw 2–3 mm long; medial petal boat or cup-shaped, 5–6(–7.5) × 2–4.5(–6) mm, 2.8–3.7(–3.9) mm deep, whitish except lilac base; staminodes 3, yellow with pinkish-purple bases, antherode bilobed; lateral stamens with filaments (5–)6–8.5 mm long, S-shaped, glabrous, anthers 1.4–1.7 × 0.8–1 mm, pollen white; medial stamen filament 4.5–6.3 mm long, anther 1.6–2 × 1–1.4 mm, pollen white. Ovary 1.5–1.8 × 1–1.2 mm, with patent glandular hairs; style 7–8 mm long; stigma capitate. Capsules substipitate, tan with dark brown spots, lustrous, obovoid to obovoid-ellipsoid, bivalved, trilocular, 3.6–4.3(–5) × 1.9–2.1(–2.6) mm, puberulous; dorsal locule very prominent, 1-seeded (or occasionally empty), ventral locules each 2-(or, by abortion, 1-)seeded. Dorsal locule seed broadly ellipsoid, 1.65–2.15 × 1.35–1.55 × ± 1 mm, testa orange-buff, spotted and striped with dark brown; ventral locule seeds ovoid to subtriangular, (1.3–)1.5–1.7(–1.8) × 1.4–1.8 × 0.9–1.2 mm, testa tan, orange-buff or orange-brown, spotted and striped with dark brown (or maroon?).

KENYA. Lamu District: Kitwa Pembe Hill and vicinity, 15–16 July 1974, *Faden & Faden* 74/1083! & Manda (East) Island, 5 Aug. 1982, 1982 *Brathay Expedition* 106! & Lamu Island, 1 Aug. 1971, *Schlieben* 12121!

DISTR. **K** 7; Somalia

HAB. Coastal sand dunes and other sandy habitats; 0–50 m
Flowering specimens have been seen from July and August; in the field the flowers open 09:00–09:30 and fade ± 13:00.

NOTE. *Schlieben's* specimens differ from the *Fadens'* in having shorter internodes and apparently somewhat fleshy leaves. They were probably growing more exposed to the effects of salt-laden sea breezes than were the Fadens' plants which came from the sheltered, leeward side of 45–60 m high sand dunes.

Plants of *A. lamuense* show great variation in habit, ranging from unbranched, erect individuals a few centimeters tall to much-branched ones up to about 40 cm tall and 50 cm in diameter (*Brathay Expedition* 106 refers to plants of 1 m, presumably in length). In general, the shoots are erect to ascending; only in the largest plants do they become somewhat decumbent.

The stamen retention mechanism is poorly developed in *A. lamuense.* In the field the period for which the lateral stamens were held by the medial petal could not be determined because visiting bees appeared to be instrumental in their early release. However, in cultivation the stamens were found to be released in less than an hour and often after only a few minutes, suggesting that the pollinators were not the cause of the early release. The anthers escape from a very narrow gap between the still inrolled, lateral margins of the medial petal.

Aneilema lamuense is highly distinctive because of its inflorescence structure and fruiting pedicel curvature, characters it shares only with *A. clarkei* and *A. succulentum*, its closest relatives. *Aneilema lamuense* differs from the former by its smaller flowers and narrower capsules, from the latter by its petiolate leaves and larger capsules and seeds, and from both by its annual habit, long, uniseriate hairs on the pedicels and lateral sepals, petal color, staminode lobes with pinkish purple or violet spots at their bases, and white medial anther with white pollen. It is also ecologically distinct from both.

21. **Aneilema succulentum** *Faden* in Smithsonian Contr. Bot. 76: 136, fig. 55, t. 4e, f, t. 6j (1991). Type: Kenya, Tana River District, Garsen–Witu road, near Nyangoro Bridge, *Faden & Faden* 74/1152 (US!, holo.; BR!, EA!, FT, K!, MO!, PRE, iso.)

Perennial herbs; roots fibrous; vegetative shoots repent, much branched, often forming mats; flowering shoots ascending, unbranched or little branched, to 30 cm tall, puberulous, very succulent, becoming swollen in the rainy season, the nodes then appearing as constrictions. Leaves distichous; sheaths 2–6(–10) mm long, pilose-puberulous, ciliate at apex; lamina succulent, often slightly falcate, sessile, narrowly lanceolate-elliptic to elliptic, 1.3–4(–6) × (0.4–)0.7–1.3(–1.9) cm, base slightly asymmetric and cuneate, margin often maroon, scabrid, sparsely ciliate near base, apex acute; both surfaces slightly lustrous, adaxial frequently mottled with maroon, puberulous or pilose-puberulous. Inflorescences terminal and on short shoots from the uppermost 1–4 nodes or on longer shoots produced from the lower nodes of the main flowering shoot, each inflorescence consisting of a short solitary cincinnus partially enclosed by a pair bracts; peduncles ± 0–4 cm long puberulous, with terminal bract; cincinni to ± 5 mm long and 10-flowered; bracteoles eccentrically cup-shaped, perfoliate, 3.2–7 × 1–3 mm, slightly thickened apically but eglandular, sparsely puberulous, rarely glabrous. Flowers bisexual and male, 9–14(–17) mm wide; pedicels 7–10(–13) mm long, apically recurved in fruit. Sepals with the laterals larger than the medial, ovate to lanceolate-elliptic, 3.5–4.5(–5.5) × 1.8–3.2 mm, apparently eglandular, glabrous, or puberulous basally; paired petals white to pale lilac, ovate-elliptic to reniform, 7–8(–10.5) × 5–7(–9) mm of which the whitish claw 1.5–2.5(–3.5) mm; medial petal white or pale lilac, boat- or cup-shaped, 5.5–7(–9.5) × 3–5(–7) × (2–)2.5–3(–5) mm; staminodes 3, yellow, antherode bilobed; lateral stamens with filaments 5–6(–7) mm long, S-shaped, anthers (0.7–)0.9–1.4 × (0.45–)0.6–1 mm, pollen white or sometimes orange-yellow; medial stamen filament 4–5(–5.5) mm long, anther (1.2–)1.5–2.3 × 1–1.2 mm, pollen orange-yellow. Ovary 1.3–1.6 × 0.9–1.1 mm, with patent, glandular hairs; style 5.5–7 mm long, gently arcuate-decurved; stigma capitate. Capsules shortly stipitate, greyish tan with dark brown spots, ellipsoid to obovoid-ellipsoid, keeled, bivalved, trilocular,

(2.8–)3.2–3.7(–4.3) × (1.5–)2.1–2.3(–2.5) mm, puberulous; dorsal locule prominent, 1-seeded (rarely empty), ventral locules each 2-(or, by abortion, 1-) seeded. Seeds broadly ellipsoid to broadly ovoid, 1.3–1.6 × 1.2–1.3 mm (dorsal) or subtriangular to ovoid, 1.3–1.5(–1.6) × 1.3–1.5 mm (ventral), 0.8–1 mm thick, testa tan or greyish tan.

KENYA. Kilifi District: Kakoneni–Jilore Forest Station road km 2.5, 10 Jan. 1972, *Faden et al.* 72/39!; Tana River District: Garsen–Witu road, near Nyangoro Bridge, 20–21 July 1974, *Faden & Faden* 74/1152!; Teita District: Nairobi–Mombasa road, 11.7 km towards Mombasa past Maungu Station, 30 Apr. 1974, *Faden & Faden* 74/524!

DISTR. **K** 7; not known elsewhere

HAB. Dry deciduous bushland and bushland thicket dominated by species of *Commiphora*, *Euphorbia*, *Lannea* &c. or, less commonly, *Diospyros* or *Acacia*, also in *Brachystegia* forest or woodland and along roadsides; sandy or clayey soils, occasionally on rocks; usually in partial shade or full sun; 10–600 m

Flowering specimens have been seen from January, April, May, July, October and December; in the field the flowers open 11:30–12:30 and fade 14:45–15:30.

NOTE. The stamen retention mechanism of the medial petal is poorly developed in *A. succulentum*. It has been noted both in field populations and cultivated plants that frequently some flowers have one or both stamens free when the flowers open. Furthermore, in no case are the stamens held for more than half an hour.

The occurrence of *A. succulentum* in *Brachystegia* woodland in the Kilifi District is interesting, because the closely related *A. clarkei* also frequently grows under *Brachystegia* in the same district. However, the soils and understory shrubs and herbs are quite different where each of the two occurs. Most notable is the presence of numerous succulents – e.g., *Euphorbia breviarticulata* [= *grandicornis* of authors] is the dominant understory shrub – in those *Brachystegia* habitats where *A. succulentum* is present. Succulents are completely lacking in the *Brachystegia* woodland containing *A. clarkei*. The two species have not been found together.

In the inland localities (Teita and Kwale Districts) and in Kilifi District, *A. succulentum* occurs only in sandy soil or occasionally in well drained rocky habitats. In Tana River District it grows mostly on a clayey alluvium mixed with sand.

Aneilema succulentum appears to have two flowering periods inland, April to June and October to December or January, corresponding to the two rainy seasons in this region. In the Tana River District it may have one long flowering season (April to November?) or perhaps two.

22. **Aneilema zebrinum** *Chiov.* in Webbia 8: 38, fig. 12 (1951), as *zebrina*; Obermeyer & Faden in F.S.A. 4(2): 40, figs. 8, 9–3a, b (1985); Faden in Smithsonian Contr. Bot. 76: 140, t. 4g, h, t. 7a (1991) & in U.K.W.F. 2nd ed.: 308 (1994); Ensermu and Faden in Fl. Eth. 6: 357, fig. 207.12 (1997). Types: Ethiopia, Gemu-Gofa Prov., Caschei R. bank, *Corradi* 2154 (FT!, syn.; photo K) & same locality, *Corradi* 2163 (FT!, syn.; photo K); Cashei, *Corradi* 2155 (FT!, syn., photo K: lectotype of Brenan in K.B. 19: 67, 1964)

Perennial herbs; roots fibrous; shoots repent, sparsely branched, puberulous. Leaves distichous; sheaths 3–9 mm long, puberulous and sparsely pilose, ciliate at apex; lamina somewhat succulent, sessile, ovate to ovate-elliptic, 1–3.5(–4) × 0.5–2 cm, base symmetric, usually amplexicaul, margin ciliolate, scabrid in dried specimens but not in life, apex acute or occasionally obtuse, thickened and usually mucronate; both surfaces lustrous, puberulous, often mottled with maroon between the veins, when not so mottled often grey-green. Inflorescences axillary, perforating the leaf sheaths, generally consisting of one (rarely 2), short cincinnus subtended by 3 bracts, frequently a secondary inflorescence developing from the axil of the inflorescence bract; peduncles up to 8 mm long (often not developed), puberulous with a basal to apical bract; cincinni to ± 2 cm long and 4(–6)-flowered; bracteoles eccentrically cup-shaped, perfoliate, 1.2–2(–2.2) × 0.3–1 mm, eglandular, densely puberulous. Flowers bisexual and (very rarely) male, 7–10 mm wide; pedicels

(0.5–)1.5–3.5 mm long, to 5 mm long in fruit, puberulous. Sepals with the lateral slightly larger than the medial, ovate-elliptic to ovate, 2–3 × (1–)1.5–1.9 mm, apparently eglandular, puberulous; paired petals pale lilac, ovate to broadly ovate, 4–5.7 × 3.1–4.5 mm of which the whitish claw 1.25–2 mm; medial petal same colour, cup-shaped, 3.5–4.7 × 2.7–4 × 1.7–2 mm; staminodes 3, the medial hidden in the sepal, yellow, antherode bilobed; lateral stamens with filaments 3–3.5(–4.5) mm long, ± S-shaped, glabrous, anthers 0.5–0.55 × 0.45–0.6 mm, pollen yellow; medial stamen filament 2.5–2.8(–3.6) mm long, anther 0.6–0.85 × 0.6–0.8 mm, pollen yellow. Ovary (1–)1.2–1.5 × (0.7–)0.9–1 mm with a mixture of glandular hairs and hook-hairs; style 2.7–3.2(–3.6) mm long, arcuate-decurved; stigma capitate. Capsules substipitate, pale yellow or rarely chocolate brown, lustrous, obovoid-elliptic, bivalved (when partially dehiscent), trilocular, 3–4 × 2.1–2.5(–3) mm, puberulous; dorsal locule 1-seeded (or occasionally empty), ventral locules each 2-(or, by abortion, 1-)seeded. Seeds broadly ellipsoid to reniform, 1.4–1.8 × 1–1.25 mm (dorsal), or ovoid to ovoid-cordate, 1.2–1.6 × 1.1–1.5 mm (ventral), 0.5–0.6 mm thick, testa grey or greyish brown. Fig. 13/6–7, p. 120.

KENYA. Elgeyo District: 2.1 km S of Tot, 23 Nov. 1970, *Faden et al.* 70/888!; Tana River District: 13 km from Galole on Galole–Garsen road, 10 July 1974, *Faden & Faden* 74/1052!; Teita District: 18 km on Voi–Taveta road from Nairobi–Mombasa road turnoff, 5 May 1974, *Faden & Faden* 74/533!

TANZANIA. Pare District: between N Pare Hill & N end of Lake Jipe, 17 Mar. 1960, *Bally* 12145!; Tanga District: 3 km towards Maramba from turnoff to Lelwa on Maramba (Malamba)–Korogwe road, 27 Mar. 1974, *Faden et al.* 74/330!; Bagamoyo District: Dar es Salaam–Morogoro road, ± 10 km E of Chalinze, 11 Dec. 1976, *Mhoro & Wingfield* BM2411!

DISTR. **K** 3, 7; **T** 3, 6; Ethiopia, South Africa

HAB. Deciduous or semi-evergreen bushland and bushland thickets often dominated by species of *Acacia* and/or *Commiphora*, also in woodland, woodland thickets and light forest, sometimes in rocky areas, occasionally on steep slopes; sandy or clayey soils; partial or dense shade; 10–1150 m

Flowering specimens have been seen from January, February, April, May, July and December; in the field the flowers open 10:30–11:15 and fade 13:15–14:30.

SYN. *Ballya zebrina* (Chiov.) Brenan in K.B. 19: 64, map 1, fig. 1 (1964); E.P.A.: 1519 (1971); Ross, Fl. Natal, 117 (1972); Faden in U.K.W.F.: 653 (1974)

NOTE. The monospecific genus *Ballya*, based on *A. zebrinum*, was shown to belong to *Aneilema* section *Lamprodityros* by Faden (1991) because of its large, cup-shaped lower petal that is concolorous with the paired petals, its fused filament bases and five-seeded capsules. Nevertheless, it does have some unusual features, such as the usually strictly axillary inflorescences, eglandular bracteoles, abundant hook-hairs on the ovaries and capsules, and indehiscent or partially dehiscent capsules.

Within section *Lamprodithyros*, *A. zebrinum* is most closely related to *A. clarkei*, *A. lamuense* and *A. succulentum* and especially to the last species. The repent flowering shoots and strictly lateral inflorescences of *A. zebrinum* readily distinguish it from these species.

This species is very distinctive but easily overlooked because of its repent habit and small flowers that are borne beneath the leaves. The major gaps in its range are likely the result of under-collection rather than disjunct distribution.

23. **Aneilema beniniense** (*P. Beauv.*) *Kunth*, Enum. Pl. 4: 73 (1843); C.B. Clarke in DC., Monogr. Phan. 3: 224 (1881) & in F.T.A. 8: 68 (1901); Morton in J.L.S. 59: 464–468, fig. 12, 13 (1966), including subsp. *sessilifolium*, & in J.L.S. 60: 169 (1967); Brenan in F.W.T.A., 2nd ed.: 3: 31 (1968); Faden in Smithsonian Contr. Bot. 76: 145, t. 4, I (1991) & in U.K.W.F. 2nd ed.: 308 (1994); Ensermu & Faden in Fl. Eth. 6: 357 (1997). Type: Nigeria, vicinity of Benin, *P. de Beauvois* (G!, holo.)

Perennial, decumbent, rooting at the nodes; roots thin, fibrous; shoots erect to ascending, (15–)40–130 cm tall, densely branched below, usually unbranched distally, glabrous. Leaves spirally arranged; sheaths 1–4 cm long, glabrous or very sparsely puberulous along the fused edge distally, sparsely ciliolate or eciliolate at apex; lamina (sub-)petiolate, narrowly lanceolate to elliptic, (5–)7–15(–18.5) × (1.2–)2–5(–6.7) cm, base cuneate to ± rounded, margins scabrid distally, apex acuminate; both surfaces glabrous or the abaxial sparsely puberulous. Thyrses terminal and occasionally on a short shoot from a distal leaf or inflorescence bract, very dense, ovoid or cylindric, 2–6 × 1.5–6 cm, of (10–)18–55 ascending cincinni (or the lowest patent); inflorescence axes glabrous or sparsely puberulous; peduncles 2–4 cm long with a bract halfway; cincinni to 3 cm long and 9-flowered; bracteoles cup-shaped, perfoliate, prominently glandular subapically, glabrous. Flower bisexual and male, 7–10(–13) mm wide; pedicels 3–6 mm long in flower, 4–7 mm long and erect to strongly recurved in fruit, glabrous. Sepals with the medial slightly larger, green or greenish white, ovate or elliptic, 2–3 × 2–2.5 mm, convexo-concave, hooded apically, glandular subapically, glabrous; paired petals white or pale lilac, lavender or violet, 4–5 × 2.5–4 mm of which the claw 1–1.5 mm long; medial petal white or greenish white, ovate or broadly ovate, 3–4 × 2–3 mm; staminodes 2–3, yellow, antherode bilobed; lateral stamens with filaments 3.5–5 mm long, sigmoid, sparsely bearded, anthers 0.8–1 mm long; medial stamen filament 3–3.5 mm long, anther 0.8–0.9 mm. Ovary 1.5–2 × 1 mm, glabrous; style arcuate-descending, ± 4 mm long; stigma capitellate. Capsules dark brown or grey-brown, oblong-ellipsoid, (4.5–)5–7 × 2.5–3.5 mm, bi- or trilocular, bivalved, glabrous; dorsal locule 0–1-seeded, ventral locules (1–)2–3-seeded. Ventral locule seeds broadly ovoid to reniform, 1.3–2.4 × 1.2–1.8 mm, testa pinkish brown, orange-brown or grey.

UGANDA. Ankole District: Kalinzu Forest, 4 km NW of Saw Mill, W of Rubuzigye, 19 Sep. 1969, *Faden et al.* 69/1132!; Toro District: Labuongwe Forest, 27 W of Fort Portal, 19 Dec. 1949, *Dawkins* D486!; Masaka District: Malabigambo Forest, 6.4 km SSW of Katera, 2 Oct. 1953, *Drummond & Hemsley* 4547!

KENYA. North Kavirondo District: Malava (Kabras) Forest, along Kakamega–Broderick Falls Road, 27 Nov. 1969, *Faden & Evans* 69/2060!; Kakamega Forest, along Kubiranga Stream at N end of forest, 16 Mar. 1977, *Faden & Faden* 77/883!; Kericho District: Lugusida Stream near Kericho, 3 Oct. 1979, *Lavranos & Newton* 17746!

TANZANIA. Bukoba District: Minziro Forest Reserve, Kinwa Kyaishemweru forest area SE of Minziro Village, 16 July 2001, *Festo, Bayona & Wilbard* 1633!; Buha District: Kasakala Chimpanzee Reserve, 25.6 km N of Kigoma, by Gombe Stream, 17 Nov. 1962, *Verdcourt* 3337!; Mpanda District: Ntakatta Forest, 11 June 2000, *Bidgood, Leliyo & Vollesen* 4647!

DISTR. **U** 2, 4; **K** 5; **T** 1, 4; Senegal, Guinea Bissau, Guinea, Sierra Leone, Liberia, Ivory Coast, Ghana, Benin, Nigeria, Cameroon, Equatorial Guinea (including Bioko), Príncipe, Gabon, Chad, Central African Republic, Congo Brazzaville, Congo Kinshasa, Burundi, Sudan, Ethiopia, Angola, Zambia

HAB. Evergreen forest, especially along forest edges and gaps, riverine and swamp forest; 700–1600 m

Flowering specimens have been seen from all months except May; the plants flower in the morning, although one collection from Uganda was recorded as open at 13:30.

SYN. *Commelina beniniensis* P. Beauv. in Fl. Oware 2: 49, t. 87 (1816)

Aneilema lujai De Wild. & T. Durand in Ann. Mus. Congo. Bot., Sér. 2, 2: 63 (1900), as *lujaei*. Type: Congo–Kinshasa, Stanley-Pool, between Sabuka & Kinshasa [Léopoldville], *Luja* 46 (BR!, holo.) [Collection number not cited in the publication; flowers recorded on type as "blue"]

A. mortehanii De Wild. in B.J.B.B. 5: 84 (1915); Brenan in K.B. 7: 197 (1951). Syntypes: Congo–Kinshasa, Dundusana, *Mortehan* 1017 (BR!, syn.) & Yangambi, *Michiels* 16 (BR!, syn.)

NOTE. This species is common in the forests of southern and southwestern Uganda where it can have white, pale lilac, lavender or violet flowers. Morton (1966) separated subsp. *sessilifolium* (Benth.) J.K. Morton from subsp. *beniniense* by the former's smaller white flowers and glabrous or nearly glabrous vegetative parts (among other characters). The distributions of the two subspecies were not clearly stated within West Africa and no mention was made of the distributions of these taxa E or S of Cameroon. Brenan (1968) found it impossible to distinguish these taxa in West Tropical Africa, and I also do not find them recognizable in our flora.

Three collections from **T** 4 (*Verdcourt* 3337; *Pirozynski* P47 & P101) are unusual in having narrowly lanceolate to lanceolate-oblong leaves that are subpetiolate, thus giving them more the aspect of *A. lanceolatum* than of typical *A. beniniense*. Moreover, *Pirozynski* P101 is the only specimen seen from our area that has long uniseriate hairs on the adaxial leaf surface. There is nothing noted on the labels to suggest that these plants were growing in a more open, sunnier locality than usual, so their appearance in unexplained. Their glabrous inflorescences and fruits and moist, apparently shady habitats clearly indicate that they are *A. beniniense*. *Bidgood et al.* 2838 from **T** 4 is unusual is having a row of minute hook-hairs on the leaf margins proximally.

Similarly *Aneilema lujai* was considered a synonym of *A. lanceolatum* in Faden (1991) because of its sessile, lanceolate leaves, despite what would be an unlikely habitat ('marsh') and location for that species. Having restudied the type, its glabrous capsules and reticulate seeds, somewhat covered by a whitish papery layer, clearly place it in *A. beniniense*. The authors' spelling for the specific name ('*lujaei*') has been corrected because the collector was a man.

Morton in J.L.S. 60: 169 (1967) referred to the middle stamen as staminodial. However, its anther is about the same size as the others, and almost every time that I have observed it, it produces pollen. Morton does not mention testing the viability of the pollen, and neither have I, but in *A. umbrosum* and *A. lanceolatum*, in which Morton also refers to the medial stamen as staminodial, I have found that the pollen produced was indeed viable. Therefore I suspect that Morton is mistaken about *A. beniniense* as well.

24. **Aneilema lanceolatum** *Benth.* in Hooker, Fl. Nigrit.: 546 (1849); C.B. Clarke in DC., Monogr. Phan. 3: 227 (1881) & in F.T.A. 8: 72 (1901); Brenan in K.B. 7: 202 (1952); Morton in J.L.S. 59: 453 (1966); Brenan in F.W.T.A., 2nd ed.: 3: 31 (1968); Faden in Smithsonian Contrib. Bot. 76: 145 (1991). Syntypes: Nigeria, confluence of the Niger and Benue rivers at Stirling Hill, *Ansell* & *Vogel* s.n. (K!, syn.); same data, 1842, *Ansell* s.n. (K!, syn.); Niger R. [Quorra River], *Vogel* s.n. (K!, syn.) [same sheet as previous] [two sheets at Kew, one with a single collection, the other with two collections. Both sheets have "Type" labels on them]

Geophytic perennials; roots tuberous, narrowly stipitate; shoots solitary or several, erect to ascending, to 60 cm tall, not rooting at the nodes, puberulous. Leaves spirally arranged; sheaths to 2 cm long, puberulous or pilose-puberulous, ciliate or eciliate at apex; lamina sessile, narrowly lanceolate to oblong-lanceolate, (2.5–)4–12.5 × 0.8–1.7 cm, base cuneate to rounded, margins scabrid, sometimes ciliate at base, apex acuminate (or acute), sometimes mucronulate; surfaces scabrid, usually with a mixture of long uniseriate hairs, and short hook-hairs and prickle-hairs, occasionally the long hairs lacking. Thyrses terminal on the main shoots, dense, ovoid or narrowly ovoid, 4–5 × 2.3–3 cm, of numerous ascending cincinni; all axes puberulous with minute hook-hairs; peduncles 2.5–5 cm long, with a bract ± halfway; cincinni to 1.7 cm long and 6-flowered; bracteoles symmetrically or asymmetrically cup-shaped, perfoliate (sometimes split to the base by pedicels), 1.5–2 mm long, prominently glandular subapically, glabrous or puberulous proximally. Flowers bisexual and male, 6.5–8.5 mm wide; pedicels erect in flower and fruit, 4–6 mm long, puberulous with hook-hairs its whole length or only distally. Sepals subequal, ovate to elliptic, ± 2.5 × 1.5 mm, convexo-concave, glabrous or puberulous with hook-hairs, the upper hooded and glandular at apex; paired petals white or mauve, broadly ovate, ± 5.5 × 4–5 mm of which the claw ± 1.5 mm; medial petal ovate, cup-shaped, 2.5–3 × 1.75–2.5 mm; staminodes 3 (rarely the medial one lacking?), with bilobed, yellow antherodes; lateral stamens with filaments (4–)5–6 mm long, sigmoid, finely bearded near the middle, anthers ± 1 × 1 mm, pollen dirty yellow; medial stamen filament ± (2.2–)3.5–4 mm long, anther 0.6–1 mm long and wide, pollen bright yellow. Ovary ± 1.7 × 1.3 mm, densely glandular-puberulous; style arcuate-decurved, 4.5–6 mm long; stigma capitate. Capsules medium brown, bilocular, bivalved (the valves persistent), 5–7 × 3–4 mm, puberulous, locules 2-seeded. Seeds broadly ovoid to transversely elongate, 2–2.9 × 1.5–2 mm, testa foveolate-reticulate to scrobiculate, tan to pinkish tan or reddish tan.

UGANDA. Acholi District: Kitgum Matidi, Apr. 1943, *Purseglove* 1509!; Karamoja District: Kidepo Valley National Park, E. B. area, 27 May 1969, *Harrington* 520!; Teso District: Serere, May 1932, *Chandler* 548!

KENYA. Northern Frontier District: Lake Turkana [Rudolf], rec'd 14 Sep. 1899, *Wellby* s.n.!

DISTR. **U** 1, 3; **K** 1; Mali, Burkina Faso, Ghana, Togo, Benin, Nigeria, Cameroon, Congo–Brazzaville, Chad, Central African Republic, Congo–Kinshasa, Sudan

HAB. Scattered tree grassland; 950–1200 m

Flowering specimens have been seen from March to May.

SYN. *Lamprodithyros lanceolatus* (Benth.) Hassk. in Schweinfurth, Beitrag Fl. Aethiop. 211 (1867)

L. gracilis Kotschy & Peyr., Pl. Tinn.: 47, t. 23A (1867). Type: Sudan, Djur near Bongo, *Heuglin in Exped. Tinn.* 4 (W†, holo.)

Aneilema gracile (Kotschy & Peyr.) C.B. Clarke in DC., Monogr. Phan. 3: 228 (1881) & in F.T.A. 8: 73 (1901)

A. schweinfurthii C.B. Clarke in DC., Monogr. Phan. 3: 227 (1881) & in F.T.A. 8:71 (1901). Syntypes: Sudan, Bongo, Gir, *Schweinfurth* 1886 (K!, syn, P!, isosyn.); Djur, Seriba Ghattas, *Schweinfurth* 1332 (K!, syn.); Djur, Seriba Ghattas, *Schweinfurth*, ser. III, 214 (K!, syn.)

NOTE. Our plants belong to *A. lanceolatum* subsp. *lanceolatum*. Two subspecies are recognized in West Africa, subsp. *lanceolatum*, in which the leaves are fully developed at flowering time and subsp. *subnudum* (A. Chev.) J.K. Morton, in which the leaves are not fully expanded when flowering begins. Although the leaves of *Purseglove* 1509 are small and the plant would thereby fall into the size range of subsp. *subnudum*, the leaves appear to be mature in the single, rather scrappy specimen examined, and no other characters of that subspecies are distinctly present. Both Morton (1966) and Brenan (1968) consider subsp. *subnudum* to be restricted to West Africa, although there are a series of specimens from Parc National de la Garamba, in NE Congo–Kinshasa, e.g. *Troupin* 43, that clearly belong to this subspecies, and that locality is not very far from NW Uganda. Thus the presence subsp. *subnudum* in our area would not be surprising.

I have not seen this species growing in the field, and the only living plants that I have are from seed collected in Togo. The major difference from the five Ugandan specimens that I have studied is the more prominently glandular sepals in the West African plant.

The floral description has been taken mainly from pressed flowers and dissected buds from the Ugandan collections. Because none of the flowers was carefully pressed, some details, such as the petal sizes and shapes, were added from liquid preserved flowers from *Lock* 84/93 from Togo.

In the limited material available much variation was found in the degree of development of the medial staminode. Morton (loc. cit., text-fig. 8) described it as small and depicted the antherode as unlobed. In the material that I examined I found the medial staminode to be typically much smaller than the lateral staminodes, with a bilobed antherode with much smaller lobes than in the lateral staminodes. There were two exceptions: in one pressed flower of *Purseglove* 1509 the three staminodes were clearly subequal in size and apparently in development, and in a carefully dissected mature bud of *Synnott* 1035 – the only bud examined on this specimen – no trace could be found of the medial staminode.

White flowers were mentioned on the labels of two the three Ugandan collections that indicated flower colour. If they are indeed frequent in our area, it would represent a difference from West African plants.

Pollen viability was tested in the lateral and medial anthers of three flowers of a cultivated plant of the *Lock* collection from Togo. Approximately 90% (86–94%) of the pollen in both anther types was found to stain deeply, indicating viability. Morton's description of the medial stamen in this species as staminodial (Morton, loc. cit.) is not supported.

25. **Aneilema dispermum** *Brenan* in K.B. 7: 198 (1952); Faden in Smithsonian Contrib. Bot. 76: 145 (1991). Type: Malawi, Cholo Mt, *Brass* 17763 (K!, holo.)

Decumbent perennial rooting at the nodes; roots thin, fibrous; shoots erect to ascending, 30–150 cm tall, usually densely branched, puberulous or glabrescent. Leaves spirally arranged; sheaths 1–2.5 cm long, puberulous with minute hook-hairs, long-ciliate at apex, long hairs usually also present distally on the line of fusion; lamina usually petiolate (except sometimes in the distalmost leaves), narrowly lanceolate to elliptic, 7–18 × (1.5–)2–6.5 cm, base cuneate (rarely rounded to subcordate in distal leaves), margins scabrid, apex acuminate, adaxial surface ± scabrid, often with sparse short or long uniseriate hairs, rarely uniformly puberulous; abaxial surface pilose mainly or exclusively on the veins, sometimes mixed with sparse hook-hairs. Thyrses terminal on the main shoots, very dense, ovoid to narrowly ovoid, 2.5–6 × (1.5–)2–4 cm, of numerous ascending (or the lowermost patent) cincinni; inflorescence axes puberulous; peduncle 1.6–3 cm long, with bract halfway; cincinni to 2 cm long and 4-flowered; bracteoles asymmetrically perfoliate, sometimes split by recurved fruiting pedicels, glandular subapically, glabrous. Flowers bisexual and male, 7–12 mm wide; pedicels 2–5.5 mm long, erect or strongly recurved in fruit, glabrous or sparsely puberulous near apex. Sepals white, violet apically and marginally, with a green subapical gland, subequal, ovate to ovate-elliptic 3–3.5 × 2–3 mm, glabrous; paired petals lilac, blue-violet, blue-purple, pale lilac, mauve or violet, 5.5–6 × 4–6 mm of which the white claw ± 2 mm long; medial petal obovate, 2.5–4 × 2.5–2.7 mm, white tinged with the same colour as the paired petals; staminodes 3, antherodes bilobed, yellow; lateral stamens with filaments (4–)5–6 mm long, sigmoid, minutely bearded near apex, anthers (0.7–)1.1–1.3 mm long, pollen dull yellow; medial stamen filament 3–4 mm long, anther ± 1 mm long, pollen golden yellow. Ovary 1.5–1.8 × 1.2–1.5, with short glandular hairs; style arcuate-decurved, 4–6 mm long; stigma slightly enlarged. Capsules medium brown, bilocular, bivalved, obovoid to obovoid-orbicular, 3.7–5 × 3.9–5.2 mm, glabrous, valves persistent, locules 1-seeded. Seeds shortly ellipsoid, 2.3–2.7 × 1.85–2.3 mm, testa rugose and finely pitted, dull, dark brown, not farinose.

TANZANIA. Morogoro District: Uluguru Mountains, Morningside to Bondwa, 3–4 July 1970, *Faden, Evans & Kabuye* 70/311!; Iringa District: Udzungwa Mountain National Park, Mt Luhomero, 28 Sep. 2000, *Luke et al.* 6729!; Mbeya District: Ngozi Forest, 26 June 1996, *Faden et al.* 96/409!;

DISTR. **T** 6, 7; Malawi; not known elsewhere (see notes)

HAB. Moist evergreen forest, sometimes near streams or waterfalls, less often in the bamboo zone, among shrubs and in grassland at timberline, or in cultivation; 1450–2600 m

Flowering specimens have been seen from all months except January, February and April.

NOTE. This species was described to include white-flowered plants from the Cameroon highlands and Bioko. Studies of plants using a relative abundance of more recent collections have led me to conclude that the East and West African plants are not conspecific. References to *A. dispermum* in F.W.T.A. (Brenan, 1968) and by Morton (in J.L.S. 59: 445, 1966) refer only to the West African plants. To his credit, Morton (loc. cit.) did note that the eastern African – referred by him to as "Central African"– plants differed from the Cameroonian plants by having "pinkish blue or pinkish" instead of exclusively white flowers. He suggested that further differences might be manifest in living plants, but concluded that at the then state of knowledge the plants from opposite sides of the continent did not "warrant taxonomic separation." In addition to the difference in flower color, the Central African plants have leaves with ciliolate margins (at least proximally) and seeds with a different testa pattern. Indeed the testa of the West African plants closely resembles that of *A. beniniense* where as in true *A. dispermum* the testa more closely resembles that of the Mozambiquan and South African species *A. brunneospermum*.

This species can be confused only with *A. beniniense* but is easily separated by its 2-seeded capsules that are about as broad as long. Plants of *A. dispermum* also have much more pubescent leaves and inflorescences than *A. beniniense* in our area. The two species do not overlap in their distributions, with *A. dispermum* confined to the Eastern Arc mountains and *A. beniniense* in Tanzania occurring only west of Lake Victoria and in the vicinity of Lake Tanganyika.

26. **Aneilema umbrosum** (*Vahl*) *Kunth*, Enum. Pl. 4: 71 (1843); Brenan in K.B. 7: 199 (1952); Morton in J.L.S. 59: 459 (1966) & in J.L.S. 60: 168 (1967); Brenan in F.W.T.A., 2nd ed.: 3: 30 (1968); Faden in Smithsonian Contrib. Bot. 76: 145 (1991). Type: Guinea, Asohia ["Assajanae" in Vahl], *Isert* s.n. (C!, holo)

BASIONYM: *Commelina umbrosa* Vahl, Enum. Pl. 2: 179 (1806)

subsp. **ovato-oblongum** (*P. Beauv.*) *J.K. Morton* in J.L.S. 59: 461, fig. 11 (1966); Brenan in F.W.T.A., 2nd ed.: 3: 30 (1968); Faden in Smithsonian Contrib. Bot. 76: 145, fig. 4, t. 7g (1991). Type: Nigeria, *P. de Beauvois* s.n. (G!, holo.)

Decumbent perennial rooting at the nodes, usually lacking a definite base; roots thin, fibrous; shoots much branched, ascending to ± 30 cm, puberulous with minute hook-hairs. Leaves spirally arranged; sheaths to 1.2 cm long, commonly splitting longitudinally, puberulous with hook-hairs, ciliate at apex; lamina petiolate, ovate to ovate-elliptic, 1.5–6 (including petiole) × 1–2.5(–3) cm, base cuneate to rounded, margins ± undulate, scabrid distally, apex acute; adaxial surface sparsely puberulous with minute hook-hairs or pilose-puberulous, abaxial surface scabrid with prickle-hairs and hook-hairs. Thyrses terminal and lateral from the axils of the inflorescence bract and the uppermost foliage leaf, broadly ovoid, ± lax, 1.5–3 × 2–4.5 cm, of 1–6 cincinni; inflorescence axes puberulous or glabrous; peduncles 1–2.5 cm long, with bract halfway or medial or basal; cincinni up to 2.5 cm long and 11-flowered; bracteoles cup-shaped, perfoliate, 1–1.5 mm long, glandular subapically, glabrous or with sparse minute hook-hairs at the base. Flowers bisexual, ± 6.5 mm wide; pedicels 3–7.5 mm long, declinate in fruit, puberulous with hook-hairs at distal end. Sepals green with reddish purple margins proximally, convexo-concave, ovate to elliptic, 2–3 mm long, hooded at apex, and glandular subapically, usually sparsely puberulous; paired petals white, ovate, ± 4.5 × 2.5 mm of which the claw ± 1.5 mm; medial petal greenish white, ovate-elliptic, slightly convexo-concave, ± 3 × 1.5 mm, apex slightly hooded; staminodes 2–3, the medial lacking or vestigial, filaments ± 2 mm long, antherodes bilobed, yellow; lateral stamens with filament ± 2 mm long, gently sigmoid, bearded above the middle, anther ± 0.4 × 0.4 mm, pollen yellow; medial stamen filament ± 1.8 mm long, anther 0.4 × 0.5 mm, pollen yellow. Ovary ± 1 mm long and wide, glabrous; style arcuate-decurved, ± 3 mm long; stigma capitate. Capsules light brown or greyish tan to grey-green, bilocular, bivalved (the valves persistent), oblong-ellipsoid, 3.5–4.5 × 2.5–3 mm, glabrous, locules (1–)2-seeded. Seeds ovoid to transversely ellipsoid, 1.5–1.9 × 1.3–1.5 mm, testa coarsely foveate-reticulate, light brown, orange tan, pinkish grey or pale grey. Fig. 12, p. 111.

UGANDA. Kigezi District: South Maramagambo Central Forest Reserve, ± 9 km up Kaizi–Biterako road, off Katunguru–Ishasha (Congo) road, 18 Sep. 1969, *Faden et al.* 69/1109!; Mengo District: Kipayo, Apr. 1914, *Dummer* 769! & Mpanga Forest Research Station, 5 km E of Mgigi, 9 Sep. 1969, *Faden, Evans & Lye* 69/997!

TANZANIA. Bukoba District: Minziro Forest Reserve, 4 July 2000, *Bidgood et al.* 4844!

DISTR. **U** 2, 4; **T** 1; Senegal, Sierra Leone, Liberia, Ivory Coast, Ghana, Benin, Nigeria, Cameroon, Bioko, Gabon, Congo–Brazzaville, Central African Republic, Congo–Kinshasa, Sudan; also in the neotropics

HAB. Evergreen forest and swamp forest, in moist soil; 1050–1500 m

Flowering specimens have been seen from February, April, May, September and December. The flowers have been observed closing in the field at 14:00.

SYN. *Aneilema ovato-oblongum* P. Beauv., Fl. Oware 2: 72 (1818), fig. 1 (1819); Kunth, Enum. Pl. 4: 72 (1843); Benth., Fl. Nigrit. 545 (1849); C.B. Clarke in DC., Monogr. Phan. 3: 226 (1881), excl. var. *nigritanum* C.B. Clarke, & in F.T.A. 8: 69 (1901), excl. var. *nigritanum*

Commelina ovato-oblonga (P. Beauv.) J.A. Schult., Syst. Veg., Mant. 1: Addit. 1, 376 (1822)

Aneilema umbrosum (Vahl) Kunth var. *ovato-oblongum* (P. Beauv.) Brenan in K.B. 7: 200 (1952)

NOTE. *Aneilema umbrosum* is extremely variable in its morphology and chromosome numbers in West and Central Africa. Two subspecies were recognized by Morton (1967) and in F.W.T.A. (Brenan 1968). Subsp. *umbrosum* is distinguished by its larger inflorescences with more numerous cincinni, often larger leaves, the flowers sometimes mauve, and the presence of long, reddish hairs on the summit and sometimes also on the surface of the leaf sheaths. Subsp. *umbrosum* is very variable and perhaps comprises more than one taxon, but it does not occur in our area. Subsp. *ovato-oblongum* is distinguished by its smaller inflorescences and leaves, the flowers always white, and the occurrence of long, colorless hairs only at the summit of the leaf sheaths (if anywhere), not on the surface. *Aneilema umbrosum* subsp. *ovato-*

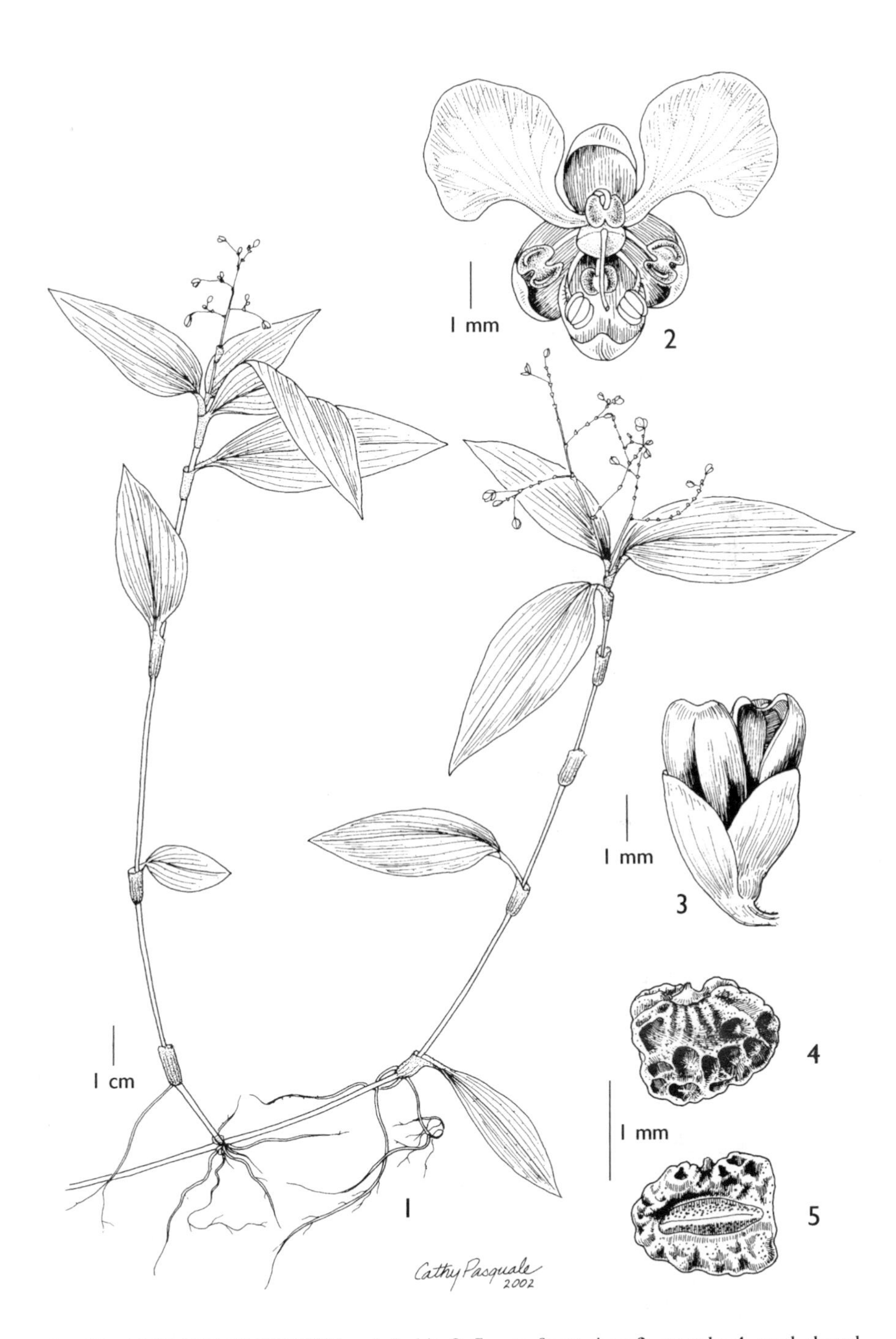

FIG. 12. *ANEILEMA UMBROSUM* — **1**, habit; **2**, flower, front view; **3**, capsule; **4**, seed, dorsal view; **5**, seed, ventral view. 1 from I.N.P.A. 48.584; 2–5 from *Gentry* 12021. Drawn by Cathy Pasquale.

oblongum is uncommon in our area but quite uniform. Although not identical to our plants, plants of subsp. *ovato-oblongum* from the neotropics show very little variation throughout their extensive range, suggesting a single introduction from Africa in the not very distant past. Hybrids between the two subspecies are recorded as frequent in West Africa (Morton 1966).

Aneilema umbrosum has several unusual features. It is unique among the species in our area in having the pedicels curve forward and downwards in fruit. In all other species, except *A. nyasense*, fruiting pedicels are typically erect or curve backwards to varying degrees. In at least some West and Central African populations of *A. umbrosum*, the capsules (plus pedicels) have been observed to fall off the plant when the seeds ripen, a feature not seen in any other African species. It is not clear whether this happens in plants from our area: some capsules have been found with mature seeds on herbarium specimens, but mature capsules are infrequent on them. The shortly pedunculate, small, terminal inflorescences usually have axillary inflorescences closely associated with them. The very large antherodes in such a small flower and the form of the antherode lobes are both unusual. The ends of the lobes of the antherode curve downwards, i.e. towards the ends of the other lobe, and are so long that the ends of the two lobes overlap, giving the whole antherode a circular appearance.

Floral measurements have been recorded largely from cultivated material from a single population (*Lye & Katende* 6235 from Uganda), so further variation is to be expected.

Morton (1966: 464) referred to an apparent new species from the Kenya coast that was very similar to small plants of *A. umbrosum*. That species is *A. taylorii* (see above).

27. **Aneilema welwitschii** *C.B. Clarke* in DC., Monogr. Phan. 3: 229 (1881) & in F.T.A. 8: 71 (1901); Faden in Smithsonian Contr. Bot. 76: 145, t. 4j, t. 7h (1991). Syntypes: Angola, Huilla, Lake Ivantâla, *Welwitsch* 6596 (BM!, syn.); Golungo Alto, between Catumba and Ohai, *Welwitsch* 6597 (BM!, K!, syn.)

Perennial geophyte; roots stipitate, fusiform tubers to ± 10 × 0.6 cm; shoots erect to ascending, sometimes decumbent but not rooting, ± 15–50 cm tall, pilose-puberulous with hook-hairs and long uniseriate hairs. Leaves spirally arranged, consisting of a series of bladeless sheaths to 2 cm long at the base, the normal foliage leaves with sheaths to 3 cm long, pilose-puberulous, ciliate at apex; lamina sessile, narrowly lanceolate to ovate, 2–10 × 0.5–2.4 cm, base rounded (to cuneate), margins ciliolate, scabrid distally, apex acute (to acuminate); adaxial surface pilose-puberulous; abaxial surface pilose- to hirsute-puberulous, the long hairs often confined to the major veins. Inflorescence a terminal, solitary very dense thyrse, usually ovoid to broadly ovoid, less commonly cylindrical, 1.5–5(–7) × 1–4(–6) cm; inflorescence axes puberulous with hook-hairs, often mixed with long, hairs; peduncle (2–)4–10 cm long with a bract halfway; cincinni up to 3 cm long and 5-flowered; bracteoles ovate (to ovate-lanceolate), very asymmetrically cup-shaped, perfoliate or not perfoliate, 1.5–3 mm long, glandular subapically, puberulous, sometimes with one or more long hairs on the margin. Flowers female, bisexual and male, 9–14 mm wide; pedicels 3–7(–10) mm long, erect to strongly recurved in fruit, puberulous. Sepals convexo-concave, ovate to elliptic, 2–5–4.5 × 2–3 mm, hooded at apex, glandular subapically, sparsely puberulous (or glabrous); paired petals orange, orange-yellow, yellow or pale apricot, broadly ovate, 6–8 × 5–6.5 of which the claw 2–3 mm; lower petal green tinged with orange, ovate, 2.5–4.5 × 2.5–3.5 mm long; staminodes 3, with yellow, bilobed antherodes; lateral stamens with filaments 4.5–8 mm long, sigmoid, bearded with minute white hairs, anthers 0.7–1.6 mm, pollen pale yellow; medial stamen filament 2.5–5 mm long, anther 0.5–1 × 1–1.6 mm, pollen orange-yellow. Ovary 1.5–3 × 1.7–2 mm, densely glandular-puberulous; style arcuate-decurved, 5–9 mm long; stigma capitate. Capsules lustrous light to medium brown, marked with violet distally and midventrally, bilocular, bivalved (the valves persistent), oblong-elliptic, (4–)6.5–9.5 × 4–6 mm, glandular-puberulous, locules (1–)2-seeded. Seeds broadly ovoid to oblong or reniform, 2.6–4.3 × (2–)2.2–2.8 mm, testa orange-brown, reddish purple or greyish purple dotted with paler tan, reticulate-alveolate and irregularly ribbed.

TANZANIA. Ufipa District: 8–9 km off road from Sumbawanga to Mbala (Zambia) on road to Safu, 3 Nov. 1992, *Gereau et al.* 4985!; Njombe District: Livingstone Mountains, road from Mile Mission turning to Madunda Mission, 1 Feb. 1961, *Richards* 14066!; Songea District: 19 km W of Songea near Likuyu River, 30 Dec. 1955, *Milne-Redhead & Taylor* 7792!

DISTR. **T** 4, 6–8; Congo–Kinshasa, Angola, Zambia, Malawi, Mozambique, Zimbabwe

HAB. *Brachystegia* woodland (often secondary), grassland, seepage areas and stream banks; 300–2000 m

Flowering plants have been seen from October to February.

SYN. *Aneilema erectum* De Wild. in F.R. 12: 289 (1913). Type: Congo (Kinshasha), Haut-Katanga, Lubumbashi [Élisabethville], Oct. 1911, *Hock* s.n. (BR!, holo.)

A. florentii De Wild. in F.R. 12: 289 (1913). Congo (Kinshasha), Haut-Katanga, Welgelegen, *Florent in Homblé* 640 (BR!, holo.)

A. katangense De Wild. in F.R. 12: 290 (1913). Congo (Kinshasha), Haut-Katanga, Shilongo [Chilongo], Sep. 1911, *Hock* s.n. (BR!, holo.)

A. densum T.C.E. Fr. in R.E. Fr., Schwed. Rhodesia–Kongo Exped. 1: 222, t. 16, fig. 7, text fig. 19 (1914). Type: Zambia, Msisi near Mbala [Abercorn], *Fries* 1306 (UPS!, holo.)

NOTES. On the BM sheet of *Welwitsch* 6597, 'Golungo Alto' was crossed out by Hiern and 'Huilla' was written in its place. On both syntypes Welwitsch records the flowers as blue, which would be an uncharacteristic error by him. There are no verified records of blue-flowered individuals or populations in any yellow- to orange-flowered species of *Aneilema*, and I believe that *C. welwitschii* is no exception. Therefore I expect that Welwitsch's notes were erroneous. *Leedal* 509 from **T** 7 also records the flowers as blue, but I believe that this too was in error.

This species is very variable in its morphology. Outside our area plants may be much more robust and taller. For example, the type of *A. katangense* is a plant 70 cm tall with leaves up to10.5 cm long and 3.7 cm wide. However, the largest inflorescence seen anywhere (7 × 6 cm) is on *Schmidt et al.* 1240 from **T** 4. Thus size of all parts may be variable, but it does not appear to have a consistent geographic basis.

Pubescence is also variable, but all collections seen from our area have long uniseriate hairs, in addition to short hook-hairs, on the internodes, surface of the leaf sheaths and lower leaf surface. Most collections also have such hairs on the upper leaf surface, in the inflorescences and on the peduncle (at least on the internode below the inflorescence bract). Although long, uniseriate hairs may not be present on all collections of *A. welwitschii* throughout its range – they have been noted to sparse or lacking in many Angolan collections, including the Kew sheet of the syntype of *Welwitsch* 6597 – their presence does characterize the collections from our area and will distinguish *A. welwitschii* from *A. macrorrhizum*, a species known from Zambia and Congo–Kinshasa. *A. welwitschii* further differs from that species by the veins prominently raised on the lower leaf surface, larger flowers, broader capsules and larger seeds in which the hilum is in the center of the ventral groove, not attached to one side of the groove. *Watermeyer* 189 from Sao Hill (**T** 7) was originally thought to be *A. macrorrhizum* but, upon re-examination, its vegetative pubescence, raised veins on the lower leaf surface, and the position of the hilum of an immature seed in the center of a broad, shallow groove clearly demonstrate that it is *A. welwitschii*.

Specimens with mature capsules are uncommon in our area, and I have not found any seeds on them. Therefore I have used the collections made in Zambia for these details, as well as for many of the floral details. Certainly some of the capsules measured in these Zambian collections exceeded any found in our area, but because the capsules from our area were so few and difficult to measure, it is impossible to know whether such large capsules may actually be present in our flora. It is also impossible to know whether the sizes of some of the floral parts in the spirit collections, which exceeded those measured from the floral remains present on some herbarium specimens from our area, were larger simply because of the different preservation methods.

On *Milne-Redhead & Taylor* 8261 two immature capsules were measured at 4 × 4 and 5 × 4 mm respectively. They were obviously not fully developed and they were well outside the size range of the other capsules from our area.

28. **Aneilema spekei** *C.B. Clarke* in F.T.A. 8: 72 (1901); F.P.S. 3: 239 (1956); Faden in U.K.W.F.: 667 (1974); Blundell, Wild Fl. E. Afr.: 412, fig. 630 (1987); Faden in Smithsonian Contr. Bot. 76: 147 (1991) & in U.K.W.F. 2nd ed.: 308 (1994); Ensermu & Faden in Fl. Eth. 6: 357 (1997). Type: Tanzania, Kahama/Tabora District: Unyamwezi, Mininga, [Speke and] *Grant* 165 (K!, syn.) &, near Tabora, 1860, [Speke and] *Grant* s.n. (K!, syn.); Malawi, North Nyasaland, *Whyte* s.n. (BM?, syn., not seen)

Annuals; roots thin, fibrous; shoots erect or straggling through other vegetation, sometimes decumbent and rooting at the nodes, to 15–45(–60) cm tall when erect, to 1 m long when straggling, puberulous, rarely pilose-puberulous. Leaves spirally arranged; sheaths commonly split longitudinally, to 1.5 cm long, pubescent and sometimes also pilose with long hairs, white-ciliate at apex; lamina sessile (sometimes shortly petiolate in proximal leaves), lanceolate to elliptic, 3–7.5(–12) × 0.7–3.5(–5) cm, base rounded to cuneate, margins scabrid, sometimes also ciliate proximally, apex acute (to acuminate); surfaces puberulous, sometimes scabrid adaxially, occasionally also pilose on one or both surfaces. Thyrses terminal and frequently axillary, sometimes (in old plants) arising from nearly all the nodes, ± dense, cylindrical to ovoid, 2–6 × 1–2.5 cm, of up to 40 cincinni; inflorescence axes puberulous; peduncle 3.5–9 cm long, with a bract halfway; cincinni to 1.5 cm long and 7-flowered; bracteoles ovate to ovate-lanceolate, 0.5–1.5 mm long, not perfoliate, with or without a subterminal or terminal gland, glabrous. Flowers bisexual and male, 8–12 mm wide; pedicels erect in flower and fruit, (2–)4–7 mm long, puberulous. Sepals convexo-concave, hooded at the apex, lanceolate-elliptic to ovate, 2.5–4.5 × 1.8–2 mm, glandular subapically, puberulous; paired petals blue to blue-violet or bluish purple, ovate, 5–9 × 4.5–7 mm of which the claw ± 1.5–2 mm, concolorous or contrasting; medial petal green or hyaline white, sometimes with reddish margins, ovate or spatulate, hooded and cup-shaped distally, 3–4.5 × ± 2.5 mm; staminodes 3, with yellow, bilobed antherodes; lateral stamen filaments, 4–6 mm long, S-shaped, densely bearded distally with bluish-purple hairs, anther blue and yellow, 1–1.3 mm long, pollen golden yellow; medial stamen filament 2–3 mm long, anther 0.3 × 0.4–0.6 mm, sterile or producing a small amount of yellow pollen. Ovary 1–1.5 × 0.6–1 mm, densely glandular-puberulous; style arcuate-descending, 4.5–7 mm long; stigma capitate. Capsules lustrous pale brown or pale yellow, bilocular, bivalved (valves persistent), oblong to oblong-ellipsoid (or subquadrate), 3.5–5(–6) × 2.5–3.5 mm, puberulous, locules (1–)2-seeded. Seeds ovoid to trapezoidal or transversely ellipsoid, (1.4–)1.8–2.3 × 1.4–1.8 mm, testa beige to orange-tan, scrobiculate, sparsely whitish or light brown farinose.

UGANDA. West Nile District: small valley 0.4 km S of Maracha Rest Camp, 27 July 1953, *Chancellor* 58!; Bunyoro District: 37 km N of Butiaba turnoff on Butiaba–Murchison Falls National Park road, 9.6 km N of Sambiye River crossing, near Kisansya, 16 Sep. 1969, *Faden, Evans & Lye* 69/1069!; Teso District: Soroti–Moroto, km 24, 13 Oct. 1952, *Verdcourt* 823!

KENYA. Turkana District: near Kacheliba, 11 Oct. 1964, *Leippert* 5123!; Masai District: road from Keekorok gate to Narok, 16 Aug. 1971, *Kokwaro & Mathenge* 2710! & Masai Mara Game Reserve, 7 km S of Keekorok, 25 May 1977, *Kuchar & Msafiri* 6282!

TANZANIA. Musoma District: Plains N of the Balangeti River, 9 Mar. 1962, *Greenway* 10509!; Dodoma District: km 40 S of Itigi Station, 20 Apr. 1964, *Greenway & Polhill* 11650!; Mbeya District: Madibira–Igawa track, 14 km SW of Madibira, 12 June 1996, *Faden et al.* 96/174!

DISTR. **U** 1–4; **K** 2, 3, 5, 6; **T** 1, 4, 5, 7; Congo–Kinshasa, Burundi, Rwanda, Sudan, Ethiopia, Zambia

HAB. Thickets, swamps, marshes, seepage areas, river and stream margins, rocky places, grassland, bushland, woodland, forest and forest edges, cultivation and other disturbed habitats; rocky, sandy or clayey soil; sun or shade; 700–1800 m

Flowering specimens have been seen from all months.

SYN. *Aneilema* sp. nov., Thomson in Speke, Nile, Append., 650 (1864)
A. tacazzeanum Baker in Trans. Linn. Soc. 29: 163 (1875), *non* A. Rich. (1850)
A. lanceolatum of Clarke in DC., Monogr. Phan. 3: 227 (1881) pro parte, *non* Benth. (1849)

NOTE. Plants of *A. spekei* have sometimes been described as short-lived perennials, perhaps because of their decumbent habit. However, *Faden et al.* 96/174, a sprawling specimen of this species, which was collected in a seepage area in southern Tanzania after the rainy season had ended, had completely gone to fruit and was clearly an annual. I believe that all plants of *A. spekei* are annuals.

This species is easily recognized by the presence of a bladeless sheath typically near the middle of the inflorescence peduncles, relatively large inflorescences, bracteoles usually lacking a linear apex terminating in a capitate gland, stamen filaments densely bearded with bluish purple hairs; and 4-seeded capsules with a truncate apex. The fusion of the stamen filament bases, noted in three populations, might serve as a distinguishing character from related species, but this character needs to be further evaluated. The sepal glands may be prominent or not very conspicuous and those of the lateral sepals are usually unlobed.

Specimens from Tanzania in which the cincinnus bracts and bracteoles have a ± linear apex terminating in a capitate gland have sometimes been separated from typical *A. spekei* as *A.* sp. aff. *spekei*. Such bracts and bracteoles resemble those found in most species of *Aneilema* sect. *Pedunculosa* [species 29–37] (all except *A. richardsae* and typical *A. spekei*), although they are much less extreme than those found in *A. hirtum*, with which *A. spekei* can occur. These atypical specimens of *A. spekei*, e.g. *Renvoize & Abdallah* 2251 from Iringa District may share additional characters with *A. hirtum*, such as a scarcity of bladeless sheaths near the middle of the peduncles, an occasional long uniseriate hair in the inflorescence, all the cincinni 1-flowered, all three sepals with prominent bilobed glands, and the capsule apex of the immature fruits rounded and dark blue. *Greenway & Kanuri* 14338, from the same locality as *Renvoize & Abdallah* 2251, has all of its inflorescence bracts basal instead of medial and agrees with that collection in all other characters just listed. These collections are not typical of *A. spekei*, but they also strongly depart from *A. hirtum* by their much larger inflorescences, shorter linear apices of the bracteoles, much longer pedicels, capsules with a truncate to emarginate apex, and locules always 2-seeded except by abortion. Although possible hybridization between the two species might explain these two collections, there are others specimens with similar bracteoles that have fewer *A. hirtum* characters but instead connect these collections with more typical *A. spekei*. For example, *Mdelwa* 23 has mostly medial inflorescence bracts, lacks all uniseriate hairs, and has elongate cincinni with up to six flowers per cincinnus. Although the possibility of hybridization between *A. spekei* and *A. hirtum* and back-crossing cannot be eliminated as a possible explanation for the morphological observations, I think it best to treat such collections as variants of *A. spekei* until proven otherwise. These unusual specimens come from scattered localities, are few in number, and no consistent differences other than the form of the bracts and bracteoles have been observed in dried specimens to separate them from *A. spekei* specimens with more typical bracts and bracteoles. Judging by the abundance of fruits and seeds, these unusual specimens appear to be fully fertile. Living plants have not been observed, so the possible occurrence of useful floral traits cannot be determined. These specimens are best treated as atypical specimens of *A. spekei*, rather than as a distinct taxon. In the event that further collecting and the availability of living plants should lead to these specimens being recognized as a distinct taxon, some examples may be cited.

TANZANIA. Buha District: 144 km SW of Kibondo, Rondobera swamp near Buhoro village, 11 Mar. 1973, *Mutch* 247!; Iringa District: Rangers Post, Magangwe, 13 Apr. 1970, *Greenway & Kanuri* 14338! & same location, 19 May 1968, *Renvoize & Abdallah* 2251!

DISTR. **T** 4, 5, 7; Congo–Kinshasa (*Bequaert* 3456), Burundi (*Lewalle* 5649)

HAB. Woodland, grassy areas, and edge of permanent swamp at base of small shrubs; 1200–1600 m

Flowering specimens have been seen from March to May.

29. **Aneilema richardsae** *Brenan* in K.B. 15: 219 (1961); Faden in Smithsonian Contr. Bot. 76: 147, t. 7, e (seeds) (1991). Type: Zambia, Mbala District, Kalambo Farm, Saisi Valley, *Richards* 1789 (K!, holo.; K!, iso.)

Annuals; roots thin, fibrous; shoots erect to ascending, to about 30 cm tall, puberulous. Leaves spirally arranged, usually strongly ascending; sheaths ± 0.4–1 cm long, puberulous, glabrous or ciliolate at the mouth, long hairs entirely absent; lamina sessile, linear to linear-lanceolate, 2.5–11.5 × 0.2–0.6(–1) cm, base very gradually narrowed or not narrowed, margins scabrid, apex acute or subacute to acuminate; both surfaces puberulous and scabrid. Thyrses terminal and often axillary from the distal leaves; dense to moderately dense, ovoid, (0.7–)1.5–3 × 0.5–2.5 cm; inflorescence axes puberulous; peduncles 2.5–14 cm long, with a bract halfway or higher; cincinni to 1.5 cm long and 10-flowered; bracteoles ovate to ovate-triangular, scarious, 1–2 mm long, sparsely puberulous, often distinctly glandular subapically. Flower pedicels usually blue, 1.5–4 mm long, puberulous at apex. Sepals strongly convexo-concave, elliptic, 2–3 mm long, hooded-obtuse at the apex, puberulous, glandular subapically; paired petals blue, bluish purple or mauve-blue, obovate-orbicular, ± 5 × 3–3.5 mm of which the claw (1.5–)2 mm; medial petal narrowly obovate, (2.5–)3.5 × 1.5 mm, apex obtuse; staminodes 3, antherodes yellow, bilobed; lateral stamens with filaments ± 4 mm long, densely bearded with blue to lavender hairs near the apex, anthers ± 0.7 × 1 mm; medial stamen filament 1.5 mm long, 0.3 × 0.4 mm. Ovary 1 × 0.8 mm, glandular-puberulous; style glabrous, ± 2–2.5 mm long. Capsule pale yellow, bilocular, bivalved (the valves persistent), oblong to obovoid, (3–)3.5–4.5 × 2.5–3 mm, shortly glandular-pubescent. Seeds (1–)2 per locule, ovoid (to subreniform), 1.4–1.8(–2) × 1.2–1.4(–1.5) mm, testa with light brown or light grey, 10–12 inconspicuous ribs, with a prominent midventral pit.

TANZANIA. Ufipa District: 2 km on Tatanda–Mbala road, 24 Apr. 1997, *Bidgood et al.* 3379! & 5 km N of Namanyere, 2 May 1997, *Bidgood et al.* 3628! & M'wimbe Dambo, Old Sumbwanga road, 21 Apr. 1962, *Richards* 16358!

DISTR. **T** 4; Congo–Kinshasa, Zambia

HAB. Damp soil and drainage ditches in grassland; 1450–1700 m

All collections from our area flowering in April; in Zambia from March to May and July.

NOTES. The narrow, sessile, linear to linear-lanceolate leaves, make this a very easy species to recognize. Also distinctive are the absence of long hairs from the inflorescence and from the summits of the leaf-sheaths, and the absence of elongate, clavate ends on the cincinnus bracts and bracteoles. The bracts and bracteoles may be puberulous, which is an unusual feature in this species, but they can also be glabrous. The seeds are also highly unusual, with the testa ribbed and the hilum borne in an oblong-elliptic pit that is closed off at both ends instead of the more usual groove that has the ends open. The sessile leaves, presence of a conspicuous inflorescence bract ± midway on the peduncle, and absence of elongate ends on the bracts and bracteoles in *A. richardsae* all agree with *A. spekei*, to which *A. richardsae* is probably most closely related. *A. spekei* differs by its broader leaves, presence of long hairs on the leaf-sheath margins, usually longer pedicels and larger capsules. Its seeds lack the unusual characters of those of *A. richardsae*.

There are two sheets at K of the type collection, each with numerous plants, and both are labeled as holotypes. Sheet 1 (K000345730) has Mrs. Richards' original label and extensive notes by Brenan and is best considered the holotype, whereas Sheet 2 (K000345731) seems to be just the extra plants, and it is best treated as an isotype. If this constitutes lectotypification, then I select *Richards* 1789, sheet 1 (K) as the lectotype of *A. richardsae* Brenan.

The seeds of *A. richardsae* are very distinctive because of their midventral pit (see Faden, 1991, Pl.7, e) – as opposed to a groove in *A. hirtum* and *A. chrysopogon* – and their weakly ribbed testa. The hilum is very short and the absence of farinose granules and larger particles from the dorsal surface is also unusual. I have not examined all of the collections cited by Brenan, but I have studied eight in total, and I have been unable to find seeds as large as he records. Indeed his parenthetical lower dimensions are more the norm among the collections that I measured. Therefore I have placed his largest dimensions in parentheses. It is possible that a different subset of specimens will yield different results, and that the seeds of this species are more variable than I have seen.

30. **Aneilema hirtum** *A. Rich.*, Tent. Fl. Abyss. 2: 343 (1850), as *hirta*; C.B. Clarke in DC., Monogr. Phan. 3: 228 (1881) & in F.T.A. 8: 74 (1901); Brenan in K.B. 15: 220 (1961); Faden in Smithsonian Contr. Bot. 76: 147, t. 4o & 7d (1991) & in U.K.W.F. 2nd ed.: 308, t. 138 (1994); Ensermu & Faden in Fl. Eth. 6: 360 (1997). Type: Ethiopia, Kouaieta, *Quartin Dillon & Petit* s.n. (P herb. Richard!, lecto in Fl. Eth. 6: 360 (1997), P!, iso.)

Annuals; roots thin, fibrous; shoots erect to ascending, straggling or decumbent, sometimes rooting at the proximal nodes, to ± 40 cm tall (or 1 m long), puberulous or hirsute-puberulous, with long and short hairs. Leaves spirally arranged; sheaths 0.5–1.3 cm long, puberulous or hirsute-puberulous, whitish-ciliate at the apex; lamina sessile (to shortly petiolate), ovate-elliptic to ovate or lanceolate, 2–8 × 0.5–2.5 cm, base cuneate to rounded, margins scabrid, apex acute to acuminate; surfaces scabrid and hirsute-puberulous. Thyrses terminal and axillary, ± dense, ovoid to subspherical (rarely narrowly cylindrical), 1–2.7 × 0.7–1.7 cm; inflorescence axes puberulous, mixed with long hairs; peduncles 1.5–7.5 cm long with a basal or rarely subbasal, bract; cincinnus to 4 mm long and 2-flowered, usually most or all 1-flowered; bracteoles linear to linear-lanceolate, 1–2.2 mm long, with a terminal gland, sometimes puberulous. Flowers bisexual and male, 8–10(–12) mm wide; pedicels erect in flower and fruit, ± 1–2.5 mm long, to 3.3 mm long in fruit, puberulous, rarely glabrous. Sepals convexo-concave, prominently glandular subapically, glabrous to sparsely puberulous, upper sepal lanceolate-elliptic, 2.7–4 × 2–2.3 mm, paired sepals ovate to ovate-elliptic, 2.5–3 × 2 mm; paired petals blue to lavender, blue-violet or lilac, 4.5–5.5 × 3.5–4 mm of which the claw 2 mm; medial petal spatulate, 3 × 2.5 mm, with a large maroon spot; staminodes 3, antherodes bilobed, yellow; lateral stamens with filaments 3.7–5 mm long, S-shaped, densely bearded above the middle with maroon to purple or reddish purple hairs, anthers 0.4–0.5 × 0.5 mm, pollen golden yellow; medial stamen filament ± 3.5 mm long, anther 0.3 × 0.5–0.7 mm, pollen yellow. Ovary ± 1.7 × 1.5 mm, glandular-puberulous; style sharply curved downwards, 3–4 mm long; stigma capitate. Capsules bilocular, bivalved, nearly circular or broadly ellipsoid to obovoid, substipitate, 2.5–5 × 3–3.5 mm, sparsely glandular-puberulous, capsule valves persistent, locules 1–2-seeded. Seeds ovoid to transversely ellipsoid, (1.6–)2–2.5 × 1.4–1.7 mm, testa tan to brown or grey-brown, scrobiculate to foveate-reticulate, sometimes shallowly furrowed with dense tan, white or brown particles in the depressions.

UGANDA. Mbale District: Kapchorwa, 11 Sep. 1954, *Lind* 343

KENYA. Trans-Nzoia District: Hoey's Bridge, ± 20 km S of Kitale on Kitale–Eldoret road, 25 Sep. 1971, *Faden, Faden & Tweedie* 71/853! & Caves of Elgon, Kitale, rec. 25 Jan. 1943, *Tweedie* 76B! & SE Elgon, rec. May, 1946, *Tweedie* 175!

TANZANIA. Iringa District: Udzungwa Mountain National Park, Camp 281 – pt 282, 7°41'S, 36°31'E, 25 May 2002, *Luke et al.* 8442!; Mbeya District: 14 km on Tunduma–Sumbawange road, 21 Apr. 1997, *Bidgood et al.* 3348!; Songea District: ± 3 km NE of Kogonsera, 14 Apr. 1956, *Milne-Redhead & Taylor* 9640!

DISTR. **U** 3; **K** 3; **T** 4, 5, 7, 8; Congo–Kinshasa, S Sudan, Ethiopia, Zambia, Malawi

HAB. Grassland and woodland, rarely thicket edges, roadside banks or in cultivation; sandy or lateritic soils; 950–2050 m

Flowering specimens have been seen from March to July and September.

SYN. *Lamprodithyros hirtus* (A. Rich.) Hassk. in Schweinfurth, Beitr. Fl. Aethiop.: 295 (1867)

Aneilema whytei C.B. Clarke in F.T.A. 8: 72 (1901); Brenan in K.B. 15: 220 (1961). Type: Malawi, *Whyte* s.n. (K!, holo.)

A. ringoetii De Wild. in F.R. 12: 291 (1913). Type: Congo–Kinshasa, Katanga, Shinsenda, *Ringoet in Homblé* 545 (BR!, holo.; BR!, *Ringoet* 545, lacking '*Homblé*', iso.)

NOTE. Clarke (1901) described *A. whytei* when he concluded that Richard's *A. hirtum* was a *Floscopa*. Clarke based that decision on Richard's description of the inflorescence of *A. hirtum* and on its 2-seeded capsule, whereas *A. whytei* had a 4-seeded capsule. It is curious that Clarke did not refer to having seen the type of *A. hirtum*, but only to Richard's description, because in his monograph, Clarke (1881) lists Paris as one of the sets of collections that he had studied. *Aneilema hirtum* can have either 2- or 4-seeded capsules, which is essentially the only difference between *A. hirtum* and *A. whytei*, as Brenan (1961) pointed out.

Specimens from N of our area, including the type of *A. hirtum*, consistently have capsules nearly circular in outline with 1-seeded capsule locules. In our area and further S both oblong-elliptic capsules with 2-seeded locules occur in Kenya and Tanzania as do ones with one-seeded locules.

The flower description was taken mainly from *Faden, Evans & Tweedie* 71/853 from **K** 3 which has a separate label with a very detailed description made in the field, including measurements. These notes were supplemented with details from two Zambian collections.

This species is one of the more difficult in this group to identify, especially when its lower leaves are petiolate. It is often determined by elimination of all other possible species. The more distinctive features are the usually purple to maroon hairs on the stamen filaments, contrasting with the bluer petals, the maroon spot on the lower petal – described on *Milne-Redhead & Taylor* 9640 as "lower petal reddish purple with pale green margin" – and the often 2-seeded capsules. Small plants with petiolate lower leaves are most readily confused with *A. termitarium*, which differs by its smaller flowers, lack of a medial staminode and, especially, by its 3-lobed seeds. The usual absence of a bract – the inflorescence bract – near the midpoint on the peduncle, smaller inflorescences, the usual presence of long, uniseriate hairs in the inflorescence, and shorter pedicels distinguishes *A. hirtum* from *A. spekei*, with which it may occur.

Plants of *A. hirtum* closely resemble those of *A. chrysopogon* and non-flowering plants are difficult to separate. *A. hirtum* usually lacks long, uniseriate hairs on the sepals. When present, such hairs are usually confined to the bases of the sepals (rarely more widespread), whereas in *C. chrysopogon* they are consistently scattered all over the sepals.

Two very different chromosome numbers have been obtained for this species, a tetraploid based on 9 (2n = 36) and a tetraploid based on 13 (2n =52) (Faden, unpublished). This suggests that the species as currently understood may need further revision. Herbarium specimens do not show a similar dichotomy.

31. **Aneilema chrysopogon** *Brenan* in K.B. 15: 222 (1961); Faden in Smithsonian Contr. Bot. 76: 147, t. 4, n (1994). Type: Zambia, Mbala District, near Mpanda, *Richards* 4471 (K!, holo., SRGH!, iso.)

Annuals; roots thin, fibrous, produced only at the base; shoots erect to ascending, 10–30(–50) cm tall, puberulous, occasionally mixed with sparse long uniseriate hairs. Leaves spirally arranged; sheaths 0.3–1 cm long, puberulous, often mixed with long hairs, white-ciliate at apex; lamina lanceolate to lanceolate-elliptic (to ovate), 1.5–7.5 × (0.5–)1–2(–3.5) cm, base rounded (to cuneate), margins scabrid, apex acute to acuminate; surfaces puberulous or hirsute-puberulous, often mixed with longer uniseriate hairs. Thyrses terminal and axillary, dense, ovoid, 0.5–2.5 × 1–1.7 cm; inflorescence axes puberulous, mixed with a few long hairs; peduncles 1.5–6.5(–10) cm long, with a basal (very rarely medial) bract; cincinni all 1-flowered; bracteoles linear or abruptly expanded at the base, 1–1.5 mm long, glandular-capitate at apex. Flowers bisexual and male, (9.5–)13–15 mm wide; pedicels often blue, erect in fruit, 2–3.5 mm long, puberulous. Sepals convexo-concave, (3.5–)4–5 × 2–2.5 mm, hooded at apex, glandular subapically, puberulous, upper sepal lanceolate-ovate, paired sepals oblong to lanceolate-elliptic; paired petals blue to lavender or violet, 7.5–8 × 6–7.5 mm of which the white claw 3–3.5 mm; medial petal greenish white with a maroon spot, boat-shaped-saccate, ± 4.5 × 1.5 mm, hooded at the apex; staminodes 3, yellow, the antherodes bilobed; lateral stamens with filaments 6.5–8.5 mm long, S-shaped, densely yellow-bearded distally, anthers (0.6–)0.9–1.3 mm long, pollen orange; medial stamen filament 4–4.5 mm long, anther bearing 2 distal thecae ± 0.5 mm long. Ovary 2 × 1 mm, glandular puberulous; style 5.5–7.5 mm long; stigma capitate. Capsule pale yellow, bilocular, bivalved (the valves persistent), oblong-ellipsoid, 4–4.5 × 3–3.5 mm, sparsely glandular-pubescent, not at all stipitate. Seeds 2 per locule, ± trilobed, 1.6–1.8 × 1.5–1.7 mm, testa scrobiculate, pale brown, sometimes with darker brown flecks, with dense white or tan particles in the depressions.

TANZANIA. Ufipa District: 2 km on Tatanda–Mbala road, 24 Apr. 1997, *Bidgood et al.* 3376! & 36 km from Sumbawanga on Sopa [Isopa] road, 17 June 1996, *Faden et al.* 96/269! & Sumbwanga–Chapota road, Kasesi River, 5 Mar. 1957, *Richards* 8462!

DISTR. **T** 4; Zambia

HAB. Grassland, grassy depressions, swampy ground, wet roadside ditches and banks; 1650–1750 m

Flowering plants have been seen from March, April and June.

NOTE. This species bears a strong resemblance and is probably most closely related to *A. hirtum*. Among the similarities are the lower petal with a central maroon spot and the leaf margins with hook-hairs in addition to the typical prickle-hairs. *A. chrysopogon* was distinguished from *A. hirtum* by Brenan by its larger flowers, differently shaped capsules, and yellow stamen filaments densely bearded with yellow hairs. The yellow-bearded filaments are an obvious difference and the flowers of *A. chrysopogon* are distinctly larger than those of *A. hirtum*, but the reported difference in capsule shape is useful only for plants of *A. hirtum* with one-seeded capsules locules. Plants of *A. hirtum* with strictly two-seeded capsule locules have capsules similar in shape and size to those of *A. chrysopogon*.

Two additional characters have been found that can be used to separate non-flowering specimens of *A. chrysopogon* and *A. hirtum*. In *A. chrysopogon* uniseriate hairs (in addition to shorter, hook-hairs) are always present on the sepals and are typically scattered all over them. In *A. hirtum* the sepals usually lack uniseriate hairs, but when present, such hairs are typically confined to the sepal bases. The narrowing of the seeds of *A. chrysopogon* at the level of the hilum makes them ± 3-lobed, whereas the seeds in *A. hirtum* have no corresponding narrowing and are not 3-lobed.

In our area *A. chrysopogon* is found in moist, open places. In Zambia it has also been collected in disturbed places around cultivation and can occur in partial shade.

32. **Aneilema nicholsonii** *C.B. Clarke* in F.T.A. 8: 70 (1901); Obermeyer & Faden in F.S.A. 4, 2: 46, fig. 9–1 (1985); Faden in Smithsonian Contrib. Bot. 76: 147, t. 4, p, t. 7, b (1991) & in U.K.W.F. 2nd ed.: 308 (1994). Type: Mozambique(?) ["North Nyasaland and Upper Loangwa River"]: road from Missala [= Missale?] to Luenha [Luia] R., Jan. 1897, *Nicholson* s.n. (K!, holo.)

Annual; roots thin, fibrous; internodes to 12 cm long, pubescent; shoots erect to ascending or scrambling, often decumbent at the base, 10–45 cm high, to 80 cm long, unbranched to much-branched. Leaves spirally arranged; sheaths 0.5–1.7 cm long, pubescent, sometimes with a line of long hairs along the fused edge or scattered over the surface, white-ciliate at apex; lamina petiolate, lanceolate-elliptic to ovate, 2.5–10.5 × 1–3.5 cm, base cuneate (to rounded), margins scabrid, apex acuminate to acute; surfaces pilose-puberulous. Thyrses terminal and axillary from the distal leaves, ± dense, ovoid to ellipsoid, (1–)2–3.5 × 1.5–2.5 cm; inflorescence axes puberulous; peduncles (1–)2–4 cm long, with a basal bract; cincinni to 8 mm long and 4-flowered; bracteoles ovate, 0.5–0.8 mm long, usually glandular at apex. Flowers perfect and staminate, (8–)10–15 mm wide; pedicels 2–5.5 mm long, erect to recurved in fruit, puberulous. Sepals convexo-concave, ovate to lanceolate-elliptic, 2.7–3 × 2–2.3 mm, hooded at apex and glandular subapically, glabrous or sparsely puberulous, the upper sepal slightly narrower; paired petals blue to blue-purple or lavender, 6–7.5 × 5–6 mm of which the lavender or whitish claw 2–2.5 mm; medial petal convexo-concave, 3.5–4 × 2–2.7 mm, white with a medial green or pale lilac stripe; staminodes 3, antherodes bilobed, yellow; lateral stamen filaments 5.5–7 mm long, geniculate, densely bearded with reddish purple, blue or blue-purple (rarely white) hairs above the middle, anthers 0.4–0.7 mm long, pollen yellow; medial stamen filament 4–5 mm long, anther small, pollen golden yellow. Ovary ± stipitate, 1.2–2 × 0.9–1 mm, glandular-puberulous; style 4–6 mm long; stigma subcapitate. Capsules pale yellow, sometimes with a blue patch (or 2 patches) at the distal end, tri- or bilocular, bivalved (dorsal valve often deciduous), oblong-ellipsoid, 4.5–6 × 2–3 mm, substipitate, sparsely glandular-pubescent; dorsal locule sometimes not developed, when present indehiscent, (0–)1-seeded, ventral locules (2–)3-seeded. Ventral locule seeds 3-lobed, 1.2–2.1 × 1.2–1.7 mm, testa foveolate-scrobiculate, tan to light orange-tan, sometimes with dark brown flecks. Fig. 13/1–3, p. 120.

KENYA. Masai District: W side of old alignment of Namanga–Kajiado road, ± 0.5 km from Namanga, foothills of Ol Doinyo Orok, 20 June 1971, *Faden & Evans* 71/504! & same locality, 6 June 1974, *Faden, Faden & Ng'weno* 74/811!

TANZANIA. Kondoa District: Great North Road, Kolo, 26.7 km N of Kondoa, 10 Jan. 1962, *Polhill & Paulo* 1125!; Mpwapwa District: Mpwapwa, 22 Jan. 1935, *Hornby & Hornby* 588!; Iringa District: Msembe–Kimiramatonge Circuit, 6.7 km from Msembe, 24 Feb. 1970, *Greenway & Kanuri* 13944!

DISTR. **K** 6; **T** 2, 5, 7; Zambia, Malawi, Mozambique, Zimbabwe, Namibia (Caprivi Strip)

HAB. Bushland, thickets, woodland (miombo, often degraded), grassland, often associated with rocks and rock outcrops; sandy soils; 700–1700 m

Flowering specimens have been seen from January through June.

NOTE. The country of the type collection is uncertain. 'Missala' on the label appears to be 'Missale,' according to Pope & Pope (1998). It is on the border of Zambia and Mozambique. The Luia [= Luenha] River is in Mozambique, so most likely the plant was collected in Mozambique.

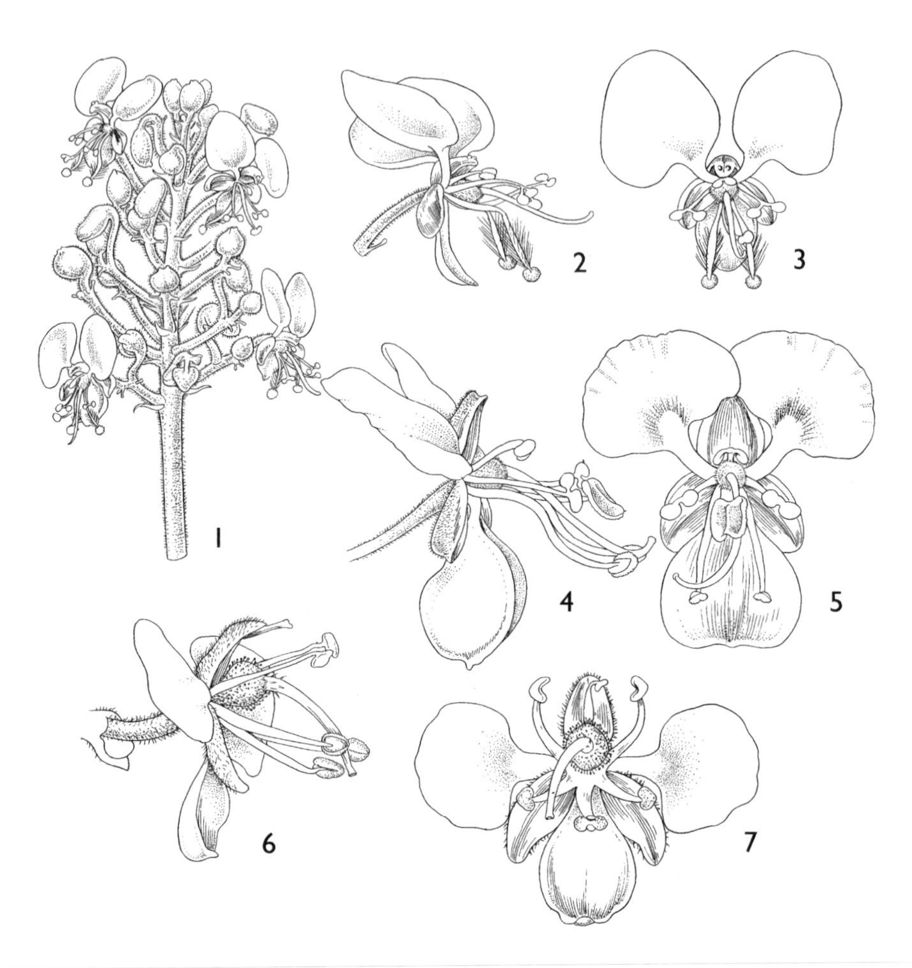

FIG. 13. *ANEILEMA NICHOLSONII* — **1**, inflorescence, × 2; **2** & **3**, flower, side and front view, × 2. *ANEILEMA INDEHISCENS* subsp. *LILACINUM* (Note: subspecies does not occur in our area) **4** & **5**, flower, side and front view, × 2. *ANEILEMA ZEBRINUM* — **6** & **7**, flower, side and front view, × 2. 1–3 from *Pawek* 12545; 4–5 from *Faden & Faden* 74/208; 6–7 from *Faden, Faden & Faulkner* 74/330. Drawn by R. Holcroft and reproduced with permission from F.S.A. 4, 2 fig. 9 (1985).

The dorsal valve appears to be persistent when the dorsal locule seed is not produced. Most likely when the dorsal valve is lacking in specimens, it has been shed with an enclosed dorsal locule seed.

The capsules with three 3-lobed seeds per ventral locule and a dorsal locule that often produces a seed readily distinguish this species from all other similar species. The only other species with similar, trilobed seeds is *A. termitarium*, which has much smaller inflorescences, flowers and capsules, the capsules at most 4-seeded, the valves always persistent, the dorsal locule not developed, and the medial staminode lacking. *Aneilema termitarium* is found mainly in Zambia and Congo–Kinshasa. It is restricted to **T** 4 in our area, whereas *A. nicholsonii* is much more widespread but is not recorded from **T** 4. Another taxon that can be confused with *A. nicholsonii* is *A.* sp. aff. *termitarium* (see discussion under that species). Like *A. nicholsonii* it can develop a dorsal locule seed. It differs from *A. nicholsonii* by its smaller, more compact inflorescence (cincinnus peduncles to 4 mm long), pedicels puberulous only distally, consistently glabrous sepals, smaller capsules with at most two seeds per ventral locule, and seeds neither trilobed nor with a tranverse dorsal groove.

Hammond 262 from **T** 5 (Iramba Plateau) consists of three small plants with small inflorescences. The seeds are somewhat 3-lobed, but less strongly so than in more typical collections, and, except in one seed, which has a partial groove, they lack a dorsal groove. Perhaps this is just part of the normal variation in *A. nicholsonii* seeds.

The Kenyan population is unusual is having very small seeds that lack a dorsal groove and have a ± smooth dorsal surface. The seeds are 3-lobed and there can be three seeds per ventral locule, so there is no question that this population belongs in *A. nicholsonii*. Representing a peripheral distribution of the species, it is not surprising that it shows some unusual features. *Richards* 25428, from Tarangire National Park (**T** 2), the closest locality to the Kenyan plants, has a single mature seed, which also has a smooth dorsal surface, but it has a partial dorsal groove, linking it with more typical specimens from further south.

Lovett 1621 from **T** 7, Iringa District, is either an extreme form of *A. nicholsonii* or else a distinct species related to *A. nicholsonii*. It comes from a suitable habitat for *A. nicholsonii* ('*Acacia-Commiphora-Combretum* bush') and agrees with this species in having a capsule with 3-seeded ventral locules, a deciduous dorsal capsule valve and often a dorsal locule seed; but differs by the presence of numerous long white uniseriate hairs on the sepals, in addition to more numerous than usual hook-hairs on the sepals. Seeds present in a packet on the sheet are very distinct from those of *A. nicholsolii*, but it is unclear whether they could have come from any of the nine plants or shoots on the sheet, only one of which is old enough to have had mature seeds. I suspect that the seeds in the packet were in the fine sand and other material left in the newsprint after the specimens were mounted, not from any particular plant on the sheet. The four seeds in the packet are scrobiculate all over, orange-tan and completely or almost completely lacking in granules in the depressions on the dorsal surface. They also lack a transverse dorsal groove. Two of the seeds are ellipsoid and are probably dorsal locule seeds, which have not been observed in *A. nicholsonii* in our area. The other two, clearly ventral locule seeds are ovate and scarcely three-lobed. They probably come from the same collection as the ellipsoid seeds and definitely do not match typical seeds of *A. nicholsonii*.

Brenan (in K.B. 15: 220, 1961) intentionally omitted *A. nicholsonii* from a table comparing five species in this group (*A. richardsae*, *A. hirtum*, *A. spekei*, *A. pedunculosum* and *A. chrysopogon*) as doubtfully distinct from *A. pedunculosum*. He wrote that extreme plants of *A. nicholsonii* seemed recognizable but concluded that the most distinctive characters were inconstant. In this account we not only recognize *A. nicholsonii* as distinct from *A. pedunculosum*, but also accept *A. leiocaule*, which Brenan also included in *A. pedunculosum*, as a distinct species. We also describe two new additional species, *A. termitarium* (species 34) and *A. minutiflorum* (species 37). The three species that Brenan included *in A. pedunculosum* (*A. pedunculosum*, *A. leiocaule* and *A. nicholsonii*) are readily separable on habit (*A. leiocaule* is a perennial, the others annuals), habitat (*A. pedunculosum* and *A. leiocaule* grow in moist habitats and *A. nicholsonii* in dry habitats), geography (*A. pedunculosum* and *A. leiocaule* do not overlap in their distributions) and capsule and seed characters (see key and discussions under the species). This is a very difficult group, but fortunately most of the taxa are annuals that readily produce capsules and seeds, so there is often good material available for comparison.

Aneilema nicholsonii occurs in drier habitats than any related species. Where it occurs it is likely to be the only species of sect. *Pedunculosa* present.

33. **Aneilema termitarium** *Faden* **sp. nov**. *A. nicholsonii* affinis sed inflorescentiis et floribus parvioribus staminodio mediano absenti vel vestigiali capsula loculis ventralibus 2-seminalibus loculo dorsali haud evoluto recedit. Type: Zambia, Kitwe–Ndola road km 28, near source of the Misaka River, Misaka Forest Reserve, *R.B. & A.J. Faden, Handlos, Laing & Fanshawe* 74/173 (US!, holo.; BR!, K!, MO!, NHT!, WAG!, iso.)

Annual herbs 6–33 cm tall; roots fibrous; shoots erect to ascending, densely puberulous. Leaves spirally arranged, not clustered distally, except occasionally the distalmost 2; sheaths green, 0.3–1 cm long, puberulous or pilose-puberulous, ciliate at apex; lamina petiolate (except in the distalmost leaves), lanceolate to narrowly lanceolate-elliptic, 1.5–6 × 0.3–2.2 cm (including the petiole), base cuneate to rounded, margins scabrid but ciliate proximally, apex acute to acuminate; surfaces discolorous, the adaxial dull, dark green, hirsute-puberulous, abaxial surface lustrous paler green, pilose-puberulous. Thyrses terminal and axillary from the distal 1–3 leaves, dense, ovoid to obovoid or subglobose, 0.5–1.5 × 0.5–1.3 cm, with erect to ascending (or the lowermost patent)cincinni; inflorescence axes pilose-puberulous; peduncles 1.7–2.7 cm long, with basal or medial bract; cincinni to 6 mm long, 1–3(–4)-flowered; bracteoles trough-shaped, ovate to linear, amplexicaul but not perfoliate, 0.5–1 mm long, glabrous, apex glandular-capitate. Flowers bisexual and male, 4–7.5 mm wide; pedicels 1 mm long in flower, to 2.5 mm and erect or recurved in fruit, puberulous distally. Sepals convexo-concave, ovate, ± 2 × 1.3–2 mm, the medial broader than the lateral, hooded distally, glandular subapically, puberulous; paired petals bluish purple, blue violet, pale violet, blue, pale blue, mauve, lilac, cream or white, ovate to suborbicular, 3–3.5 × 2 mm of which the white claw ± 1 mm; medial petal slightly convexo-concave, ovate, 2 × 1.6 mm, white to greenish white, sometimes tipped with green or with a green midrib; staminodes 3, antherodes yellow, bilobed; lateral stamens with filaments ± 2.3 mm long, decurved after the proximal $^1/_3$, densely bearded distally with maroon hairs, anther 0.3 × 0.3 mm, pollen yellow; medial stamen filament ± 1.6 mm long, anther 0.2 × 0.3 mm, pollen yellow. Ovary ± 1.3 × 0.9 mm, with glandular-capitate hairs; style sharply decurved; stigma capitate. Capsules sessile, pale yellow, lustrous, usually tipped with violet, bilocular, bivalved, oblong-ellipsoid to obovoid, 2.5–3.8 × 2.2–2.8 mm, sparsely puberulous, valves persistent, spreading ± 90°, locules 2-(or, by abortion, 1-)seeded, cells of the capsule wall transversely elongate. Seeds trilobed, 1.5–2.1 × 1.3–1.6 mm, dorsiventrally compressed, testa tan or light brown to orange-brown, finely ribbed or foveate-reticulate.

TANZANIA. Ufipa District: Kasanga–Sumbawanga road, 1 Apr. 1959, *Richards* 11052! & Kalambo Falls, 28 Mar. 1947, *Van Meel* 1390!
DISTR. **T** 4; Congo–Kinshasa, Zambia
HAB. In woodland in shade; 1525 m
Flowering specimens have been seen from March and April; in the field the flowers open 07:30–08:30 and fade 11:00–12:00.

SYN. *Aneilema termitarium* Faden in Smithsonian Contrib. Bot. 76: 147 (1991), *nom nud.*

NOTE. Twenty-six collections have been seen, all but four from Zambia. We collected *A. termitarium* six times there in 1974 and found it strongly associated with termite mounds (termitaria or anthills), hence the species name. The termite mounds on which it occurs are chiefly in woodland in sandy soil. The species also grows on laterite and we recorded it twice as a low epiphyte. Its altitude range in Zambia is 1250–1740 m and it flowers there from January to April. It likely has been overlooked by collectors in southern Tanzania

Aneilema termitarium is distinctive because of its small stature, small inflorescences and especially small flowers, which are smaller than those of any similar species in its range. It is probably most closely related to *A. nicholsonii* with which it shares 3-lobed seeds. *Aneilema nicholsonii* has larger inflorescences and flowers, distinctly bilobed (vs. unlobed) glands on the sepals, a fully developed (vs. absent or vestigial) medial staminode, usually bluish purple (vs. reddish purple to maroon) hairs on the stamen filaments, 3 (vs. 2) seeds per ventral capsule locule and often a seed produced in the dorsal locule (vs. dorsal locule not developed). *Aneilema richardsae, A. hirtum* (usually) and *A. chrysopogon* all have sessile leaves and only seeds of *A. chrysopogon* are ± 3-lobed, but it is easily distinguished by its much larger flowers and yellow-bearded lateral stamen filaments.

Friend 49 and *Richards* 19812 represent a distinct species (*Aneilema* sp. aff. *termitarium*) that resembles *A. termitarium*. It is a much branched annual 20–25 tall with lanceolate to ovate, petiolate leaves 2.5–5 × 1–2 cm. The shoots have a small terminal inflorescence (and sometimes also a lateral inflorescence next to it) 1–2 cm long and wide. The main differences from *A. termitarium* are: 1) the inflorescences lack or have very few long, uniseriate hairs; 2) the sepals are glabrous or have at most a few hook-hairs at the base; 3) the distal glands on the sepals are more distinctly bilobed; 4) all three staminodes appear to be fully developed; 5) fruiting pedicels are up to 3.5 mm long; 6) the dorsal locule often forms a seed; 7) the dorsal valve of the capsule is often deciduous; and 8) the ventral locule seeds are unlobed. The flowers are recorded as violet or pale blue.

TANZANIA. Tabora District: Tabora, 25 Mar. 1959, *Friend* 49!; Chunya District: Kipembawe [Kepembawe], base of Igila Hill, 22 Mar. 1965, *Richards* 19812!
DISTR. **T** 4, 7; known only from the two collections cited.
HAB. Woodland loam, sandy soil; 1500 m

NOTE. *Aneilema* sp. aff. *termitarium* resembles *A.* sp. aff. *leiocaule* in its small dense inflorescences and petiolate leaves. It differs by its annual habit, fewer long hairs on the inflorescences, glabrous sepals, deciduous dorsal capsule valve and often presence of dorsal locule seed.

Aneilema sp. aff. *termitarium* can also be confused with some specimens of *A. hirtum* which may sometimes have the proximal leaves ± petiolate. Those specimens differ from *A.* sp. aff. *termitarium* by their pubescent sepals, the dorsal locule never developed and the dorsal capsule valve never deciduous. Typical specimens of *A. hirtum* also can be distinguished by the presence of numerous, long uniseriate hairs in the inflorescence. For comparison of *A.* sp. aff. *termitarium* with *A. nicholsonii* see notes under that species.

34. **Aneilema pedunculosum** *C.B. Clarke* in DC., Monogr. Phan. 3: 228 (1881), excl. var. *lutea* C.B. Clarke; C.B. Clarke in F.T.A. 8: 73 (1901) pro parte; Faden in Smithsonian Contrib. Bot. 76: 147 (1991). Type: Mozambique, Lower Zambesi, Shiramba, between Tete [Tette] and the coast, *Kirk* 265 (K!, holo.)

Annual; roots thin, fibrous; shoots much branched, trailing or tufted, often decumbent and rooting at the nodes and ascending distally, to 30(–50) cm tall or 60 cm long, sparsely puberulous. Leaves spirally arranged, crowded distally; sheaths to 1 cm long, often splitting longitudinally, sparsely puberulous, white-ciliate at apex; lamina petiolate (except sometimes the distalmost), ovate-elliptic to lanceolate, 3.5–10 × 1–4 cm (including petiole to 2 cm long), base cuneate, margins scabrid-ciliolate, apex acuminate; both surfaces scabrid. Thyrses terminal and often from upper leaves, in old plants produced from many nodes, ± dense, ovoid, 1–2 × 1–1.5 cm; inflorescence axes densely puberulous; peduncle 2–4 cm long, with basal bract; cincinni to 7 mm long and 4-flowered; bracteoles linear or with a broad base and linear apex ending in a capitate gland, not perfoliate, 0.5–1.2 mm long, glabrous. Flowers bisexual and male, 9–12 mm wide; pedicels (1.5–)2–4.5 mm long in flower, to 6 mm long and erect in fruit, densely puberulous. Sepals strongly convexo-concave, ovate-elliptic, 2.5–3 × 1.8–2 mm, ± glabrous, the upper hooded at apex, with a bilobed gland, the laterals eglandular; paired petals pale lilac, blue or violet or 'blue and white', broadly ovate, 4.5–5 × 3.8–4 mm of which the whitish claw 1.5–2 mm; medial petal cup-shaped, ovate to obovate, 3 × 1.5–2 mm, hooded at the apex; staminodes 3, antherodes bilobed, yellow; lateral stamens with filaments 4–4.5 mm long, sigmoid, sparsely bearded distally with long purple to bluish purple hairs, anthers ± 0.8 mm long, pollen orange-yellow; medial stamen filament 3–4 mm long, anther 0.5–0.6 × 1 mm, pollen sparse, yellowish. Ovary 1.7–2.2 × 1.2–1.5 mm, densely glandular-pubescent; style arcuate-decurved, 3–3.5 mm long; stigma capitate. Capsules light to medium brown, bi- to trilocular, bivalved, oblong-ellipsoid, 4.5–5.5 × 2.5–3 mm, sparsely glandular-puberulous, 4–5-seeded (or fewer by abortion); dorsal locule, when developed, 0–1-seeded, ventral locules (1–)2-seeded. Ventral locule seeds ovoid to broadly ovoid, 1.8–2.3 × 1.6–1.8 mm, testa foveolate-reticulate, tan or orange-tan, whitish or tan-farinose in the depressions (or not); dorsal locule seed ellipsoid, 2.7–2.8 × 1.7–1.8 mm, testa smooth to alveolate, orange-tan.

TANZANIA. Kilwa District: Selous Game Reserve, Tundu Hills – Lung'onyo River (airstrip thicket), 17 May 1968, *Rodgers* 320! & Selous Game Reserve, Kingupira, Longonya plain, 10 Mar. 1975, *Vollesen* 1905!; Ulanga District: Mahenge area, Mbangala, 21 Feb. 1932, *Schlieben* 1810!

DISTR. **T** 6, 7?, 8; Zambia, Malawi, Mozambique, Zimbabwe

HAB. Thickets or forest, sometimes near streams, growing in shade; 15–800 m

Flowering specimens have been seen from February, March and May.

SYN. *Floscopa beirensis* Kuntze, Rev. Gen. 3: 319 (1898). Type: Mozambique, Beira, 1 Apr. 1894, *Kuntze* s.n. (NY!, iso?, K!, holo., US!, iso.)

NOTE. The type of *A. pedunculosum* has attached to it watercolour sketches by John Kirk of a flowering shoot and details of the flower, floral parts and fruit. On the back of the sketches are copious descriptive notes by Kirk. Despite these details this species has been confused with others right from Clarke's original description. At the end of that description Clarke

(1881) recognized var. *luteum*, based on a *Welwitsch* collection from Angola, but that taxon is *Aneilema angolense* C.B. Clarke var. *luteum* (C.B. Clarke) Faden. In F.T.A. Clarke (1901) excluded var. *luteum* from *A. pedunculosum* but added further collections, including the type of *A. leiocaule* (*Volkens* 1436), which I consider a distinct species (see next species).

The most distinctive features of *A. pedunculosum*, and those that distinguish it from *A. leiocaule* (also see notes at end of that species), are its annual habit, usually puberulous internodes (at least distally), less regularly spaced leaves that are mostly clustered below the inflorescences, the upper sepal with a much more prominently bilobed gland, and capsules often with a well-developed dorsal locule, sometimes 5-seeded, and with the dorsal valve often deciduous. In our area it occurs at low to mid elevations and is restricted to southern Tanzania, extending from there further south and southwest.

Schlieben 1810 is unusual in seemingly having only 4-seeded capsules. However, its almost horn-like, prominent, upper sepal glands, and the presence of only one capsule valve in some old capsules in the BR and P specimens – suggesting that the dorsal valve containing a dorsal locule seed may have been shed – all place this plant in *A. pedunculosum*.

The record from **T** 7 is based on *Batty* 924 from Kidatu (Iringa District). My notes indicate that the plant is an annual with a 5-seeded capsule, but the habitat is described on the label as a "steep bank of [a] recently made road," which does not necessarily suggest a moist situation. Its flowers are recorded as 'white', a colour otherwise not reported in this species. This would represent the sole collection seen from **T** 7.

South of our area specimens of *A. pedunculosum*, including the type of the species, have less prominent glands on the upper sepal and more consistently puberulous internodes. Some specimens, especially from Zimbabwe, may have the sepals distinctly puberulous beyond their bases, unlike any collections seen from our area.

Floral details in the description were taken from the only spirit collection available, *Faden & Faden* 74/196 from Malawi. Its upper sepal glands were not very prominent, either in the pressed specimens or the spirit material, compared with specimens from Tanzania, so that point was omitted from the description.

Blundell, Wild Fl. E. Afr.: 412, fig. 856 (1987) confuses *A. pedunculosum* and *A. leiocaule*. His illustration is of *A. leiocaule*.

35. **Aneilema leiocaule** *K. Schum.* in P.O.A. C: 136 (1895); Blundell, Wild Fl. E. Afr.: 412, fig. 856 (1987) as "*A. pedunculosum*"; Faden in Smithsonian Contrib. Bot. 76: 147, t. 4,m (1991); & in U.K.W.F. 2nd ed.: 308 (1994); Ensermu & Faden in Fl. Eth. 6: 360 (1997). Type: Kilimanjaro, *Kersten* s.n. (B†, syn.) & Kilimanjaro [Kilimandscharo], *Volkens* 1436 (B, syn†, BM!, G!, PRE!, iso.)

Perennial; roots thin, fibrous; shoots ascending to decumbent and rooting at the nodes, sometimes straggling, much branched, to 30(–60) cm tall, glabrous or with sparse hook-hairs below the nodes. Leaves spirally arranged; sheaths to 1.5(–2) cm long, sparsely puberulous, sometimes pilose-puberulous, white-ciliate at apex; petiole to 2.5 cm long, lamina ovate to ovate-elliptic, 2–6.5(–11) × 1.5–4(–5) cm, base symmetric, cuneate to rounded, margins scabrid, rarely ciliate (see discussion), apex acute to acuminate; adaxial surface puberulous and sometimes scabrid or pilose-puberulous, abaxial surface usually more densely pubescent. Thyrses terminal and usually solitary, sometimes also axillary from the uppermost leaf, ± dense, ovoid to subspherical, 1.5–3 × 1.5–2 cm; inflorescence axes densely (or sparsely) puberulous; peduncle 2.5–5 cm long, with basal (rarely halfway) bract; cincinni up to ± 7 mm long and 4-flowered; bracteoles often overlapping, cup-shaped, not perfoliate, 1–2.5 mm long, the apex often drawn out into a linear tip with a terminal, capitate gland. Flowers bisexual, male and sometimes female, 9–13 mm wide; pedicels 2.5–4.5 mm long in flower, to 5.5 mm long and erect to recurved in fruit. Sepals strongly convexo-concave, 2.5–3.5 × 2–2.5 mm, hooded at apex, glandular subapically, glabrous or minutely puberulous; paired petals lavender, mauve, blue, purple, blue-violet, lilac or rarely white, broadly ovate, 5–6.5 × 3.5–4.5 mm of which the white claw 1.5–2 mm; medial petal cup-shaped, 3–3.8 × 2–2.5 mm, greenish white; staminodes 3, antherodes bilobed, yellow; lateral stamens with filaments 5–7 mm long, sigmoid, densely bearded distally with bluish purple hairs, anthers 0.7–1 mm long, pollen yellow; medial stamen filament 4–4.2 mm long, anther ± 0.7 mm long and producing a small amount of yellow pollen or sterile. Ovary 1.3–2 × 1.1–1.3 mm, glandular-puberulous; style arcuate-deflexed, 3.5–6 mm long; stigma capitate. Capsules lustrous grey-tan or brown, bilocular, bivalved (the valves persistent), 4-seeded, 4–5 × 3–3.5 mm, with scattered glandular hairs, locules (1–)2-seeded. Seeds ovoid to transversely elliptic, (1.4–)1.9–2.4(–2.9) × 1.4–1.8(–2) mm, testa foveolate-reticulate to foveolate-scrobiculate, orange-tan (to orange-brown).

UGANDA. Toro District: Ruimi [Wimi] River, 13 Jan. 1896, *Scott Elliot* 7335!; Kigezi District: 4 km N of Lake Mulehe, 11 Dec. 1968, *Lye & Lester* 953!

KENYA. Northern Frontier District: Mt Kulal top, 25 July 1958, *Verdcourt* 2255!; Machakos District: Donyo Sabuk, 1 Feb. 1970, *Faden & Evans* 70/49!; Teita District: Vuria Mt, N face, 9 Feb. 1966, *Gillett, Burtt & Osborn* 17077!

TANZANIA. Arusha District: Ngurdoto National Park, Ngurdoto Crater, Leopard Point, 5 Oct. 1965, *Greenway & Kanuri* 11961!; Moshi District: Mt Kilimanjaro, W slope, E of Lemosho Glades, 13 Jan. 1970, *Thulin* 305!; Lushoto District: West Usambaras, Mkusu Valley between Mkuzi and Kifungilo, 23 Apr. 1953, *Drummond & Hemsley* 2199!

DISTR. **U** 2; **K** 1, 3, 4–7; **T** 2, 3, 6, 7; Ethiopia, Congo–Kinshasa, Burundi, Zambia? (see notes)

HAB. Upland forest & forest edges, clearings and tracks, sometimes growing near water; often in shade; 1000–2750 m

Flowering specimens have been seen from all months.

SYN. *Aneilema pedunculosum* C.B. Clarke in F.T.A. 8: 73 (1901) pro parte; Faden in U.K.W.F.: 667 (1974)

NOTE. This is a common, moist forest understory species at middle to high elevations in N Tanzania and the S half of Kenya. Although it has been confused with *A. pedunculosum* ever since Clarke in F.T.A. treated it as a synonym of that species, the two species are quite distinct and occupy entirely different ranges. *A. leiocaule* differs from *A. pedunculosum* by a perennial habit, leaves more evenly spread out on the shoots, inflorescences usually one (or occasionally two, rarely three) per shoot; internodes typically glabrous; much less prominent glands on the upper sepal; and the 4-seeded (vs. 5-seeded) capsule with both capsule valves persistent (vs. the dorsal valve deciduous).

In two spirit collections, *Faden & Evans* 70/49 and *Faden & Evans* 69/786 (both from **K** 4), the stamen filaments were observed to be fused at the base (as in *A. spekei*). I expect that this will prove typical for the species, but until it is found in populations from elsewhere, it requires confirmation.

Mature seeds are very scarce in herbarium specimens, so the lowest length and width (in parentheses) are taken from Ethiopian specimens. The largest dimensions are from a single seed in a packet on *Archer* 360 from **K** 1. Although 1-seeded locules are frequent they always result from abortion of the second ovule and do not arise from single ovulate locules, in contrast with *A. minutiflorum* and some populations of *A. hirtum*.

Luke et al. 8499 from Udzungwa Mountain National Park, the southernmost collection and sole one from **T** 7, is unusual in having white flowers (the only such collection). It also has several unusual or unique leaf pubescence characters: red uniseriate hairs abundant at the summit of the sheaths, on the margins of the petioles and on the adaxial leaf surface; hook-hairs abundant on both surfaces; and leaf margins ciliate with several-celled uniseriate hairs. Were there additional specimens that matched it, they could easily be treated as a distinct variant of *A. leiocaule*. *Verdcourt* 4024 has leaf sheaths with uniseriate red hairs on the surface, as well as at the summit.

Thomas Th 2767 from Kitojo (**U** 2, Toro District, 1830 m) is only questionably placed in this species because of several unusual features: the bracteoles are not at all elongated at the apex, and the sepals bear an occasional, scattered uniseriate hair and even rarer hook-hair; the pedicels are puberulous only at the summit, and the inflorescence is denser than in typical *A. leiocaule*. It is similar to *A. leiocaule* in stem and leaf characters, basal inflorescence bract, long uniseriate hairs in the inflorescence, and inconspicuous sepaloid glands. The pressed flowers clearly show long-bearded stamen filaments and the bracteoles are not perfoliate, so it must belong in sect. Pedunculosa. The immature fruits are clearly 4-seeded. Although the bracteoles more closely resemble those of *A. spekei*, it is unlikely to be that species because the pubescence of the internodes, inflorescence axes, pedicels and sepals is wrong, as are the petiolate leaves and, especially, the position of the inflorescence bract. The forest edge habitat would be unusual for *A. spekei* and the elevation of *Thomas* Th 2767 is higher than any recorded for that species, but both would be typical for *A. leiocaule*. For the time being, this specimen is best treated as an atypical specimen of *A. leiocaule*.

Aneilema leiocaule barely overlaps in central Kenya (**K** 3) with *A. minutiflorum*, another forest species, but the two have not been found in the same locality. *A. leiocaule* differs by it more robust habit, usually glabrous internodes, larger inflorescences on longer peduncles, much larger flowers with the paired petals much broader than the lower petal and always differing in colour from it, longer pedicels, well developed medial staminode, and much larger capsules with typically 2-seeded locules.

Dale 2667 records the Kikuyu name for this species as *Mukengeria* and gives the following use: "the leaves are wrapped round the injured member of Kikuyu boys after circumcision."

Azuma in Kyoto University Expedition 474 and *Bidgood et al.* 4196, both from **T** 4, seem to match three collections from N Zambia (*Richards* 4791, 1531 & 21359) and to represent a distinct taxon. The plant is described as a perennial in *Bidgood et al.* 4196, and none of the specimens is distinctly annual. In addition to their perennial habit, these plants also agree with *A. leiocaule* and differ from typical *A. pedunculosum* by the capsules apparently 4-seeded with the valves persistent, but only *Richards* 4791 has mature capsules and seeds. The gland on the medial sepal is bilobed, as in both species, but much less prominently so than in *A. pedunculosum* but more so than in *A. leiocaule*. The distinctive features of this taxon are: sheaths and often the internodes ± densely pilose-puberulous (or the internodes only densely puberulous), the bracteoles ovate with a shortly drawn out glandular apex, and the sepals pubescent. The collections seen from our area are:

TANZANIA. Kigoma District: Kabago Mts, 7 May 1963, *Azuma in Kyoto University Expedition* 474!; Mpanda District: Livandale Mt, 30 May 1997, *Bidgood et al.* 4196!

DISTR. **T** 4; Zambia

HAB. Tall closed forest along stream and riverine forest; 790–1200 m

Flowering specimens have been seen from May.

NOTE. The Zambian collections were made around waterfalls and in woodland and were collected in flower in March and April; they have the distinctive character of the few lowermost cincinnus bracts per inflorescence each very broad, with 2–3 separate, elongate, glandular apices. This is not present in the Azuma collection and it was not checked in *Bidgood et al.* 4196, so it is not certain whether either of the Tanzanian plants agree with those from Zambia. However, they do not match either *A. leiocaule* or *A. pedunculosum* and are allopatric with them.

36. **Aneilema minutiflorum** *Faden* **sp. nov**. a *A. leiocaule* habitu annuo inflorescentiis et floribus parvioribus cincinno infirmo a ceteris inflorescentiae separato interdum basali saepe evoluto petalo inferiore petalo laterali latiuse staminodio mediano absenti vel vestigiali capsula parviora plerumque 2-seminali differt. Type: Kenya, Mt Elgon, Suam Saw Mill road off Suam–Endebess road, *R.B. Faden & A. Evans* 70/899 (US!, holo.; BR!, EA!, G!, K!, MO!, NHT!, P!, WAG!, iso.)

Gracile, decumbent annual to 45(–60) cm high; roots thin, fibrous; shoots much-branched puberulous or pilose. Leaves spirally arranged; sheaths 0.5–1 cm long, puberulous, ciliate at apex; lamina petiolate, ovate, 2–6.5 × 1–3 cm, base rounded or abruptly cuneate, margins scabrid, apex acute to acuminate; both surfaces pilose-puberulous with long and minute hook-hairs; petiole to 1.2 cm long. Thyrses terminal and sometimes axillary from upper leaves, usually dense, ovoid to spherical, 0.8–1.5 × 0.5–1.3 cm, of few to many, mainly 2-flowered cincinni; inflorescence axis puberulous; peduncle 0.5–2 cm long; bracteoles 0.5–1 mm long, narrowed distally into a glandular, capitate tip. Flowers bisexual and male, 3.5–5 mm wide; pedicels 1–2 mm long, puberulous. Sepals glandular, sparsely pubescent or glabrous, blue to white or violet, tipped with green, the upper ovate to oblong, the paired narrowly elliptic to obovate-elliptic, 2.2–2.8 × 1–2 mm; paired petals blue, mauve, pale violet, pale lavender, lilac, pink or white, broadly elliptic to orbicular or obovate-cuneate, 2.7–3.3 × 1–1.7 mm of which the white claw 1–1.6 mm; medial petal same colour except a green keel or apex, boat- or cup-shaped, 2–3 × 1.3–2 mm; staminodes 2, the medial lacking or vestigial; antherodes bilobed, yellow; lateral stamens (1.7–)2.5–4 mm long, usually S-shaped, densely bearded with blue-violet, purple, mauve or lavender-tinged white hairs, anther 0.3–0.8 × 0.4–1.2 mm, pollen yellow; medial stamen filament 2–2.5 mm long, anther ± 0.3 mm long, pollen yellow. Ovary 0.7–1.1 × 0.8–1 mm, with appressed glandular hairs; style straight or arcuate-decurved, 1.2–2.5 mm long; stigma large, capitate-deltate. Capsules pale yellow with darker margins, bilocular, bivalved, broadly ellipsoid to obovoid, 2.5–3.5 × (2.1–)2.5–3 mm, puberulous, capsule valves persistent, spreading dorsal locule not developed, ventral locules usually 1-seeded (rarely 2-seeded). Seeds ellipsoid, 1.6–2.1 × 1.25–1.4 mm, testa foveolate-reticulate, light orangish brown, sparsely whitish or tan farinose in the depressions.

UGANDA. Mbale District: Suam, along the Suam River, ± 1 km upstream from Suam Bridge on Endebess–Bukwa road, 26 Sep. 1971, *Faden & Tweedie* 71/860B! & Suam River, 0.8 km above bridge on Mgishu Mountain Road, Aug. 1971, *Tweedie* 4090!

KENYA. Laikipia District: North Laikipia, no date, *Gardner* 3514!; Trans-Nzoia District: Suam Saw Mill road off Endebess–Suam road, along forest reserve boundary cut line, 24 Nov. 1970, *Faden & Evans* 70/899! & Cherangani foothills, 23 Sep. 1949, *Maas Geesteranus* 6373!

TANZANIA. Iringa District: Udzungwa Mountain National Park, 7°42'S, 36°33'E, *Luke et al.* 8469 & Mufindi, 26 Mar. 1970, *Paget-Wilkes* 793! & Lupeme Estate, 16 km NE of Mufindi, pressed from cult. 29 Jan. 1974, *Paget-Wilkes* 'A'!

DISTR. **U** 3; **K** 3; **T** 7; not known elsewhere

HAB. Forests and forest clearings, sometimes near streams, weed in maize; 1800–2350(–2500) m Flowering specimens have been seen from March, July to September and November.

SYN. *Aneilema* sp. aff. *pedunculosum* of Lewis in Sida 1: 279 (1964)
A. sp. *B*, Faden in U.K.W.F.: 665, 667 (1974)
A. minutiflorum Faden in Smithsonian Contrib. Bot. 76: 147 (1991), *nom nud.* & in U.K.W.F. 2nd ed.: 308, t. 138 (1994), *nom. nud.*

NOTE. Sixteen collections have been seen, eleven from Kenya, two from Uganda and three from Tanzania. There is some variation, particularly in flower colour and in the position of the solitary cincinnus that is separate from the rest of the terminal thyrse. When present, it is most often basal on the peduncle, but sometimes it is borne higher, even above the middle.

A chromosome number of $2n = 18$ was obtained from cultivated material of *Paget-Wilkes* A.

Field notes from *Faden & Tweedie* 71/860 indicated that plants were growing in and at the edge of maize cultivation in recently cleared forest. The plants were so vigorous that it was speculated that they might be perennials. More likely they were just annuals as were plants of *Paget-Wilkes* A in greenhouse cultivation.

Notes made from the field collection of *Faden & Evans* 71/463 indicate that an inflorescence, when viewed from the top, looked like a composite flower with white ray florets and blue disc florets. The white was from the larger buds on the outside at the base of the inflorescence, and the blue from the more central, smaller buds higher up.

Typically the ovary has two one-ovulate and the capsule two one-seeded locules. An ovary in *Faden & Evans* 70/899 was found to have a single large ovule in one locule and two small ovules in the other locule. Field notes from *Faden & Tweedie* 71/860 record the observation of some four-seeded capsules, but these appear to be uncommon in the species. Seeds from two-seeded locules are considerably smaller and have a different shape from seeds from one-seeded locules, and therefore they have been described separately.

This species is most distinctive because of its tiny flowers, the medial petal broader than a lateral petal and often concolorous, absent or vestigial medial staminode, very small inflorescence with the lowermost cincinnus widely separated from the others and often basal, and very small capsules with usually 1-seeded locules.

11. **COMMELINA**

L., Sp. Pl. 40 (1753); Clarke in DC., Monogr. Phan. 3: 138–194 (1881) & in F.T.A. 8: 33–61(1901); Morton in J.L.S. 55: 507–531 (1956); Faden in Smithsonian Contr. Bot. 1–166 (1991); Ogwal in Mitt. Inst. Allg. Bot. Hamburg 23b: 573–592 (1990); Kayemba-Ogwal in Proc. XIIIth Plenary Meeting AETFAT Malawi 1: 415–420 (1994); Faden in Kubitzki, Fam. Gen. Vasc. Pl. 4: 126 (1998)

Perennial or annual herbs, occasionally rhizomatous; roots fibrous or occasionally tuberous; flowering shoots sympodial. Leaves spirally arranged or distichous, sessile or petiolate, base commonly oblique or sometimes symmetric. Inflorescences terminal and leaf-opposed, solitary or clustered, each inflorescence composed of 1–2 contracted, pedunculate cincinni subtended by and partly or wholly enclosed in a folded bract (spathe), the spathes with margins free or fused proximally; upper (distal) cincinnus commonly exserted from the spathe and producing a single male (rarely bisexual) flower, occasionally multi-flowered (the flowers usually then usually all male), sometimes vestigial or lacking; lower (proximal) cincinnus wholly enclosed within the spathe, usually several- to many-flowered, with some or all of the flowers bisexual. Flowers bisexual and male, strongly zygomorphic, pedicellate. Sepals free or the lower (paired) 2 partly to wholly fused. Petals free, unequal, upper 2 (paired) petals larger and clawed, always conspicuous, lower (outer) petal usually very reduced, not clawed, concolorous with the paired petals to strongly discolorous, sometimes colourless. Androecium composed of 3(–2) shorter, posterior staminodes and 3 longer, anterior stamens, filaments glabrous, usually free, antherodes (staminode anthers) usually 6-lobed (the 2 medial lobes often producing a minute amount of pollen), sometimes reduced, usually yellow (occasionally with a dark central spot), lateral stamens usually longer than the medial stamen with the anthers different in shape and often colour and sometimes bearing pollen of a different colour, the medial anther typically saddle-shaped, pollensacs longitudinally dehiscent. Ovary sessile, bi- or trilocular, 2–5(very rarely 6)-ovulate, dorsal locule 0–1(very rarely 2)-ovulate or suppressed, ventral locules 1–2 ovulate; style slender; stigma usually capitate or deltate. Capsules bi- or trilocular, bi- or trivalved, 1–5(very rarely 6)-seeded, usually glabrous, dorsal capsule valve sometimes deciduous, dorsal locule, when present, 0–1-seeded, dehiscent or indehiscent, when indehiscent, its wall ± fused to the seed or not fused, ventral locules usually 1–2-seeded (rarely empty), dehiscent. Seeds uniseriate, with a linear hilum and lateral embryotega, sometimes appendaged or apparently arillate.

A pantropical and pan-temperate (except Europe) genus of about 170 species of which ± 100 are African, with 51 in the Flora; occurring in almost all habitats.

The species are difficult to identify even in the field because the distinguishing characters are often subtle, the flowers are ephemeral, capsules and seeds may be lacking, and several similar species may occur together. The most important characters to note are whether the spathe margins are free or fused, and the flower color. The latter can be tricky, especially for many of the species in the yellow to orange or buff colour range. Even in some of the species that regularly produce sky blue flowers there may be mauve-flowered individuals or very rarely a white-flowered plant. Mature capsules, when present, should be checked to see whether they split into two or three valves, and how many seeds there are in each locule. The latter usually can also be determined from immature fruits.

For species identification pubescence can be important, both that of the vegetative parts and of the spathes. Several species or pairs of species can be distinguished by the type of hairs present on the upper side of the leaf midrib, e.g. small, acicular (needle-like) white hairs in *C. latifolia* and *C. lukei*, but even smaller hook-hairs (hairs that terminate in a hook) in *C. kotschyi*, *C. imberbis* and *C. mascarenica*. The presence or absence of long, uniseriate hairs on the spathes, in addition to shorter hook-hairs, can help separate some species with sky blue flowers.

In species 22–42 hook-hairs are generally present on the surface of the spathes (except 30. *C. foliacea*, 35. *C. congesta*, 39. *C. melanorrhiza* and 41. *C. disperma*). The hook-hairs may be uniform in size, of two different sizes or variable in length. In species 1–4, 39 and 41 hook-hairs may be present on other parts, such as the peduncles of the cincinni, but they are lacking from the spathe surfaces. In all other species (5–21, 30 and 35) hook-hairs appear to be lacking everywhere.

When flowers are present, a number of characters should be noted and recorded. The most obvious one is the flower, i.e. paired petal, colour and also the shape and colour of the lower petal. The paired (or lower) sepals should be checked to see whether they are free or fused. Whether there are two or three staminodes and whether the antherodes are pure yellow or have a central dark spot should be noted. The pollen in the three fertile stamens should be observed to see whether it is all the same color. Finally, it should be noted whether the flowers are male – the ovary and style would be lacking or very reduced – and/or bisexual.

Commelina seeds are often dimorphic within a single capsule. The dorsal locule seed, which is often retained within an indehiscent locule, is commonly different in form and testa pattern from the ventral locule seeds. The dorsal capsule valve, containing the dorsal locule seed, is often deciduous and can serve as a disseminule or seed-dispersal structure.

Seeds may have specific structures that are related to or probably related to seed dispersal. In 22. *C. capitata* the seed is covered by an orange-yellow or dull yellow material that is fleshy in fresh seeds and appears to be an aril. In 37. *C. erecta*, 39. *C. melanorrhiza* and 40. *C. albescens* the seeds are surrounded by a ring (or sometimes two patches in *C. erecta*) of soft, whitish (when fresh) material that functions as an elaiosome to attact ants, which have been observed carrying seeds of *C. erecta* and probably also do so in the other two species. Species 28. *C. kotschyi* and 29. *C. lukei* have the hilum raised and extended at both ends into appendages that presumably also are for seed-dispersal. Finally, in 51. *C. aspera* the hilum bears a creamy white ridge that is sometimes extended beyond the seed at one or both ends.

A number of *Commelina* species show so much variation that it has not been possible to divide them satisfactorily into infraspecific taxa. Some variable species, e.g. *C. eckloniana*, have been studied in detail, and a good but provisional treatment has been included. For other species, e.g. *C. benghalensis*, in which the great variation observed in the field cannot be fully recognized in herbarium specimens, even such a provisional account has been impossible. Clearly much more study is required to understand fully the morphological variability in some of the most common and widespread *Commelina* species. Adding to the complexity in the genus, recent collections from central Tanzania, especially by Bidgood and Vollesen, have shown that there likely are new species of *Commelina* from that area yet to be discovered and described.

The arrangement of the species follows no particular classification but places together the species that I believe are closely related. Species 1–21 have spathes with free margins, whereas species 22–51 have fused spathe margins. 22. *Commelina capitata* will key out both ways because the fused part of the margin may tear open during fruit development or with age. 17. *Commelina nyasensis* has spathes that may be very shortly and inconspicuously fused at the base, so it too is keyed out both ways. Species 10–21, 37–39, 44 and 48–51 have trivalved capsules, whereas all of the others and some populations of species 37 and 38 have bivalved capsules. Capsules with the ventral locules one-seeded characterize species 6, 35–42 and 49–51 and some plants or populations of 4. *C. africana*. In 32. *Commelina forskaolii* and some plants or populations of 4. *C. africana* only the dorsal locule seed matures, the four ventral locule ovules/seeds usually aborting. All other species have two-seeded ventral locules, although irregularly only a single seed will develop due to abortion or lack of pollination of one of the ovules.

Among the species with one-seeded ventral locules the ventral locule seeds are borne a little distal to the dorsal locule seed in the capsule in species 4, 6, 35 and 36 because the basal ventral locule ovules or developing seeds regularly abort. In all other species with one-seeded ventral locules, the ventral locule seeds are basal in the locules and are borne at the same level as the dorsal locule seed, if one is produced. Several *Commelina* species, e.g. *C. africana*, *C. benghalensis* and *C. forskaolii*, are significant weeds. Some are used as fodder for domestic animals or medicinally. These are mentioned under the individual species.

Note: the second dimension given for the spathe is the height of the folded spathe, which is half of the actual width of the spathe, were it unfolded and laid flat. The actual width is the dimension used in some Floras. Similarly the shape of the spathe apex is that shown by the folded, not the flattened spathe.

1. Spathes with margins free to the base 2
   Spathes with margins fused at least proximally 25
2. Flowering shoots leafless at anthesis; flowers usually white . . . . 9. *C. scaposa* (p. 159)
   Flowering shoots leafy at anthesis; flower colour various 3
3. Leaves with red hairs at the summit of the sheaths; flowers yellow; seeds covered by a yellowish aril . . . . 22. *C. capitata* (p. 178)
   Leaves with white or colorless hairs at the summit of the sheaths; flower colour various; seeds exarillate 4
4. Capsules trilocular, trivalved (the dorsal locule sometimes tardily dehiscent); flowers buff to orange, pink-orange, apricot or reddish brown (seldom yellow; lilac, pale blue or purple in *C. fluviatilis*; if flowers sky blue, see notes under 11. *C. subulata*) 5
   Capsules bi- to trilocular, bivalved; flower colour various but rarely buff to orange, pink-orange or reddish brown 17
5. Flowers lilac, pale blue or purple; plants often growing in deep water, with floating stems . 10. *C. fluviatilis* (p. 161)
   Flowers buff to orange, pink-orange, apricot or reddish brown (rarely yellow); habitat various, but plants usually not growing in deep water 6
6. Flowers pale yellow; spathes mainly subsessile, clustered on short lateral shoots; testa foveolate to foveolate-alveolate, the depressions usually surrounded by a raised reticulum . . . . 5. *C. pycnospatha* (p. 153)
   Flowers buff to orange, pink-orange, apricot or reddish brown; spathes pedunculate (sometimes very shortly), usually not clustered, terminal and leaf-opposed on elongate shoots; testa various 7
7. Perennials 8
   Annuals 10
8. Spathe peduncles all or mostly >10 mm long 9
   Spathe peduncles all or mostly <10 mm long . 14. *C. purpurea* (p. 168)
9. Stems commonly looping along the ground and rooting at the nodes, occasionally erect; leaves usually twisted, margins strongly undulate (widespread but not in **T** 4–8) . . . 15. *C. reptans* (p. 169)
   Stems mostly erect, seldom looping; leaves not twisted, margins not strongly undulate (**T** 4, 5, 6/8, 7) . . . . 16. *C. pseudopurpurea* (p. 171)

10. Spathes completely glabrous; lower internodes thick and spongy; lake margins (if spathes glabrous and seeds are transversely grooved, see 11. *C. subulata*) .... 21. *C. chayaensis* (p. 178)
    Spathes with margins ciliate or ciliolate, at least proximally, often surface pubescent as well; lower internodes not thick and spongy; habitat various 11
11. Seeds with two dorsal grooves, making them 3-lobed, strongly dorsiventrally compressed, with a grain of sand-like appearance; spathes strongly curved downward 12
    Seeds not 3-lobed, strongly dorsiventrally compressed or not, sand grain-like or not; spathes curved downward or not 13
12. Proximal (lower) leaves 5–11 mm wide; surface of spathe glabrous; lower petal strongly contrasting in colour with the paired petals 18. *C. trilobosperma* (p. 174)
    Proximal leaves 3–7.5 mm wide; surface of spathe usually hirsute, occasionally glabrous; all 3 petals ± concolourous .... 19. *C. merkeri* (p. 175)
13. Seeds with one deep circular pit or V- or U-shaped groove or with an ovate-deltate pit containing a central boss of similar shape, but smaller, leaving a furrow all around it 14
    Seeds various but not as above 15
14. Spathes with glabrous surfaces; dorsal seed surface (the surface opposite the one bearing the hilum) with a deep circular pit or V- or U-shaped groove .... 12. *C. polhillii* (p. 164)
    Spathes densely hirsute; ventral seed surface (the surface bearing the hilum) with an ovate-deltate pit containing a central boss of similar shape, but smaller, leaving a furrow all around it .... 13. *C. sulcatisperma* (p. 167)
15. Spathes subsessile, the peduncles up to 5 mm long; seeds with testa transversely grooved (or with elongate pits) and ridged on the dorsal surface, with 3–5 ridges (or elongate pits) and 2–4 grooves .... 11. *C. subulata* (p. 162)
    Spathes mostly distinctly pedunculate, the peduncles mostly 3–15 mm long; seeds with testa usually alveolate, foveolate-alveolate, foveolate-reticulate, or sparsely and irregularly finely pitted 16
16. Spathes 5–8(–10) mm high; upper cincinnus usually lacking or vestigial; lower cincinnus (1–) 2–4-flowered; seeds with testa usually alveolate, foveolate-alveolate or foveolate-reticulate, not sand-grain like .... 17. *C. nyasensis* (p. 173)
    Spathes (2–)2.5–5 mm high; upper cincinnus well-developed and 1-flowered; lower cincinnus 3–8-flowered; testa sparsely and irregularly finely pitted, sand-grain like ... 20. *C. pallidispatha* (p. 176)

17(4) Flowers yellow, yellow-orange, apricot to brownish yellow or nearly white 18
Flowers sky-blue, lavender to purple or pale lilac, pink, mauve, purple or violet to white 20

18. Perennials; spathes solitary, peduncles 0.8–4 cm long .... 4. *C. africana* (p. 146)
Annuals; spathes often clustered, peduncles mostly 0.3–1.4 cm long 19

19. Plants decumbent to repent or subscandent; lamina lanceolate to elliptic or ovate, 0.7–2 cm wide; flowers ± 7.5 mm wide, pale yellow ... 5. *C. pycnospatha* (p. 153)
Plants usually erect to ascending; leaves linear to linear-lanceolate, 0.1–0.6 cm wide; flowers 10–16 mm wide, yellow-orange or apricot to brownish yellow or nearly white .... 17. *C. nyasensis* (p. 173)

20. Flowers sky blue, rarely mauve to lilac 21
Flowers lavender to purple or pale lilac, pink, mauve, purple or violet to white 23

21. Plants decumbent to repent, sprawling or straggling, usually without a definite base; roots thin, fibrous; capsules 5-seeded; plants of wet or moist habitats; leaves completely green (widespread) 22
Plants with erect or ascending stems from a definite base; roots tuberous; capsules 3-seeded; plants of dry habitats; leaves sometimes with purple spots or chevrons (**K** 1) .... 6. *C. stefaniniana* (p. 154)

22. Spathes 0.4–4 cm long, peduncles 0.7–2.5(–3) cm long; capsules (4.7–)5–6(–7) mm long; ventral locule seeds 1.8–3.1 mm long, lacking a middorsal ridge; pedicels and sepals glabrous; antherodes completely yellow (widespread) .... 1. *C. diffusa* (p. 140)
Spathes (2.5–)3–6.5(–8) cm long, peduncles (1–)1.6–5(7.5) cm long; capsules 7–9 mm long; ventral locule seeds 2.7–3.2 mm long, with a middorsal ridge (occasionally not very prominent); pedicels puberulous distally or sepals puberulous proximally; antherodes with a central dark spot and yellow lobes (**T** 8) 2. *C. scandens* (p. 144)

23. Spathe peduncles 1.5–3 cm long; peduncles of both cincinni puberulous with hook-hairs; seeds smooth to alveolate; plants of forest and forest edge (**U** 2) .... 3. *C. acutispatha* (p. 145)
Spathe peduncles 2.5–11(–18) cm long; peduncle of upper cincinnus glabrous or puberulous with short, uniseriate hairs, peduncle of lower cincinnus glabrous; seeds foveolate to foveolate-reticulate; plants of grassland or *Brachystegia* woodland (**T** 4, 7) 24

24. Spathes broadly striped or flushed with red or reddish purple, densely to sparsely pubescent; upper cincinnus usually 1-flowered, its peduncle 10–22 mm long; capsules 5–6 mm long; flowers usually pale blue to lavender or purple, the lower petal strongly contrasting with the paired petals; plants mainly of montane grassland at (1600–)1830–2880 m elevation .... 8. *C. kituloensis* (p. 158)
Spathes green or sometimes purple, purple-veined or purple-margined, usually glabrous, occasionally pubescent; upper cincinnus (1–)7–14-flowered, its peduncle 18–30 mm long; capsules 8–9 mm long; flowers usually white to pink, all 3 petals concolorous; plants of *Brachystegia* woodland and grassland up to 1980 m elevation .... 7. *C. hockii* (p. 157)
25(1) Spathes with both halves folded outward, forming a flat, leaf-like surface when viewed from above; flowers sky-blue, open mainly in the afternoon .... 30. *C. foliacea* (p. 198)
Spathes with both halves closely appressed, not leaf-like; flower colour various, flowers open mainly or exclusively in the morning 26
26. Flowers yellow to orange, sometimes mixed with pink or brown 27
Flowers blue, lavender, lilac, mauve, purple or white 36
27. Leaves with red hairs at the summit of the sheaths; seeds entirely covered by a yellowish aril; forest plants .... 22. *C. capitata* (p. 178)
Leaves with white or colorless hairs at the summit of the sheaths or hairs lacking; seeds exarillate but sometimes appendaged; plants of various habitats 28
28. Spathe margins nearly free, fused up to 1 mm 17. *C. nyasensis* (p. 173)
Spathe margins manifestly fused for at least 2 mm 29
29. Capsules trilocular, trivalved, 3- or 5-seeded 30
Capsules bilocular or rarely trilocular, bivalved, 4-seeded 35
30. Perennials; spathes with upper cincinnus 1-flowered, lower cincinnus ± 9–10-flowered; capsule 5-seeded; seeds smooth (**K** 3) ..... 44. *C. kitaleensis* (p. 226)
Annuals; spathes with upper cincinnus lacking or vestigial, lower cincinnus 1–4-flowered; capsule 3- or 5-seeded; testa various (not in Kenya) 31
31. Capsules 5-seeded; seeds sparsely warty; spathes always solitary .... 48. *C. aurantiiflora* (p. 231)
Capsules 3-seeded; seeds various; spathes solitary or densely clustered 32
32. Spathes densely clustered 33
Spathes ± solitary 34

33. Seeds muricate or echinulate, with 2 transverse or transverse-diagonal pits on the dorsal surface; minutely puberulous (at high magnification), hilum lacking a raised, usually creamy white ridge; spathe margins fused for 4–8 mm .... 49. *C. nigritana* (p. 233)
   Seeds ± smooth, lacking dorsal pits, glabrous, hilum with a raised, usually creamy white ridge; spathe margins fused for 5–19 mm .. 51. *C. aspera* (p. 236)
34. Seeds ellipsoid, longer than broad, 2.35–3.7 mm long, smooth to echinulate or muricate (rarely ribbed), with 2 transverse or transverse-diagonal pits on the dorsal surface; minutely puberulous (at high magnification); capsules broadly ellipsoid, 3–4.5 mm long . 49. *C. nigritana* (p. 233)
   Seeds ± dumbbell-shaped, about as broad as long, 1.5–2 mm long, smooth, lacking dorsal pits, very rarely puberulous; capsules ± obovoid, 2.5–3.5 mm long .... 50. *C. saxosa* (p. 235)
35. Plants usually ± perennial (rarely distinctly annual); spathes usually solitary, rarely subclustered, 1.4–2.7 cm long, (0.9–)1.1–1.6 cm high, surface glabrous or puberulous, spathe peduncles (1–)1.5–4(–9) cm long, conspicuous; flowers 15–22 mm wide; upper sepal 3–5.5 mm long .... 46. *C. neurophylla* (p. 229)
   Plants consistently annual; spathes clustered, (0.5–)0.8–1.7 cm long, 0.4–1.3 cm high, surface pubescent, rarely nearly glabrous, spathe peduncles, 0.3–1.1 cm long, mainly hidden within the sheaths; flowers ± 10 wide; upper sepal 2–2.5 mm long .... 47. *C. triangulispatha* (p. 230)
36. Spathes, peduncles and leaves completely lacking hook-hairs, all hairs uniseriate; flowers lavender to blue violet and similar hues, but never sky blue; antherodes with a central dark spot and yellow lobes; paired (lower) sepals completely fused into a cup; medial petal minute; ventral capsule locules 2-seeded 37
   Spathes usually puberulous with hook-hairs (requires at least 20× magnification) (when hook-hairs are lacking the ventral capsule locules are 1-seeded), sometimes also pilose or hirsute with longer uniseriate hairs, leaves and peduncles sometimes with hook-hairs; flowers often sky blue, but sometimes lilac, lavender, purple, mauve or white; antherodes entirely yellow; paired sepals and medial petal various; ventral capsule locules 1–2-seeded 39

37. Leaves all or mostly basal, linear, 8–40(–50) cm long; spathes 2.7–4.7 cm long, 1.5–2.6 cm high, glabrous(except ciliate or ciliolate along the fused edge) .... 45. *C. grossa* (p. 227)
Leaves all cauline, linear to oblong, lanceolate, ovate or elliptic, 2–20(–27) cm long; spathes 1.4–3.7(–4.2) cm long, 0.9–1.7(–1.8) cm high, glabrous to densely pilose or hirsute 38
38. Roots thin; leaves linear, linear-oblong, linear-lanceolate to lanceolate-oblong, 2–9.5 cm long, 0.1–1(–1.4) cm wide; fused portion of spathe margin reflexed at the distal end of the fusion; seeds muricate to echinate .... 46. *C. neurophylla* (p. 229)
Roots thick and sometimes tuberous; leaves linear to oblong, lanceolate, ovate or elliptic, 2–20(–27) cm long, 0.3–2.5(–3) cm wide; fused portion of spathe margin not reflexed; seeds smooth .... 43. *C. schweinfurthii* (p. 222)
39. Leaves usually with long, red hairs at the summit of the sheath (sometimes also on the sheath surface, proximal leaf margins and on the spathes); cleistogamous flowers and fruits sometimes present on short subterranean shoots from the base of the plant; ventral capsule locules 2-seeded .... 31. *C. benghalensis* (p. 200)
Red hairs completely lacking; cleistogamous flowers and fruits usually lacking (if present, borne from rooted nodes of decumbent shoots, see 32. *C. forskaolii*); ventral locules 1- or 2- seeded 40
40. Spathes clustered at the ends of the shoots 41
Spathes solitary or two together 45
41. Cleistogamous flowers and fruits often present on subterranean short shoots from the base of the plant; ventral capsule locules 2-seeded 31. *C. benghalensis* (p. 200)
Cleistogamous flowers and fruits usually lacking (if present, borne on short shoots from rooted nodes, see 32. *C. forskaolii*); ventral capsule locules 1-seeded 42
42. Flowers white or whitish (pale pink, lilac mauve or blue); shoots rooting at the nodes; ventral locule seeds borne slightly distal to the dorsal locule seed; spathes glabrous (in our area); seeds lacking a peripheral ring or two patches of soft white to brown material (**U** 2, 4) .... 35. *C. congesta* (p. 208)
Flowers blue, blue-mauve, mauve, lilac, lavender, or pale purple (rarely white); shoots not rooting at the nodes; ventral locule seeds borne at the same level as the dorsal locule seeds (when present); spathes densely pubescent to glabrous; seeds with a peripheral ring (or sometimes two patches) of soft white to brown material (widespread) 43

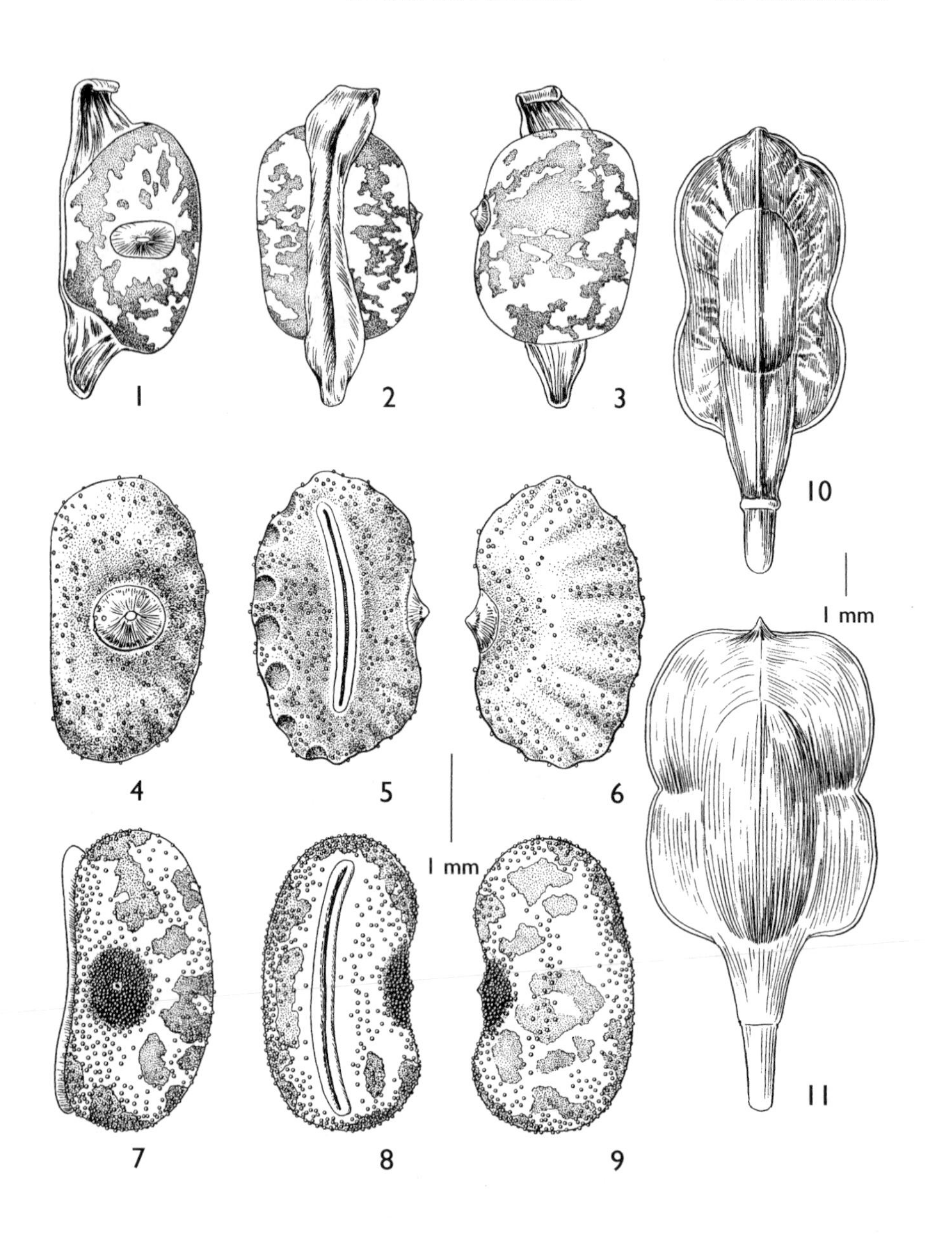

FIG. 14. *COMMELINA* seeds: *C. LUKEI* — **1–3**, ventral locule seed, lateral/ventral/dorsal view; **10**, capsule, dorsal view. *C. IMBERBIS* — **4–6**, ventral locule seed, lateral/ventral/dorsal view; **11**, capsule, dorsal view. *C. MASCARENICA* — **7–9**, ventral locule seed, lateral/ventral/dorsal view. 1–3, 9 from *Luke et al.* 7080; 4–6, 11 from cultivated material of *Gillespie* 4; 7–9 from *Faden & Faden* 77362b. Drawn by A.R. Tangerini, and reproduced with permission from Novon 18: 471 fig. 2 (2008).

43. Leaves lanceolate-elliptic to ovate, to 7 cm long and 2.5 cm wide; spathe surface subglabrous to sparingly pilose, lacking hook-hairs; upper cincinnus often well developed and 1-flowered, its peduncle exserted from the spathe; capsules trilocular, trivalved, with 3 smooth, dehiscent locules . . . . 39. *C. melanorrhiza* (p. 217)
Leaves linear or linear-lanceolate (to elliptic or ovate), to 17 cm long and 4 cm wide; spathe surface usually hirsute-puberulous, including hook-hairs, occasionally glabrous; upper cincinnus usually lacking or vestigial, the peduncle lacking or enclosed in the spathe (very rarely 1-flowered and the peduncle shortly exserted); capsules very variable, trilocular-trivalved, trilocular-bivalved or bilocular-bivalved, the dorsal locule, when present, smooth and dehiscent or muricate and indehiscent 44

44. Plants glaucous; spathes with grey-green or green margins and a white or whitish base with contrasting dark green veins; flowers mauve (rarely white); paired petal limb suborbicular; lateral stamen filaments strongly compressed laterally and ± winged distally; capsules trilocular-trivalved; the dorsal locule always muricate and indehiscent 40. *C. albescens* (p. 218)
Plants not glaucous; spathes green, without contrasting veins; flowers light or dark blue; paired petal limb broadly ovate to ovate-reniform or ovate-deltate; lateral stamen filaments neither laterally compressed nor winged; capsules bi- or trilocular, bi- or trivalved, when trilocular the dorsal cell either smooth and dehiscent or muricate and indehiscent . . . . 37. *C. erecta* (p. 210)

45. Flowers pure white; leaves sometimes maroon-spotted above; capsules 3-seeded, with one dorsal and 2 ventral locule seeds, the ventral locule seeds borne slightly distal to the dorsal locule seed (**K** 5, **Kakamega Forest and vicinity**) 36. *C. albiflora* (p. 209)
Flowers usually blue; leaves entirely green; capsules 1–5-seeded; widespread 46

46. Capsules 1–3-seeded; ventral locules (0–)1-seeded; paired sepals long-fused from the base 47
Capsules 4–5-seeded (occasionally less through abortion); ventral locules 2-seeded; paired sepals usually free or nearly so (except 34. *C. zambesica*) 52

47. Capsules usually 1-seeded, with only the dorsal locule seed developing, the ventral locules usually empty or rarely forming a single seed; dorsal capsule valve ornamented on the sides and often proximally and distally with a variety of tubercles and raised ridges, these often forming longitudinal rows; lateral stamen filaments winged distally; cleistogamous flowers sometimes borne on small shoots from rooted nodes .... 32. *C. forskaolii* (p. 203)
    Capsules 2–3-seeded (1-seeded only by abortion), usually forming 2 ventral locule seeds and sometimes also a dorsal locule seed; dorsal capsule valve, when present, smooth or muricate; lateral stamen filaments not winged; cleistogamous flowers lacking 48
48. Capsules 2-seeded, the dorsal locule not developed 49
    Capsules 3-seeded, the dorsal locule 1-seeded 51
49. Leaves linear to linear-lanceolate, up to 1.7 cm wide, often auriculate at the summit of the sheaths; spathes sometimes hirsute-puberulous; seeds smooth, with a peripheral ring (or sometimes two patches) of soft white to brown material .... 37. *C. erecta* (p. 210)
    Leaves lanceolate-oblong to lanceolate-elliptic or ovate, (1–)2–3.9 cm wide, not auriculate; spathes puberulous proximally or glabrous, never hirsute; seeds smooth or reticulate, lacking a peripheral ring or two patches of soft white to brown material 50
50. Roots thick and fleshy; upper cincinnus lacking or vestigial; seeds with a low, raised orange reticulum (**T** 4) .... 41. *C. disperma* (p. 220)
    Roots thin and fibrous; upper cincinnus often well developed and 1-flowered; seeds uniformly brown, smooth (**U** 2, 4) .... 42. *C. zenkeri* (p. 221)
51. Spathes with margins fused for 2–6 mm; paired petals held in a lateral position, i.e. facing one another; dorsal capsule locule striate when indehiscent ; seeds lacking a peripheral ring or two patches of soft white to brown material, embryotega with a long, pointed apicule .. 38. *C. bracteosa* (p. 214)
    Spathes with margins fused for 7–19 (or more) mm; paired petals held in a dorsal position, i.e. both in the back of the flower; dorsal capsule locule muricate when indehiscent; seeds with a peripheral ring (or sometimes two patches) of soft white to brown material, embryotega with a short conical apicule ... 37. *C. erecta* (p. 210)

52. Spathes subsessile, usually hirsute-puberulous; peduncles 0.2–0.5(–0.7) cm long; seeds foveolate-reticulate, sometimes the depressions radially aligned; cleistogamous flowers sometimes borne on short shoots from the base of the plant . . . . 31. *C. benghalensis* (p. 200)
Spathes pedunculate, puberulous or hirsute-puberulous; peduncles (0.5–)0.7–4(–8) cm long; seeds various but not foveolate-reticulate; cleistogamous flowers lacking 53
53. Sheaths auriculate at the summit; leaf midrib glabrous above; roots thick, sometimes tuberous; seeds brown to nearly black, usually with slightly raised, irregular, orange or orange-yellow (rarely red) lines sometimes forming an irregular reticulum . . . . 34. *C. zambesica* (p. 205)
Sheaths not auriculate; leaf midrib usually variously puberulous above; roots mostly thin and fibrous, rarely thick (in *C. schliebenii*) or tuberous; seeds variable but not as above 54
54. Capsules ± square; seeds subspherical to very shortly ellipsoid 55
Capsules oblong; seeds ellipsoid 57
55. Seeds with testa verrucose to echinulate; spathes tending to spread at right angles to the stem in fruit; lower petal blue, concolorous with the paired petals . . . . 24. *C. eckloniana* (p. 182)
Seeds with testa nearly smooth to alveolate; spathes not spreading at right angles to the stem in fruit; lower petal blue or white, slightly to strongly contrasting with the paired petals 56
56. Spathes hirsute-puberulous, 1–1.7 cm high; margins fused for 5–12 mm; internodes and sheaths hirsute; lower petal blue, slightly paler than the paired petals . . . . 33. *C. schliebenii* (p. 204)
Spathes puberulous, 0.5–1.4 cm high; margins fused for 2–5 mm; internodes glabrous, sheaths puberulous; lower petal white or very pale blue, strongly contrasting with the paired petals . . . . 23. *C. latifolia* (p. 180)
57. Capsules with an apical extension, rounded to truncate; seeds appendaged at both ends 58
Capsules without an apical extension, truncate to emarginate or retuse; seeds not appendaged 59
58. Annuals; spathes 1–1.9 cm long, peduncles 0.5–2.1(–2.5) cm long; upper cincinnus lacking; leaf margin strongly undulate; leaves puberulous on both surfaces, including only colourless hook-hairs on the adaxial leaf midrib (20× magnification required); flowers 8–15 mm wide . . . . 28. *C. kotschyi* (p. 194)
Perennials; spathes (1.4–)1.8–2.5 cm long, peduncles (1.6–)2–4(–5.5) cm long; upper cincinnus 1–3-flowered; leaf margin slightly undulate; leaves puberulous only on the abaxial surface, adaxial leaf midrib with whitish acicular hairs only, otherwise glabrous; flowers (13–)15–21 mm wide . . . . 29. *C. lukei* (p. 195)

59. Lower leaves petiolate; upper cincinnus 1(–2)-flowered; seeds with 4–5(–8) transverse tuberculate ribs; not farinose .... 27. *C. petersii* (p. 192)
Lower leaves with bases cuneate to rounded or cordate, not petiolate; upper cincinnus lacking, vestigial or 1(–2)-flowered; seeds smooth to faintly alveolate or weakly to strongly radially ridged, not tuberculate; white-farinose 60

60. Upper cincinnus usually lacking (rarely present and 1-flowered in **U** 1 & 3 and **T** 2); seeds weakly to strongly radially ridged, uniformly dark brown to nearly black (or rarely medium brown); upper sepal 2.7–3 mm long; lower petal white or bluish white, contrasting with the paired petals; filaments yellow or yellowish white .... 25. *C. imberbis* (p. 187)
Upper cincinnus 1(–2)-flowered; seeds smooth to faintly alveolate, mottled different shades of grey and/or brown; upper sepal 3.5–5 mm long; lower petal concolorous with the paired petals; filaments wholly or partly lavender to blue .... 26. *C. mascarenica* (p. 189)

1. **Commelina diffusa** *Burm. f.*, Fl. Ind. 18, t. 7, fig. 2 (1768); Merrill in J. Arnold Arb. 18: 64 (1937); Morton in J.L.S. 55: 521, fig. 18 (1956) & in J.L.S. 60: 181, fig. 3 (1967); Brenan in F.W.T.A., 2nd ed.: 3: 47, fig. 332 (1968); Obermeyer & Faden in F.S.A. 4(2): 26, fig. 5 (1985) pro parte; Ogwal in Mitt. Inst. Allg. Bot. Hamburg. 23b: 577 (1990); Faden in U.K.W.F. 2nd ed.: 304 (1994); Kayemba–Ogwal in Proc. X11th AETFAT, Malawi 1: 416 (1994); Ensermu & Faden in Fl. Eth. 6: 362, fig. 207.13 (1997); Wood, Handbook Yemen Fl.: 317 (1997); Faden in Fl. N. Amer. 22: 194 (2000) & in Taxon 52: 831 (2003). Type: India, Coromandel, *Burman* s.n. (G!, holo.)

Perennials or annuals; roots thin, fibrous; shoots decumbent to repent, regularly rooting at the nodes, much branched, with a line of very short uniseriate hairs descending from the line of fusion of the distal sheath or ± glabrous. Leaves distichous; sheaths to 2.5 cm long, with a line of uniseriate hairs along the fused edge and elsewhere glabrous or pilose, white-ciliate at apex; lamina (sub-)sessile, ovate to narrowly lanceolate, 1.5–8.5 × 0.6–2.7 cm, base oblique, cuneate to rounded, margins scabrid distally, apex acute to acuminate or obtuse, sometimes mucronulate; surfaces glabrous to pilose or rarely puberulous, occasionally scabrid adaxially. Spathes solitary; peduncles 0.7–2.5(–3) cm long, glabrous or with a line of hairs; spathes 0.8–4 × (0.3–)0.4–1.3 cm, not or slightly falcate, rarely strongly falcate, base cordate, margins free, apex acute to acuminate; usually completely glabrous, occasionally ciliolate and/or puberulous or pilose; upper cincinnus often exserted from the spathe, producing 1–5 male (occasionally 1 bisexual) flowers; lower cincinnus 2–7(–10)-flowered, the flowers mostly bisexual, occasionally functionally male; bracteoles lacking. Flowers bisexual and male, 10–15 mm wide. Sepals hyaline white, 2.5–3.7 mm long, paired sepals fused proximally for ± $^1/_3$ their length, 2.5–4 × 1.5–2.5 mm; all 3 petals usually concolorous and blue, or occasionally mauve to lilac, rarely the lower petal somewhat discolorous, paired petals 4.5–9 × 4–7 mm of which the claw 2–3 mm long, medial petal somewhat cup-shaped, 3–4.5 × 3–6 mm; staminodes 2 or 3, the medial often vestigial or lacking; lateral stamens with filaments 4–7.5 mm long, anthers 0.7–1.7 mm long, pollen golden yellow, medial stamen with filament 3.5–5.5 mm long, anther 1–2.1 mm long, pollen golden yellow. Ovary 1–1.6 mm long; style 4–5 mm long; stigma capitate-deltate. Capsules ± pale yellow or

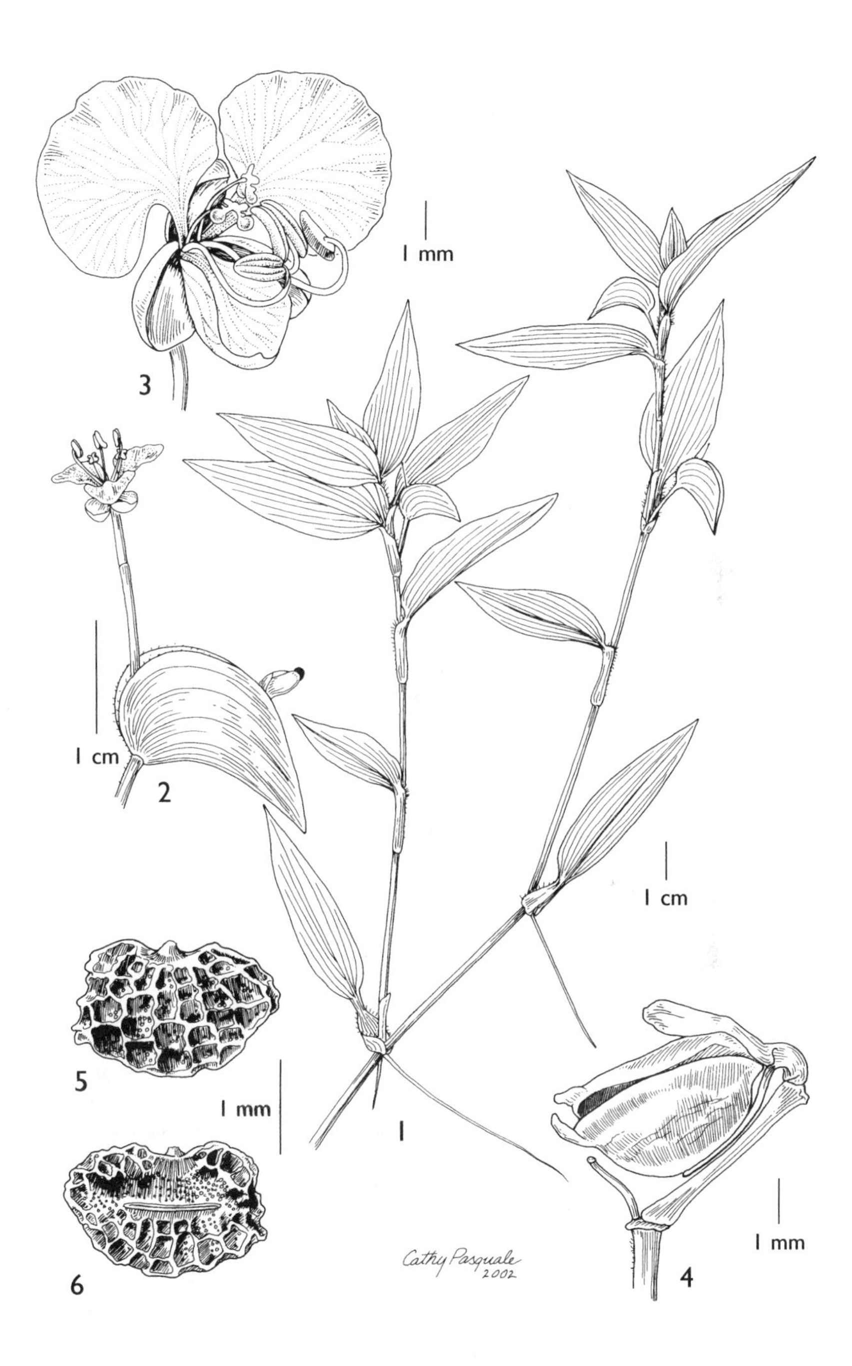

FIG. 15. *COMMELINA DIFFUSA* — **1**, habit; **2**, spathe with male flower in upper cincinnus; **3**, male flower; **4**, capsule, side view; **5** & **6**, ventral locule seed, dorsal and ventral view. 1 from *Prévost* 2714; 2, 4–6 from *Faden* 76/1; 3 from *Faden* 76/325. Drawn by Cathy Pasquale.

heavily spotted with dark brown, trilocular, bivalved, the valves persistent, 5-seeded, oblong-ellipsoid in dorsal view, (4.7–)5–6(–7) × 2.5–3 mm, slightly contracted between the seeds, rostrate or not rostrate, dorsal locule 1-seeded, indehiscent, ventral locules 2-seeded. Ventral locule seeds broadly ovate to elliptic, dorsiventrally compressed, 1.8–3.1 × 1.5–2 mm, testa medium to dark brown, reticulate or rarely foveolate-reticulate or smooth to alveolate; dorsal locule seed dorsiventrally compressed or elongate-hemispheric, oblong-ellipitc to elliptic or reniform, (2.45–)3–4.1 × 1.8–2.2 mm. Fig. 14, p. 141.

subsp. **diffusa**

Leaves ovate to linear, 1.5–8.5 cm long; upper cincinnus often well developed and long-exserted from the spathe, 1–5-flowered; flowers blue or occasionally mauve to lilac; seeds reticulate or rarely foveolate-reticulate; plants usually associated with water or a weed in cultivation; perennials.

UGANDA. Bunyoro District: 20 km S of Victoria Nile on road to Masindi, 2 Oct. 1962, *Lewis* 6004!; Kigezi District: Kachwekano Farm, Jan. 1950, *Purseglove* 3229!; Mengo District: near Kitamiro, 26 Sep. 1949, *Dawkins* D. 405!

KENYA. Northern Frontier District: South Horr, beside Korungwe River, 8 km N of town, 14 Nov. 1978, *Hepper & Jaeger* 6796!; Nairobi District: Nairobi, Langata, Forest Edge Road, 6 Feb. 1977, *Faden, Faden & Ng'weno* 77/286!; Central Kavirondo District: 1 km SE of Ukwala, 8 Oct. 1971, *Gillett* 19347!

TANZANIA. Lushoto District: Magoma, NE of Korogwe, 2 Aug. 1978, *Archbold* 2392!; Iringa District: Dabaga Highlands, near Idete School, 88 km SE of Iringa, 31 Jan. 1971, *Mabberley* 634!; Rungwe District: Itungi Port, Lake Nyasa, ferry terminal, 28 June 1996, *Faden et al.* 96/450!

DISTR. **U** 1–4; **K** 1, 3–7; **T** 1–4, 6–8; **Z**, **P**; pantropical and warm temperate

HAB. Wet places such as swamps, marshes, edges of pools, streams, rivers, lakes and irrigation channels, flood plains, moist grassy areas, banks and secondary vegetation, also in forest clearings, edges, paths and roadsides; often a weed in cultivation, including rice, bananas and tree plantations; clayey or sandy soils; 40–2100(–3400) m

Flowering specimens have been seen from all months.

SYN. *Commelina nudiflora* L., Sp. Pl. 41 (1753) pro parte, not as to lectotype (see Merrill in J. Arnold Arb. 18: 64 (1937)); C.B. Clarke in DC., Monogr. Phan. 3: 144 (1881); Schumann in P.O.A. C: 134 (1895); Rendle in Cat. Afr. Pl. Welw. 74 (1899); C.B. Clarke in F.T.A. 8: 36 (1901)

C. sabatieri C.B. Clarke in DC., Monogr. Phan. 3: 146 (1881) & in F.T.A. 8: 37 (1901). Type: Sudan? ("Voyage aux sources du Nil-Blanc"), 1842, *Sabatier* s.n. (P!, holo., K, photo!)

C. aquatica Morton in J.L.S. 55: 515, fig. 12 (1956). Type: Ghana, by Ada road, 25 km from Accra, *Morton* A30 (holotype not designated; K!)

C. diffusa Burm. f. subsp. *scandens* (C.B. Clarke) Obermeyer in Bothalia 13:437 (1981); *sensu* Obermeyer & Faden in F.S.A. 4(2): 26, (1985) pro parte maj., not as to type

C. diffusa Burm. f. subsp. *aquatica* (J.K. Morton) Ogwal in Proc. XIIIth Plenary Meeting AETFAT, Malawi, 417 (1994)

NOTE. The occurrence of this species at 3400 m on Mt Elgon (Kenya, Kimili track, open *Erica* stand, *Wesche* 1819), some 1300 m above the next highest elevation, is remarkable. I know of no other location in its worldwide distribution where it approaches this altitude. Normally, the only *Commelina* species to be expected at very high elevations in Africa is a variety of *C. africana*, which this specimen certainly is not. The only unusual morphological feature of *Wesche* 1819 is the ciliolate spathe margin, which is uncommon in our area.

subsp. **montana** *J.K. Morton* in J.L.S. 60: 181 (1967); Brenan in F.W.T.A., 2nd ed.: 3: 47 (1968); Kayemba-Ogwal in Proc. X111th AETFAT, Malawi 1: 417 (1994); Faden in U.K.W.F. 2nd ed.: 304 (1994). Type: Cameroon, Bamenda area, path to Lake Bambelue from Santa road, *Morton* K 281 (K!, lecto., selected here)

Leaves mostly ovate, 1.5–4 cm long; upper cincinnus often abortive or, at most, shortly exserted from the spathe and 1-flowered; flowers mauve to lilac; seeds smooth or alveolate; plants of forest edges and paths; annuals?

UGANDA. Mengo District: Mabira Forest, 1–2 km E of Kiwala, 13 Apr. 1969, *Lye* 2476!

KENYA. North Kavirondo District: Kakamega Forest, along Kubiranga Stream at N end of forest, 16 Mar. 1977, *Faden & Faden* 77/850! & Kakamega Forest, 9 Dec. 1956, *Verdcourt* 1650!
DISTR. **U** 4; **K** 5; Nigeria, Cameroon (incompletely known, see note)
HAB. Forest paths and roadside at forest edge; 1200–1550 m
Flowering specimens have been seen from March, April and December.

SYN. *Commelina sabatieri* of Ogwal, Taxon. study Commelina in Uganda, M.Sc. Thesis: 38, fig. 5 (1977), not of Clarke (1881)

NOTE. Morton (1967) defined this taxon on the basis of the following characters: long hairs on the sheaths, spathes and peduncles, peduncles longer than the spathes and leaves ovate. Brenan (1968) came up with additional characters: plants annual, leaf bases sessile and cordate or subcordate, and seeds smooth (as opposed to reticulate). In his key he chose to deemphasize the sheath pubescence of Morton, indicating that subsp. *montana* typically did not have sheaths that were pubescent all over.

Ogwal (1977) recognized *C. sabatieri* as a distinct species, basing it on a plant with smooth seeds, which agrees with *C. diffusa* subsp. *montana*, but not with Clarke's original description of *C. sabatieri*. She based it on *Chandler* 462, which I have not examined. In the type specimen of *C. satabieri*, the seeds have a reticulate testa, as in typical subsp. *diffusa*.

In our area plants from the Kakamega Forest in W Kenya and from the forested parts of **U** 2 and **U** 4 bear a strong resemblance to West African *C. diffusa* subsp. *montana*, but only three such collections have been observed with seeds. The seeds vary from smooth to alveolate. Plants very similar to these have also been encountered in the forest area of SW Ethiopia (*Faden & Faden* 2003/32). These however had typical *C. diffusa* reticulate seeds. Therefore, plants from our area that lack seeds are being treated as *C. diffusa* subsp. *diffusa* in the absence of more conclusive evidence that they are subsp. *montana*.

The three collections that we are treating as subsp. *montana* have mauve to lilac flowers, so it is unknown whether blue-flowered plants of this variety occur in our area, but it would not be unexpected because the plants from the Cameroon highlands, that include the type, all had blue flowers.

Purseglove 3229 from **U** 2 has foveolate-reticulate seeds. While the testa pattern is somewhat intermediate between the reticulate seeds of typical plants of subsp. *diffusa* and the alveolate seeds in some plants of subsp. *montana* the overall morphology of subsp. *montana* is ill-defined so as to make it impossible to state whether *Purseglove* 3229 is intermediate in other characters too. The capsule of *Pursglove* 3229 was measured at 7 mm in length, the longest found, but that included a 1 mm long beak.

NOTE (on the species as a whole). The dorsal locule seed is normally not attached to the capsule wall and is easily removed. Indeed in some old spathes the dorsal locule may even give the appearance of being dehiscent, which might account for Ogwal (1990) having described all three locules as dehiscent in *C. diffusa*. However, I have only seen one capsule in which dehiscence may have occurred. At the other extreme, *Bax* H (from **T** 1) has the capsule wall closely adherent to the dorsal locule seed so that it has to be removed in small pieces and it would be extremely time consuming to remove in its entirety. Concomitant with this, the seed in that locule is completely smooth, in contrast with the normal reticulate seeds in the ventral locules. In populations in which the dorsal locule seed is easily removable, its testa has the same pattern as the ventral locule seeds. At this time we know too little about the distribution of these different patterns of seed development and release in the dorsal locule to attempt to use it taxonomically, but it could make an interesting study in the field. *Bax* H is a very large, robust specimen of *C. diffusa* subsp. *diffusa*. Unfortunately this species appears to have one of the lowest percentages of specimens with seeds among the *Commelina* species that I have studied.

From a series of well pressed flowers in a packet on *Hepper & Jaeger* 6796 (from **K** 1) it was clear that, at least for this population, the male flowers in the upper cincinnus had much longer pedicels and stamen filaments than the bisexual flowers of the lower cincinnus. The spirit collection for *Verdcourt* 1650 initially seemed to demonstrate the same type of dimorphism, but on closer examination the spathes with the male flowers in the upper cincinnus were likely collected at a later time – they had started to fade whereas the bisexual flowers are still at anthesis – and from a different population – the specimen does not show such a well developed upper cincinnus – and thus a fair comparison cannot be made. All of the *Hepper & Jaeger* pressed flowers are consistent in having only two fully developed staminodes. How frequent three fully developed staminodes are in our area remains to be determined, but in *Faden, Faden & Ng'weno* 77/286 from **K** 4 it is recorded that the staminodes were either 2 or 3 in the flowers and *Archbold* 2392 from **T** 3 records 3 staminodes in this population.

Plants of *C. diffusa* with long, narrow spathes and narrow leaves, which occur throughout Africa and well beyond, present a special problem. In West Africa they were described as *Commelina aquatica* J.K. Morton by Morton (1956) and later changed to *Commelina diffusa* subsp. *aquatica* (J.K. Morton) Ogwal (1994). Ogwal (1994) accepted subsp. *aquatica* for the flora of Uganda. In F.W.T.A. *C. aquatica* was considered part *of C. diffusa* subsp. *diffusa*. In F.S.A such plants were treated as *C. diffusa* subsp. *scandens* (C.B. Clarke) Oberm., which was based on *Commelina scandens* C.B. Clarke. Here we are following F.W.T.A. not because we believe it is necessarily correct but because we are unhappy with the alternatives. *C. aquatica* does not separate from *C. diffusa* on any important characters, and subsp. *scandens* is based on a misunderstanding of *C. scandens* (cf. Faden in Taxon 52: 831 [2003]). Non-African material has not been studied, with the exception of some specimens from Madagascar (Faden, loc. cit.).

In our area, plants with the narrowest leaves and long, narrow spathes. e.g. *Kabuye et al.* in TPR 735 (**K** 7), *Archbold* 1211 (**T** 3) and *Vollesen* 2672 (**T** 8), sometimes stand out, and are all from low elevations. Other collections, however, tend to blur the lines with subsp. *diffusa*, so that the separation of it from a narrow spathe form would seem entirely arbitrary and without sufficient justification.

Two specimens from our area, *B.D. Burtt* 1731 [Kondoa District, Simbo Hills, between Jogose and Mangoloma; flowers clear blue] and *Pirozynski* P217 [Kigoma District, Gombe Stream Reserve, Kakombe Valley; flowers violet blue] with large leaves with cordate bases and free-margined spathes at first appeared to represent a distinct species related to *C. diffusa*. In addition to their large, unusual looking leaves – to 11 cm long and 2 cm wide in the *Burtt* collection and up to 10.5 cm long and 3.2 cm wide in the *Pirozynski* collection, both longer than in any collection of *C. diffusa* – the leaf sheaths were also longer than those in *C. diffusa* (up to 3 cm long in both collections). Moreover, in both collections the spathes were glabrous outside but pilose inside – a combination of characters not seen in *C. diffusa* – and the spathe margins in both collections were ciliolate proximally, a rare character for *C. diffusa* in our area. Both collections lack capsules and seeds and useful floral remains. Upon further reflection these characters all make sense if it is the flower colour on the label that is incorrect and should have been yellow. All the unusual characters for *C. diffusa*, including the habitat, make sense if the species is instead *C. africana*. In that case the plants would clearly be *Commelina africana* var. *lancispatha*. In all likelihood, the collectors misremembered the flower colour when they wrote up their notes.

2. **Commelina scandens** *C.B. Clarke* in DC., Monogr. Phan. 3: 146 (1881); Rendle, Cat. Afr. Pl. Welw. 2: 75 (1899); C.B. Clarke in F.T.A. 8: 38 (1901); Faden in Taxon 52: 831 (2003). Syntypes: Angola, Pungo Andongo, banks of the Cuanza River, near Nbilla, *Welwitsch* 6642 (BM!, lectotype of Faden, loc. cit., 832) & Madagascar, Nosy Be [Nossi-Bé], *Boivin* 2009 (P!, syn.)

Robust, straggling, sprawling or creeping perennial, rooting at the lower nodes; roots thin, fibrous; shoots 1.2–3 m long, much branched, to 8 mm thick, glabrous. Leaves apparently spirally arranged; sheaths sometimes suffused with purple, mostly 1.5–4 cm long, sparsely puberulous along the line of fusion, ciliate at apex, otherwise glabrous; lamina sessile, linear to linear-lanceolate, 3–14 × (2.5)4–12(–18) mm, base cuneate to attenuate, margins scabrid, apex acuminate to attenuate; surfaces glabrous or rarely the upper scabrid or the lower pilose on the midrib. Spathes solitary, peduncles (1–)1.6–5(–7.5) cm long, puberulous or occasionally glabrous; spathes green or occasionally purple-veined or suffused with purple, (2.5–)3–6.5(–8) × 4–10(–13) mm, falcate or not, base rounded to cordate, margins free, scabrid distally, rarely ciliate proximally, apex attenuate, surfaces glabrous, rarely sparsely puberulous; upper cincinnus long-exserted, 1–4-flowered, the flowers bisexual or male, lower cincinnus 5–7-flowered. Flowers bisexual and male, pedicels 2.5–5.5(–6.5) mm long, puberulous at least at the apex. Sepals free, ± glabrous, upper sepal 4–5.5 mm long, lower sepals 3.5–5.5 mm long; all 3 petals concolorous, bright blue, rarely pale blue, the paired petals 10.5–11 × 9.5 mm of which the claw 3.5–5 mm; lower petal broadly ovate, ± 5.5–7 mm long; staminodes 3 or 2 and one vestigial; lateral stamens with filaments ± 6–8 mm long, anthers (1–)1.5–1.9 mm long; medial stamen with filament ± 5.5–7 mm long, anther (1.6–)2–2.5 mm long, connective purple or black, sometimes with yellow basal lobes. Ovary ± 2 mm long;

style 6–10 mm long. Capsules ± grey-green, trilocular, bivalved, oblong-ellipsoid, 7–9 × 3.5–4 mm, apiculate, 5-seeded, dorsal valve sometimes deciduous, dorsal locule indehiscent (although the capsule may split both above and below it), 1-seeded, ventral locules 2-seeded. Dorsal locule seed fused to the capsule wall, testa (when capsule wall is removed) ± smooth, brown; ventral locule seeds transversely ellipsoid or oblong-ellipsoid, 2.7–3.2(–3.7) × 1.7–2 mm, testa dark brown with a low, raised, lighter brown reticulum and often a suggestion of a finer reticulum within each areole, also a longitudinal, middorsal keel present.

TANZANIA. Songea District: Johannsbruck, at Rovuma River, 10 Mar. 1956, *Milne-Redhead & Taylor* 9083!
DISTR. **T** 8; Congo–Kinshasa, Angola, Zambia, Botswana, Namibia
HAB. Wet riverside vegetation; 900 m
Flowering in March.

SYN. *Commelina vanderystii* De Wild. in B.J.B.B. 3: 268 (1911). Type: Congo–Kinshasa: Wombali, Nov. 1910, *Vanderyst* s.n. (BR!, holo.)
C. diffusa Burm. f. var. *scandens* (C.B. Clarke) Oberm. in Bothalia 13:437 (1981) pro parte, as to type, but not as to a majority of the collections; Obermeyer & Faden in F.S.A. 4, 2, 26 (1985) pro parte, as to type, but not as to a majority of the collections

NOTE. South of our area this species grows in swamps and wet ditches, but mainly along rivers, sometimes in deep water. Whether this species is always perennial in uncertain. No definite base has been collected or remarked upon, but *C. scandens* is likely to be a diffuse spreader. In view of its growing in places that would tend to be permanently wet, it is likely to be perennial all or most of the time.

This species closely resembles and was confused with a narrow-leaved and narrow-spathed form of *C. diffusa* in Flora of Southern Africa. All of the specimens seen at Kew from South Africa were *C. diffusa*, and only two collections from Namibia and one from Botswana in that Flora were *C. scandens*. *Commelina scandens* may be separated from *C. diffusa* by the former's larger capsules and seeds, the seeds with a middorsal keel and much less farinose, the antherodes with blue to purple or maroon connectives, and the pubescent pedicels.

Commelina scandens has sometimes been confused with *C. macrospatha* Mildbr., which was reported from Uganda and Kenya by Morton (in J.L.S. 60: 188 (1967)) and Brenan (F.W.T.A. 2, 3: 47 (1968)) but omitted from the account of the genus in Uganda by Ogwal (Mitt. Inst. Allg. Bot. Hamburg 23b: 573–592 (1990)). No specimen so-named from Kenya could be located at Kew, but of the three collections filed as *C. macrospatha*, and so annotated by Brenan, from Uganda, two were found to be narrow-spathed, narrow-leaved forms of *C. diffusa* and the third was a narrow-leaved form of *C. africana*, with the flower colour wrongly stated by the collector. Although *C. macrospatha* closely resembles *C. scandens* in herbarium specimens, the former is a grassland or woodland species, occasionally occurring in moister situations or in cultivation, that ranges from Sierra Leone to Cameroon and possibly Congo–Kinshasa, if *C. vanderystii* is correctly placed in synonymy. It differs from *C. scandens* by its rhizomatous base, shoots that do not root at the nodes, usually pubescent spathes (at least in Nigeria and Cameroon), usually a solitary male flower in the upper cincinnus and three bisexual flowers in the lower cincinnus, subequal petals, very reduced, white antherodes, and especially by its seeds, which are completely and densely covered by white farinaceous material, even in immature seeds.

Madagascan plants that have been called *C. scandens*, including the syntype *Boivin* 2009, differ in having smaller spathes on shorter peduncles, smaller capsules and seeds, and seeds that lack a middorsal ridge. They are *Commelina diffusa*, not *C. scandens* as it is being used here.

3. **Commelina acutispatha** *De Wild.*, Pl. Bequaert. 5: 164 (1931); Faden in K.B. 62: 139 (2007). Types: Congo–Kinshasa, near Mobwasa, *Reygaert* 538 (BR!; lectotype of Faden, loc. cit.) & Yambuya, *Bequaert* 1337 (BR, syn.)

Scrambling, straggling or sprawling perennial, rooting from the nodes; roots thin, fibrous, very long when produced from stems well above the ground; shoots to 1.5 m tall and 2 m or more in length, glabrous to puberulous. Leaves distichous; sheaths mostly 1.5–2.5 cm long, with a line of hairs along the fused edge, otherwise glabrous to puberulous or patently pilose, ciliate or ciliolate at apex; lamina sessile, linear-lanceolate to oblong-lanceolate, (3.5–)5.5–10.5(–13) × (0.9–)1.2–1.7(–2) cm, base oblique and rounded to cuneate, margins scabrid, sometimes ciliate at base, apex attenuate; surfaces pilose, at least on the nerves, or the upper glabrous. Spathes solitary; peduncles 1.5–3 cm long, puberulous at least in a longitudinal line; spathes green, (2.5–)3–6.2 × 1–1.6 cm, usually slightly falcate, base cordate, margins free, scabrid and ciliate, apex attenuate, surfaces pilose outside and inside; upper cincinnus usually present, included (in our area) or shortly exserted, 1–2(–3)-flowered, the flowers sometimes bisexual, lower cincinnus 4–7-flowered. Flowers bisexual and male, ± 15 mm wide. Sepals 3.5–4.5 mm long, the paired sepals apparently free; paired petals white to very pale lilac, mauve, purple or violet, 7–9 × 5 mm of which the claw ± 3–4 mm, lower petal broadly ovate, 3–4 × 2.5 mm, concolorous; staminodes 2, the medial apparently lacking; lateral stamens with filaments 5–5.5 mm long, anthers 1.2–1.5 mm long; medial stamen with filament ± 4.5–5 mm long, anther 1.8–2.3 mm long, connective apparently with a large, blue patch; style ± 5–7 mm long. Capsules tan or greenish tan, often with a few, scattered, maroon flecks, trilocular, bivalved, oblong-ellipsoid, 5.5–6 × 4 mm, 5-seeded (or fewer through abortion), rostrate, dorsal locule indehiscent, 1-seeded, ventral locules dehiscent (1–)2-seeded. Seeds transversely oblong-elliptic in outline, 2–2.5 × 1.5–1.9 mm, ± dorsiventrally compressed, testa smooth to alveolate, dark brown and faintly to distinctly mottled with lighter brown, or reddish brown and not mottled (the latter not fully mature?), not farinose.

UGANDA. Toro District: near Kirmia, W of Ntandi in Semliki Forest, *Faden* 69/1263!
DISTR. U 2; Sierra Leone, Liberia, Ghana, Togo, Nigeria, Cameroon, Gabon, Congo–Kinshasa
HAB. Forest edge and forest with *Cynometra*; 700 m

SYN. *Commelina thomasii* Hutch. in K.B. 1939: 243 (1939) & in F.W.T.A. 1, 2: 318 (1936), English descr. only; Morton in J.L.S. 55: 528, fig. 26 (1956) & in J.L.S. 60: 189 (1967); Brenan in F.W.T.A. 2, 3: 47 (1968); Ogwal, Taxon. study Commelina in Uganda: 36, fig. 4 (1977). Type: Sierra Leone, Yonibana, *Thomas* 5205 (K!, holo.)

NOTE. This is a common, weedy species in the forested regions of Cameroon and parts of West Africa. It should be widespread in Central Africa, but few specimens have been seen other than from Cameroon. The species occurs chiefly in disturbed, secondary vegetation, often straggling through other plants in moist situations. In West African specimens the flowers have sometimes been reported as blue, but I think that this is in error, and Brenan in F.W.T.A. appropriately ignored them. Morton (1956) reported unusual antherodes with reduced lobes in plants from Ghana. I have been unable to confirm that for plants from other areas, including FTEA. Morton (1967) reported hybrids between this species and *C. diffusa* from Ghana and Cameroon, but they would seem to be rare.

Living material of the single collection from our area was cultivated in Nairobi until it produced very pale pink flowers and smooth seeds, so the initial determination was confirmed.

4. **Commelina africana** *L.*, Sp. Pl.: 41 (1753); C.B. Clarke in DC., Monogr. Phan. 3: 164 (1881); Schumann in P.O.A. C: 135 (1895); Rendle in Cat. Afr. Pl. Welw.: 76 (1899); C.B. Clarke in F.T.A. 8: 45 (1901); Morton in J.L.S. 55: 515, fig. 11 (1956); Brenan in Mitt. Bot. Staats. München 5: 199 (1964); Morton in J.L.S. 60: 174 (1967); Brenan in F.W.T.A., 2nd ed.: 3: 45 (1968); Schreiber et al. in F.S.W.A. 157: 6 (1969); Faden in Monocot Weeds 3: 100 (1982); Obermeyer & Faden in F.S.A. 4(2): 28, fig. 5 (1985) pro parte; Blundell, Wild Fl. E. Afr.: 412 (1987); Ogwal in Mitt. Inst. Allg. Bot. Hamburg. 23b: 577 (1990); Faden in U.K.W.F. 2nd ed.: 304, t. 135 (1994); Kayemba-Ogwal in Proc. X111th AETFAT, Malawi 1: 418 (1994); Faden in Fl. Somalia 4: 89 (1995); Ensermu & Faden in Fl. Eth. 6: 362 (1997); Wood, Handbook Yemen Fl.: 317 (1997). Type: cultivated in Hort. Upsal., Herb. Linnaeus 65.3 (LINN!, lecto. of Brenan, 1964)

Perennial herb; roots thick, fibrous; shoots prostrate to decumbent, sometimes rooting at the nodes and often forming mats, or ascending or straggling, usually much branched, with a line of pubescence or glabrous, often tinged with red. Leaves distichous; sheaths to 3 cm long, surface with a line of uniseriate hairs along the fused edge or pilose all over, apex white-ciliate; lamina sessile or rarely petiolate, linear-lanceolate to elliptic or ovate, 0.1–15 × 0.7–3.4 cm, base cuneate to rounded or cordate, margins scabrid distally, often ciliate basally, apex obtuse to acuminate; surfaces glabrous to pilose. Spathes solitary, usually slightly falcate, less commonly not falcate or strongly falcate, the spathe halves sometimes spreading at anthesis; peduncles 0.8–4 cm long; spathes rarely violet, (0.8–)1–6 × (0.3–) 0.5–1.7 cm, base rounded to cordate, margins free, often ciliate proximally, usually scabrid distally, apex acute to acuminate, outer surface glabrous to densely pilose or puberulous; upper cincinnus sometimes well developed and producing 1(–2) male flower(s), sometimes lacking or vestigial, lower cincinnus (1–)2–8-flowered; bracteoles present. Flowers bisexual and male, usually 8–10 mm wide (sometimes wider). Sepals ovate, 2–3.5 × 1.2–2 mm, glabrous or sparsely pilose, paired sepals fused at the base, shallowly cup-shaped; paired petals erect or lateral, yellow or occasionally orange or orange-yellow, rarely creamy white, broadly ovate to broadly ovate-reniform, 5–5.5(–8) × 5–5.5 mm of which the claw 2–2.5(–3) mm long; medial petal cup-shaped or flat, usually not clawed, ovate, 2.5–4 × 1.5–2 mm, concolorous with the paired petals or a little paler; staminodes 2 or 3, the medial fully developed or lacking; stamen filaments fused proximally, 4–7 mm long, anthers 0.8–1.4 mm long; medial stamen filament 3.2–5 mm long, 1.1–1.8 mm long. Ovary 1–1.5 mm long; style 4–7 mm long; stigma capitate. Capsule light or dark brown, with or without darker brown spots or reddish brown mottling, trilocular, bivalved, (1–)3-seeded, the basal ventral ovules (and sometimes the apical ventral ovules) aborting, oblong-ellipsoid to oblong, 4.5–6.5(–6.8) × 2–4 mm, rostrate or not rostrate. Dorsal locule seed fused to the capsule wall, elongate-hemispheric, sometimes slightly dorsiventrally compressed, 2.6–4.4 × 1.7–2.5 mm, testa dark brown, usually smooth (to alveolate); ventral locule seeds broadly ovoid to oblong-rectangular, dorsiventrally compressed, 1.9–2.8 × 1.2–1.8 mm, testa foveolate to foveolate-reticulate or alveolate, dark brown.

1. Leaves often petiolate or subpetiolate, 2–7.5 cm long, pubescent; spathes with a strong vein on both sides of the midrib bearing long hairs, the rest of the spathe subglabrous to densely pilose or puberulous with shorter hairs; coastal and subcoastal Kenya and Tanzania and Zanzibar .... a. subsp. *zanzibarica*
   Leaves sessile, 1.5–15 cm long, glabrous or pubescent; spathes usually lacking strong veins on either side of the midrib, glabrous or pubescent; widespread but absent from coastal and subcoastal Kenya and Zanzibar (subsp. *africana*) 2
2. Leaves lanceolate or oblong-lanceolate to elliptic or ovate, the largest ones usually 0.8–3.4 cm wide 3
   Leaves linear-lanceolate, up to (0.8–)1 cm wide 5
3. Leaves normally up to 6(–7) cm or less in length; spathes mostly 1.5–3.5 times as long as wide, 1.2–4 cm long; lower cincinnus (1–)2–3(–4)-flowered; pedicels 1.5–3 mm long 4
   Leaves (at least the larger ones) 7–15 cm long; spathes 1.5–5 times as long as wide, 2–6 cm long; lower cincinnus (3–)4–8-flowered; pedicels 3.5–6 mm long . b3. var. *lancispatha*
4. Lamina glabrous or nearly so (excluding marginal cilia and sometimes a line of hairs on the abaxial midrib) b1. var. *africana*
   Lamina ± densely pubescent .... b2. var. *villosior*
5. Lamina glabrous or nearly so (excluding marginal cilia) b4. var. *glabriuscula*
   Lamina ± densely pubescent .... b5. var. *milleri*

subsp. **zanzibarica** *Faden* **subsp. nov.** a subsp. *africana* foliis lamina saepe petiolata vel subpetiolata apice longiacuminata spathis nervo prominenti longitudinali trichomata longa ferenti utroque costae posito differt. Type: Kenya, Kwale District: Jadini Hotel, 30 km S of Mombasa, *R.B. Faden & A. Evans* 70/442 (K!, holo.; EA!, iso.)

Trailing herb rooting at the nodes; leaves often petiolate to subpetiolate, narrowly lanceolate to lanceolate (or the smaller leaves ovate), 2–7.5 × 0.7–1.8 cm, apex acuminate to long-acuminate (acute in the smaller leaves), both surfaces pilose; spathes (1.6–)2–3(4.6) × (0.6–)0.8–1.1(1.5) cm, with a strong vein on both sides of the midvein forming a pouch at the base of the spathe, the strong veins with very long hairs, normally much longer than those elsewhere on the surface; spathe also densely pilose inside in a characteristic pattern of a band on each side starting at the base and curving forward and downward until they meet at the midvein; upper cincinnus peduncle 4–10 mm long, glabrous; lower cincinnus 1–3-flowered, its peduncle 6–8 mm long, glabrous; flowers yellow; pedicels ± 2 mm long, glabrous; staminodes 2; capsule usually 3-seeded or sometimes 1-seeded.

KENYA. Lamu District: Track from Mpekatoni to Kitwa Pembe Hill, 16 July 1974, *Faden & Faden* 74/1123! & Mambasasa, Utwani Forest Reserve, 16 Oct. 1957, *Greenway & Rawlins* 9343!; Tana River District: Tana River National Primate Reserve, Mchelelo Forest, 11 Mar. 1990, *Kabuye et al.* in TPR 14!

TANZANIA. Tanga District: Magila near Muheza [Muhesa], 8 Nov. 1970, *Archbold* 1292; Zanzibar: Mwera Swamp, 19 Aug. 1960, *Faulkner* 2695!; Mtoni, 16 Oct. 1950, *Robinson* 77!

DISTR. **K** 7; **T** 3, 6; **Z**; not known elsewhere

HAB. Forest (including riverine and semi-deciduous), evergreen thickets, grassland, cultivation and other disturbed situations such as roadsides and pastures, often in damp places; 0–250 m

Flowering specimens have been seen from all months except January and May.

SYN. *Commelina africana* L. var. *villosior* of Brenan in Mitt. Bot. Staats. München 5: 207 (1964) pro parte, as to specimens from Zanzibar

NOTE. This is the first infraspecific taxon in *Commelina africana* that combines a distinctive morphology with a clear geographic distribution and without any evidence of hybridization or intermediates with other forms. The scarcity of collections from Tanga Province, Tanzania is surprising because Mrs. Faulkner lived and collected plants there for many years. On the label of her collection 2695 from Zanzibar she recorded that the plant also occurs on the mainland, but she apparently made no collections of it there.

The label of *Faden & Evans* 70/442 records the sometime presence of cleistogamous flowers which would appear to be the only record of them in *C. africana*. However, they are not clearly present on the holotype and they have not been noted in any other collection of var. *zanzibarica*. They should be looked for in the field.

Field notes for the same collection indicate that the sepals were free. No preserved flowers of this subspecies have been studied, so that observation remains unconfirmed.

b. subsp. **africana**

Leaves sessile, 1.5–15 cm long, glabrous or pubescent; spathes usually lacking distinctive strong veins on either side of the midrib, glabrous or pubescent; widespread but lacking from coastal and subcoastal Kenya and Zanzibar.

b1. var. **africana**; Brenan in Mitt. Bot. Staats. München 5: 203 (1964) & in F.W.T.A., 2nd ed.: 3: 45 (1968); Obermeyer & Faden in F.S.A. 4(2): 28 (1985); Kayemba-Ogwal in Proc. X111th AETFAT, Malawi 1: 418 (1994)

Generally mat-forming plants to 60 cm (1 m) across, the shoots usually decumbent or procumbent, uncommonly erect to ascending; leaves 1.5–6.5 × 0.7–2.5 cm, glabrous except for marginal cilia proximally; spathes 0.8–3.2 × 0.5–1.2 cm, glabrous (except for ciliate margins) to sparsely pubescent outside, usually pubescent inside; upper cincinnus usually abortive, lower cincinnus 3-flowered, its peduncle glabrous or puberulous; sepals glabrous to sparsely pubescent; capsule 3-seeded, the basal ventral locule seeds aborting; ventral capsule locule sometimes pilose; the flowers are usually recorded as yellow, but also orange-yellow, orange, yellow-brown, apricot, pinkish yellow, cream, or white.

UGANDA. Karamoja District: Lodoketemit [Lodokeminet], 20 May 1963, *Kerfoot* 4937!; West Nile District: 1.3 km SE of Metu rest camp, 15 Sep. 1953, *Chancellor* 277!; Busoga District: Bugonzo, 7 km W of Kaliro, just to N of road to Kamuli, 30 May 1953, *Wood* 718!

KENYA. Northern Frontier District: Mt Kulal, Gatab, 17 Nov. 1978l *Hepper & Jaeger* 6869!; South Nyeri District: below Castle Forest Station, 4 Apr. 1970, *Gillett & Mathew* 19114!; Nairobi District: Eastleigh Estate, S of Eastleigh Community Centre, 19 May 1971, *Mwangangi & Mukenya* 1587!

TANZANIA. Arusha District: Momella, Park Headquarters, 1 Apr. 1968, *Greenway & Kanuri* 13286!; Moshi District: Moshi, Oct. 1927, *Haarer* 949!; Lushoto District: Korogwe, 10 May 1965, *Archbold* 441!

DISTR. **U** 1–4; **K** 1, 3–5; **T** 1–4, 6–8; Sierra Leone, Ghana, Nigeria, Cameroon, Congo–Kinshasa, Sudan, Eritrea, Ethiopia, Djibouti, Somalia, Angola, Zambia, Malawi, Mozambique, Zimbabwe, Namibia, South Africa; also in Madagascar, Yemen, and naturalized around Sydney, Australia

HAB. Grassland, especially with short grasses and other short plants, rocky places, including rock crevices, woodland, thickets, light forest, open damp places such as marshes and near ponds and lakes, disturbed places, such as roadsides, termite mounds and moorland, also as a weed in cultivation (fide Kayemba-Ogwal, 1994); 300–2900 m

Flowering specimens have been seen from all months except August.

SYN. *Commelina mannii* C.B. Clarke in DC., Monogr. Phan. 3: 167 (1881); Morton in J.L.S. 55: 318 (1955) & in J.L.S. 60: 188 (1967). Type: Cameroon, Mt Cameroon, *Mann* 2136 (K, holo.)

C. mannii C.B. Clarke var. *lyallii* C.B. Clarke in DC., Monogr. Phan. 3: 168 (1881). Type: Madagascar, *Lyall* 106 (K!, holo.)

C. lyallii (C.B. Clarke) H. Perrier in Notul. Syst. 5: 179 (1936) & in Fl. Mad. 37: 13 (1938)

C. africana L. var. *mannii* (C.B. Clarke) Brenan in Mitt. Bot. Staats. München 5: 206 (1964) & in F.W.T.A., 2nd ed.: 3: 45 (1968)

NOTE. Kayemba-Ogwal (1994) reported that var. *africana* differed from var. *villosior* by having two seeds in each ventral locule. I am unable to confirm that unless she was referring to the large white basal ventral locule seeds that can be full-sized when the capsule dehisces but very rapidly shrivel (see discussion of the species).

Morton (1967) reported *C. africana* var. *mannii* (C.B. Clarke) Brenan (as *C. mannii* C.B. Clarke) from Mt Kilimanjaro, Tanzania based on a collection that he had made. I have not seen that collection, and Brenan (1964) did not report this variety from our area. *Kindeketa et al.* 779 from Ketumbeine Forest Reserve (**T** 2) would key out to var. *mannii* in Brenan (1964), but its spathes are relatively large (up to 2 cm long) on peduncles up to 2 cm long, and I think it is no more than a reduced var. *africana*. Indeed, I think that var. *mannii* is likely no more than an extreme form of var. *africana*.

I have examined the type of *C. mannii* var. *lyallii*, which in the *Flore de Madagascar* is accepted as a full species, and I can see no basis for separating it from *C. africana* var. *africana*.

b2. var. **villosior** (*C.B. Clarke*) *Brenan* in Mitt. Bot. Staats. München 5: 207 (1964) & in F.W.T.A., 2nd ed.: 3: 45 (1968); Kayemba-Ogwal in Proc. X111th AETFAT, Malawi 1: 418 (1994). Type: South Africa, Natal, 30°S, 1855, *Sutherland* s.n. (K, lecto. of Brenan, 1964)

Mostly low, spreading plants, occasionally erect, not often rooting except at or near the base; leaves 1.5–7 × (0.6–)1–2.5(–2.8) cm, pubescent; spathes (0.8–)1–3.5(–4) × (0.3–)0.5–1.2 cm, usually pubescent outside and inside; upper cincinnus usually abortive, lower cincinnus (1–)2–3(–4)-flowered, its peduncle glabrous or puberulous; pedicels and sepals usually sparsely pubescent; capsule 3-seeded, the basal ventral locule seeds aborting; capsule usually pilose; the flowers are usually recorded as yellow, but also orange-yellow, orange, pale apricot and white.

UGANDA. Karamoja District: Amudat, 11 June 1959, *Symes* 607!; Kigezi District: Kachwekano Farm, Dec. 1949, *Purseglove* P.3161!; Toro District: Rwenzoli National Park [Queen Elizabeth National Park], 5 Oct. 1962, *Lewis* 6011!

KENYA. Northern Frontier District: Moyale, 16 Apr. 1952, *Gillett* 12802! & Marsabit, 28 Jan. 1961, *Polhill* 346!; Kisumu-Londiani District: Tinderet Forest Reserve, Camp 2, 22 June 1949, *Maas Geesteranus* 5166!

TANZANIA. Lushoto District: Western Usambara Mts, Magamba, near Magamba Secondary School, 2 June 1996, *Faden et al.* 96/18!; Lushoto/Tanga Districts: Ndola, 24 May 1950, *Verdcourt & Greenway* 217!; Njombe District: Kitulo [Elton] Plateau, 6 May 1975, *Hepper, Field & Mhoro* 5359!

DISTR. **U** 1–4; **K** 1–7; **T** 1–8; Sierra Leone, Ivory Coast, Ghana, Nigeria, Cameroon, Congo–Kinshasa, Sudan, Eritrea, Ethiopia, Angola, Zambia, Mozambique, Zimbabwe, Botswana, Namibia, South Africa

HAB. Grassland, sometimes with scattered trees or shrubs, woodland, rocky outcrops, shallow soil over rocks, rock crevices, forest undergrowth, montane forest, dry forest, forest edges and glades, thickets, edges of shrubs, also a weed in disturbed or waste ground, plantations, gardens, old or abandoned cultivation, roadsides and pathsides; (300–)500–2800(–3250) m Flowering specimens have been seen from all months except July.

SYN. *Commelina barbata* Lam. var. *villosior* C.B. Clarke in DC., Monogr. Phan. 3: 167 (1881)
C. krebsiana Kunth var. *villosior* C.B. Clarke in F.T.A. 8: 47 (1901)

NOTE. This is the pubescent counterpart of var. *africana* and it was tempting to lump them. Although I could not corroborate the differences in numbers of ventral locule seeds that Kayemba–Ogwal (loc. cit.) reported for the these two varieties in Uganda – 4 ventral locule seeds in var *africana* and only 2 in var. *villosior* – I have not seen enough fruiting specimens of both to contradict her description. Var. *villosior* tends not to occur in wet habitats the way that var. *africana* frequently does, but their ecology overlaps a great deal.

b3. var. **lancispatha** *C.B. Clarke* in Fl. Cap. 7: 10 (1897); Brenan in Mitt. Bot. Staats. München 5: 211 (1964); Brenan in F.W.T.A., 2nd ed.: 3: 45 (1968); Obermeyer & Faden in F.S.A. 4(2): 28 (1985). Type: South Africa, Alexandria Division, Zuurberg Range, *Drège* 8779 (K, lecto. of Brenan, 1964)

Plants straggling or sprawling, to 60 cm (1 m); leaves 6–15 × 1–3.4 cm, glabrous or pubescent, apex usually acuminate, base rounded to cordate; spathes 2–6 × 0.6–1.7 cm, glabrous or pilose outside, almost always pilose inside; upper cincinnus usually well developed, its peduncle puberulous with hook-hairs; lower cincinnus (3–)4–8 flowered, its peduncle usually puberulous with hook-hairs; pedicels 4.5–6 mm long; pedicels and sepals usually puberulous with hook-hairs or the upper (outer) sepal pilose; flowers yellow (including pale or deep yellow), also recorded as orange or rarely white; staminodes 3.

UGANDA. Karamoja District: Lodoketemit [Lodokeminet], 20 May 1963, *Kerfoot* 4934!

KENYA. Northern Frontier: near Wamba on edge of Wamba Mt, Lekanto Lugga, 25 June 1987, *A. Faden* 87/40!

TANZANIA. Ufipa District: Ilemba Gap, road to Rukwa, 12 Mar. 1959, *Richards* 11167!; Dodoma District: Berekoa–Babati road, 15 Jan. 1974, *Richards & Arasululu* 28696!; Iringa District: Ngwazi [Ngowazi] Lake dam and vicinity, 10 June 1996, *Faden et al.* 96/139!

DISTR. **U** 1; **K** 1; **T** 1, 4, 5, 7; Sierra Leone, Congo–Kinshasa, Sudan, Eritrea, Ethiopia, Angola, Zambia, Malawi, Mozambique, Zimbabwe, Botwana, Namibia, South Africa

HAB. Woodland, thickets, grassland, montane forest, forest glades and edges, shady edge of lugga, rocky slopes, roadsides, alluvial and silty washes; in clay or sandy soil; 750–2300 m Flowering specimens have been seen from January to June and September.

SYN. *Commelina boehmiana* K. Schum. in P.O.A. C: 135 (1895); C.B. Clarke in F.T.A. 8: 48 (1901). Type: Tanzania, Tabora District: Gonda, *Böhm* 12 (B, holo.; K!, photo)
C. africana L. var. *boehmiana* (K. Schum.) Brenan in Mitt. Bot. Staats. München 5: 213 (1964); Brenan in F.W.T.A., 2nd ed.: 3: 45 (1968)

NOTE. The collection *Kerfoot* 4934 from Uganda has both glabrous and hairy shoots that would be named, according to Brenan's system, var. *lancispatha* and var. *boehmiana*, respectively. Because this is the only collection seen of either variety from Uganda, and the specimens of pubescent var. *boehmiana* from Tanzania, fall completely within the range of the more common glabrous var. *lancispatha*, there seemed to be no point in maintaining them as distinct. Almost all of the plants studied with spathes glabrous outside had abundant pubescence inside, similar to that in the plants that had pubescent leaves and spathes. All had pubescent upper cincinnus peduncles and almost all had pubescent lower cincinnus peduncles, pedicels and sepals. No ventral locule seeds have been observed.

A. Faden 87/40, the sole Kenyan collection seen of this variety, more closely resembles an overly large specimen of var. *villosior* than the hairy form of var. *lancispatha*. Its upper cincinnus is abortive and its lower cincinnus only three-flowered. It was growing in a moist shady situation according to the label. However, with its leaves up to 12 cm long, short of completely redefining the taxa, there seems to be no way to exclude it from var. *lancispatha*.

Two specimens from our area, *Burtt* 1731 and *Pirozynski* P217, with descriptions of the flowers as blue or violet-blue, respectively, were first thought to be a species related to *C. diffusa*. Further study instead indicated that most likely the flower colour was wrongly recorded and that the specimens are instead *C. africana* var. *lancispatha* (see discussion under *C. diffusa* for more details).

No spirit material of this variety was available. Based on the fragmentary floral remains examined, it appears likely that the flowers in var. *lancispatha* are larger than those of the other taxa recognized here and perhaps larger than the dimensions given in the description. The large flowers are consistent with the greater number of flowers per spathe and the longer pedicels than in any other taxon.

b4. var. **glabriuscula** (*Norlindh*) *Brenan* in Mitt. Bot. Staats. München 5: 217 (1964). Syntypes: Zimbabwe, Inyanga, R. Nianoli, *Fries, Norlindh & Weimarck* 3141 (BM, LD, syn.); Inyanga, *Fries, Norlindh & Weimarck* 2446 (LD, syn.)

Erect or semi-prostrate plant to 50 cm; leaves usually complicate, linear to linear-oblong, 1.5–9 × 0.4–0.8 cm, glabrous or subglabrous except for the marginal hairs proximally, spathes (1.1–)1.6–6 × 0.3–0.8 cm, usually glabrous outside except for marginal hairs proximally, pubescent inside; upper cincinnus abortive, lower cincinnus 2–3(–4)-flowered, its peduncle with a few, minute hairs, pedicels ± 2 mm long, puberulous, upper sepal sometimes and capsules typically sparingly pilose; flowers usually described as yellow or sometimes orange-yellow, orange or pinkish yellow.

UGANDA. Karamoja District: Lodoketemit [Lodokeminet], 20 May 1963, *Kerfoot* 4938.
KENYA. Baringo District: Maji ya Moto, Kamasia, 15 July 1945, *Bally* 4539!; Naivasha District: Ol Longonot Estate, Sep. 1960, *Kerfoot* 2257!; Machakos/Masai Districts: Chyulu Hills, saddle above ODW camp, 13 Dec. 1991, *Luke* 2971!
TANZANIA. Masai District: 40 km from Arusha on Dodoma road, 13 May 1965, *Leippert* 5758!; Mbulu District: Lake Manyara National Park Headquarters, Mto wa Mbu, 23 Mar. 1964, *Greenway & Kanuri* 11,400!; Ufipa District: Mbizi Mts, Fuzu Hill, 18 June 1996, *Faden et al.* 96/295!
DISTR. **U** 1; **K** 3, 4, 6?; **T** 2, 4, 7; Eritrea, Ethiopia, Malawi, Mozambique, Zimbabwe, Botswana, Namibia, South Africa
HAB. Grassland, rocky ground, thickets, roadsides, roadside ditches and other disturbed habitats such as margins of cleared field, pasture, old quarries and pine plantation;1200–2300 m
Flowering plants have been seen from January though May, July and September through December.

SYN. *Commelina krebsiana* Kunth var. *glabriuscula* Norlindh in Bot. Not. 1948: 20 (1948)
C. africana L. var. *diffusa* Brenan in Mitt. Bot. Staats. München 5: 217 (1964). Type: Malawi, Nyika Plateau, *Richards* 14386 (K, holo.)

NOTE. Var. *glabriuscula* is considered a synonym of var. *lancispatha* in F.S.A. (Obermeyer & Faden 1985). It our area, however, it has narrower leaves and spathes, an abortive upper cincinnus, generally fewer flowers in the lower cincinnus and shorter pedicels.

b5. var. **milleri** *Brenan* in Mitt. Bot. Staats. München 5: 221 (1964). Type: Zimbabwe, Matobo, Besna Kobila, *Miller* 4061 (K, holo.)

Erect or creeping plants to 50 cm; leaves linear to linear-lanceolate,1.5–11 × 0.1–1 cm, puberulous; spathes 1.2–5 cm long, 0.35–0.8 cm high, densely puberulous outside and inside or sparsely puberulous outside and glabrous inside; upper cincinnus lacking in some populations and present in others, when present, long-exserted from the spathe; lower cincinnus 3–4-flowered, its peduncle glabrous or puberulous; pedicels and sepals puberulous or glabrous; capsule pilose distally or glabrous; mature capsule with seeds not seen; flowers usually described as yellow, rarely orange or white.

UGANDA. Karamoja District: Lodoketemit [Lodokeminet], 20 May 1963, *Kerfoot* 4935! & Kidepo Valley National Park, 23 Feb. 1976, *Ogwal* 98!; Teso District: Kyere, Jan. 1933, *Chandler* 1112!
KENYA. West Suk District: Escarpment E of Turkwell Gorge, 13 Aug. 1978, *Lye* 9077! & between escarpment and Kongelai, May 1958, *Tweedie* 1573!; Machakos District: Nairobi–Emali road, 115 km, 26 Feb. 1969, *Napper & Abdallah* 1890!
TANZANIA. Iringa District: Ngowasi [Ngwazi], 3 Feb. 1987, *Lovett* 1419!; Njombe District: Milo, 6 Oct. 1978, *Archbold* 2454! & Msima Stock Farm, 1932, *Emson* 326!

DISTR. **U** 1, 3; **K** 1, 2, 4; **T** 6, 7; Zambia, Malawi, Zimbabwe, Botswana, Namibia, South Africa
HAB. Grassland with scattered trees or tree clumps and sometimes on termite mounds, open bushland, thorn scrub, thickets, montane forest, rock crevices, alluvial fans, pasture, cultivation and old cultivation; 700–2300 m
Flowering plants have been seen from January, February, May, June, August and October to December.

SYN. *Commelina brevipila* Brenan in Mitt. Bot. Staats. München 5: 219 (1964); Kayemba-Ogwal in Proc. X111th AETFAT, Malawi 1: 418 (1994). Type: Zambia, Kapiri Mposhi, *Fanshawe* 1823 (K, holo.)

NOTE. In F.S.A. (Obermeyer & Faden 1985) all of Brenan's varieties of *C. africana* with pubescent leaves, including var. *villosior*, var. *boehmiana*, var. *milleri* and var. *brevipila*, were treated as *C. africana* var. *krebsiana* (Kunth) C.B. Clarke, a variety that Brenan considered restricted to Zimbabwe and further south. I consider that too extreme, although I too do not accept all of Brenan's varieties. Clearly the whole species is in need of further study.

NOTE (on the species as a whole). Brenan (1964) advanced our concept of *Commelina africana* by lumping a large number of separately described species into it, stopping only short of including Madagascan *C. lyallii*, which I have reduced to synonymy under *C. africana* subsp. *africana* var. *africana*. However, his system appears to be wholly artificial, using almost entirely leaf dimensions and pubescence to define the 12 varieties he recognized. Aside from the second edition of F.W.T.A, no other flora has adopted his system in its entirety. Some have ignored it completely (Ensermu & Faden 1997: Fl. Eth.) and others have reinterpreted it (Obermeyer & Faden, *loc. cit.*: F.S.A.). The treatment here is more like the latter in that some of Brenan's varieties are accepted whereas others are treated in synonymy. The final taxa, however, do not have the same circumscription as in F.S.A. The present treatment is also different in that it recognizes a new subspecies of *C. africana*, subsp. *zanzibarica* which is confined to the East African coast. This taxon, while common in Kenya and Zanzibar, is apparently scarce in mainland Tanzania, a distribution pattern that, while unpredictable, is also similar to that found in two other Commelinaceae, namely, *Aneilema clarkei* and *Murdannia axillaris*.

It is possible that a more satisfactory infraspecific classification of *Commelina africana* subsp. *africana* might result from taking into account reproductive characters of the varieties, which is impossible to accomplish from herbarium specimens alone. I have looked at the degree of development of the upper cincinnus, the number of flowers in the lower cincinnus, pedicel length and the pubescence of the cincinnus peduncles, pedicels, sepals and ovaries/capsules, and except for the more common development of an exserted upper cincinnus and the presence of longer pedicels and more numerous buds in the lower cincinnus in var. *lancispatha*, no other reproductive characters seem to separate the remaining varieties. Indeed within Brenan's var. *milleri*, even without our inclusion of his var. *brevipila*, there are some populations in which the upper cincinnus is very well developed and one-flowered and other populations in which it is lacking altogether.

Three staminodes, all equally developed, have been seen in four populations of var. *lancispatha* and only two staminodes – the medial one was lacking – in var. *villosior* and subsp. *zanzibarica*. Few populations of the latter two taxa have been checked and the number of staminodes present in the other accepted taxa is unknown for certain. Morton (1956) reported only two staminodes present in *C. africana* in Ghana.

The capsules of *C. africana*, which in the sense of Brenan (1964) comprises species 37 to 45 in Clarke's monograph (Clarke 1881) and species 22 to 30 in F.T.A., were described by Clarke (where capsules were known) as one-seeded, with all four ventral locule ovules aborting, or occasionally three-seeded, with two of the ventral locule ovules aborting. This is essentially what I have found in our area except that three-seeded capsules are more common than the one-seeded capsules. Because of limited numbers it is not clear whether a one-seeded capsule might be more common in some varieties of *C. africana* than in others. One-seeded capsules were found exclusively only in one population of var. *lancispatha*, and both capsule types have been observed in individual plants of var. *villosior*. Clarke (1901: 45) observed in *C. africana* that when the ventral capsule locules formed seeds it was the apical (upper or distal) ovules that produced them. That is indeed the pattern that I have observed in this species. What is most peculiar, however, is that the basal ovules also develop and reach full size before aborting. Indeed when the capsule dehisces it may appear to be five-seeded, because the two apical ventral seeds are dark brown and the equally large or sometimes larger basal ventral locule seeds are white or whitish. This has been noted by Ogwal (1990) although she did not identify which ovules fully matured and which produced "infertile" seeds. The white basal seeds soon shrivel such that their full stature and fresh colour can never be determined from herbarium specimens, in which they are present as small, brownish remnants. The development of the basal ventral seeds to full size before they abort has the consequence that the apical ventral locule seeds when they mature have the size and shape, including a truncate basal wall, that one would expect in a seed from a two-seeded locule. In nearly all other species with two-seeded ventral locules, abortion of a potential seed in a ventral locule is irregular and unpredictable, such that when an ovule fails to produce a seed it either does not develop at all – perhaps due to lack of fertilization – or else it does not grow very large before aborting. This reduces the constraints on the size and shape of the seed that does develop in that locule, so that it is typically larger than a seed from a two-seeded locule and rounded at both ends.

Three other species in our area have three-seeded capsules similar to those of *C. africana*: *C. stefaniniana*, *C. congesta* and *C. albiflora*. All of them have five-ovulate ovaries, a dorsal locule with a seed that is fused to the capsule wall, the basal, ventral ovules abortive, and the apical, ventral ovules usually developing into seeds. Although I have looked at a limited number of populations of these species, it appears that the ovules that abort do so much earlier in *C. albiflora* and at least somewhat earlier in the other two species than in *C. africana* because the apical ventral locule seeds that they do develop are either rounded at both ends, longer than one would expect in a two-seeded locule of that length, or both. The three seeded capsules in all four of these species should not be confused with the three-seeded capsules of such species as *C. erecta*, *C. bracteosa*, *C. nigritana* and *C. aspera* (among others) which develop from three-ovulate ovaries.

The testa of *C. africana* seeds shows much variation. The final pattern is the result of two variables: the depth of the regular depressions on the dorsal surface of the seed and the presence or absence of tiny, white or grey, bead-like particles on the high areas between the depressions. The depressions vary from deep (foveolate) to very shallow (alveolate), with the former more common. The bead-like particles can ± completely form a network around the depressions, adding a reticulation to the pattern or they can be lacking entirely. So far seeds with foveolate, foveolate-reticulate and alveolate testa patterns have been observed. I would be very surprised if alveolate-reticulate patterns did not also occur within the species. At this time no testa pattern is known to be restricted to a particular taxon (or taxa) within *C. africana* and no infraspecific taxon is sufficiently well known to be characterized by its testa pattern or patterns.

5. **Commelina pycnospatha** *Brenan* in K.B. 15: 207 (1961). Type: Zambia, Bulayo–Sambu road, Mweru-Wantipa, *Richards* 9042 (K!, holo.)

Annual; roots thin, fibrous; shoots much-branched, decumbent, repent or rarely subscandent (see note below), rooting at the nodes, to 1 m long or longer, glabrous or puberulous, at least in a line, then glabrescent, often red; Leaves distichous; sheaths often flushed with purple, 0.7–1.4 cm long, often split to the base, completely glabrous or white-ciliate at apex; lamina sessile, in proximal leaves lanceolate to ovate-elliptic, 1.5–6 × 0.7–2 cm, base cuneate to rounded, margins scabrid distally, apex acute, sometimes mucronulate; surfaces glabrous or sparsely puberulous; lamina of distal leaves, including those of short shoots, ovate, 0.8–2 cm long, apex acute or subacute to obtuse, commonly mucronulate, base rounded to cordate. Spathes solitary, leaf-opposed distally on the long shoots and also densely aggregated and nearly imbricate on short axillary shoots; spathes subsessile or pedunculate, slightly falcate, 1.1–2.3(–2.7) × (0.4–)0.6–1.2 cm, base truncate to cordate, margins free, usually ciliate or ciliolate at least proximally, sometimes scabrid, rarely glabrous, apex acute to acuminate, often with a recurved mucro, surfaces glabrous, rarely puberulous inside; upper cincinnus usually abortive or vestigial, occasionally (in larger spathes) 1-flowered, lower cincinnus (1–)2–3(–5)-flowered. Flowers bisexual, ± 7.5 mm wide. Sepals hyaline white, ± 2.5 mm long; paired petals pale yellow, limb ± reniform, lower petal ovate, cup-shaped; staminodes 2, the medial lacking; lateral stamens with filaments creamy yellow, anthers ovate, yellow with blue sutures, pollen yellow; medial stamen filament ± equalling the lateral stamen filaments in length, pollen yellow; style sigmoid, ± 5 mm long; stigma capitate. Capsules tan to greenish tan with dark brown spots, trilocular, bi- or trivalved, 3–5-seeded, oblong-ellipsoid,

4–6 × 2–3.5 mm, apiculate, dorsal valve ± deciduous, dorsal locule 1-seeded or empty, dehiscent (sometimes tardily) or indehiscent, ventral locules (0–) 1–2-seeded. Ventral locule seeds broadly ovate to trapezoidal or reniform in outline, 1.5–2.5 × 1.25–1.75 mm, dorsiventrally compressed, testa light greyish brown or medium brown, foveolate or foveolate-alveolate; dorsal locule seed free or (less commonly) fused to the capsule wall, ellipsoid, depressed hemispherical, 2.4–2.55 × 1.75–1.9 mm, colour and pattern as in ventral locule seeds or alveolate with shallower depressions.

TANZANIA. Chunya District: Mbeya–Chunya road km 50, 1 km before Salangwe village, 30 June 1996, *Faden et al.* 96/475!; Dodoma/Singida Districts: km 27 on Manyoni–Singida road, 3 July 1996, *Faden et al.* 96/540!; Ufipa District: Nkundi–Kamwanga road, km 3, by small bridge, 16 June 1996, *Faden et al.* 96/246!

DISTR. **T** 4, 5, 7; Congo–Kinshasa, Zambia

HAB. Marshy seepage areas in grassland, marshy streamsides, stream banks, disturbed river banks, wet roadside ditches and seepage, wet depressions, disturbed areas, and in cultivation; 800–2750 m

Flowering specimens have been seen from April, May and October.

NOTE. There seem to be two patterns of variation: (1) all of the spathes subsessile, the spathe margins usually scabrid, the dorsal capsule locule regularly dehiscent, the seeds with a pale grey-brown testa with a reticulum formed by irregularly arranged, whitish particles, and (2) at least the spathes along the long shoots distinctly pedunculate, the spathe margins generally ciliate to ciliolate or glabrous, the dorsal capsule locule indehiscent or tardily dehiscent, the testa darker brown, with a variable reticulum that is not like the above. The first pattern is shown by two of the three Zambian collections, including the type and by the northernmost Tanzanian collection (*Sabaya* 36) from Singida. The second pattern is shown by the seven collections from Tanzania made by *Faden et al.* and by the third Zambian collection (*Mutimushi* 3075) from the westernmost locality for the species. Since the geographic distribution does not make any sense, there is no point in considering infraspecific taxa at this time.

Young plants resemble *Commelina africana* except for their red stems and usually aquatic or semiaquatic habitat. They also produce larger, more longly pedunculate and less clustered spathes resembling those of species such as *C. africana* and *C. diffusa* than do more mature plants of *C. pycnospatha*. It appears that the short, axillary shoots bearing clusters of small, subsessile spathes are produced later. Some older specimens bear only such spathes, perhaps because the spathes of the long shoots have fallen off. Plants that resemble *C. africana* are best distinguished by their pattern of raised veins (primary and usually secondary) on the spathes, which is lacking in *C. africana*.

The colour of the paired petals is recorded as pale lemon yellow in the type collection, but on other specimens can be described as yellowish, yellow, creamy yellow or even orange. In single collections from Zambia (*Mutimushi* 3075) and Tanzania (*Richards & Arasululu* 26271) they are recorded as white, which I suspect was more of a cream, with some yellow mixed in.

Based on our observations, the species is probably much more common in central and SW Tanzania than the present collections would suggest. Its usually prostrate habit, reddish stems and preference for moist habitats make it quite distinctive.

Completely dead plants found growing well up into bushes along the Manyoni–Singida road, near a then-dry water hole, were thought possibly to belong to this species, but the characteristic arrangement of the spathes, which were almost entirely lacking, could not be confirmed, and the single seed recovered was not fully diagnostic. This was the only possible record of the subscandent habit in this species until I observed the collection *Selemani* TTSA 244 which record a plant "1 m tall". Either the plant was subscandent or else it was 1 m long.

6. **Commelina stefaniniana** *Chiov.* in Coll. Bot. Miss. Stefanini-Paoli: 169 (1916); Faden in Fl. Somalia 4: 89, fig. 56, p. 90 (1995); Ensermu & Faden in Fl. Eth. 6: 365, fig. 207.16 (1997). Types: Somalia, El Uré, *Paoli* 1079 (FT!, syn.) & El Ualac, *Paoli* 108 bis (FT!, syn.) & Baidoa forest, *Paoli* 1117 (FT!, syn.)

Perennial herb 30–60 cm tall; roots tuberous, to ± 5 mm thick; shoots annual, often much branched, erect to ascending, sometimes declinate but not rooting at the nodes, puberulous. Leaves distichous; sheaths ± 0.5–1.5 cm long, the older

FIG. 16. *COMMELINA STEFANINIANA* — **1**, habit; **2**, spathe and bisexual flower, side view; **3**, bisexual flower, partial side view, **4** & **5**, capsule, dorsal and side view; **6**, capsule, dorsal view (dorsal valve removed); **7** & **8**, seed from ventral locule, side and dorsal view. All from *Faden & Kuchar* 88/232. Drawn by Alice Tangerini, reproduced with permission from Flora of Ethiopia, Vol. 6, Fig. 207.16.

ones split to the base, surface glabrous or puberulous, with a dense line of pubescence along the fused edge, ciliate at the apex; lamina sessile or rarely shortly petiolate, narrowly lanceolate to ovate, 3–10 × 1.3–2.7 cm, base oblique, rounded to cordate, cordate-amplexicaul in the distal leaves, margins scabrid, occasionally ciliate basally, apex acute to acuminate, occasionally mucronate; both surfaces sparsely puberulous or the adaxial glabrous, adaxial surface often blotched or spotted with reddish purple or maroon. Spathes solitary; peduncles (0.6–)1–3.5 cm long, puberulous, often in a longitudinal line; spathes falcate, 2–3.7 × 0.9–1.5 cm high, base deeply cordate to sagittate, margins free, scabrid, eciliate, apex acute to obtuse, surfaces glabrous to sparsely puberulous; upper cincinnus 1-flowered, the flower male, lower cincinnus 2–8-flowered, the flowers bisexual; bracteoles present or absent. Flowers bisexual and male, 1.7–2.6 cm wide; upper sepal ovate-elliptic or ovate-lanceolate, 4–6 mm long, sparsely strigose, paired sepals fused into a cup, 4–5.5 mm long, sparsely strigose or glabrous; paired petals blue, ovate-deltate, 15–20 × 10–12 mm of which the pale violet claw ± 5–6 mm; lower petal whitish to very pale lilac, linear to linear-oblong, 6–7 mm long; staminodes 3, equal; lateral stamens with filaments ± 9–11 mm long, anthers 0.7–1.5 mm long, pollen yellow or orange-yellow; medial stamen filament ± 5.5–8.5 mm long, anther 1.5–2.5 mm long, pollen yellow or orange-yellow. Ovary ± 2 mm long; style 13–14 mm long; stigma capitate. Capsules trilocular, bivalved, 3-seeded, obovoid, 5–6 × 2.5–3 mm, dorsal valve deciduous, reddish brown, striate, ventral valve pale yellow, sometimes spotted and streaked with maroon; dorsal locule (0–)1-seeded, indehiscent, the seed fused to the capsule wall, ventral locules (0–)1-seeded, dehiscent, the basal seed aborting in each locule. Dorsal locule seeds ellipsoid, elongate-hemispheric, 3.8–3.9 × 1.95–2 mm, testa apparently dark brown and reticulate-foveolate; ventral locule seeds dorsiventrally compressed, convexo-planar, ellipsoid, 3–4.2 × 1.6–1.7 mm, dark brown with lighter brown along the ridges, testa reticulate-foveate, white-farinose in the depressions and on the ventral surface. Fig. 16, p. 155.

KENYA. Northern Frontier Province: Dandu, 14 Apr. 1952, *Gillett* 12778! & Moyale, 23 Apr. 1952, *Gillett* 12906! & same loc., 20 Oct. 1952, *Gillett* 14076!

DISTR. **K** 1, Ethiopia, Somalia

HAB. *Commiphora-Acacia* scrub on red sandy loam and degraded montane scrub with *Cussonia*, *Ficus*, &c.; 300–635 mm rainfall; 750–1100 m

Flowering specimens have been seen from April and October.

NOTE. This is a very distinctive species because of its tuberous roots, annual shoots with leaves marked with reddish purple or maroon on the upper side, simply folded spathes with deeply cordate to sagittate bases, large blue flowers with the lower sepals fused into a cup, pubescent sepals, and trilocular, 3-seeded capsules in which the basal ventral ovules typically abort. The only other species with spotted leaves in our area is *C. albiflora* from the Kakamega Forest area in western Kenya. Other species with similar 3-seeded capsules in our area are *C. albiflora*, *C. congesta* and some plants or populations of *C. africana*. The first two have fused spathe margins and are not related to *C. stefaniniana*. Yellow-flowered *C. africana* shares with *C. stefaniniana*, in addition to free-margined spathes and a 3-seeded capsule in which the basal ventral seeds abort, pubescent sepals, spathes pubescent inside and outside, and strong veins on either side of the midrib forming a pouch at the base of the spathe (present in *C. africana* subsp. *zanzibarica* and occasional plants of other taxa). It is not clear whether all of these similarities are indicative of a close relationship or are merely the result of convergence.

A collection from Somalia (*Faden & Kuchar* 88/232) has been maintained in cultivation for more than 20 years. The two halves of the spathes in these plants are not closely appressed, unlike nearly all *Commelina* species. Instead they usually spread apart at the apex and also at the base and sometimes for their entire length. It is not clear from herbarium specimens how common this character might be in the species and whether is occurs in our area.

A weak fragrance has been detected in the flowers of *C. stefaniniana* in cultivation. This is the only species of *Commelina* in which a floral scent has been recorded.

7. **Commelina hockii** *De Wild.*, Contrib. Fl. Katanga, Suppl. 3: 68 (1930). Type: Congo–Kinshasha, Katanga, Sep. 1911, *Hock* s.n. (BR!, lecto., K, photo.; selected here); Kafubu, Granat farm, *Quarré* 781 (BR!, syn.; K, photo)

Perennial; roots relatively few, thick, cord-like or occasionally tuberously thickened; shoots annual, erect to ascending, 8–25(–40) cm tall, sparsely branched, glabrous to puberulous. Leaves distichous; sheaths mostly 1.5–3.5 cm long, sometimes the basal ones longer, usually with a line of hairs along the fused edge, ciliate at the apex, and otherwise glabrous, or occasionally completely glabrous or the whole surface pubescent; lamina lanceolate-oblong to linear, usually broadest at the base, 5.5–15(–22) × (0.25–)0.8–3 cm, base rounded to cordate-amplexicaul, margins scabrid distally, occasionally also ciliate proximally, apex acuminate to attenuate; surfaces usually glabrous, occasionally puberulous. Spathes solitary; peduncles (4–)5.5–11(–18) cm long, glabrous, with a longitudinal line of hairs, or rarely more generally puberulous; spathes not at all to slightly falcate, 3–9(–11) × (0.6–)1–1.5(–1.9) cm, base cordate to deeply cordate or rarely rounded, margins usually ciliate to ciliolate at least proximally, apex usually long-attenuate, sometimes abruptly so, surfaces usually glabrous, occasionally pubescent, sometimes purple, purple-veined or narrowly to broadly purple-margined; upper cincinnus (1–)7–14-flowered, lower cincinnus (4–)10–15-flowered. Flowers bisexual and male, those of the upper cincinnus apparently all male; upper sepal tinged with reddish purple, 4.5–6 mm long, paired sepals tinged with reddish purple, essentially free, 4.5–7 mm long; paired petals usually white, occasionally pink (rarely blue?), 9–14 × 8–10 mm of which the proximally reddish purple claw 4.5–5 mm long; lower petal elliptic or ovate-elliptic, 4.5–8 mm long, concolorous with the paired petals; staminodes 3; lateral stamens with filaments ± 10 mm long, anthers (1.2–)1.6–2.2 mm long, medial stamen with filament 7–9 mm long, anther (1.8–)2–3.2 mm long. Ovary ± 1.5 mm long; style ± 12–13 mm long. Capsules bi- to trilocular, bivalved, 8–9 mm long, rostrate, up to 5-seeded, dorsal locule 1-seeded to abortive, indehiscent, ventral locules 1–2-seeded, the basal ovules sometimes aborting. Ventral locule seeds strongly dorsiventrally compressed, ± reniform to deltate, 2.5–3.5 × 1.9–2 mm, testa blackish, foveolate or foveolate-reticulate.

TANZANIA. Mpanda District: Silkcub Highlands, 6 Dec. 1956, *Richards* 7168!; Ufipa District: Chapota, 3 Dec. 1949, *Bullock* 1977! & Hill NW of Tatanda Mission, 23 June 1996, *Faden et al.* 96/362!

DISTR. **T** 4, 7; Congo–Kinshasa, Angola?, Zambia

HAB. *Brachystegia* woodland and grassland; (1250?–)1650–2000 m

Flowering in December.

SYN. *Commelina coelestis* of C.B. Clarke in F.T.A. 8: 40 (1901) pro parte, pro *Carson* 35 and *Thomson* s. n., *non* Willd.

NOTE. This is a very distinctive species because of its very long, attenuate spathes with free margins and usually white flowers. *Bullock* 1976 is unusal in having very pubescent leaves, internodes and spathes.

Both cincinni typically have numerous flowers. In smaller, gracile specimens the number of flowers per cincinnus may be reduced, to as few as one in the upper cincinnus and four in the lower cincinnus, both in *Richards* 7350 from Zambia. In dried specimens an irregular papery material is commonly found within the spathe on the pedicels of the lower cincinnus. It is likely that this is the remains of fluid that was in the spathe. It is unusual for species with spathes with free margins to produce enough liquid to leave such a residue. In the most robust specimens, the upper cincinnus may produce a secondary cincinnus. In cultivation this tertiary cincinnus may have a small spathe.

The antherodes typically are entirely yellow. In some specimens, however, the antherode connective appears to be dark, and in *Richards* 7168, from Tanzania, the antherode is clearly mainly dark with yellow lobes. This type of variation in colour is probably uncommon in *Commelina* species.

It is doubtful that fully mature seeds have been observed. The most mature ones show the foveate pattern described above.

Although neither cited sheet is designated as the type in the publication, the *Hock* collection has the label "Typus" on it, whereas the *Quarré* specimen does not. Both seem to belong to the same species, and I see nothing wrong with the *Hock* collection as the lectotype.
The single collection seen from Angola that closely resembles *C. hockii*, *Menezes* 345, clearly has blue flowers, so it very likely is a different species.

8. **Commelina kituloensis** *Faden* in Novon 11: 400, fig. 2 (2001). Type: Tanzania, Njombe District, Kitulo Plateau, 36 km E of the turnoff on Mbeya–Tukuyu road, *R.B. Faden, Phillips, Muasya & Macha* 96/438 (US!, holo.; BR!, EA!, NHT!, K!, MO!, PRE!, US!, iso.)

Perennial; roots abundant, thick, whitish, cord-like, apparently lacking distal tubers; shoots annual, tufted, arising from a subterranean base, (10–)15–45 cm tall, erect to ascending or sometimes decumbent at the base, rarely rooting, glabrous or glabrescent to densely white-puberulous or strigillose. Leaves spirally arranged or distichous; sheaths sometimes clustered at the base of the shoots, spaced at least medially, 1–3(–3.5) cm long, with a line of short or long hairs along the fused edge, otherwise the surface glabrous to puberulous or strigose, ciliate at apex; lamina sessile, oblong-lanceolate to linear-lanceolate or, rarely, linear or ovate-ellipitic, 4–12(–16) × (0.3–)0.5–2(–2.5) cm, base cuneate to rounded, margins ciliate or glabrous, not scabrid, apex acuminate to acute, surfaces densely strigose to glabrous. Spathes arising distally on the shoots, solitary; peduncles (2.5–)3–10(–15) cm long, usually densely strigose, rarely the pubescence reduced to a single longitudinal band; spathes usually not to very slightly falcate (rarely strongly falcate), 1.7–4.5(–5.3) × 0.6–1.8 cm, the folded edge often somewhat saccate proximally, base cordate to hastate, margins free, usually ciliate, occasionally glabrous, apex acute to acuminate, rarely attenuate, surfaces strigose, the the whole spathe flushed or striped with red or reddish purple (rarely entirely green); upper cincinnus usually producing 1 male flower, occasionally the flower bisexual or the cincinnus more than 1-flowered; lower cincinnus 5–12-flowered, included. Flowers bisexual and male, 25–30 mm wide; outer (upper) sepal reddish purple with green veins, ovate, 4.5–7 mm long; inner (lower) sepals apparently free, forming a broad, shallow cup, 6–8 mm long, reddish purple except green on the side where they meet or half reddish purple and half green; paired petals pale blue, lavender or purple to nearly white, 13–18 × 12 mm of which the reddish purple claw 4–6 mm long; medial petal strongly reflexed, white or cream, ovate, 6.5–9 mm long; staminodes 3; lateral stamens with filaments 7–11 mm long, anthers 1.6–3 mm long, pollen creamy yellow; medial stamen filament 6–7 mm long, anther 2.2–3.5 mm long, pollen orange-yellow. Ovary green; style 7–16 mm long; stigma small. Capsules medium brown when mature, bi- to trilocular, bivalved, 4–5-seeded, covered by the large, marcescent sepals, oblong-ellipsoid, 5–6 × 3 mm, weakly or strongly beaked, dorsal valve ± deciduous, dorsal locule 1-seeded or empty, indehiscent, ventral locules 2-seeded, dehiscent. Ventral locule seeds dorsiventrally compressed, ovoid or deltoid to transversely ellipsoid, 2–2.6 × 1.7–1.8 mm, testa brownish black, foveate-reticulate on the dorsal surface.

TANZANIA. Mbeya District: Mbeya Range, 12.9 km NE of Mbeya, 24Oct. 1962, *Lewis* 6085! & Mbeya Mt, NE of peak, 13 May 1956, *Milne-Redhead & Taylor* 10206!; Njombe District: Kitulo Plateau, 5 Jan. 1957, *Richards* 7465!

DISTR. **T** 7; also the Nyika Plateau of Malawi and Zambia

HAB. Montane grassland, sometimes in rocky areas, at roadsides or near streams, rarely in stunted *Brachystegia* woodland; (1600–)1800–2900 m

Flowering specimens have been seen from October through June, but mainly December to February; flowers were found open between 14:00 and 15:00 on a particularly cold, overcast day, so the typical flowering time is unknown.

NOTE. Although this species was considered to be related to *C. purpurea* by Brenan, based on notes in the Kew Herbarium (it was called "*Commelina* sp. 'C'. aff. *purpurea*"), its bivalved capsules and lavender (instead of orange to buff) flowers suggest that its relationships are elsewhere. I believe that it is more closely related to *Commelina hockii* because of its capsule type, in combination with its tufted perennial habit, large spathes with free margins, apparent abundant liquid in the spathes, and numerous buds in the lower cincinnus. It differs from *C. hockii* by its more numerous, whitish roots, usually densely pubescent foliage and spathes, shorter, proportionally broader spathes usually with a much less attenuate apex, lavender to purple (or blue?) paired petals with a contrasting lower petal, and typically a solitary male flower in the upper cincinnus. In addition, it occurs at generally higher elevations, mainly in montane grassland.

Fresh spathes of *C. kituloensis* are full of liquid. This is unusual for a species whose spathes do not have fused margins. The numerous flowers that we saw were not a true blue, yet most of the collectors have described them as some shade of blue or mauve-blue. A few collectors have called them violet or purple. The ruffled petals with strongly undulate margins are much more three-dimensional than in typical *Commelina* flowers. The large, contrasting lower petal is very striking.

The most atypical specimen included in *C. kituloensis* is *Richards* 18563. That specimen has the longest, most falcate spathes, with among the most attenuate apices seen in this species. The spathes also seem to be entirely green. Were it not for the plant's dense pubescence, high elevation (2100 m), and flowers described as "blue", it would be a much better match for *C. hockii* De Wild. than for *C. kituloensis*.

One of the most unusual features of *C. kituloensis*, and perhaps the most important reason why it has remained unnamed for so long, is the apparently scarcity of seed-production. Only a single specimen at Kew was found to have ripe seeds. In the field, when we checked abundant plants, albeit not at the ideal stage, intensive searching turned up very few seeds and not a great deal of evidence in the form of old capsules to demonstrate that seeds had been present. Lewis (in Sida 1: 283 (1964)) reported a chromosome number of $n = 30$ for *Lewis* 6085 (as *Commelina* sp. 10).

The type collection was made at a time of year (late June) when this species would not ordinarily have been flowering. At the previous locality where we first found the plant, the shoots of all specimens were completely dry. At the type locality, however, just four kilometers down the road, the plants were for the most part growing in a recently planted potato field and were in full flower. It is possible that either the cultivation of the ground per se and/or the addition of fertilizer, pesticide or herbicide stimulated the dormant plants of *C. kituloensis* to resume growth, providing us with a serendipitous opportunity to see, record and collect the plants in flower. The cold weather and overcast skies probably were responsible for our finding the flowers open in the middle of the afternoon.

The single record from *Brachystegia* woodland (*Lovett* 1288) seems improbable. It is the second lowest elevation (1830 m) for the species, and when we visited the locality (Lake Ngowazi Dam) in 1996, the habitat seemed quite wrong for the species. We were unable to find it, but considering how late it was in the season, the plants might well have been dormant.

9. **Commelina scaposa** *C.B. Clarke* in Dur. & De Wild. in B.S.B.B. 38, 2: 220 (1900) & in F.T.A. 8: 38 (1901). Type: Congo–Kinshasa, upper Marungu, *Debeerst* s.n. (BR!, holo.; photo, drawing K!)

Perennial scapose herb; roots tuberous, tapering distally from a narrow base; vegetative shoots typically produced after the flowering shoots and probably later in the season than them (see note), erect, unbranched, 20–30 cm tall. Leaves mainly borne on separate, usually sterile shoots, distichous; sheaths overlapping, 2.7–5.5 cm long, sometimes with a line of hairs along the fused edge, ciliate at apex, otherwise glabrous; lamina sessile to subpetiolate, oblanceolate-linear to lanceolate-oblong, 12–23 × 0.9–3.8 cm, base cuneate to attenuate-subpetiolate, margins scabrid, sometimes ciliate at base, apex attenuate; surfaces glabrous or the lower surface sparsely puberulous. Flowering shoots precocious, erect, usually sparsely branched, 20–50 cm tall, glabrous, or with a line of pubescence below the node; sheaths to 3 cm long, lamina usually a short deltate to lanceolate extension of the sheath, sometimes well developed and linear to lanceolate, 1–3.5(–10) cm long, margins ciliate proximally, scabrid distally (in well developed leaves), surfaces glabrous or adaxial surface with a few long hairs. Spathes solitary, slightly falcate, occasionally strongly or not falcate; peduncles (0.6–)2–11(–16) cm long, glabrous or sparsely puberulous in a line at the apex; spathes 1.3–2.5 × 0.6–0.9 cm, base truncate to rounded or cordate, margins free, commonly purple, ciliate proximally or glabrous, apex acuminate or occasionally acute, surfaces sometimes purple-veined or suffused with purple, glabrous; upper cincinnus 1–10-flowered, lower cincinnus ± 8–11-flowered. Flowers bisexual and male. Sepals sometimes flushed with reddish purple, upper sepal ovate to lanceolate-elliptic, (3–)3.5–6 × 2–2.5 mm, glabrous; lower sepals fused for $^1/_4$ to almost their entire length, forming a flat open cup, each sepal (2.5–)4–7 × 2.5–3.5 mm, glabrous. Paired petals white or dull mauve (or blue?, see note), broadly ovate to ovate-reniform, 8.5–18 × 7.5–10 mm of which the sometimes pinkish claw 3.5–7 mm; medial petal whitish green, broadly ovate to linear-oblanceolate, 3.5–5 × (0.5–)1–3.5 mm. Staminodes 3; lateral stamens with filaments 8–17 mm long, anthers 1.3–2.5 mm long; medial stamen filament 6–8 mm long, anther 1.9–3 mm long. Ovary 1.5–2 mm long; style 9.5–19 mm long. Capsules pale yellow, sometimes with darker flecks, trilocular, bivalved, oblong-ellipsoid, 4.5–8.5 × ± 2–3 mm, up to 5-seeded, rostrate, dorsal valve commonly deciduous, dorsal locule 0–1-seeded, indehiscent but the capsule base and apex often split on both sides of it, ventral locules (1–)2-seeded. Ventral locule seeds dorsiventrally compressed, ovoid to ovate-reniform or tranversely ellipsoid, 1.5–3.2 × 1.3–2 mm, testa dark brown or grey, radiately pitted and furrowed.

TANZANIA. Kigoma District: Ulemba–Ikola track, 97 km from Ikola, 6 Nov. 1959, *Richards* 11725!; Mpanda District: Mwese, 8 Dec. 1956, *Richards* 7155! & Ikola–Mpanda road, 32 km from Ikola, 8 Nov. 1959, *Richards* 11751!

DISTR. **T** 4; Congo–Kinshasa, Angola, NW Zambia

HAB. Woodland and grassland, sometimes on edge of swamps; 1050–1800 m

Flowering November–December.

SYN. *Commelina praecox* T. Fries in R.E. Fr., Wiss. Ergebn. Schwed. Rhod.-Kongo-Exped. 1911–1912, 1: 220 (1916). Type: Zambia, Bwana Mkubwa, *Fries* 483 (UPS!, holo., photo K, US)

NOTE. All five collections from the Flora area consistently have a one-flowered upper cincinnus, whereas all collections seen from Zambia and most from Congo–Kinshasa, including the type, have a multiple-flowered upper cincinnus. The Tanzanian specimens also include the broadest spathes and longest spathe peduncles seen. The upper cincinnus peduncles from our area are unusually short, but they are matched by a number of collections from Congo–Kinshasa but none from Zambia. The Tanzanian specimens possibly could be recognized at an infraspecific rank, but our knowledge is too incomplete to justify that.

Fully developed leaves are known from only two specimens, *Lewis* 6193 and *Milne-Redhead* 2567A, both from western Zambia. In the *Lewis* collection a vegetative shoot has arisen from the base of the plant, but it has produced a few terminal spathes. On the *Milne-Redhead* collection, the shoot is completely vegetative and was collected in the same place where a flowering specimen (*Milne-Redhead* 2567) had been collected five months previously. All five Tanzanian collections show the early development of vegetative shoots from axillary buds on the flowering shoots. The leaves so produced have a linear lamina up to 10 × 0.5 cm, with scabrid margins and sometimes scattered, long, uniseriate hairs on the adaxial surface. It is unclear whether these plants will produce more vegetative growth from the flowering shoots or even if they will produce separate vegetative shoots later. It is quite possible that their habit and leaves could differ substantially from those of Zambian and Congolese plants.

The petals are usually described as white, pale mauve or bluish but *Richards* 11700 from Tanzania describes them as blue, which Brenan, in a handwritten annotation on the label, notes as probably an error. *Schmitz* 1030 and *Quarré* 1369 record the flowers as yellow on Congolese collections, and *Quarré* 6019, on another Congolese specimen, notes them as blue or yellow, which is even more surprising. If the yellow flower records are accurate, then *C. scaposa* would be the only *Commelina* species known to have both yellow flowers and flowers in the mauve or bluish colour range. Alternatively, this species might be more complex taxonomically than can be determined from the available specimens.

Material with mature capsules and seeds is not common, so the full range of variation in this species is unknown. I have studied in detail the capsules and seeds of one of the Tanzanian collections (*Richards* 7113) and its capsules and seeds are much smaller than those of *Russell* 21 from Congo. However, I am not certain that the seeds of the latter are fully mature. In addition to the apparent size difference there is a much great tendency for seed abortion in the Congolese than the Tanzanian plant, but there are too few observations from which to generalize.

10. **Commelina fluviatilis** *Brenan* in Mitt. Bot. Staatssam. München 6: 253 (1967). Type: Zambia, Mpika District, Luitikila River, *Richards* 14984 (K!, holo.)

Ascending to decumbent perennial with shoots to ± 1 m tall, mainly unbranched, often producing fibrous roots in water or on mud from the lower nodes; roots at base of plant tuberous; internodes glabrous, the proximal ones sometimes only 1–2 cm long and covered by overlapping sheaths, the medial and distal ones elongate, to 21 cm long, sometimes thick (up to 9 mm thick) and pithy, aerenchymatous. Leaves with sheaths usually purple-striped or purplish, 0.5–4 cm long, often split to the base, completely glabrous or ciliate at apex; lamina complicate to inrolled, linear, 5–25 × 0.1–0.6 cm, base hardly broadened into the sheath, margins glabrous or sometimes ciliate at base, apex attenuate-acute; surfaces glabrous. Spathes solitary, 1–3 per stem; peduncle (1–)2.5–7 cm long, glabrous or with a fine line of pubescence; spathes horizontal or deflexed, not at all falcate to strongly falcate, 1.1–2.2 × 0.3–0.9 cm, base cuneate to truncate (cordate to sagittate), margins free, glabrous or ciliate or ciliolate proximally, apex usually acuminate to attenuate, sometimes acute, surface glabrous (in our area), usually purple-veined; upper cincinnus well-developed, 1-flowered, the flower sometimes bisexual and setting a fruit, lower cincinnus 5–11-flowered. Flowers bisexual and male, 20–30 mm wide. Sepals 3.5–5.5(–6.5) mm long, to 9 mm long in fruit, glabrous, lower sepals fused for less than half their length; paired petals lilac, pale blue or purple, 16–17 × 15–17 mm, of which the claw 6–7 mm; lower petal broadly reniform or broadly ovate, 6–7 × 8–8.5 mm, concolorous with the paired petals; staminodes 3; lateral stamens with filaments 8–10 mm long, anthers 1.4–1.8(–2.2) mm long; medial stamen filament ± 6 mm long, anther 1.8–2.5(–3) mm long, pollen in all 3 stamens orange. Ovary 5-ovulate; style 8–12 mm long. Capsules greenish tan or tan with a broad purple band along the lateral sutures, sometimes with scattered fine brown flecks, trilocular, trivalved, 1–3(–4?)-seeded, rostrate with a long (<1 mm) or short downturned beak, 4.5–7.2 × 2–2.5 mm, dorsal valves deciduous, dorsal locule dehiscent, 0–1-seeded, ventral locules 0–2-seeded. Seeds dark brown or grey-brown, dorsiventrally compressed, usually ellipsoid or ovoid, 1.9–2.7 × 1.5–2.1 mm, testa irregularly reticulate-foveolate.

TANZANIA. Buha District: Kaberi swamp, 10 Aug. 1950, *Bullock* 3129!; Dodoma District: Lake Chaya, S of Itigi–Tabora track, 16 km W of Kazikazi, 2 July 1996, *Faden et al.* 96/510!; Ufipa District: Empeta Swamp near Chapota, 8 Mar. 1957, *Richards* 8591!

DISTR. **T** 4; Congo–Kinshasa, Zambia, Namibia (Caprivi Strip)

HAB. Marshes, swamps, river and lake edges, growing in up to 60 cm of water; 1150–1650 m Flowering specimens have been seen from March, May to August and October.

NOTE. The loose growth habit, long, narrow, succulent, complicate leaves, long-pedunculate spathes which in our area are either completely glabrous or else are ciliolate proximally, the upper cincinnus sometimes producing a fruit, the lower cincinnus many-flowered, the flowers large, pale lilac to pale blue, the seeds large, dark brown, and foveolate-reticulate, and the aquatic habit, with the stems floating on the water, all make this a very distinctive species. However, because the spathes are so variable in shape and size, specimens can easily be misunderstood or misinterpreted. Thus, all five collections from Tanzania that I found in the Kew Herbarium in 1996, scattered under various epithets, were present when Brenan described *C. fluviatilis*, but were overlooked by him. Similarly, it took me more than a month to recognize that these specimens all belonged to the same named species, despite the abundant material at Kew from Zambia.

The flower of the upper cincinnus, whether bisexual or male, seems to have much larger sepals than those of the lower cincinnus. Whether one can infer that the flower is larger in all of its parts is uncertain.

Bullock 3129 has smaller spathes and, apparently, flowers than *Burtt* 3712, and the medial anther has only a suggestion of a dark band. Both specimens have similar seeds, however, and both can be matched by specimens from Zambia, so unless unless other features are discovered, these specimens must be considered as belonging to the same variable species.

Brenan's description of the capsule as 2.5–3.5 mm wide could not be confirmed. It appears that the width he gives is that of one of the deciduous dorsal valves, not of the capsule itself, which is only about 2–2.5 mm wide.

A character noted by Brenan is the strong tendency towards abortion of the ovules and the production of few-seeded capsules. This does not appear to be the predictable abortion of the basal ventral locule ovules that occurs in several species, but rather a more random pattern that I have noted previously only in *Commelina diffusa* var. *gigas* (Small) Faden from Florida, U.S.A. That variety was found be a hexaploid ($2n = 90$), so we may predict that *C. fluviatilis* will also have a high chromosome number or meiotic irregularities. Despite the aquatic habitats of this species, a lack of pollinators does not seem the likely cause of low seed set because many if not most species of *Commelina* self-pollinate as the flowers fade. Despite 15 years of cultivation and numerous pollination attempts, this species has rarely set seed.

Plants that were collected from the margins of Chaya Lake in July, 1996 (*Faden et al.* 96/510) included many that had become stranded in the mud on the receding margins of the lake, during the early part of the dry season. The plant was rooted in the ground, with tuberous roots, and was probably able to withstand total desiccation.

11. **Commelina subulata** *Roth* in Roem. & Schult., Syst. Veg. 1: 530 (1817) & in Nov. Sp.: t. 23 (1821); C.B. Clarke in DC., Monogr. Phan. 3: 148 (1881); K. Schumann in P.O.A. C: 134 (1895); Rendle in Cat. Afr. Pl. Welw. 2: 75 (1899); C.B. Clarke in F.T.A. 8: 38 (1901); Morton in J.L.S. 60: 189 (1967); Brenan in F.W.T.A., 2nd ed.: 3: 47 (1968); Schreiber et al. in F.S.W.A. 157: 9 (1969); Cribb & Leedal, Mountain Fl. S Tanzania: 169, t. 47b (1982); Obermeyer & Faden in F.S.A. 4(2): 25 (1985); Ogwal in Mitt. Inst. Allg. Bot. Hamburg. 23b: 578, fig. 2C (1990); Faden in U.K.W.F. 2nd ed.: 305, t. 135 (1994); Ensermu & Faden in Fl. Eth. 6: 364, fig. 207.14 (1997); Wood, Handbook Yemen Fl.: 318 (1997). Type: Eastern India, *Heyne* s.n. (B!, holo.; K!, iso.)

Annual herb; roots thin or somewhat thickened, fibrous; shoots erect or ascending in small plants, becoming decumbent and often rooting at the nodes in larger plants, usually much branched, to 60 cm high but often < 25 cm, with a line of short uniseriate hairs descending from the fused edge of the distal sheath, or pilose. Leaves apparently distichous on longer shoots; sheaths to 1(–1.5) cm long, frequently splitting longitudinally, puberulous or pilose along the line of fusion and sometimes all over, white-ciliate at apex; lamina ± succulent, sessile, narrowly lanceolate to linear, (1–)2–10(–20) × 0.2–0.5(–0.9) cm, base ± rounded or broadened into the sheath, margins scabrid distally or not, often ciliate proximally, apex acuminate to acute; surfaces glabrous or pilose. Spathes solitary, declinate towards the stem or not, falcate or not; subsessile, the peduncles up to 5 mm long and largely enclosed in the sheath of the subtending leaves; spathes 0.4–1.7 × 0.2–0.7 cm, base cordate or rarely sagittate, margins free, ciliate or ciliolate, rarely glabrous, apex acute to acuminate, surface glabrous or pilose-puberulous, rarely the whole spathe flushed with maroon; upper cincinnus lacking or, when fully developed, producing 1 male flower, lower cincinnus (1–)2–6-flowered. Flowers bisexual and male, often < 10 mm wide. Sepals ovate to ovate-orbicular or obovate-elliptic, 1.7–3.5 × 1–2 mm, paired sepals fused proximally for up to ± $^1/_3$ their length; paired petals apricot, buff, buff-orange, yellow or orange, rarely blue, 3–8 × 3–5 mm of which the claw 1.5–4 mm; medial petal cup-shaped, 1.1–2.2 × 1.5–3 mm, ± concolorous with the paired petals; staminodes 3; lateral stamens with filaments 1.7–5(–7) mm long, anther 0.5–1.2 mm long; medial stamen filament 1.5–4 mm long, anther 0.5–1.6 mm long. Ovary 0.7–1.2 mm long; style 1.7–8 mm long; stigma capitate. Capsules pale yellow, trilocular, trivalved (the dorsal locule sometimes tardily dehiscent), 5-seeded, oblong-ellipsoid, 2.7–4.5 × 1.5–1.7 mm, rostrate. Seeds dorsiventrally compressed, 1.1–2.3 × 0.8–1.4 mm, testa dull dark brown, white (or occasionally brown) farinose especially in the grooves, transversely grooved and ridged; dorsal locule seed ellipsoid to reniform, ventral locule seeds ovoid in outline. Fig. 18, p. 166 & seeds Fig. 18, p. 166.

UGANDA. Teso District: Serere, Dec. 1931, *Chandler* 238!; Masaka District: 4–5 km N of Lake Nabugabo, 25 Sep. 1969, *Faden* 69/1288!; Mbale District: Kapchorwa, Sep. 1954, *Lind* 451!

KENYA. Northern Frontier District: Moyale, 10 July 1952, *Gillett* 13558!; Trans-Nzoia District: Kitale Grassland Research Station, 2 Oct. 1959, *Verdcourt* 2459!; Nairobi District: Nairobi, Langata, Forest Edge Road, 6 Feb. 1977, *Faden, Faden & Ng'weno* 77/289!

TANZANIA. Ufipa District: 5 km N of Namanyere, 2 May 1997, *Bidgood et al.* 3636! & Matai–Nkowe road km 14, 22 June 1996, *Faden et al.* 96/339!; Dodoma District: 20 km E of Itigi Station, 11 Apr. 1964, *Greenway & Polhill* 11512!

DISTR. **U** 1, 3, 4; **K** 1–6; **T** 1, 3–7, 8?; widespread in tropical Africa; also in Yemen and India

HAB. Seasonally wet situations such as soil pockets in and at edge of rock outcrops, rock pools, damp grassland, swamp and waterhole edges, marshes, along rivers, streams and dams, roadside banks and ditches, murram pits, wet flushes in bushland, also in arable land and crops; in sandy or clayey soils; altitude (300–)1000–2000(–2300) m

Flowering specimens have been seen from all months.

SYN. *Commelina striata* Kunth, Enum. 4: 44 (1843); Hassk. in Schweinfurth, Beitr. Fl. Aethiop.: 207 (1867). Type: Ethopia, near Adoa, *Schimper* 360 (B, holo.; BM!, K!, iso.).

C. subaurantiaca Kunth, Enum. 4: 658 (1843); Hassk. in Schweinfurth, Beitr. Fl. Aethiop.: 208 (1867). Sudan, Kordofan, Abu-Gerad, *Kotschy* 59 (B?, holo.; BM!, K!, iso.).

C. subulata Roth var. *macrosperma* C.B. Clarke in DC., Monogr. Phan. 3: 149 (1881). Type: Sudan, Kordofan, *Kotschy* 34 (partly) (K!, holo.)

C. angustissima K. Schum. in P.O.A. C: 134 (1895); C.B. Clarke in F.T.A. 8: 39 (1901). Type: Tanzania, Mwanza District: Karumo, Uzinza [Usindschi], *Stuhlmann* 3564 (B!, holo.; K!, iso.)

NOTE. The type of *Commelina angustissima* has tiny, densely hairy spathes. The isotype seen has no upper cincinnus and the lower cincinnus is 4-flowered. The holotype has seeds that are transversely ribbed, as in C. subulata. When I examined the holotype in 1978 I noted that the seeds appeared to be larger and less deeply grooved than those of a typical specimen of *C. subulata*, namely *Gillett* 13558 from Kenya. However, my sketch of both seeds suggests that that of the *Stuhlmann* type may have been somewhat immature. In 2009 I examined specimens from **T** 1, whence the type was collected, e.g. *Rounce* 309, that were a good match for *C. angustissima* and had typical *C. subulata* seeds, so I have no doubt that *C. angustissima* is a synonym of *C. subulata*.

The questionable record from **T** 8 is a based on *Ludanga* in MRC 1299, which was collected near the **T** 6/**T** 8 boundary.

Commelina subulata is very variable in our area and at least part of the variation has a geographic basis to it. For example, plants with very large spathes (>10 mm long) are almost all confined to Tanzania (**T** 1, 2, 4, 5), with just single collections from Uganda (*Ogwal* 88, from **U** 3) and Kenya (*Bogdan* AB 1675, from **K** 3), both collected, perhaps significantly, from cultivation. My initial studies suggested that the very large- and very small-spathed plants might constitute recognizable taxa, but partly this may have been due to my selecting plants with the most extreme size differences and finding that the larger plants did indeed have larger capsules and seeds as well as significantly larger flowers, for collections of which there was spirit material. However, when a greater range of collections was studied more recently, the differences initally found were difficult to sustain, and they have been abandoned for this account. Further studies from living plants are required in order to determine whether the variation in size within *C. subulata* has any taxonomic significance.

The spathes also show much variation in pubescence. Only one collection completely lacks margins hairs, *Greenway & Polhill* 11512, but there is much greater variation in the pubescence on the surface of the spathes, largely manifested in Tanzania. Only two Kenyan collections and none from Uganda have been observed with hairs on the spathe surfaces, whereas more than half of the Tanzanian collections have pubescent spathes. At least 10 of the 23 Tanzanian collections at Kew that have some hairs on the spathes also have other plants with glabrous spathe surfaces mounted on the same sheet, so we know that this can be a variable character within populations. There is also variation, at least among populations, in the length of the spathe marginal hairs (ciliate vs. ciliolate) and rarely in the presence or absence of long, uniseriate hairs on the spathe surface, in addition to the more ubiquitous short hairs.

Variation in flower colour is more difficult to judge partly because the actual colour of the flowers can be so difficult to describe. Typically it is in the orange range, being often recorded as apricot, buff, buff-orange or yellow, in addition to orange. Occasional populations have blue flowers, e.g., *Bidgood et al.* 4374. In our area they have been recorded from **K** 1, 3, **T** 1, 4 & 7. I first thought that these plants needed to be recognized taxonomically, but I have since learned that blue flowers are of sporadic occurrence throughout the Africa distribution of the species, being recorded in F.W.T.A., F.S.A. and Fl. Eth. Thus the trait 'blue flowers' in *C. subulata* is like other noticeable characters that are rare but widespread: they are worth mentioning but are not taxonomically significant.

Flowers have been recorded as up to 10 mm wide. It is likely however that plants with petals up to 8 mm long could have flowers up to half again as wide.

Commelina subulata can be so similar to other species, e.g. *C. polhillii*, that it is best collected whenever encountered, particularly in central and southern Tanzania where these other taxa occur. Every effort should me made to collect mature capsules and seeds, for without them the species is virtually impossible to identify with certainty in that region.

12. **Commelina polhillii** *Faden & Alford* in Novon 11: 16, fig. 1–2 (2001). Type: Tanzania, Iringa District: 9 km SW of Iringa along the Mbeya Road, *R.B. Faden, Phillips, Muasya and Macha* 96/94 (US!, holo.; K!, NHT!, iso.)

Annual with erect or ascending to decumbent shoots 7–20(–45) cm long, sometimes rooting at the lower nodes; roots thin, fibrous; stems slightly flattened on one side, often maroon or with maroon stripes, especially on older parts or near the nodes, glabrous except for a line of pubescence continuous with that of the distal sheath. Leaves with sheaths 3–7 mm long, usually split to the base, sometimes purple-veined, with a line of hairs along the split edge, otherwise glabrous; lamina sessile, often falcate, linear to linear-lanceolate, 1–8.4(–11.4) × 0.3–0.7 cm, margins often ciliate proximally, scabrid distally, apex acuminate, surfaces glabrous. Spathes solitary; peduncles 0.6–5 mm long, with a line of pubescence; spathes slightly to not at all falcate, 0.4–1.1(–1.7) × 0.2–0.6 cm, base cordate or occasionally truncate to hastate, margins free, ciliate, apex acute to acuminate; surfaces entirely glabrous, green (rarely the veins faintly purple); upper cincinnus absent or vestigial, lower cincinnus 3–4-flowered. Flowers mostly bisexual, occasionally male, 7–10 mm wide. Sepals translucent, tinged pink distally, 1.4–2.1 × 1.1–2.2 mm, upper sepal cup-shaped and ovate, paired sepals basally fused for about $^1/_2$ to $^2/_3$ of their length, each broadly elliptic; paired petals buff-orange, reniform and often reflexed, 3.5–4.8 × 2.5–4 mm of which the claw 1.2–2 mm, lower petal concolorous, transversely elliptic, 1.3–2.1 × 2–2.8 mm; staminodes 3; lateral stamens with filaments 1.6–3.2 mm long, anthers 0.6–0.7 × 0.4–0.7 mm, pollen orange; medial stamen filament 1.8–2.3 mm long, anther 0.8 × 0.8–1 mm, pollen orange. Ovary 0.7–1 × 0.4–0.7 mm; style 2.5–3.2 mm long; stigma capitate. Capsules trilocular, trivalved, 2.5–4 × 1.5–2.7 mm, apiculate; dorsal locule dehiscent, 1-seeded; ventral locules each 2-seeded. Seeds circular to slightly ellipsoid, slightly dorsiventrally compressed, 0.8–1.4 mm in diameter, with a large, deep central ± circular or V- to U-shaped dorsal pit, sometimes also with two shallow ventral pits, testa dark brown, lightly covered by white farinose granules. Figs. 17 & 18, pp. 165, 166.

subsp. **polhillii**

Dorsal pit of seed ± circular in outline; ventral pits lacking; leaves 3–7 mm wide; peduncle of lower cincinnus 2.5–3.3 mm long.

TANZANIA. Mpanda District: 10 km on Mpanda–Inyanga road, 15 May 1997, *Bidgood et al.* 3972!; Iringa District: Kalenga, 15 km SW of Iringa, 7 Feb. 1962, *Polhill & Paulo* 1375!; Mbeya District: Ruaha National Park, Magangwe, 18 May 1968, *Renvoize* 2241!

DISTR. **T** 4, 7; not known elsewhere

HAB. Woodland, murram pits, edge of cultivation, shallow soil on ironstone outcrop, sandy soil; 1050–1500 m

subsp. **kucharii** *Faden* **subsp. nov.** a subspecie typica seminibus fovea dorsali V-formi vel U-formi et foveis duabus ventralibus parvioribus instructus differt. Type: Tanzania. Singida District: 21 km on Babati road from Makiungu junction, *Kuchar* 24329 (MO! holo.; US!, iso.)

Dorsal pit of seed V- or U-shaped; two shallow ventral pits filled with farinose granules present; leaves up to 3 mm wide; peduncle of lower cincinnus 1.5–2.2 mm long

TANZANIA. Singida District: Ideba Hill area of Mgori Dam, generally SW-facing hillslopes, 22 Apr. 2000, *Kuchar* 23560! & 21 km on Babati road from Makiungu junction, 7 Apr. 2001, *Kuchar* 24329!

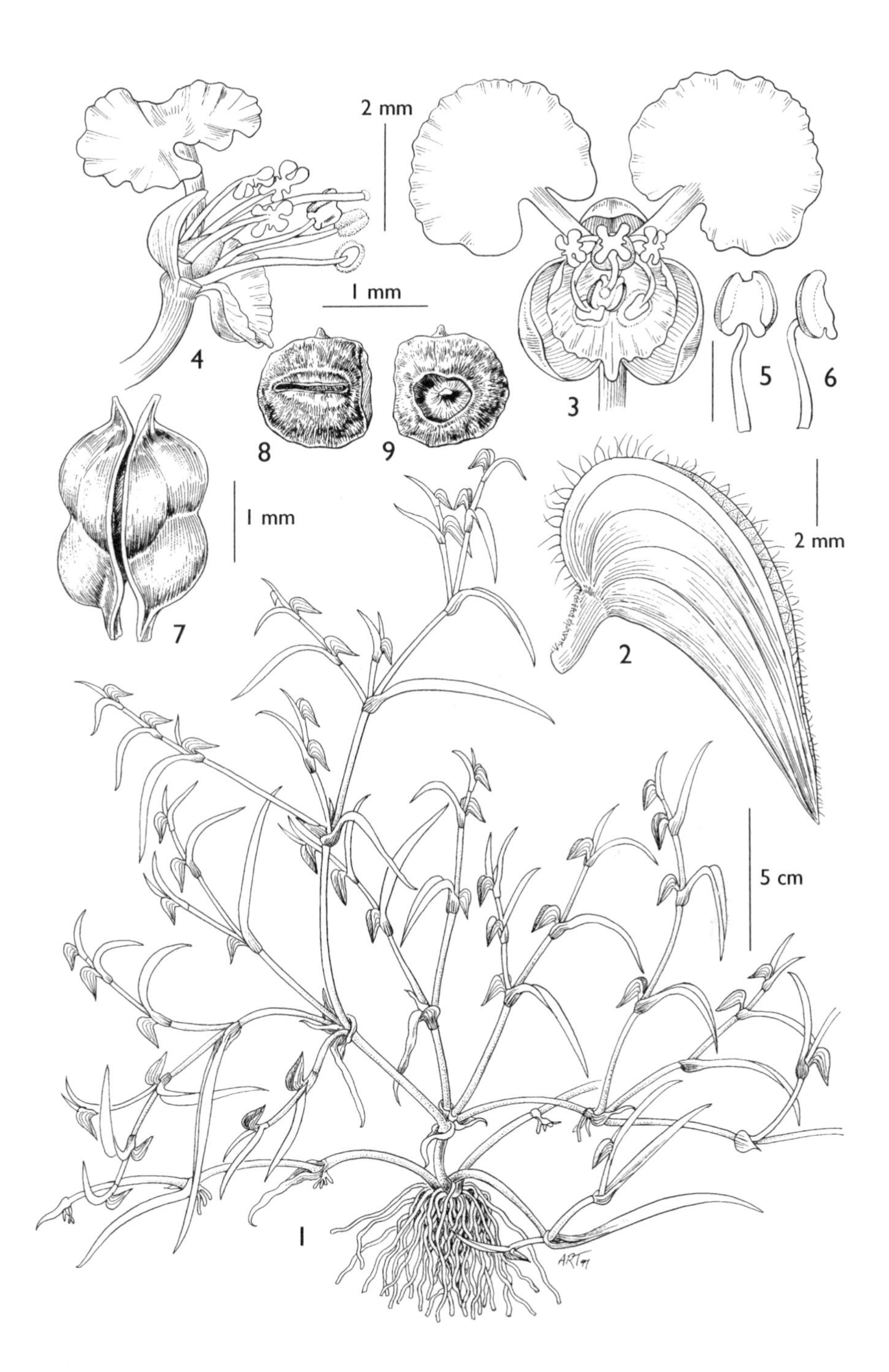

FIG. 17. *COMMELINA POLHILLII* — **1**, habit; **2**, spathe; **3**, bisexual flower, front view; **4**, idem, side view; **5** & **6**, medial stamen, front & side view; **7**, capsule; **8** & **9**, ventral locule seed, ventral & dorsal view. Reproduced with permission from Novon 11: 17 fig. 1 (2001). Drawn by A. R. Tangerini.

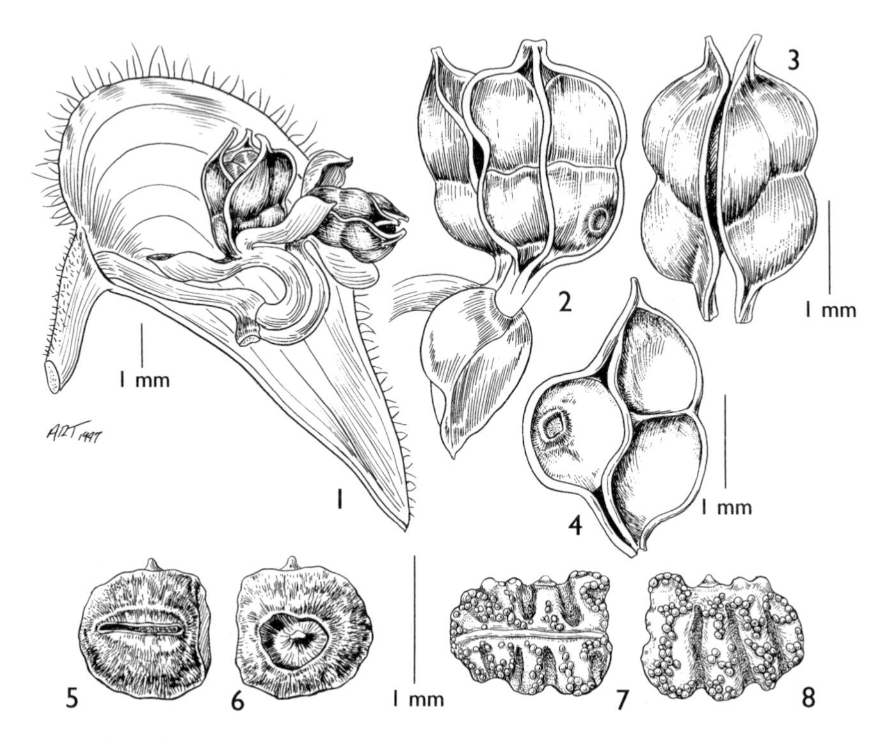

FIG. 18. *COMMELINA POLHILLII* — **1**, spathe with two dehiscing fruits, front half of spathe removed; **2** & **3**, dehiscing capsule, lateral & dorsal view; **4**, dorsilateral capsule valve, internal view; **5** & **6**, ventral locule seed, ventral & dorsal view. *COMMELINA SUBULATA* — **7** & **8**, ventral locule seed, ventral & dorsal view. 1–6 from *Faden et al.* 96/94B, ex cultivation Smithsonian Institution; 7–8 from *Faden et al.* 96/216. Drawn by A. R. Tangerini.

DISTR. **T** 5; not known elsewhere

HAB. Secondary bushland derived from miombo and *Brachystegia-Julbernardia* woodland with patchy *Combretum* thickets; 1350–1700 m

Flowering specimens have been seen from March to May.

NOTE. Both the MO and US sheets of *Kuchar* 24329 have loose spathes in a packet attached to the sheet. In the holotype 17 loose spathes were found with hairs on the surface as well as along the margin. On the whole these spathes were smaller, more falcate, and often darker in colour than the more numerous loose spathes that had only marginal hairs. In the spathes that bore capsules and seeds, it was found that the seeds present in the spathes with hairs on the surface matched those of *C. subulata* Roth, whereas the seeds from the spathes with only marginal hairs agreed with those from spathes attached to the plants mounted on the sheet. Although the majority of spathes of both types lacked seeds, it was clear that they represented two taxa, *C. subulata* and *C. polhillii* subsp. *kucharii*. It appears that the collector gathered detached spathes from the substrate below the plants and added them to the collection. Unfortunately, some of the spathes did not belong to the same species. I have treated the separate spathes of *C. subulata* as *Kuchar* 24329A, so they are excluded from the type collection.

NOTE (on the species as whole). *Commelina polhillii* is morphologically very similar to the sympatric *C. subulata* Roth (both species were collected at the type locality of *C. polhillii*). The seeds serve as the primary differentiating feature. The seeds of *C. polhillii* are round to slightly elliptic in outline (nearly isodiametric), with a single, central, round or U- to V-shaped, deep dorsal pit, while those of *C. subulata* are ovate with 3–4 deep furrows separated by warty ridges radiately arranged on the dorsal surface. The unusual seeds of *Polhill & Paulo* 1375 were initially spotted by Brenan, who separated this specimen as a distinct species related to *C. purpurea* Rendle in the Kew herbarium.

The spathes of *C. polhillii* are typically small, with glabrous surfaces and ciliate margins. Usually the veins are completely green, but they are rarely faintly purple. These spathes are identical to a subset of specimens of *C. subulata*. Additional collections of *C. polhillii* might demonstrate a greater variability in spathe characters. All specimens of *C. subulata* should be carefully checked for seeds, but especially those whose spathes match *C. polhillii* in pubescence.

From a limited amount of liquid-preserved flowers studied, plants of *C. subulata* with similar sized spathes to those of *C. polhillii* have the medial stamen anther lacking the sterile basal lobes found in *C. polhillii*, but in the single large-spathed collection of *C. subulata* studied (*Bullock* s.n.; **K** spirit No. 4568), the flowers did have sterile basal lobes. Until the significance of the variability of this character and others within *C. subulata* can be more fully evaluated, the utility of the character in possibly separating the two species remains unclear.

This species was named for Roger M. Polhill whose collections of liquid-preserved flowers of East African Commelinaceae have greatly facilitated the study of this family. Subsp. *kucharii* is named for Peter Kuchar whose collections from Singida District, Tanzania, and previously from Somalia and western Kenya, have greatly increased our knowledge of the distribution and variability in Commelinaceae species in these regions.

13. **Commelina sulcatisperma** *Faden* **sp. nov.** *C. polhillii* Faden similis seminibus fovea ad centrum sed fovea ventrali (nec dorsali) et spathis hirsutis differt. Type: Tanzania, Buha District: Moyowosi Game Controlled Area, 8 km NE of Murungu on Mpemvi–Murungu road, *Mutch* 298 (EA!, holo.)

Annual 20–35(–60?) cm high; roots thin, fibrous; shoots tufted, erect or ascending or some decumbent at the base and sometimes rooting at the proximal nodes, with a line of very fine uniseriate hairs descending from the distal node, otherwise glabrous. Leaves with sheaths to ± 1 cm long, with a band of short, uniseriate hairs along the line of fusion, white-ciliate at the summit; lamina linear, 2–14 × 0.2 cm (narrower distally), base broadened into the sheath, margin ciliate at base, otherwise glabrous, apex acuminate; surfaces apparently glabrous. Spathes solitary to subclustered at the distal nodes, decurved, falcate; subsessile, peduncles 1–4 mm long, with a band of fine pubescence; spathes 0.6–1.1 × 0.2–0.5 cm, base cordate to rounded, margins long-ciliate, sometimes blue, apex acuminate, surface densely hirsute, the veins often blue, with longitudinal rows of dark red dots between them; upper cincinnus 1-flowered (or lacking), lower cincinnus 5–6-flowered. Flowers probably bisexual and male; upper sepal ± lanceolate, ± 3 mm long, paired sepals long-fused, ovate, 3–3.5 mm long; paired petals either yellow (*Mutch* 298) or deep maroon to orange-red (*Mutch* 74); lower petal apparently lanceolate. Capsule uniformly medium brown, trilocular, trivalved (the dorsal locule splitting tardily), 5-seeded, oblong-elliptic (in dorsal view), 3 × 1.5 mm, not rostrate. Dorsal locule seed reniform-orbicular, 1.3 × 1.1 mm, testa alveolate, dark brown, with scattered irregular spots of deeper brown on the dorsal surface, ventral surface with a large, central ovate-deltate pit; ventral locule seeds ovoid-deltate, 1.1–1.2 × 1.2 mm, otherwise very similar to the dorsal locule seeds.

TANZANIA. Buha District: Moyowosi Controlled Area, 400 metres downstream from Malagarasi River bridge on Kibondo–Kasulu road, 26 Feb. 1972, *Mutch* 74! & 80 km S of Kibondo, 8 km NE of Murungu on Mpemvi–Murungu road, 27 March 1973, *Mutch* 298!

DISTR. **T** 4; not known elsewhere

HAB. Seasonally waterlogged depression in miombo woodland on lava soil; alt. 1220 m

Flowering specimens have been seen from February and March.

NOTE. This species is very similar to plants of *C. subulata* that have hirsute spathes. Its major distinction is in its seeds with a large ventral pit. The seeds somewhat resemble those of *C. polhillii*, especially subsp. *kucharii*, but in that species the large central pit is on the dorsal surface of the seed, the plants of *C. polhillii* are always much smaller and the spathe surface is always glabrous. In a dissected flower that had recently faded on *Mutch* 74 the paired sepals were fused at least two-thirds of their length and the lower petal was lanceolate and probably inconspicuous. In a dissected mature bud on *Mutch* 298, the paired sepals were completely fused and the lower petal was again lanceolate. These characters are highly unusual in the *C. subulata/C. purpurea* species group (species 10–21), but in view of all of the new taxa that are being discovered in this group, it is difficult to make generalizations.

The range of colours described for the flowers in this species is highly unusual. Because both collections were made by the same person, it is likely to be correct (unless some individual plants noted but not collected belonged to a different species).

14. **Commelina purpurea** *C.B. Clarke* in F.T.A. 8: 40 (1901); Brenan in K.B. 15: 208 (1961); Faden in U.K.W.F.: 657 (1974); Malaisse in Fl. Rwanda 4: 132, fig. 47 (1987); Ogwal in Mitt. Inst. Allg. Bot. Hamburg 23b: 577, fig. 2B (1990); Faden in Taxon 52: 832 (2003). Type: Kenya, N Nyeri District, between Ndoro and Guaso Thegu [Thego River], June 1893, *Gregory* s.n. (BM!, lecto. of Faden, 2003)

Tufted perennial; roots to 3 mm thick, not tuberous; shoots with thickened, subterranean bases, sometimes shortly rhizomatous, erect to ascending or decumbent, sometimes looping along the ground, rarely rooting at the nodes, much branched, 15–40(–85) cm long, seldom more than 30 cm high, glabrous or puberulous with a line of short uniseriate hairs below the nodes, occasionally more generally pilose, often marked with pink to maroon. Leaves with sheaths 0.3–2.7 cm long, with a line of hairs along the fused edge or occasionally pubescent throughout, ciliate at apex, the veins often purple; lamina often recurved, succulent, linear to linear-lanceolate, 1.5–16 × 0.1–0.5 cm, base broadened into the sheath, margins ciliate at base, rarely along the whole length, scabrid at apex, apex acute to acuminate; surfaces glabrous or occasionally pilose. Spathes solitary or loosely clustered distally; peduncles 2.5–8(–12) mm long, pubescent, especially in a line; spathes often strongly declinate, ± falcate, occasionally not or strongly falcate, 0.8–1.7(–2.2) × 0.3–0.8 cm, base cordate, margins free, ciliate at least proximally, apex acute to acuminate, rarely attenuate, surfaces glabrous to puberulous or pilose-puberulous, usually purple-veined; upper cincinnus 1-flowered, lower cincinnus 6–10-flowered. Flowers bisexual and male, 1.5–2.1 cm wide, upper cincinnus flower male, lower cincinnus flowers bisexual and male. Sepals free, 2–4 × 1–3 mm, the upper lanceolate-elliptic, lateral sepals broadly elliptic; paired petals ± buff to orange, salmon or apricot, broadly ovate, 10–11 × 7–9 mm of which the claw ± 3–5 mm; lower petal ovate or broadly ovate, 2–3 × 2 mm, concolorous with the paired petals; staminodes 3; lateral stamens with filaments 4.5–8 mm long, anthers 1.3–2 mm long, pollen orange; medial stamen filament 4–5 mm long, anther 1.4–2.1 mm long, pollen orange. Ovary ± 1.7 mm long; style 6–7 mm long; stigma violet. Capsules trilocular, trivalved, up to 5-seeded, 3.2–4.7 mm long, apiculate. Seeds strongly dorsiventrally compressed, transversely ellipsoid to reniform or deltate, 1.2–2.5 × 1–1.4 mm, testa pinkish brown to dark brown, radiately grooved or foveate and grooved.

UGANDA. Karamoja District: Napak, 28 May 1940, *A.S. Thomas* Th3625!; Masaka District: Lwera, 35 km on Masaka–Kampala road, 11 Feb. 1971, *Kabuye* 325!; Teso District: 24 km on Soroti–Moroto road, 13 Oct. 1952, *Verdcourt* 825!

KENYA. Uasin Gishu District: near Kaposoret [Kapsaret], 21 May 1951, *G.R. Williams* 194!; Nairobi District: behind Nairobi Golf Range, S of road to Nairobi National Park, 14 May 1974, *Faden, Faden & Ng'weno* 74/564! & N end of Kirichwa road, 1 May 1966, *Gillett* 17299!

TANZANIA. Masai District: 20 km on Kibaya–Kijungu road, 14 Jan. 1965, *Leippert* 5414!; Pare District: Mkomazi Game Reserve, below Ibaya Camp, 5 May 1995, *Abdallah & Vollesen* 95/145!; Kilosa District: Tembo, Jan. 1931, *Haarer* 2002!

DISTR. **U** 1, 2, 4; **K** 2–5; **T** 1–3, 6; Congo–Kinshasa, Rwanda, Burundi, Sudan

HAB. Seasonally moist grassland, montane grassland, open areas in scrub, wooded grassland, woodland, forest edge, shallow soil on or among rocks, edges of rock pools, swamps and marshes, roadsides, rarely a weed in cultivation; sandy or clayey soil; (750–)1050–2550 m

Flowering specimens have been seen from all months except April.

SYN. *Commelina kabarensis* De Wild., Pl. Bequaert. 5: 184 (1931). Type: Congo–Kinshasa, Kabare, *Bequaert* 3411 (BR!, holo.)

C. lugardii Bullock in K.B. 1932: 506 (1932); Faden in U.K.W.F. 2nd ed.: 305 (1994). Type: Kenya, Mt Elgon, *Lugard* 145 (K!, holo.)

NOTE. *Commelina purpurea* C.B. Clarke was based on two syntypes that represent different species (see Faden in Taxon 52: 832–835 (2003) for discussion). I had considered selecting [Speke &] *Grant* s.n. as lectotype because it was the collection on which the species name and most of the type description were based, but I was advised that, in the interest of nomenclatural stability, lectotypifying the species with *Gregory* s.n. would be the better course, which I followed. The other element in Clarke's species was described as *Commelina pseudopurpurea* Faden (see species 16).

The two collections from the lowest elevations, 760 m (*Haarer* 2002) and 875 m (*Abdallah & Vollesen* 95/145), among the easternmost localities, and one of only two specimens seen from **T** 6 as well as the sole specimen from **T** 3, respectively, are also unusual in having rhizomes that resemble a chain of beads. While this character has been noted occasionally elsewhere, particularly in western Kenya, e.g. *Maas Geesteranus* 5221, it is altogether uncommon in the species.

Lind 427 is unusual in having very long internodes (the longest seen in the species), very pubescent internodes, sheaths and laminae, and a spathe that is not at all falcate. The large size of the plant is reflective of other collections from **U** 3. The seed in the packet, which is probably a dorsal locule seed, is also unusual in being much less deeply pitted and furrowed, the particles on the ridges smaller, contrastingly darker in colour to the rest of the testa, and arranged mainly in single rows. Farinose granules are totally lacking. Unless some of these features can be found together in other examples, this must just be considered an unusual specimen. *Maitland* s.n., also from **U** 3, the longest plant measured, is also rather pubescent, although its internodes are nearly glabrous. It also has the most densely pubescent spathes seen. Seeds are lacking. *Lugard* 145, the type of *C. lugardii*, is pubescent similarly to *Lind* 427, but it is a much more dwarfed plant, perhaps reflecting its high elevation (2440 m).

This species appears intermediate between *C. reptans* and *C. subulata*, with both of which it may occur, e.g., in Nairobi. From *C. reptans* it differs by the lack of tuberous roots, the narrower leaves that are not at all twisted, spathes that are never long-pedunculate, smaller flowers and different seeds. From *C. subulata* it differs by its perennial habit, usually larger spathes, the presence of an upper cincinnus (lacking in *C. subulata* where the species overlap) and larger flowers.

15. **Commelina reptans** *Brenan* in K.B. 15: 208 (1961); Ogwal, Taxon. study Commelina in Uganda: 42, fig. 7 (1977); Blundell, Wild Fl. E. Afr.: 413, fig. 487 (1987); Malaisse in Fl. Rwanda 4: 132, fig. 48 (1987); Ogwal in Mitt. Inst. Allg. Bot. Hamburg 23b: 577, fig. 2a (190); Faden in U.K.W.F. 2nd ed.: 304 (1994); Ensermu and Faden in Fl. Eth. 6: 365, fig. 207.15 (1997). Type: Ethiopia, Mega, *Gillett* 14382 (K!, holo.)

Tufted perennial; roots tuberous, fusiform, usually stipitate; stems at first erect to ascending, to 30 cm tall, usually becoming decumbent and looping along the ground, to 50 cm long, rooting at the nodes, sometimes short, erect stems also produced from these nodes; internodes glabrous or shortly pubescent near the nodes, sometimes densely pubescent. Leaves with sheaths mostly 0.8–2 cm long, often purple or purple-striped, with a line of pubescence along the fused edge, otherwise glabrous to densely pubescent, ciliate with white hairs at apex; lamina sessile, usually linear, to 17.5 × 0.2–0.6 cm (proximal leaves) or lanceolate to linear-lanceolate and often laterally twisted, 1.5–7(–17) × 0.2–1.2 cm (medial to distal leaves), base narrowed into the sheath or rounded, margins ciliate at least proximally, scabrid at apex, apex acute to attenuate; surfaces glabous to densely pubescent. Spathes solitary, usually falcate; peduncles (0.5–)0.7–5(–9) cm long, pubescent in a line or all around; spathes (1–)1.5–2.6 × (0.4–)0.7–1.2 cm, base cordate (or truncate in very small spathes), margins free to the base, ciliate, apex acute to acuminate, surfaces green or rarely purplish, often green with purple to red veins, glabrous to densely pubescent; upper cincinnus lacking, vestigial or producing a single male flower, lower cincinnus 3–9-flowered. Flowers male and bisexual, ± 20–30 mm wide. Sepals reddish or pinkish brown, lanceolate-elliptic or ovate-elliptic, 4–5 × 2–4 mm, paired sepals fused up to half their length proximally or apparently free; paired petals buff or pinkish buff to orange buff, 15–17.5 × 9.5–17 mm of which the darker colored claw 4–6 mm long, lower petal ovate to ovate-elliptic, 4.5–5.5 × 3–4 mm, pale pink; staminodes 3; lateral stamens with filaments 7.5–9 mm long, anthers (1.3–)1.7–2.2 mm long; medial stamen filament 5–7 mm long, anther (2.2–)2.4–3.2 mm long. Ovary 1–2 mm long; style 9–12 mm long. Capsules pale yellow, sometimes marked with reddish purple flecks on the dorsal valve or with violet along the lateral sutures, trilocular, trivalved, up to 5-seeded, oblong-ellipsoid, 4–5 × 2–3 mm, dorsal locule usually present, dehiscent, deciduous, 1(–2)-seeded, ventral locules (1–)2-seeded. Seeds very variable in shape, dorsiventrally compressed, 1.6–2.5 × 1.5–1.8 mm, testa grey-brown, or mottled shades of grey, usually with light brown matted material in the depressions, irregularly foveolate to radiately ribbed, sparsely to densely white farinose over the whole surface.

UGANDA. Karamoja District: Mt Debasian, May, 1948, *Eggeling* 5817!; Ankole District: ± 26 km on Mbarara–Masaka road, 25 Sep. 1969, *Faden et al.* 69/1282!; Masaka District: ± 17 km SE of Ntusi, 19 Oct. 1969, *Lye* 4526!

KENYA. Naivasha District, near Lake Naivasha, Apr. 1932, *Napier* 1842!; Nairobi District: behind Nairobi Golf Range, 14 May 1974, *Faden, Faden & Ng'weno* 74/557!; Masai District: 8 km on Narok–Nairobi road, *Verdcourt* 3819!

TANZANIA. Musoma District: Seronera, 1 Jan. 1937, *Moore* 14!; Mbulu District: Tarangire National Park, 14 Feb. 1970, *Richards* 25435!; Moshi District: Sanya River, Moshi, Mar. 1928, *Haarer* 1223!

DISTR. **U** 1, 2, 4; **K** 3–6; **T** 1, 2; Burundi, S Ethiopia

HAB. Mainly in short seasonally moist grassland, especially in shallow soil on outcrops, occasionally in rock crevices, seepage at the base of outcrops, seasonal swamps, between tussocks in long grass, on disturbed ground, roadside ditches, open grassy patches in forest and shrubland; in clayey or sandy soil; (750–)1200–2000(–2550) m

Flowering specimens have been seen from all months, except August, but flowering is clearly bimodal in Kenya, March to July and October to January. This species flowers in the morning.

SYN. *Commelina coelestis* C.B. Clarke in F.T.A. 8: 40 (1901) pro parte, pro *Scott-Elliot* 6387 tantum, *non C. coelestis* Willd.

NOTE. This species is very distinctive because of its looping shoots, tuberous roots at the base of the plant, large, buff or pinkish buff to orange-buff flowers, large spathes on long peduncles, broad, often twisted and falcate leaves with strongly undulate margins. However, some of the most distinctive characters are often lacking. Some plants, perhaps young, may have only short, erect to ascending shoots, narrow leaves, and rather shortly pedunculate spathes. Even in these, however, the spathes still have longer peduncles and are typically larger than those in the *C. purpurea*, the only species in its range with which it can be confused. Thus, overall, even such 'atypical' plants are not difficult to identify to species.

Plants vary greatly in pubescence; some being nearly glabrous, others ± densely pubescent on all parts, including the spathes and peduncles. Spathes are sometimes pubescent inside regardless of whether they are pubescent externally, but this has not been investigated in detail.

The flower colour in *C. reptans*, as in other species in this group, has been variously described by collectors as: yellow (4 collections), yellowish, bronze-yellow, buff (6), buff-orange (3), orange-brown (4), orange-salmon, pinkish buff, apricot, buff to apricot, buff to almost brown, pale brown (2), light brownish yellow, pale rose pink, dull crimson to salmon, orange, light orange, creamy, purple to orange, dark purplish blue, and blue (3). Most of the apparent variation, I believe is due to the difficulty in describing this hue, rather than to a great variation in flower color. The records of purplish blue and blue are probably erroneous. Unfortunately, there have been no comparisons with colour charts. The lower petal always seems to contrast with the paired petals, although this is rarely commented upon and does not preserve well. In *Mathew* 6129 the carefully pressed flower shows a contrast in pollen colour between the medial (bright yellow) and lateral (creamy yellow) anthers.

Flowers of *C. reptans* throughout most of its range typically have a dark band across the medial anther, a character that is helpful in separating this species from *C. purpurea*, when they occur together, and seems to preserve well in herbarium specimens. However, the medial anther in several collections from Uganda and Tanzania (*Jarrett* 501, *Brown* 129 and *Richards* 20243) seems to be entirely yellow. If this can be confirmed, then there may be a geographic variation in this character.

Brenan distinguished *C. reptans* from *C. purpurea* by the former's stems looping along the ground and geniculate at the nodes, by its leaves usually falcate and broader, and by its larger spathes with longer peduncules. While the leaf and spathe characters work, some plants of *C. purpurea*, including the type of *C. lugardii*, have stems that loop along the ground. Unlike in *C. reptans* these stems rarely root at the nodes. *Commelina reptans* further differs from *C. purpurea* by the tuberous roots at the base of the plant and by its larger flowers. When the two species grow together, as they do in Nairobi, *C. reptans* generally grows in the drier spots and finishes flowering sooner.

Very few mature capsules and seeds have been available for study, and the seeds have been found to be rather variable. The first report of the occasional presence of two seeds in the dorsal locule was by Ensermu in Fl. Eth. Although two fully developed seeds have not been observed in our material, one maturing and one aborting seed have been seen together in a dorsal locule of *Faden et al.* 74/557, and the size of a dorsal locule in a capsule of *Dummer* 5166, also indicates that it was probably two-seeded.

Faden & Faden 71/889, collected at the highest elevation for this species (± 2530 m), is unusual is having quite small spathes, the longest of which is 1.5 cm, and seeds that have a radiately ribbed testa, which has been reported in Ethiopia and by Ogwal in Uganda, but has not otherwise been seen in the specimens examined for this study.

Koritschoner 1928 from Shinyanga is unusual in its small spathes (1.3–1.5 cm long) that are mostly not falcate (one is slightly falcate), that have an acute apex, and are very strong purple-striped. What ties it into this species are the broad leaves (to 1 cm wide) that are pubescent beneath.

Richards 25185, from Yaida Valley, Endashi, at or near the southern limits of this species, is unusual in a number of features. Its spathes are much more attenuate and more strongly falcate than those of any other specimen; the spathe margins are only ciliolate (vs. ciliate); and the seeds are deltoid or subtrilobed. The plant was collected in a habitat in which this species has otherwise not been reported, "swamp in wet clay among tussocks of long grass," which might account for its straggling habit and long leaves, the latter showing none of the characteristic features of leaves of *C. reptans*. Whether this is just an unusual form or might represent another taxon cannot be determined.

16. **Commelina pseudopurpurea** *Faden* in Taxon 52: 835 (2003). Type: Tanzania, Tabora District, [Unyamwezi district], 5°5'S, 33°E, 3600 ft, 1860–1863, *Grant* s.n. (K!, holo.)

Rhizomatous, sparsely branched perennial to 60 cm tall; roots thin or thick to ± tuberous, to 3 mm thick; shoots erect to ascending, sometimes looping and rooting at the lower nodes, glabrous except for a line of fine pubescence continuous with that of the sheath above. Leaves with sheaths ± 0.5–4 cm long, the upper ones sometimes split, purple-striped, with a line of fine pubescence along the fused edge, ciliate at apex; lamina linear to linear-lanceolate, the uppermost sometimes bract-like, 3–25 × 0.2–0.7 cm, base broadened into the sheath, margins ciliate proximally, otherwise glabrous, apex acuminate; surfaces glabrous, or the distal leaves sparsely pilose abaxially. Spathes solitary or 2–3 loosely clustered at the shoot apex, slightly falcate to not falcate; peduncles (0.3–)0.5–2.3(–3.6) cm, mostly with a longitudinal line of pubescence, occasionally densely puberulous; spathes 1–1.9(–2.4) × 0.6–1 cm, base cordate, margins free, ciliate or occasionally ciliolate, apex acuminate (to acute), surfaces usually purple-veined or striped, hirsute or puberulous or glabrous; upper cincinnus well-developed, 1-flowered, lower cincinnus 6–8-flowered. Flowers bisexual and male, ± 2 cm wide. Sepals all purple (drying reddish-maroon), upper lanceolate-elliptic, 3.5–4.5 mm long, hooded, paired sepals fused about half their length, broadly ovate, 4–5 mm long; paired petals dull orange, salmon or orange-brown (drying reddish purple), broadly ovate, 12–13 × 10 mm of which the claw ± 5 mm; lower petal purple, very broadly ovate, ± 2.5 × 4.5 mm; staminodes 3; lateral stamens with filaments 7–8.5 mm long, anthers 1.4–1.8 mm long; medial stamen filament 6–7 mm long, anther 2–2.4 mm long; style ± 10 mm long; stigma capitate. Capsules tan with dark brown flecks, sometimes with a purple band along the lateral line of dehiscence, trilocular, trivalved, 5-seeded, 4–6.2 × ± 2–2.5 mm, with an apicule, dorsal valves deciduous, dorsal locule dehiscent, 1-seeded, ventral locules 2-seeded. Ventral locule seeds ± dorsiventrally compressed, deltoid to broadly ovoid, (1.6–)1.9–2.3 × (1.4–)1.7–2 mm, testa grey or purplish grey, sometimes with scattered brownish flecks, with dense blackish or tan farinose particles in the depressions; dorsal locule seed broadly ellipsoid, 2.2–2.6 × 1.7–1.9 mm, otherwise similar to the ventral locule seeds.

TANZANIA. Dodoma District: 10 km N of Manyoni on Singida road, 15 Apr. 1988, *Bidgood, Mwasumbi & Vollesen* 1133!; Kondoa District: Great North Road, Kolo, 24 km N of Kondoa, 15 Jan. 1962, *Polhill & Paulo* 1177!; Singida District: 21 km Singida–Sepuka road, 13 Jan. 2001, *Kuchar* 23949!

DISTR. **T** 4, 5, 6/8, 7; not known elsewhere

HAB. *Acacia drepanolobium* bushland, sedge and grass marsh on alluvial flats, *Lannea-Commiphora* glade, *Lannea schimperi* woodland, riverine forest, sometimes growing near water, usually in black cotton soil; 1100–1550(–2100 m)

Flowering December to May.

SYN. *Commelina subulata* C.B. Clarke in DC., Monogr. Phan. 3: 149 (1881) pro parte, *non* Roth
C. purpurea C.B. Clarke in F.T.A. 8: 40 (1901), pro [Speke &] *Grant* s.n., not as to lectotype

NOTE. When I first studied the Grant collection, in connection with my work on U.K.W.F. Commelinaceae (Faden, 1974), I didn't realize that it was one of two syntypes for *C. purpurea*. I recognized, however, that it did not belong to the same species that we were calling *C. purpurea* in Kenya. Because there was another available name for the latter, namely *C. lugardii* Bullock, I adopted that name in the second edition of U.K.W.F. Having since confirmed that Clarke's second syntype, *Gregory* s.n., does in fact belong to the common species, I decided to lectotypify *C. purpurea* by *Gregory* s.n., thereby retaining that name for the common species, but leaving the central Tanzanian endemic without a name at that time.

The peduncles are mostly short, but, occasional long ones may occur, especially in the most proximal spathes. Rarely, plants with only long peduncles, e.g. *Kuchar* 34949 are found.

This species is geographically isolated from the two species with which it is most likely to be confused, *C. purpurea* and *C. reptans*. From the former it differs by its longer peduncles, the lower petal different in colour from the paired petals, and the seeds smaller and with a very different testa pattern. From *C. reptans* it differs by the stems rarely looping along the ground or rooting at the nodes, the leaves generally narrower, neither falcate nor twisted, smaller flowers, and a shorter upper cincinnus pedicel.

Commelina sp. nov. aff. *C. pseudopurpurea* Faden

Base lacking; shoots apparently unbranched and erect, to ± 45 cm tall. Leaves with sheaths 1–2 cm long, with a line of hairs along the fused edge, ciliate with colorless hairs at the summit; lamina sessile, linear, 5–21 × 0.4–0.6 cm, margins ciliolate throughout, ciliate proximally, slightly scabrid distally; surfaces apparently glabrous (adaxial not seen). Spathes solitary; peduncles 1–2 cm long, with a line of uniseriate hairs; spathes slightly or not falcate, 1.5–1.8 × 0.8–0.9 cm, apex acute to acuminate, base cordate, margins free, ciliate, surfaces glabrous, with abundant minute ruby-red dots forming lines parallel to the longitudinal veins; upper cincinnus shortly exserted from the spathe, 1-flowered; lower cincinnus ± 9-flowered. Flowers apparently orange (or a similar color), pedicels ± 3 mm long. Capsules pale yellow, trilocular, bivalved, (becoming trivalved by the tardy dehiscence of the dorsal locule), ± 4 × 2 mm, rostrate, apparently 1-seeded, the 4 ventral ovules all aborting and only the dorsal locule producing a seed, dorsal valve deciduous. Seed (immature) ellipsoid, ± 2.2 × 1.7 mm, testa dark brown, smooth, densely white-farinose.

TANZANIA. Tabora District: Kakoma, Feb. 1881, *Böhm* 47!

DISTR. **T** 4; known only from a single collection

HAB. In thornbush; altitude unknown

NOTE. *Böhm* 47 is an isosyntype of *Commelina echinosperma* K. Schum. (cf. *C. eckloniana* subsp. *echinosperma*). However, it completely disagrees with the type description and instead closely resembles *C. pseudopurpurea*. It agrees with the latter species in its long, linear leaves, long internodes, and solitary spathes with free, ciliate margins. The floral remains indicate an orange flower, which is also in agreement with *C. pseudopurpurea*. However, *Böhm* 47 significantly differs from *C. pseudopurpurea* by its leaves with margins ciliolate their whole length and ciliate at the base (vs. glabrous except for the ciliate base) and especially by its capsules which appear to have been derived from a trilocular, 5-ovulate ovary but are initially bivalved and one-seeded with only the dorsal locule seed maturing, this locule tardily dehiscent, containing a smooth, dark brown, white-farinose seed. Due to its unknown habit, sparse fruiting material and, especially, the lack of mature seeds, the material is too incomplete to describe.

17. **Commelina nyasensis** *C.B. Clarke* in F.T.A. 8: 40 (1901). Syntypes: Malawi, Tanganyika Plateau, Fort Hill, 3500–4000', July 1896, *Whyte* s.n. (K!, syn.); Mangonja Highlands, 4000', April 1859, *Kirk* s.n. (K!, syn.)

Sparsely branched annual (10–)20–60(–100) cm tall; roots thin or thick and cord-like, not tuberous; shoots usually erect to ascending, rarely decumbent, not rooting at the nodes, sometimes striped with reddish purple, with a line of fine pubescence continuous with that of the sheath above, otherwise glabrous. Leaves with sheaths mostly 1–2 cm long, commonly split to the base, usually purple-veined, with a line of fine pubescence along the fused edge, ciliate at the apex; lamina narrowly linear to linear-lanceolate, 1–30 × 0.1–0.6 cm, base broadened into the sheath, margins ciliate proximally, apex acuminate, surfaces apparently glabrous. Spathes solitary to densely clustered, usually slightly falcate; peduncles 0.3–1.3(–3.9) cm long, finely puberulous in a line; spathes often violet, 0.9–1.6(–2.5) × 0.5–0.8(–1) cm, base truncate, rounded or hastate, rarely cordate, margins free or nearly free, ciliate or ciliolate at least proximally, apex abruptly acute, acuminate or attenuate, the tip often turned down, surfaces often violet-veined, usually glabrous (see note); upper cincinnus usually lacking or vestigial and enclosed within the spathe, very rarely shortly exserted and 1-flowered; lower cincinnus (1–)2–4-flowered. Flowers 10–16 mm wide. Sepals brick red or maroon, the upper ovate, 2.5–3 mm long, the paired sepals ± 3–3.5 mm long, partially fused basally; upper two petals yellow-orange or apricot (rarely brownish yellow or nearly white); lower petal pink or reddish, relatively broad, ± 3 mm long; staminodes 3; lateral stamens with filaments ± 4 mm long; medial stamen filament ± 3.5 mm long, anther ± 1 mm long; style ± 4 mm long; stigma capitate. Capsules light brown, often heavily spotted with dark brown, trilocular, trivalved (or sometimes apparently bivalved), 5-seeded, oblong-ellipsoid, (4–)4.5–6 × 2–3 mm, usually contracted between the seeds, apiculate, dorsal locule dehiscent (sometimes tardily) (or indehiscent?)), 1-seeded, ventral locules 2-seeded. Ventral locule seeds dorsiventrally compressed, broadly ovoid to ovoid or deltoid, 1.5–2.2 × 1.3–1.7 mm, testa uniformly dull light or dark grey (occasionally slightly darker in the shallow depressions), usually alveolate (rarely furrowed); dorsal locule seed ellipsoid or broadly ellipsoid, 1.7–2.7 × 1.3–1.6 mm, testa similar to the ventral locule seeds but depressions often less pronounced.

TANZANIA. Iringa District: 20 km on Iringa–Mbeya road, 10 June 1996, *Faden et al.* 96/117!; Songea District: by waterfall on Luhira River near Mshangano fish pond N of Songea, 25 Apr. 1956, *Milne-Redhead & Taylor* 9805!; Tunduru District: road 100 km from Masasi, 19 Mar. 1963, *Richards* 17975!

DISTR. **T** 4–8; Congo–Kinshasa, Zambia, Malawi and Zimbabwe

HAB. Marshes, often dominated by *Leersia*, swampy or seasonally inundated grassland sometimes dominated by *Loudetia*, wet depressions and seepage areas in grassland, woodland and wooded grassland, edge of marshy areas in cultivation, wet ditches and banks, hillside seepage, and among rocks by waterfall; (250–)900–1900 m

Flowering specimens have been seen from March to June and October.

NOTE. This species was known from our area from five collections prior to 1996. Since then at least 25 additional collections have been made. In June, 1996 we found this species, generally in flower and fruit, in almost all wet places – marshes, grassy depressions, roadside ditches, etc. – that we investigated in southern Tanzania. Its abundance at that time of year, after the rains had ceased, and the absence of any collections from December through February, with only one from March, indicate this it is indeed a late-flowering species, which could account for the previous paucity of collections.

This species is very variable in spathe size, peduncle length, extent of marginal pubescence, capsule dehiscence, and seed testa. In the field the plants seemed to fall into two groups: 1) those with larger spathes, occasionally with a long peduncle, lacking violet margins, tips and veins, the capsules regularly trivalved and the seeds grey and alveolate; and 2) plants with smaller spathes that are always short-pedunculate, with violet margins and tips and sometimes veins, the capsule typically bivalved, the dorsal valve deciduous without (or before?) dehiscing, the seeds dark brown and foveolate-reticulate. The first group also appeared to grow in wetter situations, sometimes in standing water, whereas the second group on the whole occurred in somewhat drier situations, such as the margins of marshy places.

While it seems likely that some infraspecific taxa will need to be recognized in *C. nyasensis* it is not clear that the two "groups," described above exactly define all of the variation. Although the extremes in testa pattern and capsule dehiscence could be sufficient to define distinct species, if they were consistent, the testa pattern appears to be more variable than what is outlined above, and we do not have full data on capsule dehiscence. Moreover, the specimens with the largest spathes and longest peduncles mostly lack capsules and seeds, so the placement of some of them even in this species is not without doubt. One such example, *Vollesen* in MRC 4589 from the Selous Game Reserve, has much longer spathe peduncles (1.7–3.9 cm) and comes from a much lower elevation (275 m) than any other specimen included in the species and it represents the only flowering specimen from October. *Schlieben* 6357, also with long spathe peduncles, has a single foveolate-reticulate seed that does not exactly match those of any other collection. It appears that we still need additional specimens to sort out the variation within this species.

Two characters that seem to vary in this species, namely capsule dehiscence and spathe margin fusion require comment because these characters are important for determining relationships within *Commelina*. In my field notes for the collection *Faden et al.* 96/399 it was recorded that the dorsal capsule locule was indehiscent when it was shed. Upon reexamining the separate seed collection some 13 years later, it appears that nearly all of the dorsal valves still show no distinct lines of dehiscence although many of them are irregularly broken and the dorsal locule seed has been freed or nearly so. This might have resulted from the packaging and handling, however. When the plant was cultivated from field-collected seeds, the dorsal locules did not dehisce at all, but it appeared that those capsules might not have fully developed, so a direct comparison with the field-collected capsules might not be informative. The upshot of the probability that at least some populations of *C. nyasensis* do not have trivalved capsules is that this character defines the *C. subulata/purpurea* species group (species 10–21 of this account), to which *C. nyasensis* would otherwise appear to belong.

Even more unexpected is the apparent fusion for a very short distance (up to ± 1 mm) of the spathe margins at the base, e.g. in *Bidgood et al.* 3958. It may not be conspicuous or even present in all specimens of *C. nyasensis*, but species 1–21 should have completely free margins, whereas species 22–51 have the spathe margins fused for a greater distance than what is shown in this species. Perhaps the typical lack of a cordate spathe base in this species has led to the short fusion of the spathe base. An apparently undescribed *Commelina* species from southwestern Ethiopia with a spathe very similar to that of *C. nyasensis* also shows a short basal fusion of the spathe margins.

Four (out of 30) collections seen from our area have at least sparsely hirsute spathes. Their seeds have been found not to be uniform, however, so most probably this is just a minor variation. Two such collections, *Faden et al.* 96/267 & 96/324, were made from among populations with spathes with typically glabrous surfaces. *Faden et al.* 96/267 was a single plant whose seeds differed markedly from those of other plants in the population that lacked hirsute hairs (collected as *Faden et al.* 96/266), whereas both *Faden et al.* 96/324 (hirsute spathes) and *Faden et al.* 96/323 (glabrous surface spathes), from the same population, had identical seeds.

18. **Commelina trilobosperma** *K. Schum.* in P.O.A. C: 134 (1895); C.B. Clarke in F.T.A. 8: 39 (1901); Ogwal in Mitt. Inst. Allg. Bot. Hamburg 23a: 578 (1990); Faden in U.K.W.F. 2nd ed.: 305 (1994). Type: Tanzania, Mwanza District: Karumo, Usindji, *Stuhlmann* 3566 (B!, holo.)

Sparsely- or densely-branched annual 12–30(–60)cm tall; roots thin; shoots erect or, in larger plants, often decumbent and rooting at the base, then ascending, with a line of fine pubescence continuous with that of the sheath above, otherwise glabrous. Leaves with sheaths to 1.5 cm long, commonly split to the base, with a line of hairs along the line of fusion, rarely hirsute all over, ciliate or eciliate at the apex; lamina of proximal leaves linear-lanceolate to lanceolate-oblong, 3–10 × 0.5–1.1 cm, base tapered into the sheath to rounded, margins strongly undulate, ciliate proximally, apex acute to acuminate or cuspidate; surfaces usually glabrous, rarely the abaxial hirsute; distal leaves strongly falcate, 0.8–5 × 0.5–1.5 cm, base broadly cuneate to cordate-amplexicaul, often partially enclosing the spathe. Spathes solitary, often strongly curved downward, not to slightly falcate; peduncles usually completely hidden, ± 2–7 mm long, with a line of pubescence; spathes 0.6–1(–1.8) × 0.3–0.6(–0.9) cm, base cordate to hastate, margins free or very shortly fused at the base, scabrid to ciliolate or ciliate, apex acute to acuminate, surface glabrous; upper cincinnus vestigial and included or 1-flowered and exserted; lower cincinnus 3–6-flowered. Flowers bisexual and male, probably 1.5–2 cm wide; upper sepal 2.5–3 mm long, hooded; paired sepals shallowly cup-shaped, fused proximally for ± half (to $^{4}/_{5}$) their length, 3–3.5 mm long; paired petals apricot, yellow-orange, orange, apricot-brown, buff or pink-brown, reniform to broadly cordate, 10–11 × 9 mm of which claw 4–5 mm; lower petal pale pink, magenta pink or rose-purple, broadly orbicular, 3 × 4 mm; staminodes 3; lateral stamens with filaments 5.5–6 mm long, anthers 1–1.5 mm long; medial stamen filament ± 5 mm long, anther 1.5–1.8 mm long; style 6–6.5 mm long; stigma capitate. Capsules tan with numerous, large or small, dark brown spots, trilocular, trivalved, 5-seeded, with a very short downturned apicule, 3.5–4.7 × ± 2 mm, dorsal valves deciduous, dorsal locule dehiscent, 1-seeded, ventral locules 2-seeded. Ventral locule seeds dorsiventrally compressed, deltoid in outline, trilobed, 1.6–2.3 × 1.4–1.7 mm, testa beige or greyish tan, with two diagonal or vertical grooves, sometimes with a dark brown wash or streaks, very finely pitted or rugulose on the central boss and sometimes on the lateral lobes, in general white-farinose.

UGANDA. Teso District: Kapiri Ferry, 16 Sep. 1954, *Lind* 425!

KENYA. Central Kavirondo District: Paponditi, 10 Dec. 1968, *Kokwaro* 1643! & Kabotho near Nyalunga River, 2.3 km beyond Paponditi on the Nyakwere road, 18 Oct. 1970, *Evans & Maikweki* 13!

TANZANIA. Musoma District: Serengeti, Seronera, 23 Apr. 1961, *Greenway* 9879!; Mbulu District: near Great North Road, 1.6 km N of Magugu, 20 Apr. 1964, *Welch* 565!; Shinyanga District: Shinyanga new aerodrome, 6 May 1931, *B.D. Burtt* 2425!

DISTR. **U** 3; **K** 5; **T** 1, 2; not known elsewhere

HAB. Open grassland especially in places that are seasonally wet or marshy, roadsides, sandy washes or open sandy area with sparse vegetation; 1000–1550 m

Flowering specimens have been seen from March to May, September and October.

NOTE. This is a very distinctive species because of its three-lobed seeds and short, broad, falcate, amplexicaul upper leaves that partially conceal the base of the spathe *Greenway* 9879 is unusual in a number of characters: it is the most robust, although not the tallest, specimen; the upper leaves are the least reduced, and its spathes are by far the largest seen in any specimen. Furthermore, it is the only specimen that has hirsute sheaths and abaxial leaf surfaces in the lower leaves. The flowers appear typical for this species, however, and the seeds differ from other specimens mainly in being slightly smaller and in having the grooves more vertical than diagonal. Overall, it is not distinct enough to warrant separation at any taxonomic level.

19. **Commelina merkeri** *K. Schum.* in E.J. 36: 207 (1905). Type: Tanzania, Masai steppe, S, SW and W of Kilimanjaro [Kilimanscharo] and Meru, 1902, *Merker* s.n. (B!, holo.)

Annual 10–60 cm tall, usually sparsely branched; roots thin, fibrous; shoots ascending, rarely decumbent and rooting at the nodes, with a line of fine pubescence continuous with that of the sheath above, otherwise glabrous. Leaves with sheaths 0.7–2 cm long, commonly split to the base, with a line of fine pubescence along the sometime fused edge, ciliate to glabrous at apex; lamina sessile, linear to linear-oblong or narrowly lanceolate to ovate in distal leaves, 2.5–14 × 0.3–0.8 cm, base gradually narrowed into the sheath (larger leaves), sometimes rounded, margins glabrous or ciliate proximally (in larger leaves), or scabrid distally, apex acute to acuminate; surfaces usually glabrous, rarely the abaxial sparsely pubescent. Spathes solitary to clustered, subsessile, mostly strongly curved downward against the stem, the base sometimes hidden by the base of the subtending leaf; peduncles 2.5–5(–7) mm long, sparsely puberulous in a vertical line; spathes falcate, (0.8–)1.2–1.5(–1.8) × 0.4–0.8 cm, base cordate, margins free, ciliolate to ciliate, apex acute to acuminate, surface hirsute (occasionally glabrous); upper cincinnus often well developed, included to shortly exserted from the spathe, producing a single male flower, lower cincinnus 4–6-flowered. Flower bisexual and male, 11–17 mm wide. Sepals white, upper sepal lanceolate-elliptic, ± 3 × 1 mm, paired sepals free, ovate, 3–4 × 2–2.5 mm. Petals orange-buff (also recorded as metallic bronze, apricot-yellow, dull orange, pale brown, or blue), the upper ovate-reniform, 9–10 × 7–8 mm of which the claw 4–5 mm; lower petal broadly ovate, 2.5–3.5 × 3–4.5 mm, concolorous; staminodes 3; lateral stamens with filaments ± 6 mm long, anthers ± 1 mm long, pollen orange; medial stamen filament ± 4 mm long, anther ± 1.5 mm long, pollen orange; style ± 5 mm long; stigma capitate. Capsules light brown, with maroon flecks, trilocular, trivalved, 5-seeded, 3.5–5 × 1.5–2 mm, apiculate, dorsal valves deciduous, dorsal locule dehiscent, 1-seeded, ventral locules 2-seeded. Ventral locule seeds dorsiventrally compressed, deltoid in outline, trilobed, 1.7–2.3 × 1.3–1.7 mm, testa usually greyish tan, often with fine darker brown streaks, with two diagonal grooves, very finely pitted dorsally on the boss and sometimes on the lobes; dorsal locule seed similar to the ventral locule seeds, but more ovoid and less trilobed, 1.8–2.3 × 1.6–1.8 mm.

TANZANIA. Dodoma District: Km 37 on Dodoma–Manyoni road, between Issanga and Mlejeou, 24 Apr. 1962, *Polhill & Paulo* 2122!; Iringa District: Ruaha National Park, 1.5 km SW of Msembe, 3 May 1971, *Bjørnstad* 962!; Mbeya District; Usangu Plain, near Utemngule [Utencile], 29 Jan. 1963, *Richards* 17599!

DISTR. **T** 2, 5, 7; Zambia? (see note)

HAB. Dry or moist grassland, lake margins, sometimes on soils with impeded drainage, on flat granite rocks with shallow pools [with] damp marsh in between; 800–1350 m

Flowering specimens have been seen from January to April.

SYN. *C.* sp. aff. *trilobosperma* of Bjørnstad, Vegetation Ruaha National Park: 33 (1976)

NOTE. This species is distinctive because of its trilobed seeds and subsessile, strongly downward curved spathes. It can be confused only with *C. trilobosperma* which has similar seeds and downward curved spathes. *C. merkeri* can usually be distinguished by the absence of short, broad, falcate amplexicaul bract-like upper leaves partially concealing the bases of the spathes (not clear in some specimens), by the spathes usually hirsute, and by having all 3 petals ± concolorous. Plants of *Faden et al.* 96/503 grown from seed at US eventually produced glabrous spathes although the field collection had hirsute spathes, so this character may be influenced by environmental conditions.

Anderson 368 is somewhat unusual in having all of the leaves complicate and sparsely pubescent abaxially.

Plants similar to *C. merkeri* have been seen from the Luangwa Valley, Zambia, e.g. *Astle* 5034. They differ in having smaller, glabrous to sparsely hirsute spathes and narrower, complicate leaves. Depending upon their flowers, which are unknown, they likely would have to be considered a new species or a subspecies of *C. merkeri*.

20. **Commelina pallidispatha** *Faden* **sp. nov.** a *Commelina chayaensi* Faden internodiis inferioribus haud spongiosis spathis margine ciliato et pagina plerumque hirsuta (nec spathis omnino glabris) pedunculis spatharum brevioribus et linea distali trichomatum instructis cincinno inferno 3–8 (nec 9–11) flores efferenti sepalis infernis breviusibus (2.5–3.5 vs. 4–4.5 mm longis) capsulis angustioribus (1.5–2 vs. 2.3–2.6 mm latis) differt; a *C. trilobosperma* K. Schum. et *C. merkeri* K. Schum. pedunculis spatharum longioribus seminibus nec trilobatibus differt. Type: Tanzania, Iringa District: Trekimboga Track down towards the Igawira Drainage Line, *Greenway & Kanuri* 14475 (K!, holo., EA! (3 sheets), iso.)

Much-branched, tufted annual herb 20–75 cm tall; roots thin, fibrous; shoots much branched, erect to ascending, not rooting at the nodes, with a longitudinal line of fine pubescence continuous with that of the sheath above, otherwise glabrous. Leaves with sheaths to 1(–1.5) cm long, commonly split to the base, crimson-veined, with a line of fine pubescence along the fused edge, ciliate at the apex; lamina linear, ± inrolled, 1.5–13 × 0.1–0.2 cm, base broadened into the sheath, margins ciliate proximally, apex acuminate, surfaces apparently glabrous. Spathes mostly solitary, occasionally two together subterminally, not at all to strongly falcate, usually not curved downward; peduncles (2–)5–15 mm long, with a line of fine uniseriate hairs distally; spathes 0.7–1.3 × 0.2–0.5 cm, base rounded to truncate or cordate, margins free, long-ciliate, green or purple, apex acuminate to attenuate, surface usually densely long-hirsute, occasionally sparsely puberulous or glabrous, whitish with contrasting green or purple veins; upper cincinnus well developed, one-flowered; lower cincinnus 3–8-flowered. Flowers bisexual and male; upper sepal white or tinged with mauve, lanceolate, 2.5–3.2 mm long; paired sepals fused proximally for about × to $^2/_3$ their length, ovate, 2.5–3.5 × 1.5–2 mm, apparently lavender, mauve, reddish purple or violet tinged; paired petals drying reddish (sienna red or yellow brown on labels), broadly ovate, ± 7.5 × 6 mm of which the concolorous claw ± 2.5 mm; lower petal drying the same colour as the paired petals or pale lavender and possibly contrasting with them, broadly ovate to ovate-reniform, 2–2.5 × 3.8 mm; staminodes 3; lateral stamens with filaments (4–)5.5 mm long, anthers 0.7–1.2 mm long, medial stamen filament 3–3.3 mm long, anther 0.9–1.2 mm long. Ovary ± 1 mm long; style (4.5–)5–6.5 mm long; stigma capitate. Capsule light brown, sometimes with dark brown flecks, sometimes pale violet along the lateral suture, trilocular, trivalved, 5-seeded, 3–4.8 × 1.5–2 mm, including a long downward curved beak; dorsal locule dehiscent, 1-seeded, ventral locules 2-seeded. Ventral locule seeds dorsiventrally compressed, ovoid to deltoid, 1.3–1.7 × 1.1–1.6 mm, testa pale grey-tan, sometimes with very fine dark brown streaks, irregularly finely pitted; dorsal locule seed broadly ovoid to ± hexagonal, 1.3–1.8 × 1.2–1.5 mm, otherwise similar to the ventral locule seeds.

TANZANIA. Tabora District: Tabora, 5 Mar. 1924, *Durham* s.n.!; Dodoma District: Kazikazi, 17 May 1932, *B.D. Burtt* 3653!; Mbeya District: Madibira–Igawa track, 14 km SW of Madibira, 12 June 1996, *Faden et al.* 96/175!

DISTR. **T** 4, 5, 7; not known elsewhere

HAB. Spring head, seepage area in woodland, seasonally wet grassy area in woodland, *Lannea* glades or fringing seasonal rain-ponds; on sand and sandy clay loam; 850–1300 m

Flowering in March, May and June.

NOTE. The flower descriptions and dimensions come from *Durham* s.n. and *Burtt* 3653 (cited above). Overall, the flowers are rather similar in their dried appearance, although the flower colours on the labels, "yellow-brown" and "sienna red" make them sound very different in hue. In the *Durham* collection, in which some flowers were carefully pressed, the lower petal appears concolorous with the paired petals, whereas in the *Burtt* specimen it appears somewhat discolorous. No other collections record the flower colour or have useful floral remains. Based on the pressed flowers in the *Durham* collection, the upper cincinnus flower is always male, whereas the lower cincinnus flowers are mostly bisexual. A single, functionally male flower with an underdeveloped gynoecium was seen. It most likely had been on a lower cincinnus because of its short pedicel.

Commelina pallidispatha resembles *C. chayaensis* in its pale, sand grain-like seeds and general aspect. It differs by its lower internodes not spongy, the spathes always ciliate marginally and usually hirsute (vs. completely glabrous), spathe peduncles shorter and with a distal line of pubescence, the lower cincinnus fewer flowered, smaller sepals and narrower capsules. It differs from *C. trilobosperma* and *C. merkeri*, whose pale seeds also resemble sand grains, by longer spathe peduncles and its seeds not trilobed.

The spathes of *C. pallidispatha* are almost always hirsute, in addition to having ciliate margins. The exceptions are *Faden et al.* 96/175, in which the spathes have a completely glabrous surface, and *Carter, Abdallah & Newton* 2336, from nearly the same locality in **T** 7, in which the single spathe on one shoot has a glabrous surface whereas the two spathes on the second shoot have numerous short, uniseriate hairs on the surface. Perhaps these are just local variants, but there are too few good specimens to be certain.

Eight collections of this species have been seen, several of which are quite poor and only questionably placed here.

21. **Commelina chayaensis** *Faden* **sp. nov.** a *C. pallidispatha* Faden internodiis inferioribus spongiosis spathis omnino glabris pedunculis longioribus glabris cincinno inferno 9–11 flores (nec 3–8 flores) efferenti sepalis infernos longioribus (4–4.5 vs. 2.5–3.5 mm longis) capsulis latioribus (2.3–2.6 vs. 1.5–2 mm latis) differt. Type: Tanzania, Dodoma District, Lake Chaya, *B.D. Burtt* 3801 (K!, sheet II, holo.; K!, sheet I, EA!, iso.)

Probably annual herb to ± 45 cm tall; roots relatively thick and cord-like, fibrous; shoots decumbent at the base, then erect to ascending, the lower internodes to ± 5 mm thick and spongy, glabrous. Leaves nearly vertical; sheaths to ± 3 cm long, commonly split to the base, sometimes not well differentiated from the lamina, glabrous; lamina linear, 2.5–12 × 0.1–0.5 cm, base broadened into the sheath, margins glabrous, apex acute; surfaces glabrous. Spathes solitary to subclustered, not to slightly falcate; peduncles (0.7–)1.1–4(–5.5) cm long, glabrous; spathes 0.7–1.1 × 0.3–0.5 cm, base cordate, margins free, glabrous, sometimes purple, apex acute to acuminate, surfaces sometimes purple-veined, glabrous; upper cincinnus well developed, 1-flowered; lower cincinnus ± 9–11-flowered. Flowers bisexual and male; upper sepal 3–4 mm long, probably greenish white; paired sepals 4–4.5 mm long, fused proximally, evidently flushed with purple at least along the margins. Petals orange (fide Burtt). Capsules light brown, often with numerous maroon spots, sometimes the ventral valve with a longitudunal purple band, trilocular, trivalved, 5-seeded, 3.5–5 × 2.3–2.6 mm, apiculate with a 1 mm long downturned apicule, dorsal valves deciduous, dorsal locule dehiscent, 1-seeded, ventral locules 2-seeded. Ventral locule seeds dorsiventrally compressed, deltoid to ovoid, 1.5–1.8 × 1.5–1.7 mm, testa grey or grey-tan, alveolate-scrobiculate to very shallowly reticulate-foveolate; dorsal locule seed broadly ovoid, 2–2.1 × 1.7–1.8 mm, the pits on the dorsal surface more radiately aligned and groove-like, otherwise similar to the ventral locule seeds.

TANZANIA. Dodoma District: Lake Chaya, 12 July 1932, *B.D. Burtt* 3801! & same locality, 27 Apr. 1964, *Greenway & Polhill* 11743!

DISTR. **T** 5; known only from Lake Chaya

HAB. Mud flats along the lake margin; 1200–1250 m

Flowering specimens have been seen from April and July.

NOTE. This species is very distinctive because of its swollen shoot bases, small, long-pedunculate, completely glabrous spathes, and sand grain-like seeds that are not trilobed. The capsules are broader and the seeds less dorsiventrally compressed than some other similar-appearing species. The large number of flowers per spathe is also distinctive.

22. **Commelina capitata** *Benth.* in Fl. Nigrit.: 541 (1849); C.B. Clarke in DC., Monogr. Phan. 3: 176 (1881) & in F.T.A. 8: 54 (1901); Morton in J.L.S. 55: 519 (1956) & in J.L.S. 60: 179 (1967); Brenan in F.W.T.A., 2nd ed.: 3: 47 (1968); Malaisse in Fl. Rwanda 4: 129, fig. 45 (1988); Ogwal in Mitt. Inst. Allg. Bot. Hamburg 23b: 579 (1990); Faden in U.K.W.F. 2nd ed.: 305 (1994). Type: Liberia, Cape Palmas, *Vogel* 52 (K!, holo.)

Perennial; roots thin, fibrous; shoots single or tufted, erect to ascending, decumbent or scrambling, usually 20–60 cm tall, rarely to 245 cm (in scrambling plants), sometimes densely branched, glabrous, glabrescent or puberulous (especially distally). Leaves distichous; sheaths fused, except when split by lateral shoots, sometimes reddish purple, 1–3 cm long, puberulous, ciliate with long red hairs at apex; lamina shortly petiolate, lanceolate-oblong to ovate, 4.5–15.5 ×

(1.5–)2–5 cm, base strongly oblique with one side cuneate and the other side rounded, margins glabrous or rarely ciliate at base, often scabrid distally, apex acuminate to attenuate or occasionally acute; adaxial surface glabrous or sparsely puberulous along the midrib, abaxial surface puberulous, sometimes restricted to along the midrib. Spathes densely clustered (solitary in early flowering), subsessile; peduncles largely hidden, 3.5–7 mm long, glabrous or puberulous; spathes not or only slightly falcate at apex, 1.3–2.6 × 0.8–1.3(–1.7) cm, folded edge often somewhat saccate basally, base cordate, margins free or fused for up to 7 mm basally, ciliate, apex acute to obtuse; surfaces puberulous, also often sparsely hirsute with long red hairs, pale green, whitish basally, strongly contrasting with the dark green leaves; upper cincinnus 1-flowered, its flower male or bisexual, lower cincinnus 3–6-flowered. Flowers bisexual and male, 1.5–1.7 cm wide; upper sepal greenish white, ± 4 mm long, paired sepals greenish white, forming a cup, fused for ± $^3/_4$ of their length, each ± 5 × 3 mm; paired petals yellow, ovate-cordate, 8–14 × 9–11 mm of which the claw ± 4 mm; medial petal yellow at base, white distally, ± 6 × 1 mm; staminodes 3; lateral stamens with filaments ± 6 mm long, anthers ± 1 mm long, pollen yellow; medial stamen filament ± 5 mm long, anther ± 1.5 mm long, pollen yellow; style ± 7 mm long; stigma capitate. Capsules white when immature, light brown at maturity, bilocular, bivalved, 4-seeded, oblong-elliptic to elliptic, (6–)8–9.3 × (3.5–)4.8–6.5 mm, apex rounded to ± truncate, rarely slightly emarginate, valves persistent, locules 2-seeded or fewer by abortion. Seeds transversely ellipsoid, dorsiventrally compressed, 3.2–4.5 × (1.7–)1.8–2.3 mm, testa medium brown, the dorsal surface foveate or foveolate, completely covered by a dull orange-yellow to creamy yellow material that is fleshy when fresh and apparently an aril, this material sometimes forming a longitudinal middorsal crest, or the material clearly pitted, the ventral surface not covered, smooth.

UGANDA. Bunyoro District: Budongo Forest, 28 Nov. 1971, *Synnott* 759!; Kigezi District: Ishasha Gorge, 6 km SW of Kirima, 21 Sep. 1969, *Faden* 69/1190!; Masaka District: Malaigambo forest, 6.4 km SSW of Katera, 2 Oct. 1953, *Drummond & Hemsley* 4538!

KENYA. North Kavirondo District: Kakamega Forest, Kibiri Block, 21 Jan. 1970, *Faden* 70/31!

TANZANIA. Bukoba District: Minziro Forest Reserve, Bulembe, 11 Aug. 2000, *Festo* 730!; Kigoma District: Tubira Forest, 1 Apr. 1994, *Bidgood & Vollesen* 3022!

DISTR. **U** 2, 4; **K** 5; **T** 1, 4; Senegal, Guinea Bissau, Guinea, Sierra Leone, Liberia, Ivory Coast, Ghana, Benin, Nigeria, Cameroon, Equatorial Guinea, Central African Republic, Gabon, Congo–Brazzaville, Congo–Kinshasa, Rwanda, Burundi, Angola

HAB. Evergreen forest, forest edges and paths, swamp forest, ground water forest, secondary forest, rarely scrub and dry forest; 1050–1850(–2600) m

Flowering specimens have been seen from all months except October and December.

NOTE. Neither Bentham (1849) nor Clarke (1881) mention a collection number for the type specimen of *C. capitata*, although Clarke normally cites collection numbers in his monograph. Clarke in F.T.A. cites *Vogel* 52, so perhaps the collection numbers were added subsequently to the original collection. Nevertheless the type specimen is clear.

Commelina capitata is the only yellow-flowered *Commelina* in African rainforests (although collections of *C. africana* subsp. *zanzibarica* may occur in forests along the Kenyan and Tanzanian coast). When not in flower, it is still easily recognized by its densely clustered, subsessile, pale green spathes that usually have long red hairs on the margins and commonly on the surface. The leaves almost always have similar hairs at the summit of the sheath and sometimes on the petiole. The occasional absence of red hairs from the spathes has led some authors, such as Morton (1956, 1967), to describe such spathes as glabrous. However, all spathes in our area, and virtually all that I have examined from other parts of the range of this species, are puberulous with minute hook-hairs that are readily apparent with magnification. Such hairs also occur on the spathe peduncles and on the peduncle of the upper cincinnus.

The flower of the upper cincinnus is commonly bisexual, judging by the frequency with which fruits are found in this position. This is highly unusual for a species that has only one flower in this cincinnus.

This is one of the few species in which the spathe margins may be either fused or free. The margins appear to start off fused basally in young spathes, but they either tear or split to the base as the fruits develop and/or the spathe ages.

The seeds of *C. capitata* are unique in the genus in that the dorsal surface is covered by a somewhat fleshy, orange-yellow to creamy yellow material. I believe that this material is an aril to serving to attract birds and other animals for seed dispersal, but its development and therefore its true structure has not been determined

The flowers are typically bright lemon yellow, but six collections from Uganda refer to them as white (five collections) or whitish, and Ogwal (1990) states that they may be yellow or white in Uganda. I would suspect that sometimes the flowers are a creamy yellow, but not a pure white. It may be significant that white flowers are almost never mentioned by collectors in West Tropical Africa, so white or whitish flowers may be much rarer or lacking there.

Cunningham 3160 has mostly fawn-colored, rather than red hairs. The few red hairs at the summits of the sheaths agree with this species, and all other characters are also in agreement.

23. **Commelina latifolia** *A. Rich.*, Tent. Fl. Abyss. 2: 340 (1850); C.B. Clarke in DC., Monogr. Phan. 3: 173 (1881) pro parte, & in F.T.A. 8: 50 (1901) pro parte; Malaisse in Fl. Rwanda Sperm. 4: 129, fig. 47.2 (1988); Ogwal in Mitt. Inst. Allg. Bot. Hamburg 23b: 578, 584, 585, fig. 3a (1990); Faden in U.K.W.F. 2nd ed.: 305, t. 136 (1994); Ensermu & Faden in Fl. Eth. 6: 368, fig. 207.17 (1997); Wood, Handbook Yemen Fl.: 318 (1977). Type: Ethiopia, without precise locality, *Schimper* 1686 (P, holo.; BM!, BR!, K!, L!, iso.)

Annuals or perennials; roots thin, fibrous; shoots usually 1–2 at the base, sometimes many, tufted, usually creeping, trailing, straggling, scrambling to semiscandent, sometimes erect to ascending, commonly decumbent and rooting at the nodes, 20–100(–150) cm long, much branched, internodes glabrous. Leaves distichous; sheaths to 2.5 cm long, sparsely puberulous, eciliate at the apex, sometimes heavily spotted with maroon; lamina linear-lanceolate to lanceolate-elliptic, (3–)5–15(–17) × (0.3–)0.5–3(–4) cm, base cordate-amplexicaul to sagittate-amplexicaul, margins scabrid distally, apex acuminate; adaxial surface glabrous or scabrid with a line of white, acicular hairs along the midrib, abaxial surface puberulous. Spathes solitary or a few loosely clustered, usually falcate; peduncle 0.7–3.5(–5) cm long, puberulous; spathes 1–2.3(–2.8) × 0.5–1.4 cm, base rounded or truncate to cordate, margin fused proximally for 2–5 mm, ciliolate along the fused part, apex acute to acuminate, sometimes mucronate; surfaces puberulous, green contrasting dark veins; upper cincinnus usually lacking, rarely 1(–2)-flowered, lower cincinnus 2–8-flowered. Flowers bisexual and male. Sepals hyaline, upper sepal 3 mm long; paired sepals free, ovate-elliptic, 3.5–4.5 × 2.5–3 mm; paired petals light sky blue to dark blue, ovate-reniform, 12.5–14 × 8.5–10 mm of which the claw 5–6 mm, lower petal white to very pale blue, ovate to ovate-elliptic, 3.5–4.5 × 2–2.3 mm; staminodes 3; lateral stamens with filaments 9–9.5 mm long, anthers 1.2–1.7 mm long, pollen dirty yellow; medial stamen filament 6–7 mm long, anther 2–2.5 mm long, pollen golden yellow. Ovary ± 0.8 mm long; style 10–12 mm long; stigma enlarged. Capsule pale yellow, sometimes with dark brown flecks, bilocular (to trilocular), bivalved, ± square, 5.2–7 × 4.2–6 mm, contracted between the seeds, very shortly or not apiculate, dorsal locule, when present, 1-seeded (or empty?), indehiscent, ventral locules (1–)2-seeded. Seeds ± spherical, 2–2.5 × 1.9–2.4 mm, testa very dark brown to black, with many wart-like bumps, also densely white-farinose.

var. **latifolia**

Annuals (or perennials?) usually with 1–2 shoots from the base, shoot bases not thickened at the base, shoots creeping, trailing, straggling, scrambling to semiscandent, often decumbent and rooting at the lower nodes, mostly 50–100 cm long or longer; lamina (1.5–)3–15(–17) × (0.3–)0.5–3(–4) cm, margins not or scarcely undulate

UGANDA. Toro District: Kilembe, 12 Apr. 1976, *Ogwal* 124!; Mengo District: Kampala, July 1938, *Chandler* 2437! & Lutembe beach, 30 Sep. 1976, *Ogwal* 126!

KENYA. Trans-Nzoia District: Kitale, 10 Jan. 1964, *Bogdan* 5667!; Nairobi District: Herbarium grounds, 25 May 1966, *Gillett* 17311!; North Kavirondo District: Kakamega Forest, along Yala River, ± 4.8 km SE of Forest Station, 25 Nov. 1969, *Faden, Evans & Bally* 69/2020!

TANZANIA. Ngara District: Murgwanza, Bugufi, Ngara, 29 Dec. 1960, *Tanner* 5556!; Lushoto District: Jaegertal, 10 Jan. 1966, *Archbold* 588! & Soni, 8 June 1974, *Faulkner* 4862!

DISTR. **U** 2, 4; **K** 1, 3–7; **T** 1–3; Rwanda, Burundi, Ethiopia, South Africa; Yemen

HAB. Weed in crops, gardens, disturbed places, roadsides, less commonly near streams, among rocks, and on forest and thicket edges; 1100–2100 m

Flowering specimens have been seen from all months except February.

SYN. *Commelina sagittifolia* Hassk. in Schweinfurth, Beitr. Fl. Aethiop.: 206 (1867). Type: Ethiopia, without precise locality, *Schimper* 1686 (K!, L!, iso.)

C. madagascarica of C.B. Clarke in F.T.A. 8: 52 (1901), not as to type

var. **undulatifolia** *Faden* **var. nov.** a var. *latifolia* basibus surculorum tumidis et marginibus foliorum valde undulatis differt. Type: Kenya, Kitui [Machakos] District: Thika–Garissa road, 2.6 km towards Garissa after junction with Kitui road, near Kangonde, *Faden & Faden* 74/733 (K!, holo.; EA!, MO!, iso.)

Perennials with several to many tufted shoots from the base, shoot bases distinctly thickened, shoots generally erect to ascending, usually not rooting, except at the base, occasionally decumbent and rooting at the nodes, mostly 20–50 cm long (rarely to 100 cm); lamina 1.5–8.5 × 0.3–2 cm wide, margins strongly and finely undulate.

KENYA. North Nyeri District: Naro Moru River Lodge, Naro Moru, 5 Sep. 2003, *Faden & Faden* 2003/003!; Masai District: Aitong Airstrip, Mara Area, 5 June 1961, *Glover, Gwynne & Samuel* 1560! & near Ngong, 28 Feb. 1961, *Stewart per McCabe* 631!

TANZANIA. Arusha District: 27 km from Makuyuni on Makuyuni–Arusha road, 31 May 1996, *Faden et al.* 96/5!; Masai District: Ngorongoro Crater, near Siedertopf's, 9 Apr. 1941, *Bally* 2269!

DISTR. **K** 3, 4, 6; **T** 2; Ethiopia

HAB. Grassland, sometimes with scattered *Acacia drepanolobium*, on black cotton soil, and roadside ditches in a similar habitat, once reported on rocks; 1250–1750 m

Flowering specimens have been seen from [February], April, June and September.

NOTE (on the species as a whole). Some reported distributions of *C. latifolia* are misleading, such as the record of *C. latifolia* from Sudan in Fl. Eth. which is based on specimens that are probably *C. imberbis*. The citation of this species from Cameroon in Fl. Eth. was also based on a wrongly identified specimen at Kew. The record of this species from Selous Game Reserve (**T** 8) (Vollesen, Opera Bot. 59: 97 (1980)) is based on two collections of *Commelina erecta*. The specimens attributed to *C. latifolia* in Flore de Madagascar have not been examined, but from the description they are unlikely to be this species.

In contrast, the improbable record from South Africa (*Bos* 1236, K) is indeed *C. latifolia*. In view of its omission from F.S.A. (Obermeyer & Faden, 1985) and the considerable distance from the nearest locality, this may well represent a local introduction from elsewhere or a mislabeled specimen. A second specimen of this number from WAG was found to be *C. ecklonia*na, so it is even less likely that the Kew sheet represents a natural distribution.

Particularly narrow-leaved forms (all leaves <1 cm wide) occur occasionally in Uganda (e.g. *Ogwal* 112), and include both collections seen from Burundi. Such forms also have been collected in Eritrea, so there does not seen to be a geographic basis for creating a subspecies. Clarke in F.T.A. treated two collections of the narrowed-leaved form of *C. latifolia*, *Scott Elliot* 7588 from Uganda and *Schweinfurth* 131 from Eritrea, as *C. madagascarica*. I know of no records for that species' occurrence on mainland Africa.

Commelina latifolia characteristically has the leaf bases all amplexicaul and the midrib on the leaf above with only acicular hairs. The latter character has been checked in every specimen at Kew from throughout the range of the species and has been found to be consistent. In *Kibue* 105 from Masai District, Kenya only the distal leaves have amplexicaul bases and the lowermost are actually petiolate, which suggests *C. petersii*. Fruits are lacking for a definitive determination, but the hairs on the adaxial midrib, which is exposed in only one leaf, agree with *C. latifolia*, not *C. petersii*, although the shoot that has the adaxial midrib exposed in not the one with the petiolate leaves, so this might represent a mixed collection. The combination of amplexicaul leaf bases and only needle-shaped hairs on the adaxial midrib of the leaves is elsewhere found only in *C. lukei*, a mainly coastal species. Both species occur in the Taita Hills (**K** 7) and Western Usambara Mts (**T** 3). *C. lukei* differs by its elongate capsule, ellipsoid seeds and its blue (vs. white or very pale blue) middle petal. In the small area of overlap *C. lukei* tends to occur in more natural habitats whereas *C. latifolia* is much more weedy and is perhaps introduced.

Ensermu (pers. comm.) writes that in Ethiopia there are plants that are intermediate between the two varieties that we accept here. Nevertheless, var. *undulatifolia* appears to be a distinct ecotype and a characteristic plant of black cotton soils in northern Tanzania and central Kenya.

Ogwal (1990) reported the capsules of *C. latifolia* in Uganda as 5-seeded. As she was a very careful worker I accept her record.

The collection *Wallace & Forlonge* 246/87, which I am treating as var. *undulatifolia* because of its relatively small leaves with strongly undulate margins and its small, shortly pedunculate spathes is from Ol Doinyo Sabuk, near Thika. The habitat is carefully described as "skeletal soil on schists, growing in full sun", which is totally wrong for this variety. It is unknown whether this extreme environment can produce at least some of the features of var. *undulatifolia*, in which case it would undermine its validity and value as an ecological indicator taxon.

The seeds of *C. latifolia* have been described either as smooth (Ogwal 1990), nearly smooth (Clarke, 1901) or tuberculate (Ensermu & Faden 1997). It is not a matter of variability but rather at what distance and with what magnification one wishes to examine them. They do in fact have some texture, due to the slightly raised, wart-like bumps, but they are not tuberculate. I think in this case Clarke got it right. In any event, the seeds are highly distinctive and should not be confused with those of any other species.

Blundell's photograph (Wild Fl. E. Afr.: 413, fig. 861 (1987)) labeled *C. latifolia* is actually *C. lukei* and his account of the species includes at least these two species, and possibly others.

24. **Commelina eckloniana** *Kunth*, Enum. 4: 57 (1843); C.B. Clarke in DC., Monogr. Phan. 3: 174 (1881) & Clarke in Fl. Cap. 7: 11 (1897); Phillips in Fl. Pl. S. Afr. 9: t. 326 (1929); Obermeyer & Faden in F.S.A. 4, 2: 32 (1985); Ogwal in Mitt. Inst. Allg. Bot. Hamburg 23b: 578 (1990) pro parte; Faden in U.K.W.F. 2nd ed.: 305 (1994) pro parte; Ensermu & Faden in Fl. Eth. 6: 371 (1997) pro parte. Type: South Africa, Cape, *Ecklon & Zeyher* s.n. (B, K!, holo., TCD!, right hand shoot only, iso.)

Perennials or annuals; roots thin and fibrous, sometimes cord-like, or consisting of stalked fusiform tubers clustered at the base of the plant; shoots erect to decumbent and rooting at the nodes, internodes puberulous. Leaves with sheaths puberulous or hirsute-puberulous, often with long hairs along the fused edge, ciliate at the apex with white hairs; lamina sessile or petiolate, sometimes only the proximal leaves petiolate, narrowly lanceolate to ovate-elliptic, 2–15 × 0.4–2.5 cm, margins scabrid distally, surfaces puberulous or pubescent. Spathes solitary or more commonly loosely clustered; peduncles puberulous; spathe margins fused basally for 3–9 mm, puberulous along the fused margins, otherwise glabrous, surface puberuous and/or sparsely hirsute; upper cincinnus vestigial and enclosed in the spathe or well developed and producing 1 (rarely to 6) male flower(s). Flowers bisexual and male. Sepals free or nearly so. Petals pale blue to sky blue, rarely white, the lower one much reduced or nearly the same size as the paired petals. Staminodes 3, sometimes reduced to tiny knobs. Ovary bilocular, locules 2-ovulate; stigma capitate. Capsules bilocular, bivalved, 4-seeded, nearly square, constricted between the seeds, apex truncate to retuse, rarely rounded. Seeds subspherical, testa dark brown, verrucose to echinulate or echinate.

1. Plants decumbent or erect; roots fibrous, thin or cord-like 2
   Plants ± erect, roots tuberous 4
2. Plants erect (although older shoots may sometimes flop), with a definite base; shoot bases swollen; central Kenya .... a. subsp. *nairobiensis*
   Plants decumbent, with or without a definite base; shoot bases not swollen 3

3. Leaves sessile, adaxial midrib with acicular hairs only; spathes with usually a few to many long, uniseriate hairs; W Uganda .... b. subsp. *claessensii*
   Leaves petiolate, adaxial midrib with hook-hairs only; spathes lacking long, uniseriate hairs; central Kenya .. c. subsp. *thikaënsis*
4. Staminodes with antherodes 4–6-lobed; lower petal distinctly smaller than the paired petals; S Tanzania d. subsp. *critica*
   Staminodes with antherodes capitate, unlobed; lower petal subequal to the paired petals; E Uganda, W Kenya, Tanzania .... e. subsp. *echinosperma*

a. subsp. **nairobiensis** (*Faden*) *Faden* **comb. nov**. Type: Kenya, Kiambu District: Nairobi–Thika road, ± 200 m Thika side of Nairobi City boundary, *Faden, Evans & Msafiri* 70/901 (US!, holo.; BR!, EA!, K!, MO!, UPS!, iso.)

Perennial herb with swollen, moniliform, subterranean stem bases somewhat resembling orchid pseudobulbs; roots cord-like, not tuberous; shoots erect to ascending, 20–60 cm tall, sparsely to densely hirsute-puberulous, the proximal internodes glabrescent. Leaves usually not petiolate except sometimes the lower ones, 2–8.5 × 1–2.5 cm, base cuneate and often distinctly oblique (proximal leaves) to cordate and ± amplexicaul (distal leaves), apex acute to acuminate; both surfaces densely puberulous, usually also with a few long uniseriate hairs. Spathes solitary, usually somewhat falcate, 1.5–2.6 × 0.9–1.4 cm, peduncle 0.9–3.5(–6) cm long, base subcordate to truncate, margins ciliolate along the line of fusion, apex acute; upper cincinnus well developed (rarely absent), long-exserted from the spathe, producing one male flower; lower cincinnus 3–6-flowered. Flowers 1.5–2.5 cm wide. Upper sepal lanceolate to ovate-elliptic, 5 × 2–2.5 mm, paired sepals lanceolate-elliptic to oblong-elliptic, 5–5.5 × 3–4 mm, glabrous. Paired petals sky blue to dark blue, rarely white, broadly ovate to ovate-reniform, 10–15 × 9–14 mm, of which the claw 3–4 mm, medial petal ovate, concolorous, 5–6 × 4–5 mm. Staminodes subequal; lateral stamens with filaments ± 8 mm long, anthers 1.5–1.9 mm long, pollen dirty yellow, medial stamen with filament 6–7 mm long, anther 2.5–3 mm long, pollen yellow. Style 8–9 mm long. Capsules 5–6 mm long and wide, locules 2-seeded. Seeds spherical or subspherical, 2–3 mm in diameter, with raised pale warts and sometimes short ridges, farinose.

KENYA. Kiambu District: 0.4 km towards Nairobi from Thika Road Baptist Church, W side of road, 12 Dec. 1968, *Faden* 68/885!; Nairobi District: Eastleigh Estate, S of Eastleigh Community Centre, 19 May 1971, *Mwangangi & Mukenya* 1586; Masai District: ± 4 km towards Nairobi from Kiserian on Magadi–Nairobi road, crossing on Nol Chora stream, 27 June 1971, *Faden & Evans* 71/509!

DISTR. **K** 3, 4, 6; endemic to central Kenya
HAB. Seasonally wet grassland, especially on black cotton soil; 1450–2000 m
Flowering specimens have been seen from May, June and December.

SYN. *Commelina nairobiensis* Faden in Novon 4: 226 (1994) & in U.K.W.F. 2nd ed.: 305 (1994) & in Krupnick & Kress, Plant Conservation, a Natural History Approach: 108 (2005)
C. sp. "A" of Faden in U.K.W.F.: 660 (1974)

NOTE. Faden (2005) considered this plant threatened because of its local distribution in a densely populated area with a rapidly expanding human population.

b. subsp. **claessensii** (*De Wild.*) *Faden* **comb. nov**. Type: Congo–Kinshasa, Duota [=Duoto?], *Claessens* 1082 (BR!, holo.; K, US, photos)

Perennial with thick, fibrous roots; shoots tufted, semi-scandent to prostrate, ± rooting at the lower nodes, to 60 cm long, often much branched, puberulous to hirsute-puberulous below the nodes, otherwise glabrous. Leaves with sheaths to 3.8 cm long, often splitting longitudinally, puberulous or pilose-puberulous, ciliate with white hairs at apex; lamina sessile, flat or complicate, 3.5–14 × 0.6–1.7 cm, base cuneate, margins ciliate near point of attachment to the sheath, apex acuminate to attenuate; adaxial surface scabrid near the margins, otherwise hirsute-puberulous, abaxial surface hirsute-puberulous all over. Spathes solitary, not falcate to slightly falcate; peduncles 1.5–3 cm long, puberulous; spathes 1.6–2.3 × 0.9–1.2 cm, base truncate or rounded to cordate, margins ciliolate along the fused edge, apex acute to acuminate; upper cincinnus usually vestigial and enclosed within the spathe, when fully developed 1-flowered, its peduncle long-exserted from the spathe; lower cincinnus 4-flowered. Paired petals pale blue, rounded to slightly emarginate at the apex, base truncate to slightly cordate; medial petal ovate to ovate-lanceolate, spoon-shaped, apex recurved. Capsule ± pale yellow, 7.5 × 6 mm, apex ± truncate. Seeds (one submature) slightly ellipsoid, ± 3.2 mm long (perhaps too long due to immaturity?), testa densely covered with ridges and points, also densely white-farinose.

UGANDA. Bunyoro District: 37 km S of Hoima on Fort Portal road, N of Kabweya, 17 Sep. 1969, *Faden, Evans & Lye* 69/1091!; Toro District: Kilembe, no date, *Ogwal* s.n.!; Busoga District: Musisi [Musizi] Ridge, July 1936, *Sangster* 160!

DISTR. **U** 2, 3; Congo–Kinshasa, Cameroon?

HAB. Wet roadside banks and along streams, in full sun (incompletely known); ± 1000 m

Flowering specimens have been seen from July and September.

SYN. *Commelina claessensii* De Wild. in Pl. Bequaert. 5: 173 (1931). Type: Congo–Kinshasa: Duota, *Claessens* 1082 (BR!, holo.; photo, K, US)

C. eckloniana Kunth of Ogwal in Mitt. Inst. Allg. Bot. Hamburg 23b: 578 (1990)

NOTE. Our plants closely resemble the type of *C. claessensii* De Wild., which I have previously examined. A specimen from Tagbou, Cameroon, *Letouzey* 2274, which was identified as *C. echinulata*, seems to belong to *C. eckloniana* sensu lato. It is the only specimen seen from the western half of Africa. It looks most like subsp. *claessensii* having dense acicular hairs on the adaxial midrib, but unlike typical specimens of that subspecies the midrib also has a few hook-hairs too and spathes lack long, uniseriate hairs. Because *C. eckloniana* is so variable, it is hardly surprising that such a remote collection does not exactly match any specimens from our area.

c. subsp. **thikaënsis** *Faden* **subsp. nov.** a subsp. *nairobiensi* habitu decumbenti foliis regulatim petiolatis spathis trichomata longa uniseriata carentibus et a subsp. *claessensii* foliis petiolatis costa paginae adaxialis trichomata uncata ferenti differt. Typus: Kenya, Machakos District: Thika, along the Chania River, below the Blue Posts Hotel, *R.B. Faden, Napper & A. Evans* 69/225 (K!, holo.; EA! iso.)

Perennial with thin, fibrous roots; young shoots erect, older shoots semi-scandent to trailing and rooting at the nodes, to 1.5 m long, glabrous. Leaves with sheaths to 3.6 cm long, often splitting longitudinally, puberulous, ciliate or eciliate at apex; lamina petiolate, planar, 4.5–9 × 0.8–1.9 cm, base cuneate, apex acuminate to attenuate, surfaces including the adaxial midrib puberulous. Spathes solitary, usually slightly falcate (rarely not falcate); peduncles 1–1.5 mm long, puberulous; spathes 1.5–2 × 0.7–0.9 cm, base truncate to rounded or hastate, margins ciliolate along the fused edge, apex acute; upper cincinnus 1-flowered or abortive, when well developed the peduncle long-exserted from the spathe; lower cincinnus 2–3 flowered. Flowers with pedicels ± 3 mm long. Upper sepal hooded. Paired petals sky blue, limb 7–8 × 8.5–9.5 mm of which the claw ± 4 mm long; medial petal strongly boat-shaped, 4.5–5.5 × 2.5–4 mm, concolorous. Capsule grey-green, 5.5–6.5 × 5–5.5 mm, apex retuse, shortly apiculate. Seeds subspherical, 2–4–2.8 × 2.2–2.5 mm, testa with medium brown warts and short ridges all over, also densely white-farinose.

KENYA. Fort Hall District: bridge over Thika River on Bridges Road, which connects Kakuzi Road with the main Thika–Garissa Road, 23 Mar. 1969, *Faden et al.* 69/349!; Machakos District: Thika, along the Chania River, behind the Blue Posts Hotel, 2 Mar. 1969, *Faden, Napper & Evans* 69/225! & same locality, 10 July 1996, *Faden et al.* 96/546!

DISTR. **K** 4; not known elsewhere

HAB. Evergreen dry forest and riparian forest; 1400–1500 m

Flowering specimens have been seen from March.

SYN. *Commelina eckloniana* of Faden in U.K.W.F.: 660 (1974) & in U.K.W.F., ed 2: 305 (1994), *non* Kunth sensu stricto

NOTE. This is both very local to a small stretch of the Chania and Thika river systems, and a highly distinctive plant. It bears a resemblance to many South African plants of *C. eckloniana* subsp. *eckloniana* in its decumbent habit, but subsp. *thikaënsis* differs by its regularly petiolate leaves and its spathes lacking long, uniseriate hairs. Moreover, I have not seen any collections of similar habit in between these two countries. Similar looking but also geographically isolated subsp. *claessensii* is distinguished from subsp. *thikaënsis* by its sessile leaves, only acicular hairs (vs. hook-hairs) on the adaxial midrib, and long, uniseriate hairs on some or all of the spathes.

Genetic studies have shown that subsp. *thikaënsis* is very closely related to subsp. *nairobiensis*, which grows nearby but in a completely different habitat. However subsp. *thikaënsis* has been found to have 26 chromosomes, whereas subsp. *nairobiensis* has 24 chromosomes. Thus, despite their genetic similarity they may be functioning as distinct species. With the continuing destruction of the forest behind the Blue Posts Hotel, the main locality for this plant, subsp. *thikaënsis* must be considered endangered.

d. subsp. **critica** (*De Wild.*) *Faden* **comb. nov.** Type: Congo–Kinshasa, Katentania, plateau de la Mankia, *Homblé* 728 (BR!, holo.; K, US, photo)

Perennial; roots tuberous, fusiform, clustered at the base of the plant; shoots few, erect, sparsely branched, 15–40 cm tall, puberulous, usually also hirsute, proximal internodes glabrescent. Leaves with sheaths to 3.5 cm long, puberulous, often also hirsute; lamina sessile or in the proximal ones petiolate, planar or complicate, narrowly lanceolate to lanceolate-oblong, 4–14 × 0.8–2.1 cm, base cuneate to rounded, margins often ciliate proximally, apex acute to acuminate or attenuate; surfaces puberulous and often hirsute. Spathes solitary or subclustered distally, not falcate to slightly falcate; peduncles 1.2–4.5 cm long; spathes 1.7–3.4 × 0.9–1.7 cm, base truncate or rounded to subcordate, margins ciliate or ciliolate along the fused edge, apex acute to acuminate; upper cincinnus usually well-developed and producing a single male flower (rarely up to 6-flowered (*Bullock* 1973), peduncle long-exserted from the spathe; lower cincinnus 9–11-flowered. Flowers of the upper cincinnus male, flowers in the lower cincinnus bisexual. Upper sepal narrow, 3–5 mm long, hooded at apex; paired sepals very shortly fused at the base, rarely fused up to 3 mm, ovate to ovate-elliptic, 5–5.5 × 2.5–3 mm. Petals blue; paired petals broadly ovate to reniform, 10.5–11 × 9.5–11 mm of which the claw 3–4 mm; lower petal boat-shaped, 4–5 × 1.8–4 mm, apex recurved. Lateral stamens with filaments sigmoid, 7.5–12 mm long, anthers (0.8–)1.1–1.8 mm long; medial stamen with filament sharply recurved near the apex, 5.5–9 mm long, anther (1.3–)1.8–2.5 mm long. Ovary ± 1.5 mm long; style 10–13 mm long; stigma not enlarged. Capsule pale yellow to greenish tan, 7–8 × 5–6.5 mm, apex truncate or rarely rounded, shortly apiculate. Seeds subspherical, (2.25–)2.4–3.2 × 2–2.4 mm, testa with medium brown warts and/or points and few ridges, testa also densely white farinose.

TANZANIA. Ufipa District: Chapota, 2 Dec. 1949, *Bullock* 1973! & Sunzu Mountain, 9 Jan. 1955, *Richards* 3976! & Malonje Farm, 14 Mar. 1957, *Richards* 8734!
DISTR. **T** 4, 7; Congo–Kinshasa, Zambia, Malawi
HAB. Rocky places in grassland and near termite mounds; 1800–2250 m
Flowering specimens have been seen from November to January and March.

SYN. *Commelina critica* De Wild. in Contr. Fl. Katanga, Suppl. 3: 64 (1930)

e. subsp. **echinosperma** (*K. Schum.*) *Faden* **comb. nov.** Type: Tanzania, Tabora District: Igonda [Gonda], *Böhm* 2 (B†, syn.) & *Böhm* 8 (B†, syn., CORD!, isosyn. & lecto., selected here, photo US!); Bukoba District: Kakoma, *Böhm* 47 (B†, syn., isosyn. CORD!) & E of Lake Tanganyika, *Böhm* (*Deutsche Exped. Ost.-Afrika*?) 186 (B†, syn., isosyn. CORD!)

Perennial; roots tuberous, fusiform, clustered at the base of the plant; shoots few, erect, sparsely branched, (7–)10–30(–40) cm tall, pilose-puberulous, the proximal internodes glabrescent. Leaves with sheaths to 4 cm long, pilose-puberulous, ciliate at apex; lamina sessile or, in the proximal leaves of some plants, petiolate, complicate or planar, 2.5–15 × 0.4–2.5 cm, base cuneate or rounded, margins sparsely ciliate proximally, apex acute to acuminate; both surfaces puberulous or the adaxial subglabrous and sometimes with a few long hairs. Spathes solitary or subclustered distally, not falcate to slightly (rarely strongly) falcate; peduncles 0.8–2.5(–3.5) cm long; spathes 1.3–3 × 0.6–1.5 cm, base cordate to hastate, rounded or truncate, margins ciliolate (rarely ciliate) only in the fused part, apex acute to acuminate; upper cincinnus commonly vestigial and enclosed in the spathe, sometimes producing a single male flower, lower cincinnus 2–5(–8)-flowered. Flowers ± 2 cm wide, the one on the upper cincinnus male, the ones in the lower cincinnus bisexual or male. Sepals upper sepal boat-shaped, 3.3–5.5 mm long, hooded at the apex; paired sepals very shortly fused at the base, ovate to ovate-elliptic, 3.5–5 × 2.5–3.5 mm. Paired petals sky blue, broadly ovate or reniform, 8–11.5

× 7–10 mm of which the claw (2–)3–4 mm; lower petal subclawed, broadly ovate to ovate-deltate, 6–8 × 5–8 mm, concolorous. Staminodes 2 or 3; stamen filaments straight, lateral stamens with filament 3.5–6 mm long, anther 1.3–2.45 mm long; medial stamen with filament 4–6 mm long, anther 2.1–3.4 mm long. Ovary 1–1.5 mm long; style 4–6 mm long; stigma slightly capitate. Capsules pale yellow with or without dark flecks, 6–9 × 5.5–6 mm, apex retuse to truncate, not apiculate. Seeds subspherical, 2.4–2.9 × 2.2–2.3 mm, testa dark brown mottled with orange-brown, but largely obscured by white-farinose granules, with numerous orange-brown spines and short ridges.

UGANDA. Karamoja District: Pian County, Lodoketemit [Lodokeminet], 20 May 1968, *Kerfoot* 4936! & 4943!

KENYA. Trans-Nzoia District: Endebess–Suam road, ± 4 km from Suam Saw & Mill road towards Suam, 7 June 1969, *Faden, Evans & Tweedie* 69/719!; SE Elgon, Saboti [Seboti], July 1955, *Tweedie* 1350! & same locality, Apr. 1967, *Tweedie* 3444!

TANZANIA. Ngara District: Ngara, 12 Oct. 1959, *Tanner* 4611!; Ufipa District: Lake Sundu, 10 Dec. 1958, *Richards* 10291!; Dodoma District: Bereko–Kikare Motate, 2 Apr. 1974, *Richards & Arasululu* 29160!

DISTR. **U** 1; **K** 2, 3, 5; **T** 1, 2, 4, 5; Congo–Kinshasa, Rwanda, Burundi (see note), Ethiopia, Zambia

HAB. Grassland and open areas, often associated with rocks, riparian and roadside vegetation near cultivation; full sun or in shade; 1050–2050 m

Flowering specimens have been seen from October to January and April to July (the label of *Bax* F states that the species flowers from December to April).

SYN. *Commelina echinosperma* K. Schum. in P.O.A. C: 135 (1895); Faden in U.K.W.F. 2nd ed.: 304 (1994).

Commelina echinulata Lebrun & Taton in Explor. Parc Nat. Kagera: 28 (1948). Type: Congo–Kinshasa, Parc National de la Kagera, Gabino, *Lebrun* 9551 *bis* (BR!, holo.; K!, iso.)

Commelina sp. 'B', of Faden, U.K.W.F.: 657 (1974)

NOTE. This taxon has one of the most distinctive flowers of any *Commelina* species. The medial petal is very large, perhaps the largest in any African species. Indeed, all three petals appear to be nearly the same size, which has been noted by collectors in Burundi, Ethiopia, Kenya and Zambia. The staminodes, which may be two or three per flower, are consistently very reduced to a tiny filament with a minute, knob-shaped antherode. The stamens have very short, straight filaments that are either all the same length, or, more commonly, the medial stamen filament is slightly longer than others, a character not observed in any other *Commelina* species elsewhere. The anthers are large, entirely yellow and dry ivory in color.

The Berlin sheets of the four collections cited by Schumann with the type description of *C. echinosperma* were apparently all destroyed, but duplicates of *Böhm* 8, 47 & 186 – the last cited by Schumann as an expedition number, not a *Böhm* collection number – have been seen from CORD. Among them *Böhm* 8 shows the typical stamens, capsules and seeds of this taxon, as well as the long hairs on the spathes mainly confined to the long veins (in most spathes). It agrees perfectly with Schumann's description and therefore it has been selected as the lectotype of the name. *Böhm* (or Expedition) 186 should be from Kakoma, according to the type description, but the CORD sheet has "östlich vom Tanganyika See" as the locality and *Böhm* as the collector. Nevertheless, it appears to be the same taxon, although only the spathe pubescence can be used for confirmation. *Böhm* 47, however, has very narrow, linear leaves and glabrous spathes with free margins, and apparently a one-seeded capsule, all of which are completely at odds with the type description. It is discussed at the end of *C. pseudopurpurea*. I suspect that it was a mixed collection and that the Berlin sheet of this number probably agreed with the other collections and with Schumann's description.

Two specimens from Burundi, *Reekmans* 6751 & 7594 are a perfect match for this subspecies. What is unusual about the collections is their habit, described as "sciaphilous...prostrate or rampant at the base, the flowering shoots erect" in *Reekmans* 7594 and "prostrate to rampant" in 6751. Whether or not this habit is due to a difference in ecology, such as growing in more shade, these collections show a habit type that is lacking in the collections from our area.

Lloyd 37 from Kakoma, Tanzania had been left as undetermined because the collector indicated that the flowers were yellow. However there is a tiny remnant of a blue petal, and one of the typical anthers of subsp. *echinosperma* is embedded in thick, but transparent mounting glue, so there is no doubt that the specimen belongs here. It has the narrowest leaves (0.4 cm) of any specimen seed. *Bax* C2 is a mixed collected of this subspecies with *C. bracteosa* or *C. zambesica*.

NOTE (on the species as a whole). This is one of the most variable species of *Commelina* in Africa. The characters that hold it together are those included in the general description. Variation among the subspecies, in addition to differences in morphology, may also be geographic, e.g. the separation of subsp. *echinosperma* in western Kenya from subsp. *nairobiensis* and *thikaënsis* in central Kenya. It may also be partly ecological, such as the separation of the two central Kenyan taxa.

The major problem is distinguishing between subsp. *critica* and subsp. *echinosperma* from dried specimens. The flowers are hardly ever well preserved, collectors' notes are normally inadequate and spirit material is rare. Fortunately, it is sometimes possible to see some useful details, such as whether the staminodes are well developed, in pressed floral remnants. In the absence of any floral remains, plants of subsp. *echinosperma* tend to be smaller, with smaller spathes, but some collections are as large as any of subsp. *critica*. In subsp. *critica* the adaxial midrib often has long, uniseriate hairs mixed with hook-hairs whereas long, uniseriate hairs are very rarely present on the adaxial midrib in subsp. *echinosperma*, so this character may be helpful but it is not an absolute distinction between the two subspecies. The best, non-floral character seems to be the distribution of the long, uniseriate hairs on the spathes. In subsp. *critica* they are typically scattered all over the spathe, whereas in subsp. *echinosperma* they are mainly confined to the longer veins, especially those nearest the folded edge or midrib, and on the midrib itself. Specimens that are atypical in their pubescence and also lack flowers may be impossible to identify with certainty although it appears that both subspecies overlap in our area only in southern Tanzania.

One such problematic specimen is *Faden et al.* 96/240 from **T** 4. It is a small plant – perhaps grazed by cattle? – with small spathes that lack long, uniseriate hairs. Its leaves have sparse long uniseriate hairs scattered on both surfaces, but none on the adaxial midrib. There were no flowers in the field. The preliminary determination was *C. echinosperma*. However, a plant in cultivation – also not a natural environment – produced a flower in which the antherodes were too well-developed, and the lateral stamen filaments, although short (6.5 mm long), were considerably longer than that of the medial stamen filament (4.5 mm long), which also does not agree with subsp. *echinosperma*. But the large medial petal and short stamens do not agree with subsp. *critica* either, so this specimen is not treated under either subspecies. Its determination must await further collections from this area.

Subsp. *eckloniana*, as recognized here, is restricted to South Africa.

Finally, it must be admitted that not all of the subspecies recognized here probably merit equal taxonomic recognition. Subsp. *nairobiensis*, with its distinct habit, habitat, and chromosome number of $2n = 24$, as opposed to $2n = 26$ for all other counts of *C. eckloniana* sensu lato, clearly seems to be functioning as a distinct species, but its relationship with subsp. *thikaënsis* remains unclear. The exceeding distinct floral morphology of subsp. *echinosperma* would clearly merit recognition as a full species were there were a completely reliable way of identifying non-flowering collections and if we could be certain that there are no plants with intermediate floral morphologies, which *Faden et al.* 76/240 would seem to belie. The variation and evolution within *C. eckloniana* could make a fascinating study but clearly it cannot be done from dried specimens alone.

25. **Commelina imberbis** *Hassk.* in Schweinf. Beitr. Fl. Aethiop.: 209 (1867); C.B. Clarke in F.T.A. 8: 49 (1901), pro parte; Morton in J.L.S. 60: 185 (1967); Brenan in F.W.T.A., 2nd ed.: 3: 48 (1968); Ogwal in Mitt. Inst. Allg. Bot. Hamburg 23b: 579, fig. 3b (1990); Faden in U.K.W.F. 2nd ed.: 304 (1994); Ensermu & Faden in Fl. Eth. 6: 369 (1997); Faden in Novon 18: 471, fig. 2, D-F, K (2008). Type: Eritrea, Togodele in Shohos, *Ehrenberg* s.n. (B†, holo.)

Single-stemmed or tufted perennial; roots thin, fibrous; shoots scrambling, straggling, or climbing through other vegetation, sometimes prostrate and then ascending, rarely erect, to 2 m long; internodes puberulous or glabrous. Leaves distichous; sheaths often splitting longitudinally, to 4.5 cm long, puberulous or apparently glabrous, ciliolate at the apex or eciliolate, often spotted or tinged with purple; lamina sessile, linear-lanceolate to lanceolate-elliptic, 4–17(–23) × 0.6–4(–5) cm, base cordate-amplexicaul, occasionally sagittate-amplexicaul or those of the more proximal leaves sometimes rounded to cuneate, margins scabrid at least distally, apex acuminate; adaxial surface puberulous, often scabrid, abaxial surface puberulous or apparently glabrous. Spathes solitary or in loose terminal clusters, moderately to strongly falcate, sometimes not falcate; peduncles (0.7–)1.5–3.8 cm long, puberulous mainly distally, sometimes in a longitudinal line; spathes 1.5–3.2 × 0.9–1.5 cm high, base cordate to truncate or hastate, margins fused for 2.5–6 mm, ciliolate along the fused edge and sometimes just above, apex acute to acuminate, occasionally mucronate, surfaces puberulous; upper cincinnus usually lacking or vestigial, lower cincinnus 2–6-flowered. Flowers bisexual and male, ± 1.8 cm wide; upper sepal ovate to lanceolate-oblong, 2.7–3 × 1.5 mm, apex hooded, paired sepals free, ovate to obovate-elliptic, 4–4.5 × 3.2 mm; paired petals blue (sky, pale, bright, flax, powder or mauve-blue), rarely mauve, broadly ovate, 11–12 × 8–10 mm of which the proximally white claw 5 mm; lower petal white or bluish white, ovate, 3.5–4.5 × 2.5–4 mm; staminodes 3; lateral stamen filaments ± 8–12 mm long, ± S-shaped, anthers 1–6–1.8 mm long; medial stamen filament ± 5 mm long, anther 2.2 mm long,. Ovary ± 1.85 mm long; style ± 12 mm long. Capsules pale yellow with dark brown flecks, trilocular, bivalved, oblong, 4–5 seeded, stipitate, 7–9 × 4.2–5.8 mm, contracted between the seeds, apex truncate- or emarginate-apiculate, dorsal valve or sometimes just the portion over the dorsal locule seed striate, dorsal locule usually well-developed and 1-seeded; ventral locules 2-seeded. Dorsal locule seed embedded in the capsule wall, ellipsoid, 4–5 × 2–3 mm, testa dark brown, apparently smooth; ventral locule seeds ellipsoid to reniform-ellipsoid, 2.8–4 × 2.1–2.5 mm, testa dark brown to nearly black, rarely medium brown, radially ridged with white, light brown or orange-brown warts, entire testa white-farinose. Seeds & fruit Fig. 14, p. 136.

UGANDA. Karamoja District: Lodoketemit [Lodokeminet], *Kerfoot* 4940! & Kidepo Valley National Park, near Uganda-Sudan Border, 23 Feb. 1976, *Ogwal* 104!; Teso District: Soroti–Moroto km 24, 13 Oct. 1952, *Verdcourt* 826!

KENYA. West Pokot District: near Sebit on Sebit–Parua road, along Sebit River, 15 Mar. 1977, *Faden & Faden* 77/805!; Kiambu District: Thika, behind the Salvation Army Blind School, 12 Dec. 1968, *Faden* 68/891!; Masai District: Olenyamu, ± 58 km from Magadi on road to Nairobi, 30 June 1962, *Glover & Samuel* 2918!

TANZANIA. Mbulu District: Babati–Arusha road, stream crossing ± 12 km S of entrance to Tarangire National Park, 4 July 1996, *Faden et al.* 96/545! & Mto wa Mchanga, Lake Manyara National Park, 9 Dec. 1963, *Greenway & Kirrika* 11161!; Kigoma District: Kasoje, 16 July 1959, *Newbould & Harley* 4375!

DISTR. **U** 1–4; **K** 1, 2, 4, 6; **T** 1, 2, 4; Nigeria, Cameroon, Congo–Kinshasa, Sudan, Eritrea, Ethiopia, Zambia; Yemen

HAB. Grassland, bushland, thicket edges, rock outcrops and rocky hilllsides, roadsides and waste ground, banana plantations, rarely in forest, often near rivers and lakes, sun to part shade; (750–)1000–1750 m

Flowering specimens have been seen from all months except April.

SYN. *Commelina latifolia* of Clarke in DC., Monogr. Phan. 3: 173 (1881) pro parte, not as to type

NOTE. The holotype of this species, which was in flower according to Hasskarl, was destroyed in Berlin. Although we cannot be certain of its identity, there has been a general consensus among recent authors as to which species the specific name should be applied, and the species ought to be neotypified accordingly.

This species has caused much confusion in the past. Clarke (1881), having seen the type, considered *C. imberbis* to be a synonym of *C. latifolia*. However, in F.T.A. he restricted the name to a species with oblong capsules and ellipsoid seeds. Nevertheless, his concept of *C. imberbis* also included specimens of *C. petersii*, the recently described *C. lukei*, and at least one other species (he recorded *C. imberbis* from Madagascar, where is does not occur). The report of this species from India by Rao (Blumea 14: 352. 1967) is erroneous: the African species present in India, Burma and Sri Lanka, to which he was referring, is *C. petersii* (Faden, Revised Handbook Fl. Ceylon 14: 193, 2000). Southern African plants treated as *C. imberbis* in F.S.A. (Obermeyer & Faden, 1985) are *C. kotschyi*.

Commelina imberbis was once thought to be much more widespread in East Africa, with a coastal distribution as well as an interior one. However, the coastal specimens have all proven to be either the newly described species *C. lukei* or the Malagasy species *C. mascarenica*. The closest that *C. imberbis* gets to the coast is about Kibwezi in Kenya and Tarangire National Park in Tanzania. In contrast, *Commelina mascarenica* penetrates from the coast as far inland as the Tsavo National Park East area in Kenya and the dry country around the Pare Mountains in Tanzania. Unfortunately, none of these dry country collections has seeds, but the specimens do have an upper cincinnus, which is otherwise lacking in all Kenyan collections of *C. imberbis*. Except for some low elevation specimens of *C. imberbis* from western Tanzania, there is no overlap in elevation between these two species.

Within its range this species can be confused only with *C. kotschyi* and *C. petersii*. From the former it differs by its perennial habit, larger leaves, spathes, and flowers, the filaments yellow or yellowish white, capsules that lack an apical extension and dark brown, ± rugose seeds that lack appendages. From *C. petersii* it differs by its leaves more distinctly clasping the stem, the medial and proximal leaves not petiolate, an upper cincinnus usually lacking, and the dark brown, farinose seeds that lack the smooth ridges that have tubercles on them. From both it differs by having the lower petal white or tinged with blue, not concolorous with the paired petals. Non-fruiting specimens of *C. imberbis* may be confused with *C. latifolia*, but that species has an obvious (especially with 10 × magnification) line of white, needle-shaped hairs along the midrib on the upper side of the leaf, whereas the midrib hairs in *C. imberbis* are smaller, inconspicuous hook-hairs and not distinctly white.

The usual lack of an upper cincinnus in FTEA collections of *C. imberbis* – single collections with an exserted upper cincinnus have been seen from **U** 1, **U** 3 and **T** 2 – is surprising in view of its normal presence in collections from Ethiopia and Eritrea.

I once thought that it might be possible to separate robust specimens of *C. imberbis* with all their leaves amplexicaul from smaller specimens in which only the uppermost leaves had amplexicaul bases. The specimens with amplexicaul leaves throughout tend to have the adaxial surface scabrid with prickle-hairs, whereas smaller specimens often have hook-hairs instead. But these characters are not absolute, and it is often impossible to determine from fragmentary specimens in which of the two above categories individual specimens should be placed. If there is any basis for dividing this species in the future it will have to be determined from a study based on field work, correlated with vegetative traits and perhaps ecology.

The seeds are very variable in this species, especially in the degree of development of the radial or transverse ridges and in the amount and distribution of wart-like material on the testa. Unfortunately, there are an insufficient number of collections with seeds to determine whether there is any geographic pattern or correlation with other morphological characters.

The 9 mm long flowering pedicel (in *Bax* G1) is apparently that of a male flower, the last flower in a spathe that had already set three fruits. This could well be typical, but it is almost impossible to divine from herbarium specimens.

26. **Commelina mascarenica** *C.B. Clarke* in DC., Monogr. Phan. 3: 174 (1881); Perrier de la Bâthie in Not. Syst. 5: 189 (1936) & in Fl. Mad. 37: 22 (1938); Faden in Fl. Somalia 3: 585 (2006); Faden in Adansonia, sér. 3, 30: 47–55, fig. 1 (2008). Type: Mauritius, Madagascar and the Comoro Islands, Aug. to Oct. 1838, *MacWilliam* s.n. (G!, holo.)

Perennial herb; roots thin, fibrous, occasionally produced from decumbent stems; shoots sprawling, scrambling or ± scandent, often straggling through other herbaceous vegetation or shrubs, much branched, to 1.2 m tall or long, glabrous or with a line of hook-hairs. Leaves distichous or spirally arranged; sheaths to 4 cm long, often splitting longitudinally, sometimes tinged with red, with a line of hook-hairs along the fused edge or more widespread, sometimes subglabrous, apex cilioliate or eciliolate; lamina sessile, linear-lanceolate to ovate-elliptic, 3.5–14.5 × (0.6–)1–3(–3.5) cm, asymmetric at the base, rounded to cuneate, cordate to cordate-amplexicaul in the distalmost leaves, margins scabrid at least distally, apex acuminate to acute; adaxial surface ± puberulous, abaxial surface glabrous to sparsely puberulous. Spathes solitary, rarely 2–3 in close proximity; peduncle 1–3.8 cm long, puberulous in a line continuous with the fused spathe margins, sometimes all around; spathes usually slightly falcate, 1.8–2.9 × 0.9–1.6 cm, base cordate to hastate, margins fused for 3–6.5 mm, sparsely ciliolate along the fused edge and just distal to it, apex acute to acuminate, sometimes mucronate, surfaces puberulous; upper cincinnus 1(–2)-flowered, the flowers male; lower cincinnus 4–6-flowered. Flowers bisexual and male; upper sepal lanceolate-oblong, 4–5 × 2 mm, paired sepals ovate-elliptic, 4.5–6.2 × 3–4.4 mm; paired petals blue (pale blue, sky blue, flax blue, powder blue), occasionally mauve-blue, pale mauve or white, broadly ovate, 12–15 × 12 mm of which the claw 5–7 mm; medial petal ovate, 5–7 × 3.5 mm, concolorous; staminodes 3 or occasionally one abortive; lateral stamens with filaments (8–)9.5–14.5 mm long, anthers (1–)1.3–2 mm long; medial stamen

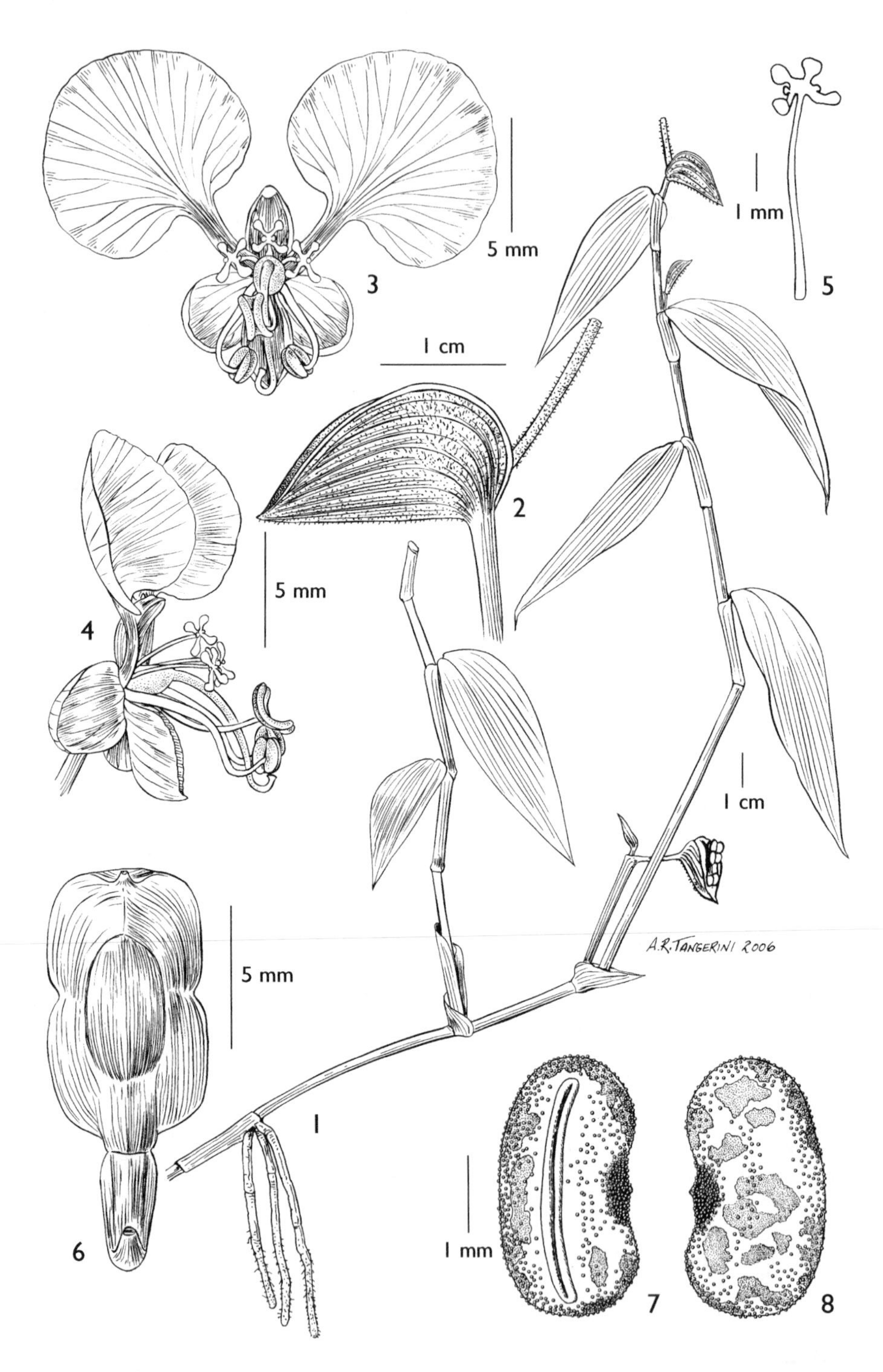

FIG. 19. *COMMELINA MASCARENICA* — **1**, habit; **2**, spathe; **3** & **4**, bisexual flower, front & lateral view; **5**, staminode; **6**, capsule, dorsal view; **7** & **8**, ventral locule seed, ventral & dorsal views. Reproduced with permission from Adansonia, ser. 3: 30: 51 fig. 1. Drawn by A. R. Tangerini.

filament (5.5–)7–8.5 mm long, anther 1.7–2.5 mm long. Ovary (1–)2–2.5 mm long; style 11–12 mm long; stigma capitate. Capsule pale yellow with dark brown flecks, trilocular, bivalved, oblong, 5-seeded, 8.3–10 × (3.6–)4.7–6 mm, constricted between the seeds, apex emarginate-apiculate or retuse-apiculate, dorsal locule one-seeded, indehiscent, striate, ventral locules 2-seeded, dehiscent. Dorsal locule seed fused to the capsule wall, dorsiventally compressed, oblong-ellipsoid to oblong, 3.8–4.3 × 2.2–2.3 mm, medium to dark brown, smooth, not farinose; ventral locule seeds cylindric to ellipsoid, not at all compressed, 3.1–4.1 × 2–2.4 mm, testa dark brown with conspicuous or inconspicuous lighter brown mottling or grey mottled with dark brown, densely white-farinose, smooth or faintly alveolate. Fig. 18, p. 190; seeds Fig. 14, p. 136.

KENYA. South Kavirondo District: Lukiri Island, Jul. 1934, *Napier* 6814!; Kilifi District: Sabaki, 6.4 km N of Malindi, 14 Nov. 1961, *Polhill & Paulo* 759!; Mombasa District: Nyali Beach Hotel, 9.6 km from Mombasa, 21 Apr. 1950, *Rayner* 294!

TANZANIA. Tanga District: Muheza, 5 Oct. 1984, *Archbold* 3027!; Bagamoyo District: 12 km on Chalinze–Korogwe road, 2 Jun. 1996, *Faden et al.* 96/26!; Morogoro District: Morogoro Agricultural College, 6 Jul. 1970, *Kabuye* 291!; Zanzibar, near Kifule, 12 Jan. 1931, *Vaughan* 1828!

DISTR. **K** 5, 7; **T** 3, 6; **Z**; Somalia, Mozambique; Madagascar, Comoro Islands

HAB. Roadsides, grassland, open bush, scattered trees and shrubs, herbaceous vegetation, wet ground, thicket edge, disturbed riverine areas; sea level to 700(–1000) m

Flowering specimens have been seen from January, March, April, June, July, and September to November.

SYN. *Commelina* sp. D in Faden, U.K.W.F.: 660 (1974) & in U.K.W.F. 2nd ed.: 305 (1994)
Commelina sp. 5 in Faden, Fl. Somalia 4: 91 (1995)

NOTE. This species has caused much confusion over the years. Plants have usually been identified as *C. imberbis* because of their elongate capsules and seeds. That species differs (in our area) by usually lacking an upper cincinnus and by having proportionally broader seeds that are shallowly ribbed, with warty material along the ribs. It also occurs further inland and does not overlap with the mainly coastal *C. mascarenica*.

The type of this species was examined in the context of determining whether the then undescribed species *Commelina lukei* could have been this Madagascan plant. While the two species proved to be distinct, some collections from the eastern African coast from southern Somalia to northern Mozambique were found to be indistinguishable from *C. mascarenica*. The more they were studied, along with the collections of *C. mascarenica* from the *Flore de Madagascar* area, the clearer it became that they were all one species, which up to then had been overlooked in Africa.

This species may grow with *C. lukei* in Kenya and Tanzania, and there at least three mixed collections, including one by this author. *Commelina lukei* is separable by its leaf bases all clasping the stem, the capsules with a short apical extension, and the seeds appendaged. With the naked eye it can be seen that *C. lukei* has a white line along the leaf midrib on the upper surface. With 10 × magnification it may be observed that this line consists of very short, needle-shaped hairs. In *C. mascarenica* the white line is lacking and the hairs along the midrib are shorter, hooked hairs that are much less conspicuous than the midrib hairs in *C. lukei*.

Mgaza 317, from the Lusunguru Forest Reserve, Tanzania, appears to be related to this species but has been excluded because some of the spathes are strongly falcate, the peduncle of the lower cincinnus is apparently glabrous, the floral parts that can be measured are much smaller than those of any other specimen, and the petals have dried a very dark blue, unlike those of other collections. Tentatively, this appears to be a different species, but in the absence of capsules and seeds it is impossible to be certain.

Kabuye 165 questionably belongs to this species. It is the only collection from **T** 3 that lacks an upper cincinnus. However, it is also from further inland than any other **T** 3 collection. Its single spathe is imperfect, so the specimen could be either *C. mascarenica* or *C. imberbis*.

Lucas, Jeffrey & Kirrika 271 is the only specimen from the Tsavo National Park area. Hence it is from further inland and possibly a drier habitat than any other **K** 7 collection. The leaves are narrower than any other collection and not even the distal ones have cordate to cordate-amplexicaul bases. The flowers are also smaller than in any other collection accepted as this species, having the shortest filaments and smallest anthers of any collection. It is not certain whether non-fruiting specimens of *C. mascarenica* can always be separated from *C. petersii*.

The occurrence of this otherwise Zanzibar-Inhambane species at Homa Point on Lake Victoria is puzzling. Most likely it got introduced from the coast, but there are collections of it over a long period of time, so it seems to be well established, at least locally.

27. **Commelina petersii** *Hassk.* in Flora 46: 385 (1863) & in Peters, Reise Mossamb. 2: 522 (1864); C.B. Clarke in DC., Monogr. Phan. 3: 169 (1881) & in F.T.A. 8: 50 (1901); Obermeyer & Faden, in F.S.A. 4, 2: 33 (1985); Faden in U.K.W.F. 2nd ed.: p. 305 (1994) & in Fl. Somalia 4: 91 (1995); Ensermu & Faden in Fl. Eth. 6: 369 (1997); Faden in K.B. 62: 139 (2007) & in Figueiredo & Smith, Pl. Angola: 176 (2008). Type: Mozambique, *Peters* s.n. (B!, holo.)

Single-stemmed or tufted perennial (rarely annual?); roots thin, fibrous, sometimes stilt-like from the lower nodes; shoots erect or scrambling, sometimes prostrate at the base and then ascending, rarely becoming decumbent; distal internodes puberulous, glabescent. Leaves distichous; sheaths often splitting to the base, to 4.3 cm long, puberulous, ciliolate at the apex or eciliolate; lamina sessile (distal leaves) to petiolate (more proximal leaves), linear to ovate-lanceolate, 5–16 × 0.4–3(–4.5) cm, base rounded to subcordate (distal leaves) or cuneate, margins scabrid with prickle-hairs, often ciliolate, apex (acute to) acuminate; adaxial surface scabrid, abaxial surface pubescent. Spathes solitary, usually falcate (sometimes not); peduncles (1–)1.4–3.4(–3.7) cm long, puberulous; spathes 1.8–3.3(–3.7) × 0.8–1.3(–1.5) cm, base cordate to truncate, margins fused proximally for 3–7 mm, ciliolate along the fused portion, apex acute to acuminate; surfaces puberulous; upper cincinnus (sometimes vestigial, rarely lacking) producing 1(–2) male flower(s); lower cincinnus 3–6-flowered. Flowers bisexual and male, ± 1.5 cm wide; upper sepal ovate-elliptic, 1.8–3.5 mm long, hooded at the apex, paired sepals ovate-elliptic to obovate, 2.5–5.3 × 2–3.7 mm. Petals all blue (bright blue, Cambridge blue or pale blue), rarely mauve, usually concolorous, paired petals broadly ovate to reniform, 7–10.5(–13) × 5–7.5(–13) mm of which the claw 2.5–5 mm; medial petal ovate, cup-shaped, 2.4–5 mm long; staminodes 3; lateral stamen filaments 5.5–9 mm long, anthers 0.7–1.5 mm long; medial stamen filament 3.5–6.5 mm long, anther 1.2–2.5 mm long. Ovary 1.5–2 mm long; style 5–7.5 mm long; stigma capitate. Capsules pale yellow with dark flecks or entirely dark brown, bivalved, trilocular, oblong, 8–10 × 4–5.5 mm, 4–5-seeded, apex emarginate, dorsal valve deciduous, strongly striate longitudinally, ventral valve weakly striate or not striate, dorsal locule 1-seeded, ventral locules 2-seeded. Dorsal locule seed embedded in the capsule wall, ellipsoid, ± 5.3 × 2–2.5 mm, testa medium brown, smooth to alveolate; ventral locule seeds reniform, 3.5–4.2(–4.8) × 2.2–2.5(–2.7) mm, testa ribbed, smooth or alveolate between the ribs, mottled black and various shades of brown or a ± uniform medium brown or grey, not farinose.

KENYA. Northern Frontier District: Modo Gash–Garissa road, 13 km S of Modo Gash, 11 Dec. 1977, *Stannard & Gilbert* 958!; Machakos District: 3 km W of the Kibwezi turnoff along the Mombasa–Nairobi road, 15 Mar. 1969, *Faden, Evans & Siggins* 69/310!; Masai District: Kajiado–Namanga road, small hill on W side of road ± 0.5 km N of Namanga, just N of turnoff to Namanga River Hotel, 16 June 1974, *Faden, Faden & Ng'weno* 74/814!

TANZANIA. Mbulu District: Great North Road, Mwembe, 125 km S of Arusha, 5 Jan. 1962, *Polhill & Paulo* 1054!; Lushoto District: Korogwe, 20 Mar. 1969, *Archbold* 998!; Iringa District: Msembi–Mbagi track, km 2.3, 19 Mar. 1970, *Greenway & Kanuri* 13136!

DISTR. **K** 1, 4, 6, 7; **T** 1–5, 7, 8; São Tomé, Congo–Kinshasa, Eritrea, Ethiopia, Angola, Zambia, Malawi, Mozambique, Zimbabwe, Namibia, Botswana, South Africa; also India, Sri Lanka, Burma

HAB. Thickets, scrub forest, woodland, grassland, often associated with rocks and rock outcrops, sometimes near rivers and lakes, rarely in disturbed areas near cultivation and on termite mounds; commonly in sandy soil; sun or shade; sea level to 1450 m

Flowering specimens have been seen from all months except July. Flowers open ± 06:00 and fade about 09:30.

SYN. *Commelina persicariaefolia* C.B. Clarke in DC., Monogr. Phan. 3: 171 (1881), *nom illeg., non* Delile (1815)

C. persicariaefolia C.B. Clarke var. *geniculata* C.B. Clarke in DC., Monogr. Phan. 3: 172 (1881). Type: Burma, Pagamew, *Wallich Cat.* 8978N (K, holo.)

C. imberbis Hassk. var. *loandensis* C.B. Clarke in F.T.A. 8: 50 (1901). Syntypes: Angola, Luanda [Loanda]: about Alto das Cruzes, *Welwitsch* 6613 (BM!, K!, syn.); at Praia de Zamba, near Loanda, *Welwitsch* 6616 (BM!, sheet with the date July 1854, lecto. of Faden, 2007; BM!, K!, iso.); near Maianga de Povo, *Welwitsch* 6618 (BM!, syn.); Ambriz: near the R. Quizembo, *Welwitsch* 6615 (BM!, syn.), mixture with *C. diffusa*)

C. trachysperma Chiov. in Ann. Bot. (Rome) 9: 148 (1911). Type: Eritrea, Beni Amer in mount Curcù, *Pappi* 7689 (FT!, holo.)

NOTE. This species has been often confused with others, especially *C. imberbis* and *C. mascarenica*. Its most characteristic features are the lower leaves distinctly petiolate, the spathes very sticky due to the abundant hook-hairs, the upper cincinni long-exserted and usually one-flowered, all three petals blue (rarely all mauve or the lower one a different colour from the upper two), and seeds with transverse tuberculate ridges and lacking farinose granules.

This species was known as *Commelina persicariaefolia* Wight ex C.B. Clarke, an illegitimate name, in India for many years, until Rao (1967) correctly concluded that the Indian plant was actually an African species. Unfortunately, he erroneously identified it as *C. imberbis* (see discussion under 25. *C. imberbis*).

African plants have for the most part narrower leaves than Asian specimens. Indeed, extremely narrow-leaved plants with the lamina less than 1 cm wide occur only in Africa. At one time I thought that the African and Asian plants might be separable as subspecies, but there is a good deal of overlap in lamina widths, and unless other characters are found, I think it best to keep them all together for the present.

The two specimens cited in F.W.T.A. (Brenan, 1968) as *C. petersii* are *C. imberbis*. The significance of the ridges on the seeds was misinterpreted. Except for the anomalous record from São Tomé, which perhaps represents a recent introduction, *C. petersii* is unknown from West Africa or the Sahel.

Richards 21018 is a very peculiar specimen, most obviously because of its long, sickle-shaped spathes, the longest and most usual shaped spathes seen in this species. It also has the longest leaves (16 cm), longest spathe peduncle (3.7 cm), longest upper cincinnus peduncle (29 mm), longest lower cincinnus peduncle (± 20 mm) and longest pedicel (9 mm). In the absence of seeds, *C. petersii* appears to be the only described species to which it could belong.

Commelina petersii generally occurs in drier habitats and at lower elevations than *C. imberbis*, but they could occur together. Similarly, although it is apparently uncommon at the coast in Kenya, *C. petersii* could occur with either or both *C. mascarenica* and *C. lukei*. I have not seen any recognizable mixed collections of *C. petersii* and other species from our area although I made a mixed collection of *C. petersii* and *C. mascarenica* in Somalia (*Faden & Kuchar* 88/269 & 88/269bis).

The spathes are commonly noted by collectors as being 'sticky' [= adhesive]. This is undoubtedly due to the hook-hairs on the surface of the spathe and on the exserted peduncle of the upper cincinnus. Similar hairs are also present in other species, so it is not clear why this species should call attention to itself in this manner whereas the other species have not. The easily detached spathes may serve for seed dispersal.

The largest paired petals were measured in the field, and none of the liquid preserved flowers in four collections examined were as large. It may be assumed that other floral parts, which were not measured in the field, would also have exceeded the dimensions given here. Similarly, the floral width given here was a field measurement from the same population as the large petals. Therefore this should be treated as the maximum size recorded for the species.

The last flowers produced in the lower cincinnus of a spathe that has set fruits may be male. Such male flowers have noticeably longer pedicels (5–9 vs. 2–4.5 mm) than either the bisexual flowers that preceded them or the male flower of the upper cincinnus. In *Mathew* 6365 from northern Kenya, each of the five spathes preserved in liquid had the upper cincinnus vestigial (or lacking) and the lower cincinnus two-flowered with one open flower. In the two spathes in which the first flower was open and the second was in bud, the open flower was bisexual. In the three spathes in which the second flower was open, one flower was bisexual and the other two were male. In the spathes with male flowers, the preceding flower had produced a fruit, whereas in the spathe that had the second flower bisexual, no fruit had been set. Thus it appears that, even when only two buds are present, fruit set by the first flower can affect the gender of the second flower. A similar pattern has been observed in other *Commelina* species, but never in spathes with just two buds.

Blundell's illustration (Wild Fl. E. Afr.: 413, fig. 862 (1987)) labelled *C. petersii* is almost certainly another species, perhaps *C. eckloniana* subsp. *nairobiensis*.

28. **Commelina kotschyi** *Hassk.* in Schweinf., Beitr. Fl. Aethiop.: 207 (1867); C.B. Clarke in DC., Monogr. Phan. 3: 173 (1881), excl. *Welwitsch* 6624; Rendle, Cat. Afr. Pl. Welwitsch 2: 77 (1899), excl. *Welwitsch* 6624; C.B. Clarke in F.T.A. 8: 49 (1901), excl. *Welwitsch* 6624; Rao in Blumea 14: 349 (1967); Ensermu & Faden in Fl. Eth. 6: 368 (1997); Faden in Friis & Ryding, Biodiversity Research Horn Africa: 223 (2001). Type: Sudan, Kordofan, Arash-Col [Arasch-Col] *Kotschy* 34 (W†, holo., E, K!, iso.)

Annual; roots thin, fibrous; shoots solitary to many tufted, ascending to decumbent, not rooting, to 1 m long, 10–60 cm high; shoot bases thickened; internodes sparsely puberulous or glabrescent. Leaves apparently spirally arranged; sheaths 1–2 cm long, very sparsely puberulous, sometimes the hairs in a line, eciliate at the apex; lamina narrowly lanceolate to lanceolate-elliptic, 2–10.5 × 0.5–2.8 cm, base usually amplexicaul to rounded, margins scabrid, apex acuminate to acute, usually mucronulate; both surfaces puberulous. Spathes solitary or loosely clustered, slightly falcate or not [rarely strongly falcate]; peduncles 0.5–2.1[–2.5] cm long, puberulous; spathes 1–1.9 × 0.5–1.1 cm, base cordate to truncate or hastate, margins fused for 2–3 mm, ciliolate along the fused portion, apex acute to acuminate, often mucron(ul)ate, surface puberulous; upper cincinnus lacking, lower cincinnus 2–4-flowered. Flowers 8–15 mm wide; upper sepal elliptic, ± 3 × 2 mm, paired sepals ovate, ± 4 × 2.7 mm, with a broad transparent margin; paired petals sky blue or dark blue, broadly ovate-reniform, 7–8 × 5–6.5 mm of which the claw ± 3 mm; medial petal broadly ovate, ± 4.5 × 4 mm, concolorous; staminodes 3; lateral stamens with filaments ± 6.5 mm long, ± sigmoid, anthers ± 1 mm long, pollen yellow, medial stamen filament straight, ± 4.5 mm long, anther ± 1.7 mm long. Ovary ± 2 × 1.2 mm; style ± 5 mm long; stigma capitate or deltate. Capsule pale yellow with brown flecks, bi- (or tri–)locular, bivalved, oblong-ellipsoid, 8–9.5(–10) × 4.5–5.5(–5.7) mm, fundamentally 4–5-seeded, but frequently less by abortion, constricted between the seeds, dorsal locule empty and flattened or occasionally 1-seeded, ventral locules 2-seeded. Dorsal locule seed fused to the capsule wall, ellipsoid to obovate-ellipsoid, 3.5–5 × 2.3–2.4 mm, testa dark brown, smooth, not farinose; ventral locule seeds not compressed, ellipsoid to quadrangular, 3–4.3 (including the appendages) × 1.9–2.4 mm, testa mottled various shades of brown, smooth, ± white-farinose.

KENYA. Northern Frontier District: Gelai, 10 km ESE of Baragoi, 18 Nov. 1977, *Carter & Stannard* 502! & Isiolo–Archer's Post road, km 2, 16 May 1971, *Faden & Evans* 71/446!; Teita District: Tsavo National Park East, Dida Harea to Ndara, 1.8 km from Sign Post 142 towards Dida Harea, 29 Jan.1972, *Faden & Faden* 72/114!

TANZANIA. Arusha District: Makuyuni–Arusha road, km 27 from Makuyuni, 31 May 1996, *Faden et al.* 96/6!; Singida District: 14 km on road W from Singida, NW end of Ititi Swamp, 26 Jan. 2001, *Kuchar* 24426!; Morogoro District: Mlali Valley/Ngerengere, near Mdanganya, 25 Feb.1970, *Pócs & Kapela* 6135/G!

DISTR. **K** 1, 4–7; **T** 1, 2, 5, 6; Ethiopia, Sudan, Angola, Mozambique, Zimbabwe, Botswana, Swaziland, South Africa; India

HAB. Seasonally wet grassland, especially on heavy black clay, near rivers and seasonal pools, in roadside ditches, as a weed in rice and maize, also in bushland and bushland thicket; 500–1600 m

Flowering specimens have been seen from January, February, May to July and November.

SYN. *Commelina imberbis* sensu Obermeyer & Faden in F.S.A. 4, 2: 32 (1985), *non* Hassk.

NOTE. Some problems of typication of this species have been discussed by Rao (1967). The holotype was apparently destroyed in Vienna. The Kew sheet of *Kotschy* 34 kept in a type folder is a good specimen of the species described above, but other specimens in various institutions under this collection number belong to other species. A specimen matching the Kew sheet and belonging to the same species is at Edinburgh, fide Rao.

Until Fl. Eth. (Ensermu & Faden, 1994) this species was overlooked (Faden in U.K.W.F. (1974) and U.K.W.F. 2nd ed. (1994)) or misinterpreted (Obermeyer & Faden in F.S.A. (1985)) in recent African floras. The very distinctive seeds were first called to my attention by Ensermu, and upon checking, I discovered that the species also occurred in FTEA, including the Upland Kenya flora area, and that I had unknowingly collected it three times in Kenya.

The seeds and capsules of *C. kotschyi* are its most distinctive features. The testa is smooth and the hilum is on a midventral ridge, producing appendages in the apical and basal concavities of the seed, one or occasionally both appendages extending beyond the rest of the seed. These appendages probably function in seed dispersal, perhaps acting as elaiosomes to attract ants. In order to accommodate these appendages, the capsule has a small, rounded to truncate apical elongation, instead of the typical emarginate apex in species such as *C. imberbis*. Other distinctive features of *C. kotschyi* are its usual annual habit, commonly small leaves with strongly and finely undulate margins, small spathes that are white or whitish basally with a contrasting green margin, green veins in the white area, and the absence of an upper cincinnus at least in our area).

Commelina kotschyi commonly grows in seasonally wet or waterlogged soils. However, specimens from "*Acacia mellifera* bushland thicket and bushland" and "*Commiphora - Grewia* bushland on gravelly soil" would suggest that it can also occur in drier habitats. It is unknown whether the plants might have been growing in seasonally wet spots within those dry habitats.

In the field and the herbarium *C. kotschyi* is most likely to be confused only with *C. imberbis* because both species have spathes with shortly fused margins, blue flowers, elongate capsules and ellipsoid seeds. Moreover, their distributions overlap in East Africa. *Commelina kotschyi* differs from *C. imberbis* by the former's annual habit, preference for seasonally wet soils; smaller leaves with margins strongly and finely undulate; smaller spathes with a whitish basal area and a contrasting green margin; smaller flowers with all three petals concolorous, and blue or bluish purple filaments; capsules truncate to rounded at the apex; and seeds smooth and appendaged.

The unusual capsule and seed characters of *C. kotschyi* occur together elsewhere only in *C. lukei* Faden of coastal and subcoastal Kenya and northeastern Tanzania. Indeed, the capsules and seeds of the two species are almost indistinguishable. However, the two species barely overlap in their distribution (only near Tsavo National Park East in Kenya), with *C. lukei* being mainly coastal and *C. kotschyi* entirely an inland species. *Commelina lukei* further differs in its ecology, never growing in seasonally waterlogged soils. It also is typically perennial, its leaves have only needle-like hairs along the adaxial midrib (vs. only minute hook-hairs – visible under high magnification), the spathes have longer peduncles, lack the proximal white area, the upper cincinnus is well developed, and the flowers are larger with distinctive, question-mark-shaped lateral stamen filaments. The two species may be closely related, but except for their capsules and seeds, they are very different in their morphology, ecology and distribution.

29. **Commelina lukei** *Faden* in Novon 18: 469 (2008); Blundell,Wild Fl. E. Afr.: fig. 861 (1987), figure only). Type: Kenya, Kwale District: Kaya Kinondo, *Luke et al.* 7040 (US!, holo.; EA, K, MO, US!, iso.)

Perennial (or annual?); roots thin, fibrous; shoots erect to ascending, becoming trailing, scrambling or straggling, to 1.8 m long, rooting only near the base or decumbent, glabrous. Leaves mostly distichous, sometimes spirally arranged; sheaths 1–3 cm long, sparsely pubescent, sometimes flushed or spotted with purple, eciliate at apex; lamina sessile, linear-lanceolate to lanceolate, 4–13 × (0.6–)1–3 cm, base cordate-amplexicaul, margins scabrid, apex acute to attenuate; adaxial surface with a line of very fine, whitish, acicular hairs along the midrib, otherwise glabrous, abaxial surface sparsely pubescent. Spathes solitary, the uppermost leaf often bract-like; peduncles (1.6–)2–4(–5.5) cm long, puberulous with hairs in a line or throughout; spathes usually slightly to strongly falcate, (1.4–)1.8–2.5 × (0.5–)0.7–1.2 cm, base cordate to truncate, margins fused for 2–5 mm, ciliate along the fused edge, otherwise glabrous, apex acute (to acuminate), surface pubescent; upper cincinnus 1–3-flowered, flowers male; lower cincinnus 5–7-flowered, the flowers mainly bisexual but sometimes male. Flowers bisexual and male, (13–)15–21 mm wide. Sepals free, upper sepal lanceolate-oblong to ovate-elliptic, 2.5–4.5 × 1.2–1.5 mm, lower sepals ovate-elliptic to obovate-elliptic, 3.5–5 × 2.2–3.2 mm; paired petals blue, sometimes pale blue, rarely pale rosy lilac, broadly ovate to ovate-reniform, (9–)12.5–16 × (6–)10–13 mm of which the claw 5–6 mm long, concolorous with the limb or paler; lower petal ovate to ovate-elliptic, (3.5–)4.5–7 × 2.5–3.5 mm, concolorous with or slightly paler than the paired petals; staminodes 3; lateral stamen filaments (7.5–)11–16 mm long, anthers (1.2–)1.5–2 mm long, pollen white; medial stamen filament

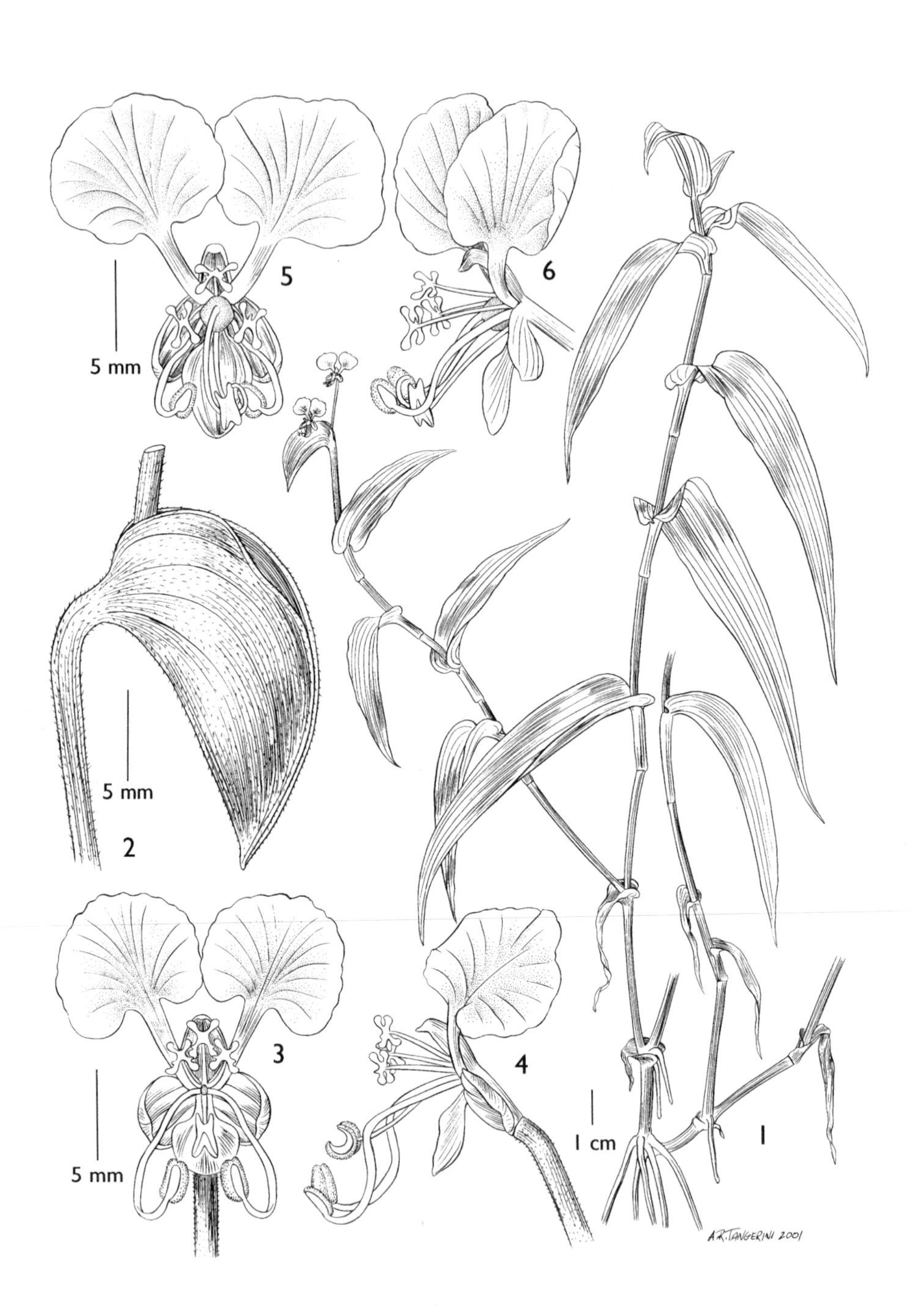

FIG. 20. *COMMELINA LUKEI* — **1**, habit; **2**, spathe; **3** & **4**, male flower, front & lateral side view; **5** & **6**, bisexual flower, dorsal and ventral view. All from cultivated material of *Luke* 7080. Drawn by A. R. Tangerini and reproduced, with permission, from Novon 18: 470 fig. 1 (2008)

(5–)5.5–8 mm long, anther (2–)2.5–3.5 mm long, pollen dull yellow. Ovary ± 2 mm long; style (8–)9–13 mm long; stigma deltate, slightly enlarged. Capsules grey-green to brown, bi- to trilocular, bivalved, 4–5 seeded, oblong, (7.5–)8–9.5 × (3.6–)4–5.5 mm, constricted between the seeds, with a rounded to truncate, obdeltate apical extension, dorsal locule 1-seeded or not developed, ventral locules 2-seeded. Dorsal locule seed fused to the capsule wall, dorsiventrally compressed, ellipsoid to obovoid-oblong, 3.3–3.7 × 1.9–2.4 mm, testa dark brown, smooth, not farinose; ventral locule seeds ± quadrangular in outline, 3–3.6 × 1.7–2.2 mm, testa smooth, dark brown or mottled various shades of brown, white- to tan-farinose. Fig. 19, p. 196; seeds & fruit Fig. 14, p. 136.

KENYA. Kilifi District: Vipingo, 36 km N of Mombasa, 16 Dec. 1953, *Verdcourt* 1056S!; Kwale District; Mrima Hill, 4 Sep. 1957, *Verdcourt* 1851S!; Teita District: ± 1.6 km S of Nairobi–Mombasa road along road from Maungu Station to Rukanga, 31 May 1970, *Faden* 70/149!

TANZANIA. Lushoto District: Western Usambara Mts, Mkusu Valley between Mkuzi and Kifungilo, 23 Apr. 1953, *Drummond & Hemsley* 2198!; Pangani District: Bushiri, 30 June 1950, *Faulkner* 629!; Tanga District: Bomalandani [Bomandani], 13.3 km S of Moa, 10 Aug. 1953, *Drummond & Hemsley* 3699!; Zanzibar, Chwaka, *Vaughan* 1745!

DISTR. **K** 7; **T** 3, 7? (see note); **Z**; Madagascar

HAB. Lowland evergreen forest, forest edges and clearings, plantations, moist thickets, bushland, grassland, roadsides, sometimes associated with rocks in moist situations, sea level to 1000(–1600) m

Flowering specimens have been seen from all months except February and November; the major flowering period is April–May.

NOTE. This species is most distinctive because of its often scrambling habit, the leaf bases all amplexicaul, the adaxial leaf midrib with only acicular hairs, the stamen filaments question mark-shaped, the capsule with a rounded to truncate apical extension, and the seeds smooth with a raised hilum that is extended as appendages at both ends. It is probably most closely related to *Commelina kotschyi*, which has similar appendaged seeds in capsules with a similar apex. *Commelina kotschyi* differs by its distinctly annual habit, mainly smaller leaves with strongly undulate margins, the presence of only or mostly hook-hairs along the adaxial leaf midrib, smaller spathes, the absence of the upper cincinnus, and by its occurrence at generally higher elevations and further inland and mainly or strictly in seasonally waterlogged soils.

In the field and in the herbarium *Commelina lukei* has been confused mainly with *C. imberbis*, with which it is allopatric, and with *C. mascarenica* C.B. Clarke, with which it is sympatric. Both *C. imberbis* and *C. mascarenica* have ± oblong capsules and ellipsoid seeds, and all collections with similar capsules and seeds, including specimens of *C. lukei* and *C. kotschyi*, were usually identified as '*C. imberbis*'. All of these species also have leaves with bases clasping the stem (at least in the most distal leaves on the flowering shoots), solitary spathes with the margins shortly fused at the base, the hairs on the spathes relatively short, and blue flowers, so it is not surprising that the species were not well understood. *Commelina imberbis* and *C. mascarenica* differ from *C. lukei* by having usually only the upper leaves amplexicaul (except in some populations of *C. imberbis*), the adaxial leaf midrib with only or mostly hook-hairs, the capsule with the apex truncate or emarginate, and the seeds not appendaged. In addition, the lateral stamen filaments in both these species are less strongly curved and not question mark-shaped, as in *C. lukei*.

Commelina lukei bears a striking resemblance to some plants of *Commelina latifolia* because they both have all of their leaves amplexicaul at the base and the adaxial midrib with only acicular white hairs. The two species only occur close to one another in the Taita Hills (**K** 7) and Western Usambara Mountains (**T** 3). In these mountains *C. lukei* is at its furthest inland extent and highest elevations. It occurs in natural habitats. In contrast, *C. latifolia* is at its closest to the Indian Ocean coast and it is limited to only disturbed situations. It may be an introduced weed in these mountains. *Commelina latifolia* can be distinguished from *C. lukei* by lacking an upper cincinnus, by its flowers having less strongly curved lateral stamen filaments, by its white (vs. blue) medial petal and by its quadrate (vs. oblong) capsules with spherical, dark brown seeds that lack appendages. The two species are probably not closely related.

The distinctive shape of the capsule apex in *Commelina lukei* and *C. kotschyi*, which can be recognized even when the capsule is not fully mature, is related to the shape of the seeds. The appendages of the hilum, which I expect function as elaiosomes for seed dispersal by ants, necessitate the extension of the capsule apex from the normal, truncate to emarginate shape found in species such as *C. imberbis* and *C. mascarenica*. In specimens of *C. lukei* with less prominently appendaged seeds the capsules have correspondingly less pronounced apical extensions and are less markedly different from those of *C. imberbis* and *C. mascarenica*.

In all three populations of *Commelina lukei* in which flowers have been photographed (*Faden, Evans & Rathbun* 69/402, *Faden & Faden* 74/513, and *Luke et al.* 7080) it has been observed that the lateral stamen filaments of the male flower in the upper cincinnus are distinctly longer than the homologous filaments of the bisexual flower in the lower cincinnus of the same spathe. Thus, it appears that stamen filament dimorphism is likely the norm in this species, although it has not been recorded in other species of *Commelina*.

The collection *Faden & Faden* 77/362, which was gathered near Shimoni, along the Kenya coast in 1977, was a major contributing factor to my early and long-term confusion between the two coastal species, *Commelina lukei* and *C. mascarenica*. It took many years to discover that the specimens collected belonged to *C. lukei*, whereas the separately collected seeds, from which living plants were grown and cultivated for more than 20 years came from plants of *C. mascarenica*.

Schlieben 1057 is from Lupembe (**T** 7), Tanzania, a great distance from the other collections of this species and much further inland. It is also unusual in some morphological characters. The upper cincinnus is mainly vestigial and enclosed within the spathe, so the specimen more closely resembles *Commelina latifolia* than typical *C. lukei*. Moreover, the capsule apex does not show distinctly the elongation that characterizes the latter species. However, the two submature capsules that are present are clearly longer than wide and the immature seeds have appendages, both characters agreeing with *C. lukei*, not *C. latifolia*. Furthermore, *C. latifolia* is found only in northern Tanzania and further north. Overall, *Schlieben* 1057 is best considered an aberrant specimen of *C. lukei*. Future collections might indicate that plants from this region should be treated as a distinct taxon.

30. **Commelina foliacea** *Chiov.*, Miss. Biol Borana, Racc. Bot: 302 (1939); Faden in U.K.W.F.: 657 (1974); Ogwal, Taxon. study Commelina in Uganda: 77, fig. 13 (1977) & in Mitt. Inst. Allg. Bot. Hamburg 23b: 580 (1990); Faden & Suda in J.L.S. 81: 304 (1980); Faden in U.K.W.F. 2nd ed.: 305 (1994) & in Novon 4: 232 (1994); Ensermu and Faden in Fl. Eth. 6: 366 (1997). Type: Ethiopia, Moyale [Mojale], *Cufodontis* 698 (FT!, holo.)

Rhizomatous perennial; roots thick, fleshy, of ± uniform diameter or slightly tapered from the base, yellow or rarely orange-yellow, arranged in two rows on the rhizome; shoots annual, arranged in two rows along the rhizome, corresponding to the roots, the subterranean bases swollen and forming the rhizome, shoots erect or ascending or occasionally trailing, to ± 30 cm high or 1 m long, rarely rooting at the nodes, glabrous or occasionally puberulous. Leaves with sheaths to 2.5 cm long, the sheaths often splitting longitudinally, glabrous or puberulous with uniseriate hairs; lamina lanceolate to ovate, 3.5–11.5 × 1.3–3 cm, base usually either cuneate and often petiolate or else sessile and cordate-amplexicaul, margins scabrid, apex acuminate (to acute); surfaces usually glabrous or occasionally puberulous. Spathes solitary, the two halves folded in the middle at right angles, forming a flat, leaf-like surface, with a pouch-like, swollen base below; peduncles 0.7–1.3 cm long, puberulous in a longitudinal, adaxial band; spathes 3.5–9.5 × 1–2 × 1 cm, base ± hastate, margins fused proximally for 3–4 mm, scabrid, sometimes ciliate proximally, apex acuminate, surfaces usually glabrous (rarely puberulous), except for an adaxial band of fine hairs around the 'mouth' of the pouch; upper cincinnus (1–)3–4-flowered; lower cincinnus 5–6 flowered. Flowers bisexual and male, ± 14 mm wide. Sepals green, glabrous, upper sepal linear-lanceolate to lanceolate-elliptic, 4–4.5 × 1.5–2.5 mm, lateral sepals connate their entire length into a cup 3.5–6 × 3 mm; paired petals dark or sky blue, rarely blue-violet, ovate-elliptic, 9–11 × 6 mm of which the green or white claw 2–3 mm; medial petal white to green, linear, often twisted, ± 3 mm × 0.3 mm; staminodes 3; lateral stamens with filaments 5–6 mm long, sigmoid; medial stamen with filament 4–5.5 mm long. Ovary ± 1 mm long; style 6–7 mm long; stigma capitate. Capsules 1 per spathe, pale yellow, sometimes with darker flecks, trilocular, bivalved, 5-seeded, oblong-ellipsoid, 8–10 × 4–4.5 mm, dorsal locule strongly humpbacked towards the capsule apex, 1-seeded, indehiscent, ventral locules (1–)2-seeded, dehiscent, dorsal valve deciduous. Seeds dark brown, often with slightly paler ridges or other raised areas, dorsiventrally compressed, reticulate, rugose-reticulate or foveolate, densely white-farinose; dorsal locule seed not fused to the capsule wall, ellipsoid, 2.4–4.2 × 2–2.6 mm; ventral locule seeds ovoid to ellipsoid, 2–4.2 × 1.7–2.6 mm.

subsp. **foliacea**

Lamina narrowed at the base, often petiolate, never cordate-amplexicaul.

UGANDA. Kigezi District, South Maramagambo Central Forest Reserve, ± 3.8 km on Kaizi–Bitereko road, 18 Sep. 1969, *Faden* 69/1114!

KENYA. Northern Frontier District: Moyale, 22 Apr. 1952; *Gillett* 12881!; Machakos District: Donyo Sabuk, 1 Feb. 1970, *Faden* 70/53!; Teita District: Taita Hills, foothills, road to Bura Mission from Voi–Taveta road, 16 Mar. 1969, *Faden, Evans & Siggins* 69/324!

TANZANIA. Moshi District: Moshi, Mar. 1929, *Haarer* 460! & Lyamungu Coffee Research Station, 17 Dec. 1943, *Wallace* 1165!; Lushoto District: Korogwe, 21 May 1969, *Archbold* 1009!

DISTR. **U** 2; **K** 1, 2, 4–7; **T** 2, 3; S Ethiopia, Congo–Kinshasa?, Malawi

HAB. Exceedingly varied: montane moist evergreen forest, forest paths, roads, edges and light gaps, dry evergreen forest, lowland wet evergreen forest, mist forest, *Acacia xanthophloea* woodland, wooded grassland, seasonally wet grassland, thickets, montane scrub, *Acacia-Commiphora* bushland, roadsides, edges of cultivation, often in riparian habitats; near sea level to 2300 m

Flowering specimens have been seen from all months except August and October.

SYN. *Commelina buchananii* C.B. Clarke in F.T.A. 8: 47 (1901), pro parte, quoad *Hildebrandt* 2643 and *Taylor* s.n. tant., *non Buchanan* 285 (lectotype of *C. buchananii* of Brenan, Bot. Staatssam. München 5: 211 (1964))

subsp. **amplexicaulis** *Faden* in Novon 4: 232 (1994) & in U.K.W.F. 2nd ed.: 305 (1994). Type: Tanzania, Morogoro District, Mkungwe Hill, NE part of the Northern Uluguru Mountains, cultivated in greenhouse and pressed from cultivation, *Faden, Evans & Pócs* 70/371 (US!, holo., C!, MO!, iso.).

Lamina base cordate-amplexicaul.

UGANDA. Karamoja District: Pian County, Lodoketemit [Lodoketeminit], 20 May 1962, *Kerfoot* 4930!; Busoga District: Iganga, 18 Dec. 1985, *Ogwal* 131!; Mubende District: Singo County, Watuba, 13 Apr. 1970, *Katende* K 120!

KENYA. West Pokot District: Sebit–Parua road, along Sebit River, 15 Mar. 1977, *Faden & Faden* 77/807! & Sebit, 31 July 1969, *Mabberley & McCall* 88! [mixed collection with *C. benghalensis*]

TANZANIA. Mbulu District: Tarangire National Park, 12.8 km from Tarangire Camp, *Richards* 24831! (in part, see notes); Morogoro District: Uluguru Mountains, road from Morogoro to Morningside, 2 Apr. 1974, *Faden & Faden* 74/384! & Northern Uluguru Mountains, N valley of Lumanga Peak, 8 June 1996, *Faden et al.* 96/76!

DISTR. **U** 1, 3, 4; **K** 2; **T** 2, 5, 6; Congo–Kinshasa?, Zambia, Malawi

HAB. Near streams, usually in shade, and in cultivation; 350–1550 m

Flowering specimens have been seen from April to July, November and December.

SYN. *Commelina foliacea* subsp. "A" in Faden and Suda in J.L.S. 81: 324 (1980)
C. foliacea var. nov. in Ogwal, Taxon. study Commelina in Uganda: 77, fig. 13 (1977)

NOTE (on the species as a whole). This species is very distinctive because of the form of its spathes, which are as long as the foliage leaves. The outwardly folded halves of the spathe, in conjunction with its size, make it more leaf-like than the spathes of any other species of *Commelina*.

Flowers open before or about noon and stay open well into the afternoon – fresh flowers have been recorded as late as 17:00.

The arrangement of the shoots, with their swollen bases, from which emerge thick yellow roots, is also highly distinctive. The rhizome grows incrementally by producing a new shoot adjacent to the most recent one, and always in a predictable place, so that the two rows of shoots and roots are maintained.

The strongly asymmetric dorsal capsule locule, which is humpbacked towards the capsule apex, is also highly distinctive. It results from the shape of the space available within the pouch and because the dorsal locule seed is positioned directly above the apical ventral locule seeds, instead of between the apical and basal ventral locules seeds, as in other species.

It is noteworthy that *C. foliacea* appears to lack hook-hairs entirely. It is thus the only sky-blue flowered species with fused spathe margins to lack such hairs.

In 1994 (Novon 4: 233) I singled out two collections from Tanzania with hairy leaves that were neither distinctly petiolate nor cordate-amplexicaul (*Richards* 24831, partly, and *Greenway & Kirrika* 11074). I suggested that they might be hybrids with another species. Further study has revealed the presence of other specimens with pubescent leaves and has made me revise my conclusion about the two cited specimens.

Gillett distinguished his 12882 from his 12881, with which it was growing, by the former's glabrous stems. Some of the internodes of 12882 are in fact puberulous, so this specimen is better differentiated by its glabrous lamina, whereas the lamina in *Gillett* 12881 is puberulous. In both collections the lamina is, at most, subpetiolate. In *Faden, Evans & Siggins* 69/324 the lamina and spathes are puberulous, and the lamina is sessile to subsessile. In *Luke* 3855B the lamina is sometimes puberulous proximally on the lower surface, and the lamina is sessile. What these three specimens have in common is that none of them comes from forest or other moist habitats. It is probable that the occasional occurrence of varying amounts of pubescence on the leaves and the lack of a distinct petiole are variants related to habitat.

Returning to the two collections that I recognized as unusual in Faden (1994), the leaves of *Greenway & Kirrika* 11074 are only sparsely pubescent beneath, but the mainly rounded leaf bases make this impossible to place in either subspecies. If it is a hybrid, then it might be between the two subspecies.

Richards 24831 is rather different. The smaller shoot is typical of subsp. *amplexicaulis*. The larger one is more densely pubescent – it is the only specimen with ciliolate leaf margins – than any other specimen with *C. foliacea*-like spathes. Its cuneate lamina base may be somewhat closer to subsp. *foliacea* than to subsp. *amplexicaulis*, but it does not match any specimen of either subspecies in leaf base or pubescence.

There are some very well pressed flowers on both shoots of *Richards* 24831, plus, in the packet, some additional flowers attached to spathes that can be assigned with assurance to one shoot or the other. All of the flowers seem to be a good match for one another and for other flowers of *C. foliacea*.

In conclusion, the pubescence and leaf base do not exclude the hairy shoot on *Richards* 24831 from *C. foliacea*, but the leaf base, in particular, does not allow its assignment to either recognized subspecies. Further collecting is necessary to determine what the significance is of the varition in *C. foliacea* in this area.

A single collection of *C. foliacea* has been seen from Congo–Kinshasa, *David* 46. Its leaves are hairy on both surfaces and their bases are rounded, so its subspecific placement is uncertain, hence the questionable listing of Congo–Kinshasa under both subspecies above. The disjunct distribution of subsp. *foliacea* in W Uganda and in Malawi from its central distribution in S Ethiopia, E Kenya and NE Tanzania is noteworthy but unexplained.

31. **Commelina benghalensis** *L.* in Sp. Pl. 41 (1753); C.B. Clarke in DC., Monogr. Phan. 3: 159 (1881); K. Schumann in P.O.A. C: 134 (1895); Rendle in Cat. Afr. Pl. Welw. 2: 76 (1899); C.B. Clarke in F.T.A. 8: 41 (1901); U.O.P.Z.: 208 (1949); Morton in J.L.S. 55: 519, fig. 15 (1956) & in J.L.S. 60: 176 (1967); F.P.U.: 195 (1962); Brenan in F.W.T.A., 2nd ed.: 3: 48 (1968); Schreiber et al. in F.S.W.A. 157: 8 (1969); Obermeyer & Faden in F.S.A. 4(2): 30, fig. 6 (1985); Blundell, Wild Fl. E. Afr.: 413 (1987); Ogwal in Mitt. Inst. Allg. Bot. Hamburg. 23b: 579, fig. 2C (1990); Faden in U.K.W.F. 2nd ed.: 305, t. 136 (1994); Kayemba-Ogwal in Proc. X111th AETFAT, Malawi 1: 416 (1994); Faden in Fl. Somalia 4: 91 (1995); Ensermu & Faden in Fl. Eth. 6: 373, fig. 207.20 (1997); Wood, Handbook Yemen Fl.: 317 (1997); Faden in Fl. North Amer. 22: 196 (2000). Type: India, *Herb. Linnaeus* No. 65.16 (LINN, conserved type)

Annuals or perennials; roots thin, fibrous; shoots ascending to decumbent, semi-erect to 60 cm or trailing or scrambling in shrubs, to 2 m long, commonly rooting at the nodes, usually much branched, glabrous or puberulous; shoot bases sometimes swollen; short, leafless shoots bearing subterranean, cleistogamous flowers and capsules often produced from the base of the plant. Leaves spirally arranged; sheaths to 3 cm long, puberulous, often also hirsute with white or red hairs, rarely subglabrous, ciliate with long red or white hairs at apex; lamina petiolate or sessile, ovate to elliptic or narrowly lanceolate, (1.5–)2.5–14.5 × 0.3–5.3 cm, base oblique, cuneate to rounded, margins scabrid at least distally, apex acute to acuminate, sometimes ± mucronulate; both surfaces puberulous, pilose or pilose-puberulous, rarely the abaxial surface glabrous. Spathes densely clustered terminally or sometimes only 1 or 2, funnel-shaped, subsessile; peduncles 2–5(–7) mm long, puberulous; spathes usually slightly falcate, 0.7–2 × 0.5–1.5 cm, base ± truncate to hastate, margins fused for 5–13 mm, ciliolate along the fused portion, apex acute, surfaces usually densely hirsute-puberulous; upper cincinnus often well-developed, producing a single male flower; lower cincinnus (1–)3–5-flowered. Flowers bisexual and male, 1–2 cm wide; upper sepal lanceolate, 3–4.5 × 1.5 mm, paired sepals usually fused proximally for up to 1 mm (rarely more), broadly ovate to obovate, 3–5 × 2.5–4 mm; paired petals usually bright blue, occasionally mauve or bluish purple, rarely white, broadly ovate to ovate-reniform, 8–14 × 7–12 mm of which the pale claw 3–5 mm; medial petal usually the same colour or bluish purple or lilac, oblanceolate-elliptic to ovate, 3–6 × 2 mm; staminodes 3; lateral stamens with filaments 7–11 mm long, anthers 1.2–2 mm long, pollen usually white, medial stamen filament 4.5–8 mm long, anther 2–2.7 mm long, pollen yellow. Ovary ± 1.5 mm long; style 9–10 mm long; stigma capitate. Capsules from chasmogamous flowers pale yellow or greyish tan, often with sparse darker spots, trilocular, bivalved (or sometimes trivalved?), 5-seeded, oblong-ellipsoid to subquadrate, 4.5–6 × 3–4.2 mm, slightly constricted between the seeds, dorsal locule indehiscent or tardily dehiscent, 1-seeded, ventral locules dehiscent, 2-seeded. Dorsal locule seed dorsiventally compressed, 2.3–3.1 × 1.6–2.25 mm, testa dark or light brown, sometimes mottled with black or various shades of brown, white-farinose all over, foveolate-reticulate; ventral locule seeds ovate, 1.6–2.1(–2.6) × (1.3–)1.6–2.1 mm, testa similar to dorsal locule seed but darker.

UGANDA. West Nile District: Mt Wati [Eti] summit, 19 July 1953, *Chancellor* 4!; Teso District: Serere, Jan. 1933, *Chandler* 1113! & km 24 on Soroti–Moroto road, 13 Oct. 1952, *Verdcourt* 827!

KENYA. Northern Frontier District: Loriu Plateau, 2 June 1970, *Mathew* 6581!; Meru District: Nyambeni Hills Tea Estate, 13 Oct. 1960, *Polhill & Verdcourt* 305!; Machakos District: ± 3 km before Simba on Nairobi–Mombasa road, 10 Feb. 1977, *Faden & Faden* 77/305!

TANZANIA. Arusha District: Ngurdoto Crater, NW slope, 21 Oct. 1965, *Greenway & Kanuri* 12186!; Tanga District: coast near Bomandani 13 km S of Moa, 10 Aug. 1953, *Drummond & Hemsley* 3696!; Songea District: between Songea airfield and Kwamponjore Valley, ± 9 km SW of Songea, 26 Apr. 1956, *Milne-Redhead & Taylor* 9826!

DISTR. **U** 1–4; **K** 1–7; **T** 1–8; **Z**; widespread in tropical and southern Africa; also in Yemen and tropical Asia; introduced and naturalized in North and South America, Indian and Pacific Ocean islands and Australia

HAB. Weed in cultivation and disturbed places such as roadsides, waste ground, cleared areas and former cultivation, also occurring in dry or moist grassland, thickets, bush and scrub, among and in shrubs, in screes and other rocky habitats, woodland, tree plantations and old nursery plots, along streams, rivers, lakes and dams, in forest, forest edges, paths and clearings, riverine forest, mist forest and ravine forest; sandy to clayey soils; 0–2350 m

Flowering specimens have been seen from all months.

SYN. *Commelina pyrrho-blepharis* Hassk. in Schweinfurth, Beitr. Fl. Aethiop.: 209 (1867); C.B. Clarke in DC., Monogr. Phan. 3: 190 (1881) & in F.T.A. 8: 60 (1901). Type: Ethiopia: Lotho, *Schimper* 1591 (K, lecto., selected here)

C. uncata C.B. Clarke in DC., Monogr. Phan. 3: 169 (1881) & in F.T.A. 8: 42 (1901). Type: Ethiopia, mountains near Gageros, *Schimper* 2269 (K!, holo.)

C. benghalensis L. var. *hirsuta* C.B. Clarke in DC., Monogr. Phan. 3: 160 (1881); Rendle in Cat. Afr. Pl. Welw. 2: 76 (1899); C.B. Clarke in F.T.A. 8: 42 (1901); Brenan in F.W.T.A., 2nd ed.: 3: 48 (1968); Kayemba-Ogwal in Proc. XIIIth Plenary Meeting AETFAT, Malawi 1: 416 (1994). Syntypes: Ethiopia, *Salt* s.n. (BM!, syn.); *Hildebrandt* 369 (not seen, syn.); Gurrsarfa, *Schimper* 1499 (K!, syn.); Madagascar & Comoros, *Boivin* 2329 (P, syn.)

C. obscura K. Schum. in P.O.A. C: 135 (1895); C.B. Clarke in F.T.A. 8: 60 (1901). Type: Tanzania, Kilimanjaro, Marangu, *Volkens* 2267 (B!, holo.; BM!, BR!, K!, iso.)
C. kilimandscharica K. Schum. in P.O.A. C: 134 (1895). Syntypes: Tanzania, Kilimanjaro, Marangu, *Volkens* 2254 (B, syn.; K!) & *Meyer* 125 (B, syn.)
C. rufo-ciliata C.B. Clarke in F.T.A. 8: 54 (1901). Syntypes: Malawi: between Mpata and the commencement of the Tanganyika Plateau, 2000–3000 ft, July 1896, *Whyte* s.n. (K!, syn.) & Kondowe to Karonga, 2000–6000 ft., July 1896, *Whyte* s.n. (K!, syn.)
C. benghalensis L. subsp. *hirsuta* (C.B. Clarke) J.K. Morton in J.L.S. 60: 215 (1967)

NOTE. This species is more diverse morphologically in our area, especially in Kenya and Tanzania, than anywhere else in its very wide range. In our area it is known to include both diploids and tetraploids and probably higher ploidies as well. There are a number of unusual forms that occur in our area, some of which might merit taxonomic recognition upon further study. *Grimshaw* 93329 and *Paulo* 404 both from **T** 2 have extremely small and narrow (up to 25 × 3–5 mm), strongly recurved, complicate leaves. Red hairs are entirely lacking but the very long, white hairs at the summits of the sheaths are typical of *C. benghalensis*. The spathes are small and not as densely clustered as in many other collections.

Archbold 560 from Kideleko (**T** 3) has long, subglabrous, narrowly lanceolate to lanceolate-oblong leaves to 11.5 cm long and 1.7 cm wide. The single spathe is also subglabrous. The flowers are recorded as white, although the remains of clearly blue petals can be seen adherent to the pressed leaves close to the spathe. The long red hairs on the apices of the sheaths clearly place this in *C. benghalensis*, but otherwise it is not at all similar to other collections.

Magor 40 from the Rumuruti area of the Laikipia Plateau (**K** 3) is a narrow-leaved plant that is common in this area. It has been found to be diploid and it grows side by side with more typical, broader-leaved plants of *C. benghalensis*. The lower surface of its leaf is glabrous. The lateral stamen anthers, observed and photographed from living plants in other populations, have yellow pollen instead of the usual white pollen.

Plants from the Chyulu Hills in Kenya, e.g. *Gilbert* 6231, often have the spathes flushed or striped with red or purple. Their relatively narrow leaves also give them a distinctive appearance.

Soleman 7480 from Amani (**T** 3) is a very robust specimen that resembles the large-leaved climbers in highland Ethiopia as exemplified by the lectotype of *Commelina pyrrhoblepharis*. A robust, large-leaved plant cultivated outside the East African Herbarium in Nairobi since it was collected as a sterile plant in the Uluguru Mountains by B.J. Harris in 1969 is likely to belong to the same taxon. So is a strikingly variegated form, as exemplified by the cultivated *Robertson & Aikman* 7484, which originally came from cultivation in Tanzania, but its original provenance is unknown.

Except for the possibility of including plants like *Soleman* 7480 in *C. benghalensis* var. *hirsuta*, none of the other variants described above can easily be accommodated in a classification that recognizes only two taxa in this species, *C. benghalensis* var. *benghalensis* and var. *hirsuta* (Morton 1967; Brenan 1968; Kayemba-Ogwal 1994). While many of the specimens of *C. benghalensis* from our can be separated using some of the characters provided by these authors, many others are not readily thus distinguished. The following characters work to some degree:

Annuals; red hairs sometimes lacking, when present, confined to the apices of the sheaths and sometimes the petiole margins; cleistogamous flowers often present at the base of the plant; diploids .... var. *benghalensis*
Perennials; red hairs always present on the surface of the leaf sheath as well as the summit, sometimes also on the spathes; cleistogamous flowers usually lacking; polyploids .... var. *hirsuta*

Var. *benghalensis* is more typical of lower rainfall areas and var. *hirsuta* of moister situations, such as forests, forest edges and clearings, but they do overlap. However, because the plants cannot always be distinguished their ecology is incompletely known. Only var. *benghalensis* is reported as a common weed in Uganda by Kayemba-Ogwal (loc. cit.).

Commelina benghalensis is an important weed wherever it has been introduced. Only the diploid form is known outside of Africa with the exception of some populations of an apparent hexaploid in southern California (Faden in Flora of North America 22: 196. 2000). Because of its ability to replant itself by means of its underground seeds, it is very difficult to eradicate. *Gillett* 13181 from northern Kenya demonstrates that even very young plants of *C. benghalensis* have already produced cleistogamous flowers to ensure that the species will return during the next rainy season.

There is some disagreement about whether the dorsal capsule valve is ever dehiscent. Most authors, including Ogwal (1990) describe it as indehiscent, but Fl. Eth. describes the dorsal locule as dehiscent or tardily dehiscent. I have found that it easily can be opened and then it splits along the dorsal suture, so I would not be surprised if it naturally split at least sometimes, but I have not actually found it already dehisced in the field or on specimens. So from my perspective its possible dehiscence is still moot.

32. **Commelina forskaolii** *Vahl*, Enum. Pl. 2: 172 (1806); C.B. Clarke in DC., Monogr. Phan. 3: 168 (1881) & in F.T.A. 8: 44 (1901); Morton in J.L.S. 55: 522, fig. 19 (1956) & 60: 185 (1967); Brenan in F.W.T.A., 2nd ed.: 3: 48 (1968); Ogwal in Mitt. Inst. Allg. Bot. Hamburg 23b: 580 (1990); Faden in U.K.W.F. 2nd ed.: 305 (1994) & Faden in Fl. Somalia 4: 92 (1995); Ensermu & Faden in Fl. Eth. 6: 374 (1997); Wood, Handbook Yemen Fl., 317 (1997). Type: Yemen, *Forsskål* s.n. (C-VAHL!, holo.)

Perennial (or annual?) herb, rarely rhizomatous; roots sometimes thick and cord-like, those from rooted nodes always thin and fibrous; shoots erect to ascending or prostrate to repent or procumbent and commonly forming mats, often rooting at the nodes and producing flowering shoots or sometimes new plants from the nodes, sometimes producing short, leafless, subterranean shoots bearing cleistogamous flowers and fruits from the rooted nodes; vegetative shoots spreading to ± 50 cm, erect flowering shoots to 15(–30) cm tall; internodes hirsutellous, puberulous or glabrous. Leaves distichous; sheaths to 1.5 cm long, puberulous or pilose-puberulous, ciliate at apex; lamina sessile to subpetiolate, often falcate, elliptic to linear or ovate, 1.5–7(–11) × 0.4–1.2(–1.8) cm, base oblique, cuneate, margins scabrid distally, ciliate proximally, apex acute to acuminate, sometimes mucronulate; adaxial surface pilose or glabrous, abaxial surface puberulous or pilose-puberulous. Spathes solitary, not to slightly falcate; peduncle 0.3–1.6 cm long, puberulous; spathes 0.9–1.9 × 0.5–1 cm, base hastate to truncate, margins fused for 2.5–5.5 mm, sometimes ciliolate along the fused part, apex acute to obtuse, surface hirsute-puberulous; upper cincinnus producing a single male flower; lower cincinnus 3–5-flowered. Flowers bisexual and male, 13–19 mm wide; upper sepal boat-shaped, 2–3 mm long, paired sepals free or shortly fused basally, ovate-elliptic to broadly elliptic, 4 × 2–3 mm; paired petals bright blue or dark blue (blue, royal blue, sky blue, cobalt blue), rarely pale blue, pale mauve, pink or white, broadly ovate, 7.5–10 × 6–9 mm of which the claw 1.5–2.5 mm long; medial petal dark blue or paler with a blue midrib, narrowly lanceolate to ovate-lanceolate, 2.2–2.5 × 0.8–1 mm; staminodes 3; lateral stamens with filaments 5–11 mm long, anthers 1.3–2 mm long, pollen white; medial stamen filament 3–5 mm long, anther 2.3–2.7 mm long, pollen yellow. Ovary 1.5–2 × 1 mm; style 5.5–10 mm long; stigma capitate. Capsules brown, often mottled different shades, rarely nearly black, trilocular, bivalved, oblong-ellipsoid to obovoid, 4.2–5.6 × 2.3–3 mm, usually 1-seeded, the dorsal locule seed developing. Dorsal locule seed fused to the wall of the indehiscent dorsal locule, oblong-ellipsoid, 3.7–3.8 × 1.9–2 mm, testa dark brown, smooth, not farinose.

UGANDA. Acholi District: Murchison Falls National Park, Paraa Lodge, 15 Sep. 1969, *Faden, Evans & Lye* 69/1054!; Karamoja District: Lodoketemit [Lodokeminet], 20 May 1963, *Kerfoot* 4939! & near Nabilatuk, 1 Jul. 1957, *Dyson-Hudson* 252!

KENYA. Northern Frontier District: Isiolo–Wajir road, 4 km W of Modo Gash, 10 Dec. 1977, *Stannard & Gilbert* 902!; Embu District: Embu–Kangonde [Kangondi] road, 11.5 km from Embu, 6 June 1974, *Faden & Faden* 74/717!; Kilifi District: Sokoke Forest, ± 3 km on track S of Gede to Jilore Forest Station, 25–27 July 1974, *Faden & Faden* 74/1249!

TANZANIA. Handeni District: Kwamkono, 29 June 1966, *Archbold* 781!; Tanga District: coast near Bomandani 13.4 km S of Moa, 18 Aug. 1953, *Drummond & Hemsley* 3697!; Uzaramo District: Dar es Salaam, 10 July 1966, *Archbold* 817!

DISTR. **U** 1; **K** 1, 2, 4, 6, 7; **T** 1–7; **Z**, **P**; Cape Verde, Senegal, Gambia, Mali, Burkina Faso, Ghana, Benin, Nigeria, Niger, Cameroon, Sudan, Djibouti, Ethiopia, Eritrea, Somalia, Zambia, Malawi, Mozambique, Zimbabwe, Namibia, Botswana, South Africa; Madagascar, Mauritius, Egypt, Arabia (including Socotra), India; naturalized in U.S.A. (Florida)

HAB. Grassland, bushland, woodland, cultivation, waste places, muddy depressions in various habitats, coastal forest, foreshore near ocean, rocky slopes, riverine, screes, sand hills, thickets, and swampy ground; most commonly in sand or sandy soil, but also in poorly drained soils or on rocks; 0–1650 m
Flowering plants have been seen from all months.

NOTE. This common species is easily recognizable because of its small spathes, small, bright blue flowers in which the lateral stamen filaments are winged, yellow sterile basal lobes on the otherwise dark medial stamen anther, and typically one-seeded capsules in which the dorsal capsule valve containing the single seed has on its sides a variety of tubercles and ridges. The plants often form mats, and it is not altogether clear even in the field which ones are annuals, but some are definitely perennials.

Some plants from the driest parts of our area, particularly in northern and eastern Kenya have particularly narrow and sometimes unusally long leaves. They match some collections from Ethiopia and Somalia. I could not find a strong character on which to separate them from the more typical forms of the species that occur throughout our area and the rest of Africa.

Faden & Faden 72/14, from the Nairobi–Mombasa road near Simba Station, is a very peculiar looking plant with strongly erect shoots, some of which may become decumbent but do not produce cleistogamous flowers, the shoots arising from thick rhizomes, and all of the leaves long and narrow. The flowers and fruits are typical of *C. forskaolii*, including a rare ventral locule seed, so there is no question about the identity of the collection. The plants were growing in clayey soil with *Pennisetum mezianum*.

The species is unusual in occurring in some of the driest habitats for any species of *Commelina*, yet it may also thrive in poorly drained soils. I suspect it would not persist in the latter, but I do not have any long-term observations.

Because of its mat-forming habit and its ability to produce underground cleistogamous flowers and fruits, this is one of the weedier species of *Commelina*. Considering its distribution and habitat preferences it is most likely to be problematic along the coast in our area. The only other species of *Commelina* that produces such flowers is *C. benghalensis*, but in that species they are produced from the base of the plant, whereas in *C. forskaolii* they are produced from rooted nodes of trailing or decument stems.

The figure labeled *Commelina forskaolii* in Blundell (fig. 860) is a different species. However his fig. 858, labeled *Commelina* sp. is *C. forskaolii*, the lyrate, winged stamen filaments being very distinctive.

33. **Commelina schliebenii** *Mildbr.* in N.B.G.B. 11: 396 (1932). Type: Tanzania, Njombe District, Lupembe area, on the Mpoponzi, *Schlieben* 1056 (B, holo., BM!, BR!, iso.)

Perennial; roots thick, but not tuberous; shoots erect to ascending or trailing, not rooting at the nodes, sparsely to densely branched, 25–40(–130) cm tall, ± densely white-hirsute. Leaves with sheaths to 3(–5) cm long, sometimes partially split, white-hirsute, sometimes dark purple-spotted, white-ciliate at apex; lamina sessile, narrowly to ovate-lanceolate, 5–12 × 1–3.5 cm, base asymmetric, rounded to cordate or one side of the lamina cuneate, not auriculate, margins scabrid, apex acute to acuminate, often mucronulate; both surfaces puberulous or hirsute-puberulous. Spathes solitary or subclustered; peduncles 1.5–6(–8) cm long, hirsute-puberulous; spathes often slightly falcate at apex, 1.5–3 × 1–1.7 cm, base subcordate or hastate to truncate, margins fused for 5–12 mm, sparsely ciliate along the fused edge, otherwise glabrous, apex acute to acuminate, surface hirsute-puberulous; upper cincinnus 1(–3)-flowered, its flower(s) male; lower cincinnus (4–)6–10 flowered. Flowers bisexual and male. Sepals free, glabrous, tinged with blue, upper sepal lanceolate- or narrowly elliptic, 4–6 × 1.7–2.5 mm, paired sepals free, obovate-oblong to broadly elliptic, ± 5.5–6.8 × 3 mm; paired petals sky blue, ± 11–15 mm long of which the concolorous claw 4–5 mm, lower petal blue but a little paler than the paired petals, oblong-elliptic to lanceolate, 5–6.5 × 3 mm; staminodes 3; lateral stamens with filaments ± 11 mm long, anthers 1.5–2 mm long, pollen dirty white to cream; medial stamen filament ± 9 mm long, anther 2.5–3.3 mm long, pollen golden yellow. Ovary ± 1.5 mm long; style ± 12 mm long. Capsules grey-green or grey-brown, nearly square, very slightly longer than wide, ± 5.5 × 4.5–5 mm, constricted between the seeds, 4-seeded (or fewer through abortion). Seeds subspherical to shortly ellipsoid, not dorsiventrally compressed, 2.1–2.6 × 1.9–2.2 mm, testa medium brown, mottled with lighter brown, smooth to partially alveolate or slightly rugose, brown-farinose in the shallow depressions.

TANZANIA. Iringa District: Ngowasi [Ngwasi] Lake dam and vicinity, 10 June 1996, *Faden et al.* 96/138!; Ngowasi [Ngwasi] pine plantation, 1 Jan. 1987, *Lovett* 1250! & Iboma Forest Reserve, Iringa Upland, 18 Nov. 1982, *Macha* 110!

DISTR. **T** 7; not known elsewhere

HAB. *Brachystegia* woodland, grassland, roadside banks, wooded termite mounds, thickets; 1750–2400 m

Flowering specimens have been seen from November to February and May.

NOTE. The label notes on the isotypes seen do not match what is given by Mildbraed along with the type description. The labels of the isotypes read: "Stromgebiet des oberen Ruhudje, Landschaft Lupembe, nördlich des Flusses, [May] 1931", while the type description includes the exact date (28 May 1931), a brief description of the plant by Schlieben (30–40 cm high, flowers sky blue) and a vernacular name (njangololo).

Both leaf surfaces always have a mixture of short hook-hairs and longer, uniseriate hairs. Mildbraed evidently overlooked the abundant hook-hairs on the leaves, because he noted only the scattered long hairs along the midrib on the type. The hook-hairs are always ± abundant on the adaxial surface, but their numbers may be greatly reduced on the abaxial surface when uniseriate hairs are abundant there. The uniseriate hairs may be equally scarce (as on the type) or equally numerous on both surfaces. When there is a difference in abundance, they are more numerous on the abaxial surface. On the adaxial surface they are sometimes ± confined to the midrib area and as a submarginal band, which is an unusual distribution pattern in *Commelina*.

The type is the only collection seen in which the sheaths are purple-spotted and the main veins on the spathes may be purple. It also has the sparsest pubescence of uniseriate hairs on the leaves.

The pubescence, especially of the internodes and sheaths, is supposed to be characteristic of the species, yet many specimens of the similar and perhaps related species *C. eckloniana*, cannot readily be distinguished by this character alone. In that species the leaves are regularly cuneate at the base (except for the uppermost ones), the roots are a cluster of fusiform tubers (in the region of overlap), and the seeds are spherical and warty or spiny. The species may also be confused with *C. zambesica*, but that species has the leaves narrowed at the base with auricles on the leaf sheaths, the spathes with margins more shortly fused, and the seeds ± reticulate.

34. **Commelina zambesica** *C.B. Clarke* in DC., Monogr. Phan. 3: 161 (1881); K. Schumann in P.O.A. C: 135 (1895), as *sambesica*; C.B. Clarke in F.T.A. 8: 43 (1901); U.O.P.Z.: 122 (1949); Brenan in K.B. 15: 209 (1961); Morton in J.L.S. 60: 190 (1967); Brenan in F.W.T.A., 2nd ed.: 3: 48 (1968); Obermeyer & Faden, F.S.A. 4, 2: 32 (1985); Malaisse in Fl. Rwanda Sperm. 4: 132, fig. 48, p. 133 (1988); Ogwal in Mitt. Inst. Allg. Bot. Hamburg 23b: 579, 585 (1990); Ensermu & Faden in Fl. Eth. 6: 370 (1997). Types: Mozambique, opposite Sena [Senna], June(?) 1860, *Kirk* s.n. (K!, syn.); banks of the River Chire [Shire] near Morrumbala Mt [Miramballas], *Kirk* 915 (K!, syn.); mouth of the Kongone [Kongoni] R., Zambesi Delta, June(?) 1861, *Kirk* s.n. (K!, syn.); Chupanga [Shupanga], between Tete [Tette] and the sea coast, 30 Mar. 1860, *Kirk* s.n. (K!, lecto., selected here); Malawi, Lake Nyasa, *Simons* s.n. (BM!, syn.)

Perennial; roots thick and fleshy, sometimes fusiform, ± tuberous, those from rooted nodes thin and fibrous; shoots 1-many, erect to decumbent and rooting at the nodes, 20–90 cm high or 1.3 m long, when decumbent, the stems often looping along the ground, ± densely hirsute-puberulous to subglabrous. Leaves distichous; sheaths to 3 cm long, with auricles at the summit, pubescent with a line hairs along the fused edge to hirsute-puberulous all over, rarely glabrous, ciliate or eciliate at apex; lamina sessile or petiolate, lanceolate to ovate-elliptic, 3.5–12(–18) × 1–4.5(–5.5) cm, base oblique, cuneate or the distalmost rounded, apex acute to acuminate; adaxial surface glabrous, scabrid or hirsute, the midrib always glabrous, abaxial surface glabrous to densely hirsute. Spathes solitary or sometimes a few loosely clustered; peduncles (0.5–)1–3.5(–5.5) cm long, puberulous mainly in a band continuous with the fused edge of the spathe; spathes falcate or not, 1.5–2.5(–3.5) × (0.8–)1–1.8 cm high, base cordate to sagittate, margins fused for 3.5–8 mm, ciliate along the fused portion, apex obtuse to acute (or rounded), surfaces hirsute-puberulous; upper cincinnus 1-(rarely up to ± 5–)flowered or vestigial; lower cincinnus 5–13-flowered. Flowers male and bisexual, up to ± 2.5 cm wide. Sepals pale blue, upper sepal ovate to ovate-elliptic, 3–5.5 × 2 mm, paired sepals fused laterally for at least half their length, ovate to obovate, 4–7 × 3–4.5(–5) mm. Paired petals blue (rarely tinged with mauve), 9–16 × 10–16 mm of which the claw 3–6 mm, medial petal blue, narrowly lanceolate to broadly ovate or obovate, cup-shaped or the margins rolled up, 5–6(–10) × 1–6(–7.5) mm. Staminodes 3; lateral stamens with filaments 7–17 mm long, anthers 1.5–2.2(–2.5) mm long, pollen white; medial stamen filament 4–8 mm long, anther 2.3–3.4(–4) mm long, pollen white. Ovary ± 2 mm long; style 8.5–15 mm long. Capsules trilocular or bilocular, tri- or bivalved, the valves persistent, fundamentally 5-seeded but commonly fewer through abortion, nearly square, 5–6.5 × 4.5–6 mm, contracted between the seeds. Seeds ellipsoid to subspherical or ovoid, testa medium brown to nearly black (rarely mottled), usually with slightly raised orange lines, also densely white-farinose; dorsal locule seed ellipsoid, 3–3.8 × 2.4–2.8 mm; ventral locule seeds subspherical or shortly ellipsoid, (2.1–)2.4–3.1 × (1.6–)2.5–2.8 mm.

UGANDA. Acholi District: NW of Chobi, Kabalega [Murchison Falls] National Park, 19 Sep. 1967, *Angus* 5994!; Bunyoro District: Rabongo Forest, 24 Apr. 1964, *H.E. Brown* 2094! & Bunyoro Ranch, near Masinde, July 1961, *Turner* 256!

KENYA. Northern Frontier District: Moyale, 7 Dec. 1952, *Woodhouse* in *Gillett* 15084!; Kwale District: Shimba Hills, Sheldrick's Falls area, 2 Aug. 1970, *Faden & Evans* 70/436! & Shimba Hills, Mkurumuji Point area, 23 Mar. 1968, *Magogo & Glover* 562!

TANZANIA. Lushoto District, Lower Sigi Valley, 30 May 1950, *Verdcourt* 236!; Morogoro District: Morogoro, Faculty of Agriculture compound, 6 July 1970, *Faden et al.* 70/383!; Tunduru District: ± 11 km E of Songea District boundary, near Libobi village, 21 Dec. 1955, *Milne-Redhead & Taylor* 7732!

DISTR. **U** 1, 2; **K** 1, 7; **T** 1, 3–8; **Z**; Nigeria, Cameroon, Central African Republic, Congo–Kinshasa, Ethiopia, Zambia, Malawi, Mozambique, Zimbabwe, Namibia, Botswana, South Africa

HAB. Grassland, woodland (especially in moist situations), streamsides, swamps, thickets and thicket edges, in and at edges of cultivation, forest and forest edges, montane scrub, roadsides; sometimes in shade or partial shade; sometimes in sandy soil; 50–1800 m
Flowering specimens have been seen from all months.

SYN. *Commelina* sp. 1, Thomson in Speke, Journal Disc. Source Nile, app. G: 650 (1863)
C. communis Baker in Trans. Linn. Soc. 29: 163 (1875), *non* L. (1753)
C. cuneata C.B. Clarke in F.T.A. 8: 51 (1901). Types: Malawi, Kondowe [Khondowe] to Karonga, 2000–6000 ft, July, 1896, *Whyte* s.n. (K!, syn.); Kavali, 1891, *Carson* s.n. (K!, syn.)

NOTE. The Kirk collection from Chupanga has an excellent watercolour by John Kirk with notes on the back of the painting, which is numbered "Fig. 223". It also has seeds. This specimen was designated as the type by Clarke on the sheet. It is the best choice, so I have lectotypified the species with it.

Commelina cuneata, which some specimens from our area had been determined, is essentially *C. zambesica* in which the dorsal locule seed did not develop. Seed abortion within the capsules of *C. zambesica* is very common, more so than in most species of *Commelina*. Although the label of *Faden* 70/383 records the basal ventral locule seeds as having a tendency to abort, producing a three-seeded capsule, in other specimens the particular seeds that abort seems to be more random. Typically, specimens with capsules in which the dorsal locule seed has aborted also have other capsules in which the dorsal locule is fully developed and one-seeded. Thus Clarke's distinction between the two does not hold and *C. cuneata* is a synonym of *C. zambesica*.

Commelina disperma Faden shares many of the significant characters of *C. zambesica*: thick roots, auriculate leaf bases, leaf pubescence type, bracteate spathes, and orange reticulum on the seeds. It differs most notably by its consistently two-seeded capsules, glabrous spathe surfaces (seen in only one specimen of *C. zambesica*: *Ogwal* s.n. from Uganda), very short spathe peduncles, fewer flowers in the lower cincinnus, smaller, less fused paired sepals, and much smaller medial petal. Whether it will prove to be merely an extreme form of *C. zambesica* is yet to be determined, but it seems best to maintain it as a separate species for the time being.

Tweedie 1674 from Morogoro has seeds that are mottled in appearance, but is otherwise a normal appearing specimen. The orange material that normally forms raised lines is instead much divided on the surface forming at most small clumps of cells.

Gillett 12800 was separated from the rest of *C. zambesica* because of its predominantly cordate leaf bases. It is at the extreme size limit in many characters, e.g. spathe length, upper cincinnus peduncle length, paired petal width, and staminode filament length, all measurements that have been matched in other, more typical collections. It also exceeds the maximum dimensions in a few cases: lower cincinnus pedicels (9 mm) and medial stamen filament (11 mm). It agrees with *C. zambesica* in some significant characters, such as auriculate apex on some of the smaller, proximal leaf sheaths, broadened lateral stamen filaments, prominent bracteoles, large medial lobes on the staminodes and the degree of fusion of the paired sepals. It lacks capsules and seeds. I believe that this is likely just an extreme form of *C. zambesica*, but I would like to see matching material with capsules and seeds to confirm that hypothesis.

Bax C1 has two very different looking plants on the sheet. One looks like typical *C. zambesica*. The other has spathes on much longer peduncles (up to 5.5 cm long), with one spathe very strongly falcate. However, a submature seed found inside this spathe agrees with *C. zambesica*, so this shoot should just be considered an atypical specimen, not a different species. The collector noted on the label that two species were mixed, which she separated as C1 and C2. C2 has not been located.

In *Gillett* 15084, the pollen-bearing medial lobes of the staminodes are unusually well developed for a species of *Commelina*. They do not however seem to have released any pollen. In some other collections these lobes are also appear well developed and have released some pollen.

The presence of bracteoles is a character of unknown significance in *Commelina*. Even its consistency within species is largely unknown. The bracteoles in this species are often quite well developed and conspicuous on the lower cincinnus and sometimes on the upper cincinnus.

The flowers seem to be very variable in size. In the larger ones, such as in *Milne-Redhead & Taylor* 7732 the lateral stamen filaments are clearly broadened, almost winged, distally. This is visible both in dried flowers on herbarium specimens and in liquid-preserved flowers. However, smaller flowers, even those preserved in liquid, do not show this character. The medial or lower petal seems to be extremely variable in shape and size. Some of the smallest widths measured may be the result of the lateral margins of this petal sometimes rolling inwards, but even when unrolled, petals as narrow as 2.5 mm have been observed. Curiously, there does not seem to be a correlation between overall size of the flower and the size of the medial petal.

This is mainly a species of low to middle elevations where the rainfall is ample or it occurs in moist situations. Only four collections have been seen from above 1300 m, and they were all from **T** 4, 5 or 7. The scarcity of this species in Kenya is curious because one would expect suitable conditions to occur near Mt Kenya and in the southwestern part of the country.

Morton (1967) mentions that in Ghana this species intergrades with *C. erecta* subsp. *livingstonii* and has a two-seeded bilocular capsule there. Brenan (1968) omits *C. zambesica* from Ghana and remains mute about any possible hybridization between these two taxa. Both these species are quite variable in our area, but no evidence of hybridization has been observed.

Richards 17759 (**T** 8, Masasi District, Masasi–Newala Road, 900 m, 6 March 1963) was identified as "*Commelina* sp. near *zambesica*" by Brenan because the dorsal capsule locule was muricate and indehiscent. I agree that this should exclude it from *C. zambesica* as we currently understand the species. However, the plant does resemble *C. zambesica* overall more than any other species, and the dark brown, subspherical seeds with raised orange "lines" forming a reticulum also strongly suggest those of that species. The spathes agree with those of *C. zambesica* in their form and pubescence, and the leaves resemble those of this species in their shape, oblique bases and basal auricles. The stems were apparently looping along the ground, which also fits *C. zambesica*. *Richards* 17759 differs from *C. zambesica* in the following characters, in addition to the one that Brenan pointed out: the upper cincinnus is 2–5-flowered (based on two spathes). Multiple-flowered upper cincinni have been observed in our area only on one other specimen, namely *Milne-Redhead & Taylor* 7732 from Songea District (which lacks fruits). The leaf pubescence is quite unusual but perhaps not totally distinct from *C. zambesica*. The adaxial surface is completely glabrous and on the abaxial surface the pubescence, consisting of hook-hairs sometimes mixed with uniseriate hairs, is confined to the midrib. Finally, although the seeds have raised orange "lines" – these are really ridges – much higher than those of *C. zambesica*, especially where they terminate at the edge of the ventral surface and around one terminal surface. The seeds further differ by having a much shorter hilum, lower apicule on the embryotega and, especially, in being puberulous with abundant minute white hairs. Finally, the dorsal locule seed is fused to the dorsal locule wall. Overall, I conclude that this is an undescribed species related to *C. zambesica* that is yet to be matched with specimens from elsewhere and any such specimens that lacked capsules and seeds would be virtually impossible to place in the correct species with certainty.

35. **Commelina congesta** *C.B. Clarke* in DC., Monogr. Phan. 3: 160 (1881) & in F.T.A. 8: 43 (1901); Morton in J.L.S. 55: 521, fig. 17 (1956) & in J.L.S. 60: 179 (1967); Brenan in F.W.T.A., 2nd ed.: 3: 49 (1968); Ogwal in Mitt. Inst. Allg. Bot. Hamburg 23b: 581 (1990). Syntypes: Gabon, Luango, Nsangaboi well, Chinchoxo, *Soyaux* 47 (K!, lecto., chosen by C.B. Clarke in F.T.A., 1901) and Cameroon, Victoria, *Kalbreyer* s.n. (B, syn.)

Decumbent or straggling perennial; roots thin, fibrous; shoots much-branched, ultimately prostrate or ascending, rooting at the nodes, glabrous or with a line of pubescence distally, usually greatly reduced and covered or nearly covered by overlapping sheaths distally on the flowering shoots. Leaves distichous; sheaths 1–2.2 cm long, with a dense line of pubescence along the fused edge, otherwise subglabrous to long-ciliate at apex; lamina sessile, lanceolate to ovate-elliptic, 4–9 × 1.5–3.5 cm, base asymmetric, cuneate-rounded, apex acute to acuminate, sometimes mucronulate; adaxial surface glabrous to shortly pilose, abaxial surface pilose on the midvein, margins scabrid distally, sometimes ciliate proximally. Spathes terminal, clustered, sometimes also appearing lateral on short shoots, 2–3(–4) together; peduncles to 7 mm long, sparsely pilose; spathes usually slightly falcate distally, 1.3–2.1 × 1–2.5 cm, base broadly cuneate to truncate, margins fused for 7–20 mm, apex acute, surfaces glabrous (in our area); upper cincinnus lacking, lower cincinnus 3–9-flowered. Flowers ± 1.1–1.3 cm wide. Sepals: upper sepal lanceolate-elliptic, paired sepals obovate, fused laterally; paired petals white or whitish to very pale pink, lilac, mauve or blue; lower petal white, linear, 1–1.5 mm wide; staminodes 3; lateral stamens with pollen yellow; medial stamen with filament slightly shorter than in lateral stamens, pollen yellow. Capsules pale yellow or grey-brown mottled with dark brown, trilocular, bivalved, obovoid to ellipoid, 5.5–7 × (2.5–)4–5 mm, 3-seeded, dorsal valve deciduous, dorsal locule indehiscent, 1-seeded, ventral locules 1-seeded, capsule constricted basally. Dorsal locule seed fused or closely adherent to the capsule wall, ellipsoid, dorsiventrally compressed, 2.3–3.8 × 2–2.5 mm, testa brown, ± smooth to alveolate; ventral locule seeds ± ellipsoid to broadly ovoid, dorsiventrally compressed, 2.55–3.3 × 1.7–2.2 mm, testa dark brown, smooth to alveolate, white farinose.

UGANDA. East Mengo District: Mabira Forest, 1–2 km E of Kiwala, 13 Apr. 1969, *Lye* 2477!; Kigezi District: South Maramagambo Central Forest Researve, ± 9 km on Kaizi–Bitereko road, 18 Sep. 1969, *Faden* 69/1118!; Toro District: Bwamba County, near Kirmia, 23 Sep. 1969, *Faden* 69/1264!

DISTR. U 2, 4; also in Guinea, Guinea Bissau, Sierra Leone, Liberia, Ivory Coast, Ghana, Togo, Nigeria, Cameroon, Central African Republic, Equatorial Guinea (Bioko), São Tomé, Gabon, Congo–Brazzaville, Congo–Kinshasa

HAB. Forest floor along tracks, forest edges, and swampy places; 700–1200 m

Flowering in September.

SYN. *Commelina heudelotii* C.B. Clarke in DC., Monogr. Phan. 3: 184 (1881). Type: Guinea, *Heudelot* 788 (K!, holo.)

C. condensata C.B. Clarke in DC., Monogr. Phan. 3: 190 (1881). Type: Bioko [Fernando Po], *Mann* 91 pro parte (K!, holo.)

NOTE. With its clustered, funnel-shaped spathes, *C. congesta* can possibly be confused only with some perennial forms of *C. benghalensis*. That species differs most obviously by its pubescent spathes, blue flowers, usually reddish hairs on the summit of the sheath, and five-seeded capsules.

The regular abortion of the basal ventral ovules, leading to a three-seeded capsule, which occurs in *C. congesta*, also occurs in other species, such as *C. albiflora* and *C. stefaniniana*. With the later it also shares a much reduced lower petal. But *C. stefaniniana* has solitary, pedunculate spathes with free margins, blue flowers and pitted seeds, so the seed reduction character is clearly convergent in these species. *Commelina congesta* is similar to *C. albiflora*, not only in capsule characters but also in its white or whitish flowers, smooth seeds and forest habitat. However, *C. albiflora* has solitary, pedunculate spathes, flowers that are always pure white, and a broader lower petal. Some plants or populations of *C. africana* also have a similar capsule, but that species has free-margined spathes and yellow flowers.

36. **Commelina albiflora** *Faden* in Novon 4: 229 (1994) & in U.K.W.F. 2nd ed.: 305 (1994). Type: Kenya, Kakamega District, Kakamega Forest, Kibiri Block, *Faden & Faden* 77/900 (US!, holo.; EA!, FT!, K!, iso.)

Decumbent perennial; roots thin, fibrous; internodes glabrescent or with a line of hook-hairs continuous with the line of fusion of the distal leaf sheath. Leaves distichous; sheaths 1–1.7 cm long, sparsely puberulous, ciliate or eciliate at apex; lamina shortly petiolate, lanceolate to lanceolate-elliptic, 3.5–8 × 1.5–2.5 cm, base usually strongly oblique, sometimes subcordate, margins undulate, scabrid, apex acuminate, adaxial surface lustrous dark green, sometimes with one or more maroon spots, usually minutely scabrid along the midrib, otherwise glabrous or with a few scattered hairs, abaxial surface puberulous. Spathes solitary; peduncles 0.8–1.3 cm long, puberulous; spathe green, gibbous, sometimes slightly falcate, 1.7–2.2 × 1(–1.5) cm, filled with mucilage when young, base truncate to subcordate, margins fused above the base for 5.5–7.5 mm, ciliolate along the fused portion, apex obtuse, ± mucronate, surfaces puberulous, with a few longer hairs along the folded edge near apex; upper cincinnus lacking; lower cincinnus 3–4-flowered. Flowers bisexual, 1.3–1.8 cm wide. Sepals ± free, upper sepal oblong to oblong-elliptic, 3.5–4 × 2.3–2.5 mm, paired sepals broadly oblong-elliptic to ovate or obovate-elliptic, 4–4.7 × 3–3.5 mm; paired petals white, broadly ovate to ovate-reniform, 9.5–11 × 8.5–9.5 mm of which the claw 3.5–4 mm; medial petal white, cup-shaped, ovate to spatulate, 4–5 × 3 mm; staminodes 3; lateral stamens with filaments 7–7.5 mm long, anthers 1.3–1.5 mm long, pollen white; medial stamen filament white, 5–6 mm long, anther 1.7–1.8 mm long, pollen white. Ovary ± 2 mm long; style 8.5–9 mm long; stigma ± capitate. Capsules greenish brown or the dorsal locule spotted with maroon, typically 3-seeded, with the one dorsal locule ovule and 2 apical ventral ovules maturing (frequently one or both of these aborting), obovoid to transversely oblong, 4–7.5 × 5–6 mm, dorsal valve deciduous, with a middorsal ridge, dorsal locule indehiscent, not striate, ventral locules typically dehiscent (tardily dehiscent or indehiscent when ventral locule empty). Dorsal locule seed fused to the capsule wall, broadly ellipsoid, 3–8–4 × 3.2–3.3 mm, testa dull greyish brown, smooth; ventral locule seeds ellipsoid, dorsiventrally compressed, 3.3–4 × 2.6–3 mm, testa smooth (to alveolate), brown, densely white-farinose.

KENYA. North Kavirondo District: Kakamega Forest, Kibiri Block, S side of Yala River, 21 Jan. 1970, *Faden* 70/20! & NW Kakamega Forest, 6 May 1971, *Mabberley* 1100!; Nandi District: Sirwa Farm, W of iron bridge on R. Yala on Kapsabet–Kisumu road, Nov. 1971, *Tweedie* 4165!

DISTR. **K** 3, 5; Congo–Kinshasa?

HAB. Moist forest in shaded situations, frequently along streams; 1500–1700 m

Flowering specimens have been seen from November to January, March and May.

SYN. *Commelina* sp. "E" of Faden in U.K.W.F.: 657 (1974); Faden and Suda in J.L.S. 81: 304 (1980)

NOTE. *Commelina albiflora* is the only *Commelina* species in FTEA that consistently has pure white flowers. Other species in the Flora that sometimes have white-flowered populations are *C. congesta* and *C. zenkeri*, forest species from Uganda, and *C. hockii*, a species of woodland from southern Tanzanzia. Another distinctive feature of *C. albiflora* is the occasional presence of one or more maroon spots on the leaves, a character otherwise found only in *C. stefaniniana* of northern Kenya. The usually 3-seeded capsule derived from a 5-ovulate ovary by the regular abortion of basal ovules of the ventral locules is also rare in the genus, having been observed in *C. congesta*, *C. stefaniniana* and *C. africana*.

Commelina albiflora can be confused only with *C. congesta* and *C. zenkeri*. *Commelina congesta* has a similar capsule and seed arrangement to *C. albiflora* and sometimes white flowers, but it differs from *C. albiflora* by its sessile leaves, funnel-shaped, subsessile spathes that are frequently clustered, and usually pink-tinged flowers. Like *C. albiflora*, *C. zenkeri* often has solitary spathes, but the spathe surfaces are entirely glabrous or puberulous only proximally, the flowers are usually blue and the capsule is bilocular and 2-seeded.

Commelina albiflora is known for certain only from the Kakamega Forest and vicinity in W Kenya, where it was locally common but by no means abundant in the 1970s. This forest is well known for its West African floristic and faunistic affinities but it also is quite low in endemism, nearly all of its species being known from Uganda and further west. Extensive field work in Ugandan forests in 1969 failed to turn up this species and Ogwal did not include it in her account (Ogwal, Taxon. study Commelina in Uganda, 1977). *Lebrun* 3939 from Djugu in E Congo–Kinshasa very closely resembles and possibly is *C. albiflora* – the collector records the flowers as white – but from one immature capsule it appears that the seed developing in one ventral locule is the basal not the apical seed as in typical *C. albiflora*. Although the pattern of seed development in *Lebrun* 3939 is not entirely clear, even if further collections from that region should confirm differences from plants from W Kenya they might still prove to be conspecific with *C. albiflora*.

37. **Commelina erecta** *L.*, Sp. Pl. 41 (1753); C.B. Clarke in DC., Monogr. Phan. 3: 181 (1881); Morton in J.L.S. 60: 183 (1967); Brenan in F.W.T.A., 2nd ed.: 3: 49 (1968); Obermeyer & Faden in F.S.A. 4, 2: 34 (1985); Blundell, Wild Fl. E. Afr.: 413 (1987); Ogwal in Mitt. Inst. Allg. Bot. Hamburg 23b: 582 (1990); Faden in U.K.W.F. 2nd ed.: 305, t. 136 (1994) & in Fl. Somalia 4: 93 (1995); Wood, Handbook Yemen Fl.: 317 (1997); Ensermu & Faden in Fl. Eth. 6: 372 (1997); Faden in Fl. North America 22: 195 (2000). Type: Dill., Hort. Eltham. 91, t. 77, fig. 88 (lecto. of Clarke (1881)), specimen in the Dillenian Herbarium (OXF!) (IDC Fiche 99/B7)

Shortly rhizomatous perennial; roots fibrous, cord-like, not tuberous (except in some plants of subsp. *erecta*); shoots tufted, erect, ascending, decumbent or sometimes scrambling, usually to 60 cm, but sometimes to 1.5 mm long, glabrous or puberulous, shoot bases often swollen. Leaves spirally arranged or distichous, often clustered terminally; sheaths to 3.5 cm long, usually auriculate at apex, usually puberulous, occasionally the hairs restricted to a band along the fused edge, white-ciliate at apex, rarely glabrous; lamina sessile, linear to ovate, 4–14 × 0.4–4 cm, base oblique, cuneate to rounded, margins scabrid distally, often ciliate proximally, apex acuminate to acute; adaxial surface pilose or hirsute, often also scabrid or just scabrid, or occasionally glabrous, abaxial surface puberulous or, less commonly, glabrous. Spathes clustered terminally, occasionally solitary, usually slightly falcate; peduncles 2–10(–15) mm long, puberulous; spathes 1.3–2.5(–3.8) × 0.8–1.9(–3) cm, base rounded to truncate, margins fused for 7–19(–30?) mm, often sparsely ciliolate along the fused edge and glabrous elsewhere, sometimes completely glabrous, apex acute to acuminate; surface usually hirsute-puberulous, less often only puberulous or glabrous; upper cincinnus usually vestigial and enclosed in the spathe or lacking, very rarely shortly exserted and 1-flowered, lower cincinnus (2–)3–5(–7)-flowered. Flowers bisexual and occasionally male, 1.5–3 cm wide. Sepals with upper sepal ovate, 2–4 × 1.5–2 mm, paired sepals shallowly cup-shaped, fused laterally, obovate to broadly ovate, 4–5 × 2.5–3(–4) mm; paired petals light to deep blue (rarely mauve?), broadly ovate to ovate-reniform or ovate-deltate, 8–11 × 8.5–10 mm of which the claw 1.5–4 mm; medial petal flat against the fused sepals, pale blue or white, linear-lanceolate to subulate, 2.5–3.5 × 0.5–1 mm; staminodes 3; lateral stamens with filaments 4.5–7 mm long, anthers 1.1–1.5(–1.9) mm long, pollen yellow or white; medial stamen filament 3–5.5 mm long, anther (1.2–)1.5–2.5 mm long, pollen usually yellow, rarely creamy white. Ovary 1–1.5 mm long; style 5–7.5 mm long. Capsules pale yellow, bilocular-bivalved, trilocular-bivalved or trilocular-trivalved, obovoid, 3–3.5 × 4.5–6 mm, 1–3-seeded, with locules 1-seeded, dorsal locule lacking or present. Dorsal locule seed similar to the ventral locule seeds when locule is dehiscent, when indehiscent the seed fused to the capsule wall, ± spherical, smooth; ventral locule seeds biconvex, usually ellipsoid, sometimes nearly circular, 2.6–4 × (2.3–)2.5–3.2 mm, testa dark brown, often mottled with lighter brown, smooth, white-farinose, with a ring of soft whitish material encircling the seed.

subsp. **erecta**

Maximum leaf width (1.8–)2–4 cm, lamina lanceolate to elliptic or ovate; spathes often glabrous (especially along the coast), occasionally sparsely hirsute-puberulous.

KENYA. Malindi District: Malindi, Robertson Plot, 23 June 2001, *Robertson* 7386!; Tana River District: Garsen–Witu road, 12.5 km towards Witu from Garsen ferry crossing, 10 July 1974, *Faden & Faden* 74/1068! & Tana River National Primate Reserve, 3 km E of Main Gate, 19 Mar. 1990, *Kabuye et al.* in TPR 638!

TANZANIA. Handeni District: Kwamkono, 29 June 1966, *Archbold* 784!; Tanga District: Coast near Bomandani, 13 km S of Moa, 10 Aug. 1953, *Drummond & Hemsley* 3680!; Kilwa District: Selous Game Reserve, ± 10 km NNW of Kingupira, 16 Jan. 1977, *Vollesen* 4331!

DISTR. **K** 7; **T** 3, 6, 8; **Z**; Senegal, Burkina Faso, Ivory Coast, Ghana, Togo, Nigeria, Equatorial Guinea (Bioko), Somalia, Zimbabwe, Namibia, South Africa; Yemen, widespread in temperate and tropical America

HAB. Lowland evergreen forest, riverine forest, dry forest, coral rag, shady thickets and thicket edges, grassland with scattered trees, pasture, dry stream bed, and in cultivation; 0–550(–1100?)m

Flowering specimens have been seen from March and May to October.

SYN. *Commelina gerrardii* C.B. Clarke in DC., Monogr. Phan. 3: 183 (1881) & in Fl. Cap. 7: 11 (1897). Syntypes: South Africa, Natal, 1838, *Gerrard* s.n. (K!, lecto., chosen by Clarke in F.T.A. 8: 58, 1901); Mozambique, Sena [Senna], 10 Apr. 1860, *Kirk* s.n. (K!, syn.) & between Lupata and Tete [Tette], Feb. 1859, *Kirk* s.n. (K!, syn.), Mozambique, *Peters* s.n. (B, syn.)

subsp. **livingstonii** (*C.B. Clarke*) *J.K. Morton* in J.L.S. 60: 184 (1967); Brenan in F.W.T.A., 2nd ed.: 3: 49 (1968). Syntypes: Mozambique, between Lupata and Sena [=Senna], Dec. 1858, *Kirk* s.n. (K!, lecto., selected here); near foot of Mt Morrumbala [Moramballa], Dec. 1858, *Kirk* s.n. (K!, syn.); between Lupata and Tete, no date, *Kirk* s.n. (K!, syn.); Tete [Tette], Feb. 1959, *Kirk* s.n. (K!, syn.) & near Sena [Senna], Jan. 1859, *Kirk* s.n. (K!, syn.)

Maximum leaf width 1.7 cm, lamina linear to lanceolate; spathes sparsely to densely puberulous or hirsute-puberulous, rarely glabrous (**T** 2, **Z**) but then leaves very narrow.

UGANDA. Karamoja District: near Monita, Aug. 1958, *Wilson* 616!; Mengo District: 9 km W of Lwampanga, 6 Sep. 1955, *Langdale-Brown* 1485!; Toro District: Rwenzoli [Queen Elizabeth] National Park, rim of Mukakunya Crater, 2 km NNW of Lake Nymumuka, 22 Sep. 1969, *Faden et al.* 69/1243!

KENYA. Kisumu–Londiani District: Kisumu, Feb. 1915, *Dummer* 1861!; Masai District: between Namanga and Kajiado, 148 km on Nairobi–Namanga road, 17 Dec. 1961, *Polhill & Paulo* 1010!; Tana River District: Garsen–Malindi road, 1.5 km towards Malindi from turnoff to Oda, 22–24 July 1974, *Faden & Faden* 74/1189!

TANZANIA. Musoma District: Musabi, 10 Feb. (?) 1968, *Greenway, Kanuri & Braun* 12163!; Pangani District: Mkwaja Ranch, 13 Sep. 1955, *Tanner* 2160!; Morogoro District: 33 km from Morogoro on Iringa road, 7 Nov. 1969, *Batty* 863!

DISTR. **U** 1–4; **K** 1, 4–7; **T** 1–3, 5?, 6; **Z**; Senegal, Gambia, Guinea Bissau, Guinea, Sierra Leone, Burkina Faso, Ivory Coast, Ghana, Togo, Nigeria, Cameroon, Central African Republic, Chad, Rwanda, Burundi, Sudan, Ethiopia, Somalia, Malawi, Mozambique, Zimbabwe, Namibia, Botswana, South Africa; Yemen

HAB. Dry or moist grassland, rocky hillsides, roadsides, woodland and open bushland, thicket or forest margins, rarely near water or in cultivation; usually in well drained soil but occasionally in black cotton soil; just above sea level to 1600 m

Flowering specimens have been seen from all months.

SYN. *Commelina livingstonii* C.B. Clarke, in DC., Monogr. Phan. 3: 190 (1881) & in F.T.A. 8: 59 (1901); Morton in J.L.S. 55: 526 (1956); Obermeyer & Faden in F.S.A. 4, 2: 34 (1985)

C. sphaerosperma C.B. Clarke in F.T.A. 8: 58 (1901). Syntypes: Mozambique, mouth of Melambe R., Zambesi Delta, 8 Feb. 1861, *Kirk* s.n. (K!, syn.); between Lupata and Tete [Tette], Feb. 1859, *Kirk* s.n. (K!, syn.); opposite Sena [Senna], 10 Apr. 1860, *Kirk* s.n. (K!, syn.). Malawi, Zomba, 2500–3500 ft., Dec. 1896, *Whyte* s.n. (K!, syn.)

C. venusta C.B. Clarke in F.T.A. 8: 58 (1901). Type: Somalia: Golis Range, March 1895, *Cole* s.n. (K!, holo.)

C. bainesii C.B. Clarke in F.T.A. 8: 57 (1901) quoad *Volkens* 2147

NOTE. The dubious record for **T** 5 is based on *Richards* 25114, which was collected on the **T** 2/**T** 5 border. This is the only collection seen from **T** 5.

NOTE (on the species as a whole). The specimen of *Commelina erecta* in LINN corresponding to Linnaeus's protologue is *Commelina communis* L. Clarke (1881) noting this, selected one of the sources cited by Linnaeus, namely Dillenius's Hortus Elthamienis, to typify the species. There is a specimen in the Dillenian Herbarium (OXF) corresponding to the plate and figure cited by Linnaeus and Clarke. It has leaves to ± 14.5 × 3 cm. No provenance is given for the specimen but there are indications from Linnaeus's other sources that the plants originated from America. At the end of the type description of *Commelina livingstonii* Clarke (1881) cited "prope flumen Zambesi (Dr. Kirk)" even though there were five different Kirk collections, mounted on three sheets at Kew, all of which Clarke had seen for the monograph. The five were subsequently separately listed in F.T.A. (Clarke, 1901). On the sheet containing the collections "between Lupata and Senna" and "foot of Moramballa" there is a red label "Type Specimen" and nearby, written in two different places by Clarke, "Type specimen" and "*Commelina Livingstoni* C.B. Clarke (type)". Although the placement of the annotations by Clarke and the Type Specimen label leave some ambiguity as to whether he is indicating the whole sheet (with both collections) or one of them in particular as the type, the only collection (of the five) that has seeds in a packet is the "between Lupata and Senna" collection. Therefore I am selecting this collection as the lectotype of the name.

Although I am using the same names for two of the three infraspecific taxa recognized in *C. erecta* in F.W.T.A. – the third one, *C. erecta* subsp. *maritima* (J.K. Morton) J.K. Morton does not occur in our area – I have found it necessary to define them differently. The delimitation of subsp. *erecta* and *livingstonii*, using the ratio of leaf length to width, which seems to work in West Africa, does not yield the same results in our area. Moreover, plants with small leaves that are proportionally broad, which F.W.T.A. would place in subsp. *erecta* and I would place in subsp. *livingstonii* (if the leaves are less than 1.7 cm broad), do not seem to be consistent with the type of the species, which has large broad leaves. In Flora of Southern Africa, *C. erecta* and *C. livingstonii* are maintained as distinct species, based on differences in capsule locule number and dehiscence. That also does not work in our area. I do not think that there is a perfect way to separate these taxa, but maximum leaf width, leaf shape and, to a lesser degree, spathe pubescence, seems to work better in our area than the characters used in F.W.T.A. or F.S.A.

Prior to Morton (1967) the African plants that we now place in *C. erecta* were usually called *C. gerrardii* C.B. Clarke in South Africa, *C. vogelii* C.B. Clarke in West Africa, or *C. livingstonii* C.B. Clarke in East, Southeast and South Africa. But Morton went further, indicating that the species not only occurred in the Americas and Africa but also in India and Australia. Brenan (1968) agreed with that circumscription, citing *C. undulata* R. Br. from Australia and *C. kurzii* C.B. Clarke from India as synonyms of *C. erecta*.

This is neither the place nor the time to present a detailed argument for a different circumscription of *C. erecta*. Suffice it to say, I am very familiar with *Commelina kurzii* in Sri Lanka which, with its consistently mauve flowers, does not agree with African or American plants of *C. erecta*. Although I am not especially knowledgeable about *C. undulata*, I am unwilling to accept its synonymy with *C. erecta* based wholly on a superficial comparison of dried herbarium specimens, so for the present I would consider *C. erecta* to be confined to the New World and Africa, with an extension into the Arabian Peninsula.

The species was characterized in West Africa as having three dehiscent locules (Morton, 1956, as *C. gerrardii*) but later it was noted that the third [dorsal] locule could be dehiscent (sometimes tardily) or indehiscent (Morton, 1967), a variability that Morton likened to the species in North America. However, that is contrary to my experience with *C. erecta* in North America where plants with three, smooth, dehiscent locules are decidedly rare, and the typical pattern is for the dorsal locule to be warty and indehiscent (Faden, 2000). In Tropical East Africa an indehiscent dorsal locule is known for certain from **K** 4, 6, 7 and **T** 2, and only in subsp. *livingstonii*, although it also occurs in subsp. *erecta* in southern Somalia

My unpublished studies of germination of dorsal locule seeds (enclosed in the indehiscent dorsal capsule locule) and ventral locule seeds from a population of *C. erecta* var. *livingstonii* from Kenya showed very strong differences in germination rates between the two seed types. These differences might be expected to be more useful to the species in lower rainfall habitats because of the less predictable rainfall. Therefore an indehiscent dorsal locule might be predicted to be more common in subsp. *livingstonii*, which occurs in drier situations than subsp. *erecta*, and that indeed is the case.

Two Kenyan populations of *C. erecta* var. *livingstonii* are known to have both capsules with three smooth dehiscent locules as well as ones with two smooth dehiscent locules and one warty, indehiscent one. *Faden & Faden* 77/303 and 77/304, were both obtained from the same population in Machakos District (**K** 4). Plants of the former had capsules with three, smooth, dehiscent locules whereas plants of the latter had two smooth, dehiscent locules and the third warty and indehiscent. These differences were maintained in later cultivation. *Polhill & Paulo* 1010, from the Kajiado–Namanga road (**K** 6), has both types of capsules in a packet on sheet 1 of this collection; the pressed specimens on both sheets have only capsules with three dehiscent locules; and the spirit collection has only capsules with two smooth locules and one muricate one. More thorough collecting and observations might yield more populations of *C. erecta* with dimorpic fruits, at least in some parts of its range.

It is very unusual to have in the same species of *Commelina* populations of plants with blue-and-white lateral stamen anthers that have white pollen and other populations in which the corresponding anthers are yellow, with yellow pollen. Although these data have been recorded in too few populations to be able to generalize, the pattern as it now presents itself is that the plants with yellow anthers and yellow pollen are from **K** 4 and **K** 6, they are all subsp. *livingstonii*, and they include both plants with two dehiscent and one indehiscent locule, e.g. *Faden & Faden* 77/586 or plants with either three dehiscent capsule locules or two dehiscent locules and one indehiscent one, e.g. *Polhill & Paulo* 1010. *Faden* 70/384 from **T** 6 records all of the anthers as yellow, but does not mention the pollen and lacks fruits. Plants recorded with blue-and-white anthers with white pollen are all from **K** 7. They are mainly subsp. *erecta*, with three dehiscent capsule locules, but there is also at least one collection of subsp. *livingstonii* with an indehiscent dorsal locule, namely *Faden & Faden* 74/1189. These sparse results suggest that fruit and floral characters are neither tracking the subspecies nor each other very well.

Faulkner 1794 from Tanga District (**T** 3) is described as "very spreading," which is generally unlike subsp. *livingstonii* (but compare *Faden* 70/384 from Morogoro District(**T** 6)). Its spathes are solitary and long-pedunculate, with peduncles up to 1.5 cm, the longest seen in this species. The lower cincinnus peduncle has sparse hook-hairs, again an unusual feature for the species. The capsule has three smooth, dehiscent locules. Overall, this collection has no match in our area and its taxonomic placement is somewhat uncertain.

Some dimensions given in the description probably do not reflect the full range of variation for the species in our area. The largest dimensions for the spathes come from collectors' notes and have not been observed in the specimens examined. Spathes that are 3 cm high would almost certainly have had fused spathe margins approximating that length. Flowers that are recorded up to 3 cm wide must have had larger paired petals and presumably other floral parts than what we have been able to document from the available pressed and spirit collections.

The two patches or a ring of soft whitish material on the ventral locule seeds of this species, *C. albescens* and *C. melanorrhiza* – they may turn brown and firm in dried specimens but they are typically (always?) white and soft in fresh seeds – have sometimes been referred to as a "ridge" or ignored altogether in the literature. In life they function as elaiosomes or fat-bodies that attract ants for seed dispersal. This has been observed in the field for *C. erecta* in Texas.

38. **Commelina bracteosa** *Hassk.* in Flora 46: 385 (1863) & in Reise Mossamb., Bot. 1, 2: 524 (1864); C.B. Clarke in DC., Monogr. Phan. 3: 180 (1881) & in F.T.A. 8: 55 (1901); Brenan in F.W.T.A., 2nd ed.: 3: 48 (1960); Ogwal in Mitt. Inst. Allg. Bot. Hamburg 23b: 581 (1990); Faden in Pl. Angola: 176 (2008). Type: Mozambique, June 1863, *Peters* s.n. (B!, holo.)

Perennial with or without a rhizomatous base; rhizome, when present, deeply buried, branched, giving rise to annual shoots; roots thin, fibrous; shoots tufted, prostrate or decumbent with short ascending stems, sometimes rooting at the nodes, to 80 cm long, usually much branched, puberulous to subglabrous with only a short line of hairs descending from the distal node, rarely pilose. Leaves distichous; sheaths auriculate at the apex, to 3 cm long, usually puberulous, occasionally only with a longitudinal band of hairs along the line of fusion, long white-ciliate at apex or eciliate; lamina sessile or petiolate, linear to ovate, 2–9(–12) × 0.4–2.7 cm, base oblique, cuneate to ± rounded, margins scabrid distally, apex acute to acuminate, sometimes mucronate; adaxial surface hirsute or less often glabrous or scabrid, abaxial surface pubescent, hirtellous or pilose, rarely glabrous. Spathes solitary or a few subclustered distally, not to slightly falcate; peduncles 0.2–1.7 cm long, puberulous mainly in a line; spathes 0.9–2.3 × 0.4–1.2 cm, base cordate to subhastate, margins fused for 2–6 mm, ciliate to ciliolate along the fused portion, apex acute to obtuse; surface puberulous or hirsute-puberulous; upper cincinnus vestigial; lower cincinnus 1–6(–14)-flowered. Flowers bisexual (and rarely male?), 12–23 mm wide. Upper sepal ovate, 2.3–3.5 × 1.5–2 mm, paired sepals fused laterally more than half their length from the base, obovate to broadly elliptic or ovate-oblong, 3–5 × 2.5–3.5 mm; paired petals held in a lateral position (that is, facing one another), with notably short claws, blue, reniform to broadly or ovate-deltate, 6–11.5 × 7–12 mm of which the claw 2–3(–4) mm; medial petal inconspicuous, usually blue, sometimes white, 1.5–4 × 0.5–1.5 mm. Staminodes 3; lateral stamens with filaments 3.5–6 mm long, anthers (1–)1.2–1.8 mm long; medial stamen filament 3–4 mm long, anther 1.5–2.6 mm long. Ovary 0.9–1.3 × 1–1.3 mm; style 4.5–8 mm long; stigma slightly 3-lobed. Capsules pale yellow or the dorsal locule contrastingly dark brown, trilocular (or bilocular though abortion), bivalved or trivalved, obovoid, 3.5–4.5 × 4.8–6 mm, stipitate, ± 3-lobed, locules all smooth, dehiscent, 1-seeded, or occasionally the dorsal locule striate, indehiscent and sometimes deciduous. Seeds subspherical to ovoid or shortly ellipsoid, 1.9–3.2 × 2–3.2 mm, testa smooth, ± uniformly dark brown or mottled various shades of brown or dark brown and orange, sparsely to densely white-farinose.

1. Plants lacking rhizomes; leaves narrowly lanceolate to lanceolate-oblong or ovate, 0.6–2.7 cm wide, leaf sheaths usually all green, or occasionally the distal ones reddish (subsp. *bracteosa*) 2
   Plants with branched rhizomes; leaves linear; 0.4–0.7 cm wide, at least the proximal leaf sheaths pink to red ... b. subsp. *rhizomifera*
2. Maximum leaf length 5–12 cm; maximum spathe peduncle length 0.8–1.7 cm; plants not rooting at the nodes, except occasionally near the base; leaves never drying bright green .... a1. var. *bracteosa*
   Maximum leaf length 4–7.5 cm; maximum spathe peduncle length 0.35–0.8(–1.1) cm; plants rooting at the nodes; leaves sometimes drying bright green .... a2. var. *lagosensis*

a. subsp. **bracteosa**

Plants lacking rhizomes; internodes sparsely to densely puberulous with hook-hairs to subglabrous with only a short line of hairs descending from the distal node, rarely densely pilose; leaves narrowly lanceolate to lanceolate-oblong or ovate, 2–9(–12) × 0.6–2.7 cm; leaf sheaths all green or occasionally the distal ones reddish.

a1. var. **bracteosa**

Maximum leaf length 5–12 cm; maximum spathe peduncle length 0.8–1.7 cm; plants not rooting at the nodes, except occasionally near the base; leaves never drying bright green.

UGANDA. Acholi/Bunyoro Districts: Kabalega [Murchison Falls] National Park, Rhino Viewpoint, Paraa, 18 June 1957, *Buechner* 70!; West Nile District: Summit of Mt Eti [= Wati], 19 July 1953, *Chancellor* 12!

KENYA. Kwale District: Cha Simba Forest, Kwale, 1 Feb. 1953, *Drummond & Hemsley* 1087! & Shimba Hills, Kwale forest area, on road to Wireless Station, 23 Mar. 1968, *Magogo & Glover* 409!; Teita District: ± 19 km from Maungu Station on Maungu Station–Rukanga road, 4 Apr. 1969, *Faden, Evans & Rathbun* 69/409!

TANZANIA. Tanga District: Eastern Usambara Mts, Lower Sigi Valley, 30 May 1950, *Verdcourt* 235!; Mpwapwa District: Mpwapwa, 7 Jan. 1942, *Hornby & Hornby* 2178!; Kilosa District: Mikumi–Kilosa Boundary Road, 36 km from Headquarters, 22 June 1973, *Greenway & Kanuri* 15211!

DISTR. **U** 1–4; **K** 7; **T** 1, 3, 4?, 5–7; Senegal, Gambia, Mali, Nigeria, Cameroon, Congo–Kinshasa, Sudan, Zambia, Malawi, Mozambique and Zimbabwe

HAB. Forest edge, moist thickets, rocky slopes and summits, woodland, grassland, bushland, cultivated fields and "drain scrapes"; usually in well-drained soil; 45–1300(–1700) m

Flowering specimens have been seen from January to July and November.

SYN. *Commelina bainesii* C.B. Clarke in DC., Monogr. Phan. 3: 184 (1881). Type: Zimbabwe, Mangwe [Mangive] River, 12 Dec. 1869, *Baines* s.n. (K!, lecto., selected here).

C. bainesii C.B. Clarke var. *glabrata* Rendle in Trans. Linn. Soc., ser. 2, Bot. 4: 52 (1894); Rendle in J.L.S. 30: 429 (1895); C.B. Clarke in F.T.A. 8: 57 (1901). Type: Malawi, Zomba, Sep. 1891, *Whyte* s.n (BM?, holo.).

a2. var. **lagosensis** (*C.B. Clarke*) *Faden* **comb. et stat nov**. Type: Nigeria, Lagos, *Millen* 21 (K!, holo.)

Maximum leaf length 4–7.5 cm; maximum spathe peduncle length 0.3–0.8(–1.1) cm; plants rooting at the nodes; leaves sometimes drying bright green.

KENYA. Kwale District: Buda Mafasini Forest, 12.8 km WSW of Gazi, 21 Aug. 1953, *Drummond & Hemsley* 3943! & Dzombo Hill, 7 Feb. 1989, *Robertson et al.* in MDE 184! & Mrima Hill, 4 Sep. 1957, *Verdcourt* 1852(S)!

TANZANIA. Handeni District: Kwamkono, 28 June 1966, *Archbold* 765!; Ufipa District(?): Casawa Sand Dunes, Lake Tanganyika, 14 June 1957, *Richards* 9240!; Rufiji District(?): Selous Game Reserve, Sand Rivers Lodge, 1 Apr. 1996, *Luke & Luke* 4436!

DISTR. **K** 7; **T** 3, 4, 6–8; **Z**; Senegal, Guinea Bissau, Sierra Leone, Ivory Coast, Ghana, Nigeria, Mozambique, Malawi, Zambia

HAB. Forest, including riverine, often on rocks, especially limestone, woodland, lake shores, sand dunes, and in short grass, often in sandy soil; 30–900 m

Flowering specimens have been seen from all months except July and December.

SYN. *Commelina lagosensis* C.B. Clarke in F.T.A. 8: 57 (1901); Morton in J.L.S. 55: 526, fig. 22 (1956) & in J.L.S. 60: 185, fig. 5 (1967); Brenan in F.W.T.A., 2nd ed.: 3: 48 (1968)

b. subsp. **rhizomifera** *Faden* **subsp. nov**. a subspecie bracteosa praesentia rhizomatis ramosi differt. Type: Tanzania, Tunduru District: ± 1.5 km E of Mawese River, *Milne-Redhead & Taylor* 7714 (K!, sheet 1, holo.; K!, sheet 2, iso.)

Plants with branched rhizomes; shoots to 30 cm long; internodes densely puberulous to densely pilose; leaves linear, 2–9 × 0.4–0.7 cm, at least the proximal leaf sheaths pink to red.

TANZANIA. Iringa District: Ruaha National Park, Mangangwe Air Strip, 17 Dec. 1972, *Bjørnstad* AB 2142!; Tunduru District: ± 1.5 km E of Mawese River, 19 Dec. 1955, *Milne-Redhead & Taylor* 7714!

DISTR. **T** 7, 8; Congo–Kinshasa, Zambia, Angola

HAB. *Brachystegia* woodland on sandy soil; 450–1350 m

Flowering plants have been collected in December.

SYN. *Commelina bainesii* C.B. Clarke in DC., Monogr. Phan. 3: 184 (1881) & in F.T.A. 8: 57 (1901) pro parte, quoad *Welwitsch* 6636 (BM!), 6641 (BM!) & 6643 (BM!), not as to lectotype

NOTE. The subterranean parts that survive the dry season and give rise to the shoots of the next growing season appear to be the thickened shoot bases of the previous year's shoots. They are thus perennating as well as storage organs, but whether they constitute true rhizomes is a matter of definition. The above-ground, new shoots that arise early in the growing season have thickened bases, and plants collected at this stage are very distinctive, particularly because the leaf sheaths that surround them are red or pink. Specimens collected later look much more like *C. bracteosa* subsp. *bracteosa*, especially when the subterranean parts are not obtained. It is not clear whether subsp. *rhizomifera* replaces subsp. *bracteosa* in part of its range or the two coexist in the same region.

NOTE (on the species as a whole). *Commelina bainesii* C.B. Clarke was based on five collections, *Baines* s.n. from Zimbabwe, three *Welwitsch* gatherings from Angola (all at BM), and a *Lemue* collection from South Africa (G, not seen). The *Baines* collection was designated as the type by Clarke on the specimen itself, and I am selecting it as the lectotype here. *C. bainesii* was treated as a synonym of *C. bracteosa* in F.W.T.A. (Brenan 1968) and as a synonym of *C. erecta* in F.S.A. (Obermeyer & Faden 1985). Aside from having an indehiscent dorsal locule, I see no strong similarity between *Baines* s.n. and some plants of *C. erecta*, as we are defining the species here. Mature fruits of specimens that are definitely subsp. *rhizomifera* have not been seen, but *Richards* 13748 from Zambia, which probably belongs to this subspecies, has a mature capsule with three equal dehiscent locules.

All plants of *C. bracteosa* sensu lato from West Tropical Africa were all called *Commelina lagosensis* until Brenan (F.W.T.A., 2nd ed., 1968) recognized two species, *C. lagosensis* and *C. bracteosa*, separating them chiefly on leaf size and shape, and spathe peduncle length. He went on to suggest that they might best be treated as subspecies. He recorded *C. lagosensis* from Kenya, Tanzania and Zanzibar. Plants from our area that agree with *C. lagosensis* occur in coastal forests in Kenya and at scattered localities in Tanzania. Extreme plants look distinctive because of their small, bright green leaves, often less pubescent that those of typical *C. bracteosa*, and small, shortly pedunculate spathes. However, there is no hard line between them and more typical specimens of *C. bracteosa*. I think that *C. lagosensis* is no more than an ecotype of *C. bracteosa*, occurring in moister habitats, often in dense shade. It deserves recognition only as a variety.

Commelina lagosensis was recognized by Clarke as a distinct species largely because of its indehiscent dorsal capsule locule. The type specimen, *Millen* 21 from Nigeria, has a striate dorsal locule with a middorsal ridge that is very similar to the capsule in the type of *C. bainesii* from Zimbabwe. In our area nearly all of the collections that have an indehiscent dorsal locule belong to var. *lagosensis* and, were this consistent, it would provide a strong argument for its recognition at a higher rank. Indeed a few specimens from Tanzanina, notably *Luke & Luke* 4436 from Selous Game Reserve and *Richards* 16086 from Mpanda District, show a dorsal valve that is not only striate and indehiscent, but also clearly deciduous and much darker brown than the two dehiscent capsule locules. Whether these specimens represent an extreme form among the indehiscent locule types or else the light-colored, striate dorsal locules seen in other collections are not fully mature has not been determined, but it seems clear that the fruits are highly variable within this variable species. Although almost all specimens seen with indehiscent dorsal locules belong to var. *lagosensis*, not all specimens of var. *lagosensis* have an indehiscent capsule locule. *Robertson et al.* in MDE 184 and *Verdcourt* 1852 clearly have capsules with three smooth, dehiscent locules. In contrast, *Faulkner* 936, which also has an indehiscent locule, clearly belongs to var. *bracteosa* (as does the type of *C. bainesii*), so the pattern of variation in the fruits does not perfectly coincide with any other characters that I have been able to detect. Relatively few specimens have mature capsules so that even if a difference in the dorsal capsule locule appearance and dehiscence were demonstrated to be important those characters would be virtually impossible to use for most dried specimens.

Spirit material of *Milne-Redhead & Taylor* 7714 includes spathes with up to 14 buds in the lower cincinnus. Whether these would all produce flowers is unknown, but that number far exceeds that of any other collection of either subspecies.

Specimens of this species can be difficult to distinguish from those of smallish specimens of *C. zambesica* and of some young shoots of *C. erecta* var. *erecta* in which only the first spathes have been produced. *Commelina zambesica* can be particularly tricky to separate because both species have auricles at the summit of the sheaths, similar-shaped spathes, similar pubescence on the leaves and spathes, and lateral sepals fused to the same extent. Moreover, when the fundamentally 5-seeded capsules in *C. zambesica* are reduced by abortion of some of the ovules/seeds, a capsule resembling that of *C. bracteosa* can be the outcome. Plants of *C. bracteosa* may best be distinguished from *C. zambesica* by the their generally smaller size, with smaller leaves and spathes, shorter internodes, the absence of thick, fleshy, sometimes fusiform, ± tuberous roots, the consistent absence of a functional upper cincinnus, 3-ovulate ovaries, the capsule locules all equal or the dorsal locule indehiscent, and the smooth seeds with an especially long apicule.

I have accepted the record of Ogwal (1990) of this species from **U** 2 and **U** 3, although I have seen specimens that are certainly determined from **U** 1, the border between **U** 1 and **U** 2, and **U** 4. The sole specimen seen of var. *bracteosa* seen from **T** 4 (*Richards* 18735) is only questionably identified as *C. bracteosa*, so this distribution record has been queried.

39. **Commelina melanorrhiza** *Faden* in Novon 4: 224, fig. 1 (1994) & in U.K.W.F. 2nd ed.: 305 (1994). Type: Kenya, Northern Frontier District, Mathews Range Forest, Ngeng River Valley and route to Leiturr, near Kitich Forest Station, *A. Faden* 87/3 (US!, holo.)

Shortly rhizomatous perennial; roots tufted, black or dark brown, tuberous, sometimes fusiform, 3–5 mm thick; shoots annual, tufted, prostrate (to ascending, in cultivation), 6–28 cm long, not rooting, with a line of pubescence continuous with the line of fusion of the sheath above, sometimes sparsely puberulous elsewhere. Leaves distichous; sheaths to 1.7 cm long, puberulous along the line of fusion and sometimes sparsely so elsewhere, white-ciliolate or ciliate at apex; lamina petiolate or sessile, lanceolate-elliptic to ovate, 1.5–7 × (0.8–)1–2.5 cm, base strongly oblique, apex acute to acuminate, both surfaces subglabrous to sparsely puberulous, the abaxial usually more pubescent, adaxial surface scabrid, margins scabrid. Spathes densely crowded in bracteate clusters, funnel-shaped, subsessile; peduncles to 5 mm long; spathes 1–1.5 × 0.8–1.3 cm, base truncate to rounded, margins fused for 5.5–8.5 mm, ciliate along the fused portion and glabrous elsewhere or completely glabrous, apex acute; surface subglabrous to sparingly pilose; upper cincinnus often well developed and exserted from the spathe, 1-flowered; lower cincinnus 2–5-flowered. Flowers bisexual (lower cincinnus) and male (upper cincinnus), ± 1.5 cm wide. Sepals glabrous, upper sepal 2–2.5 mm long, lower sepals fused for about $^2/_3$ their length, 3–4 mm long; paired petals lilac, mauve or pale purple to pale blue, ovate, 7–8.5 × 6–7.5 mm of which the white claw 3–3.5 mm; medial petal white, ± 1 mm long; staminodes 3 with the medial sometimes lacking an antherode; lateral stamens with filaments 4–4.5 mm long, anthers ± 1 mm long, pollen dirty yellow; medial stamen filament ± 4.5 mm long, anther ± 2 mm long, pollen dirty yellow. Ovary ± 1 mm long; style ± 6 mm long; stigma capitate. Capsules pale yellow, trivalved, trilocular, broadly ovoid, 2.5–3 × 2.5–3.5 mm, 3-seeded, glabrous, locules all smooth, dehiscent, 1-seeded. Seeds subglobose, 1.6–2 × 1.7–2.1 mm, with a peripheral thickened ridge of soft light brown tissue encircling the seed, testa otherwise smooth, dark brown, white-farinose.

KENYA. Meru District: Meru National Park, on the Rojwero river near the west boundary of the park, 14 May 1972, *Ament & Magogo* 210!; Machakos District: Kibwezi, 11 Dec. 1967, *Agnew* 9872!; Kitui District: Thika–Kitui road at Tiva River, 11 Nov. 1965, *Gillett* 16961!

DISTR. **K** 1, 4; not known elsewhere (but see note)

HAB. Forest, bushland and moist thickets, 750–1700 m

SYN. *Commelina* sp. "C" of Faden in U.K.W.F. p. 657 (1974)

NOTE. *Commelina melanorrhiza* is very distinctive because of its thick, black or dark brown tuberous roots. It bears a strong but superficial resemblance to small plants of *C. benghalensis* because of its short, proportionally broad leaves and small, funnel-shaped, clustered spathes. However, because of its very reduced medial petal, one-ovulate/one-seeded ovary and capsule locules, its true affinities are with *C. bracteosa* and *C. erecta*. From the former it differs by its tuberous roots, annual shoots, densely clustered, subsessile spathes, and seeds with an encircling ridge of soft material. From *C. erecta* it differs by its usually prostrate shoots, smaller and proportionally broader leaves that are mostly acute (vs. acuminate), less pubescent spathes, and all the capsule locules always smooth and dehiscent (vs. two smooth and dehiscent locules and the third locules smooth or muricate, dehiscent or indehiscent or lacking). From both it differs by its well developed upper cincinnus, lack of auricles at the summit of the sheath (auricles not always obvious in *C. bracteosa*), and usually smaller shoots. The lilac, mauve or pale purple flowers present in many populations should also serve to separate *C. melanorrhiza* from the other two species.

This species was known to me for more than two decades, but it was the 1987 collections of my wife, Audrey J. Faden, that enabled me to complete its description.

Hook-hairs are lacking from all of the vegetative parts and from the spathe surface in this species. They are restricted to the exserted portion of the upper cincinnus peduncle.

Commelina sp. 17 in Fl. Eth. (Ensermu & Faden 1997) is very similar to *C. melanorrhiza* but differs by having auricles at the summit of the leaf sheath, one capsule locule rough and indehiscent and larger seeds.

40. **Commelina albescens** *Hassk.* in Schweinfurth, Beitr. Fl. Aethiop.: 210 (1867); C.B. Clarke in DC., Monogr. Phan. 3: 189 (1881) pro parte, excl. var. *occidentalis* C.B. Clarke; Schumann in P.O.A. C:, 135 (1895); C.B. Clarke in F.T.A. 8: 59 (1901) pro parte; Morton in J.L.S. 60: 184–185 (1967), in adnot.; Blundell, Wild Fl. E. Afr.: 413, fig. 859 (1987); Ogwal in Mitt. Inst. Allg. Bot. Hamburg 23b: 582 (1990); Faden in U.K.W.F. 2nd ed.: 305 (1994) & in Fl. Somalia 4: 94 (1995); Wood, Handbook Yemen Fl.: 317 (1997); Ensermu & Faden in Fl. Eth. 6: 372 (1997). Type: Ethiopia: mountains near Gageros, 4000–5000 ft, 8 September 1854, *Schimper* s.n. (B†, syn.); Ethiopia, mountains near Gageros, *Schimper* 2268 (K!, lecto., selected here)

Perennial, shortly rhizomatous; roots fibrous, thin to ± thick; shoots tufted, sometimes purplish, sparsely to densely branched, erect or prostrate proximally and ascending distally, rarely scrambling, to ± 50 cm long or high, usually glabrous or with a short distal band of minute hairs continuous with the line of fusion in the distal leaf-sheath, occasionally entirely puberulous. Leaves spirally arranged or distichous; sheaths to 4 cm long, often splitting longitudinally, sometimes overlapping, puberulous, the hairs sometimes reduced to a band along the fused edge; lamina sessile, usually grey-green, glaucous, linear to linear-lanceolate, (2.5–)5–14(–17) × 0.3–1(–1.2) cm, base oblique, narrowly cuneate, margins scabrid distally or not, sometimes ciliate proximally, apex acuminate; adaxial surface scabrid or glabrous, abaxial surface usually puberulous, occasionally glabrous. Spathes usually densely clustered, rarely solitary, usually strongly falcate and recurved; pedicels < 5 mm long; spathes grey-green or green with a whitish base with contrasting dark green veins, 1.2–2.8 × 0.6–1.4 cm, base truncate to hastate, margins fused for 6–14 mm, occasionally ciliolate along the fused edge, apex acuminate (to acute); surface puberulous or hirsute-puberulous, rarely glabrous; upper cincinnus lacking or vestigial, very rarely (1 spathe on 1 plant) fully developed and 1-flowered, lower cincinnus (1–)2–3(–4)-flowered. Flowers bisexual or male. Sepals glabrous, upper sepal ovate, 2–2.5 × 1.5–2 mm, paired sepals obovate (to broadly ovate), shallowly cup-shaped, fused laterally except at the distal end, 3–5 × 2–3 mm. Paired petals mauve or rarely white, apparently never sky blue, suborbicular, 7.5–12 × 7.5–12 mm of which the claw 1.5–2.5 mm; medial petal colorless, often curled up, 1.8–2.5 × 0.4–1.1 mm. Staminodes 3; lateral stamens with filaments 5–5.5 mm long, anthers 1.2–1.5 mm long; medial stamen filament 3.5–5 mm long, anther 1.8–2.4 mm long. Ovary 1–1.5 mm long; style 6–7.5 mm long; stigma large, capitate-deltate. Capsules trilocular, bivalved, 3-seeded (or fewer through abortion), ± obovoid, 4–4.5 × 4–5 mm, locules 1-seeded; dorsal locule always muricate, indehiscent and deciduous, ventral locules smooth, dehiscent. Dorsal locule seed fused to the capsule wall, ± spherical, 2–2.5 mm, testa smooth, black; ventral locule seeds somewhat flattened spherical to oblong-ellipsoid, 2.7–3.5 × 2.2–2.7 mm, with a ring of soft whitish material around the outside of the seed, testa smooth, dark brown to nearly black, mottled with lighter brown, orange-brown or pinkish grey, ± white farinose.

UGANDA. Karamoja District: Lodoketemit [Lodokeminet], 20 May 1963, *Kerfoot* 4933! & Kangole, Apr. 1960, *Wilson* 863! & Kasunen Estate, May 1972, *Wilson* 2116!

KENYA. Northern Frontier District: 116 km SW of El Wak, 11 km E of Tarbaj, NE of Tarbaj Hill, 17 Dec. 1971, *Bally & Smith* B14660! & Dandu, 4 May 1952, *Gillett* 13033!; Masai District: Ol Lorgosailie [Ol Orgasaile], 9 Apr. 1960, *Verdcourt* 2632!

TANZANIA. Masai District: Olganier gorge E side of Oldoinyo Ogol, 20 Dec. 1962, *Newbould* 6389! & 72 km S of Arusha on Great North Road, 4 Jan. 1962, *Polhill & Paulo* 1039! & foot of Mt Longido, 30 Dec. 1968, *Richards* 23560!

DISTR. **U** 1; **K** 1, 2, 4, 6, 7; **T** 2; Sudan, Ethiopia, Eritrea, Djibouti, Somalia; also in Yemen (including Socotra), Saudi Arabia, Oman, Pakistan, India

HAB. Bushland with *Acacia* and/or *Commiphora*, grassland, thickets and thicket edges, rocky outcrops, swampy ground, pool and stream edges, roadsides, old cultivation; sandy or clayey (sometimes black cotton) soil; 350–1500 m

Flowering specimens have been seen from November to July, with peaks in November to January and April to May.

NOTE. This species is recognizable by its grey-green, glaucous foliage, clustered, subsessile spathes that are typically strongly falcate and have the proximal part (medial part when opened out) white or whitish with strongly contrasting green veins, the lower cincinnus peduncle puberulous, mauve petals with a ± orbicular limb, strongly laterally compressed and ± winged lateral stamen filaments, large sterile basal lobes on the medial stamen anther, and capsules that consistently have the dorsal locule, when developed, muricate and indehiscent.

Because of its obvious close resemblance to *C. erecta*, some authors have questioned whether *C. albescens* is not any more than an extreme form of that species, and even I considered it as "only moderately separable" from *C. erecta* in Flora of Somalia. Morton (in J.L.S. 60: 184, 1967) suggested that some West African collections of *C. erecta* were almost identical to Ethiopian collections, including the type of *C. albescens*. Having checked the cited specimen from Senegal, I found that Morton seemingly ignored the facts that the spathes in this specimen are hardly falcate, there is little contrast in colour between the lower and upper parts of the spathe and the flower colour is indicated as blue on the label. Brenan (1968) rightly ignored Morton's comment and omitted *C. albescens* from West Africa. Most specimens from India and Pakistan are in reasonable agreement with the African material but they have not been studied in detail. Clarke in F.T.A. reported this species from Malawi (Mlanje), but I have not been able to locate the specimen and the record is extremely dubious. Records from Mozambique, Zambia, Malawi, Zimbabwe and Botswana cited in Fl. Eth. would also seem to be erroneous.

Carter & Stannard 637, from near Marsabit in northern Kenya (**K** 1), has gigantic spathes (to 3.3 cm long and 1.8 cm high, with margins fused for up to 17 mm) that are neither as strongly recurved nor as white basally as typical *C. albescens*. However, the "bright mauve" flowers, 2-flowered lower cincinnus with a puberulous peduncle 7 mm long, all agree with *C. albescens*, suggesting that it is probably just an unusual plant of this species, perhaps representing a higher polyploid form than the tetraploid reported from Ethiopia by Lewis & Taddesse (in Kirkia 4: 215. 1964). Fruits are lacking. The extreme dimensions of this specimen have been omitted from the description above.

It has been questioned whether *C. albescens* can ever have sky blue flowers like those of *C. erecta*. Often a blue flower colour on the label is a strong indication that the specimen has been wrongly named. In other cases it may indicate either that the collector misremembered the flower colour when the label notes were written up or else that the person's vision or colour vision may have been at fault. Fortunately, the bright blue pigment that gives certain *Commelina* species their distinctive sky blue hue usually preserves very well in floral remnants on herbarium sheets, so it is sometimes possible to check whether the collector has misrepresented the colour on the specimen label. On *Newbould* 6389 and *Greenway & Kirrika* 11098, both from **T** 2, the flowers are recorded as "bright blue" and "Cambridge blue", respectively, yet the dried petals are the same brownish purple found in other collections of *C. albescens* in which the flowers are described as mauve and other similar colors, but not blue. I believe that the flower colour was just incorrectly remembered in these cases. Similarly, four collections made by Mary Richards of typical *C. albescens* from N Tanzania record the flower colour as "pale blue" or "light sky blue", whereas the floral remains, when present, indicate non-blue flowers. It is well known that Mrs. Richards' eyesight was failing in her later years, when these collections were made, so I think the colour is just incorrect. *Kabuye* 137, from Embu District, Kenya records the flower colour as light blue on a very unusual looking specimen (spathes in subsessile clusters in the leaf axils of the main shoot, in addition to subterminal) from a unique habitat for this species ("old cultivation...[on] black cotton soil"). There is no basis to exclude this specimen from *C. albescens*, but there are no floral remnants to either confirm or refute the flower color. Overall, I can find no clear case of a specimen that otherwise agrees with *C. albescens* that also has distinctly blue flowers, and thus flower colour should serve to distinguish this species from *C. erecta* as a rule.

Additional characters that can be used to separate *C. albescens* from *C. erecta* are the shape of the limb of the paired petals, the compressed lateral stamen filaments and the capsule dehiscence. In *C. albescens* the paired petal limbs are rounded with rounded bases (vs. limbs broadly ovate to ovate-reniform or ovate-deltate, with the base truncate to cordate in *C. erecta*); the lateral stamen filaments are strongly compressed laterally and ± winged distally (vs. filaments neither laterally compressed nor winged); and the dorsal capsule locule is always muricate and indehiscent (vs. smooth and dehiscent in many populations of *C. erecta*).

In the spirit collection of *Gillett* 13033, there are three bisexual flowers and three male flowers. The pedicels of the bisexual flowers range from 4 to 5 mm, whereas those of the male flowers are 7.5–9 mm long. It would appear that the flowers are dimorphic for this character. Based on the scarcity of flowers with very long pedicels I conclude that male flowers are rather infrequent in this species.

41. **Commelina disperma** *Faden* in Novon 11: 398, fig. 1 (2001). Type: Tanzania, Kigoma District, Livandabe Mountain, *Bidgood, Sitoni, Vollesen & Whitehouse* 4153 (K!, holo.; US! iso.)

Perennial herb from a short-creeping rootstock; roots dark brown, thick, fleshy but not tuberous, covered by persistent root hairs; shoots 1 or 2 per plant, erect to ascending, unusually unbranched, not rooting at the nodes, to 60 cm long, with a line of hairs, at least distally, continuous with the fused edge of the distal sheath. Leaves distichous; sheaths 1–2.3 cm long, ± auriculate at the summit, puberulous, ciliolate at apex; lamina petiolate (at least in distal leaves), lanceolate- to ovate-elliptic, 5–11 × 2–3 cm, base strongly oblique, one side rounded, the other cuneate, margins scabrid, also ciliate basally, apex acuminate to attenuate; adaxial surface scabrid, abaxial surface shortly hirsute. Spathes terminal, solitary or up to 3 loosely clustered per shoot; peduncles 6–10 mm long, puberulous with a line of hook-hairs; spathes green but whitish basally, 12–15 × 0.8–0.9 cm, base truncate to hastate, margins fused for 3–4 mm, puberulous along the fused portion, otherwise glabrous, apex acute to obtuse, mucronate, surfaces glabrous; upper cincinnus lacking or vestigial and included in the spathe, lower cincinnus 2(–4)-flowered. Flowers bisexual; upper sepal 2.6–3 mm long, hooded, lateral sepals ovate-elliptic to obovate-elliptic, ± 3.2 × 2.5 mm, completely fused in bud, fused less than half their length in flower; paired petals held erect, pale blue, claw ± 3.5 mm long, lower petal linear-lanceolate, 2–2.5 × 0.5 mm; staminodes 3; lateral stamens with filaments 6.5–7 mm long, anthers 1–1.5 mm long; medial stamen filament 5–5.5 mm long, anther 1.5–2 mm long, yellow. Ovary ± 1.5 mm long; style ± 7–8 mm long; stigma capitate. Capsules olive, bilocular, bivalved, broadly ellipsoid to obovoid, 4 × 5.2 mm, 2-seeded, apex ± truncate, locules 1-seeded. Seeds ellipsoid, 3.2–3.4 × 2.5–2.6 mm, testa dark brown, with a low raised dull orange reticulum, densely white farinose within the reticulum.

TANZANIA. Kigoma District: Livandabe Mountain, 28 May 1997, *Bidgood et al.* 4153!
DISTR. **T** 4; known only from the type collection
HAB. Tall closed forest with *Pterygota, Newtonia* &c.; 1100–1200 m
The specimen was flowering and fruiting in May.

NOTE. Among East African species *Commelina disperma* is most similar to *C. zenkeri* from Uganda. The two species have in common: a similar habit with ± tufted, ascending, typically unbranched shoots with distichous, petiolate leaves with a very oblique base; spathes shortly pedunculate, sometimes or regularly bracteate, solitary or two to three in loose clusters at the ends of the shoots; paired sepals at least partially fused; medial petal very small; and capsule two-seeded. This species differs from *C. zenkeri* by its thick roots, upper cincinnus lacking or vestigial (vs. upper cincinnus often developed and producing a male flower), sepals fused less than half their length (vs. more highly fused), and seeds with a low, raised, dull orange reticulum on the dorsal surface (vs. seeds uniformly brown, smooth).

The seeds of *C. disperma* bear a striking resemblance to those of some collections of *C. zambesica*. The thick roots and glabrous, adaxial leaf midrib also agree with this species. Although it possibly might represent an extreme form of that species, its spathes are much smaller than those of *C. zambesica* and only one of the many collections seen of that species also had glabrous spathes. *Commelina zambesica* can have up to five-seeded capsule and, although the number of seeds is frequently less than that, I have not seen a capsule that matches those of *C. disperma*.

According to one of the collectors (Vollesen, pers. com.) the type locality, near Lake Tanganyika, is very inaccessible and has yielded other interesting species. It is unclear how floristically distinct this forest patch might be from other forests near or along the lake. They would be the obvious habitats to search for further populations of *C. disperma*.

42. **Commelina zenkeri** *C.B. Clarke* in F.T.A. 8: 59 (1901); Faden in Novon 11: 404, fig. 3 (2001). Type: Cameroon, Yaoundé-Station, *Zenker & Staudt* 432 (K!, holo., BM!, iso.)

Perennial 15–40 cm tall; roots thin, fibrous; shoots ± tufted, at least in young plants, erect to ascending or decumbent, branching and rooting only near the base, flowering shoots ascending, unbranched; internodes much reduced and largely covered by sheaths distally, glabrous. Leaves distichous; sheaths to 2 cm long, green, prominently ribbed, puberulous, long white-ciliate at apex; lamina petiolate, lanceolate-oblong to ovate, 4–10.5 × (1–)2–3.9 cm, base strongly oblique, margins scabrid distally, apex acuminate, adaxial surface glabrous to puberulous, abaxial sometimes reddish, puberulous or glabrous. Spathes terminal, solitary or sometimes in pairs; peduncles 5–10 mm long, puberulous with a line of hairs; spathes not falcate to slightly falcate, ± funnel-shaped, 1.4–2 cm long, 0.8–1.3 cm high, base ± truncate, margins fused basally for 2–5 mm, ciliate or ciliolate along the fused portion, apex acute to obtuse, surfaces sparsely puberulous basally or glabrous; upper cincinnus exserted or enclosed within the spathe, to 17 mm long and puberulous with hook-hairs when exserted, producing 1 male flower, or vestigial, lower cincinnus 1–5-flowered. Flowers bisexual and male, 1.4–1.8(–2) cm wide; upper sepal ovate-elliptic to obovate-oblong, 2.5–3.5 × 1.75–3 mm, paired sepals ovate-orbicular to obovate-elliptic, 3.3–3.5 × 2.5–3.5 mm, fused laterally for most of their length into a shallow cup; paired petals blue or lavender to white, broadly ovate or ovate-reniform to ovate-deltate, 9–12 × 8–11 mm of which the white claw 3.5–5 mm long; medial petal white, lanceolate to lanceolate-oblong, 2.8–3.7 × 0.7–1.5 mm; staminodes 3; lateral stamens with filaments 5–6 mm long, anthers 0.8–1.4 mm long, pollen golden yellow; medial stamen filament 3.5–5.5 mm long, anther 1.5–2 mm long, pollen golden yellow. Ovary ± 1.3 × 1.1 mm; style 6–8 mm long; stigma capitate to slightly 3-lobed. Capsules bilocular, bivalved, 2-seeded, broadly ellipsoid, ± 4 × 5 mm, the walls thin-textured, apex emarginate, locules 1-seeded. Seeds ellipsoid, plano-convex, scarcely dorsiventrally compressed, 3.5–4 × 2.5 mm, testa smooth, dark brown, densely white- and brown-farinose.

UGANDA. Toro District: Kibale National Park, S of Ngogo Camp, 6 June 1997, *Poulsen* 1314!; Kigezi District: South Maramagambo Central Forest Reserve, ± 8 km up Kaizi–Bitereko road, 18 Sep. 1969, *Faden et al.* 69/1119!; Bunyoro District: Budongo Forest Reserve, the Nature Reserve, 30 Aug. 1995, *Poulsen, Nkuutu & Dumba* 863!

DISTR. U 2, 4; Cameroon, E Congo–Kinshasa?

HAB. Moist forest, *Cynometra alexandri/Khaya antotheca* forest, on sandy soils; 950–1300 m
Flowering specimens have been seen from February, April, August and September.

SYN. *Commelina* sp. aff. *macrocarpa* of Ogwal, Taxon. study Commelina in Uganda, M.Sc. Thesis: 116 (1977)

NOTE. When we collected *Commelina zenkeri* in Uganda in 1969 (*Faden et al.* 69/1119), I could not identify it but considered it possibly related to the West Africa *C. macrosperma* J.K. Morton because of its apparently two-seeded capsules. Ogwal (1977) did not re-collect or discover additional specimens and merely used my provisional name "*Commelina* sp. aff. *macrosperma*."

The plant was apparently not collected again until Poulsen found it in the Budongo Forest in 1995. Over the next two years he made two more fertile collections in Uganda, as well as sterile collections from plot studies in Uganda and possibly Congo–Kinshasa. At my request he also obtained living material in 1996 which he established at the University of Copenhagen Botanical Garden and later shared with me.

In the course of trying to establish the relationships of this apparent 'new species' the tropical African species of *Commelina* described with two-seeded capsules were reviewed. Clarke (1901) recognized five such species. I could readily dismiss four of them from consideration, which left the fifth species, *C. zenkeri* C.B. Clarke. When I examined the type, *Zenker & Staudt* 432, the small, sessile leaves of the specimen did not remind me of the Uganda plant.

Later I examined a collection *Commelina* collection that I had made in Cameroon in 1986 (*Faden, Satabié & Mpom* 86/1). Further study indicated that this collection was *C. zenkeri* and that it bore a striking resemblance to the 'new species' from Uganda. In the end, no significant differences between the Cameroonian *C. zenkeri* and the Ugandan plants could be found (see Faden in Novon, loc. cit., for more details), so the name *C. zenkeri* was applied to the Ugandan plants too.

Specimens that are definitely *C. zenkeri* have been seen only from Uganda and south central Cameroon. *Poulsen* 1186, from Congo–Kinshasa (Okapi Wildlife Resereve, Ituri Forest, NE of Edoro River) is sterile but closely resembles the Uganda plants. Its occurrence in this location would not be unexpected, but fertile material is essential for confirmation. A search through the herbarium of the National Botanic Garden of Belgium (BR) in September 2000 failed to turn up any specimens, but the search was hardly exhaustive. Similarly, further specimens from Cameroon may have been overlooked or misidentified as *C. bracteosa*.

The apparent disjunct distribution of *C. zenkeri* was thought to be similar to that of *Polyspatha oligospatha*. However, a thorough study of the *Polyspatha* collections from BR did reveal specimens of *P. oligospatha* from the intervening territory, so it is quite possible that a similar, detailed study of the *Commelina* collections in this herbarium would yield similar results for *C. zenkeri*.

43. **Commelina schweinfurthii** *C.B. Clarke* in DC., Monogr. Phan. 3: 158 (1881); F.P.S. 3: 240 (1956); Morton in J.L.S. 60: 189 (1967); Brenan in F.W.T.A., 2nd ed.: 3: 48 (1968); Ogwal, Taxonomic Study Commelina in Uganda, M.Sc. Thesis: 95, fig. 18 (1977); Ensermu & Faden in Fl. Eth. 6: 371 (1997). Syntypes: Sudan, Jur, Jur Ghattas, 5 July 1869, *Schweinfurth* 2022 (K!, lecto., selected here); another specimen mounted on the same sheet but with a separate label that has the same data except for the date 23 July 1969, also *Schweinfurth* 2022 (K!, syn.)

Tufted perennial; roots thick and sometimes tuberous with fusiform tubers, borne distally on the roots; shoots erect to straggling or semi-prostrate, to 60 cm long, sparsely to densely branched, not rooting, usually densely pubescent, occasionally with only a distal line of hairs. Leaves sessile; sheaths to 2.5 cm long, usually densely pubescent, sometimes reduced to a line of hairs along the fused edge of the spathe, ciliate at apex; lamina linear to ovate or elliptic, 2–20(–27) × 0.3–2.5(–3) cm, base blending into the sheath or more commonly rounded to cordate-amplexicaul, margins ciliate, sometimes only proximally, usually scabrid distally, apex acute to acuminate, sometimes mucronulate; both surfaces usually pubescent or occasionally glabrous. Spathes solitary, slightly or not falcate; peduncles (1.5–)2–7(–11) cm long, usually densely pubescent, occasionally reduced to a short line of hairs at the distal end; spathes 1.5–3.7(–4.2) × 1–1.7(–1.8) cm, base truncate to cordate, margins fused for 4–9 mm, ciliate along the line of fusion and rarely a short distance beyond the fusion, apex acute to acuminate, surface usually densely hirsute or pilose, sometimes sparsely puberulous, occasionally glabrous, veins marked with purple; upper cincinnus 1-(very rarely 2)-flowered; lower cincinnus 6–15-flowered. Flowers bisexual and male, the upper cincinnus flower always male, the lower bisexual; upper sepal 3.3–4 mm long, paired sepals completely fused into a cup 3.7–5 mm long, the margins often colored; paired petals blue (see Notes), mauve, purple, blue-mauve, violet-blue, pale lilac, pink-mauve or reddish lilac, ovate-reniform, 11–16 × 6–14 mm of which the contrastingly reddish claw 4–6 mm; medial petal white, minute, filiform; staminodes 3; lateral stamens with filaments 9–12 mm long, anthers 1–3–1.8 mm long, medial stamen filament 6–8 mm long. Ovary 1.5–2 mm long; style 9–11 mm long; stigma capitate. Capsules pale yellow or greyish brown, sometimes with darker flecks, bilocular (very rarely trilocular but the dorsal locule seed not developing), bivalved, subquadrate to oblong-elliptic, 5.5–8.3 × (3.3–)4–5 mm, 4-seeded, apiculate. Seeds subspherical to ellipsoid, usually one or sometimes both ends flattened to concave, (1.7–)2–2.5(–2.9) × 1.6–2.2 mm, testa dark brown, often with patches of lighter brown, smooth.

1. Leaves linear to narrowly lanceolate, up to 1 cm wide above the base, leaf bases mainly rounded or blending into the sheath; **T** 4, 7 . . . . b. subsp. *carsonii*
   Leaves linear to oblong, lanceolate, ovate or elliptic, (0.3–)1–2.5(–3) cm wide, bases usually rounded to cordate-amplexicaul 2
2. Leaf bases rounded to shallowly cordate-amplexicaul; plants densely or sparsely pubescent with white hairs to subglabrous; widespread . . . . a. subsp. *schweinfurthii*
   Leaf-bases deeply cordate-amplexicaul; whole plant usually cinereous and densely covered with long greyish hairs; **T** 4, 7 . . . . c. subsp. *cecilae*

a. subsp. **schweinfurthii**

Leaves usually complicate, linear to oblong, lanceolate, ovate or elliptic, 2–20(–27) × (0.3–)1–2(–3) cm, base usually rounded to cordate, occasionally narrowed into the sheath, rarely deeply cordate-amplexicaul, margins usually ciliate, at least proximally and scabrid at least distally, apex acute to acuminate, surfaces densely pubescent to glabrous; spathes 1.7–3.7(–4.2) × 1–1.7(–1.8) cm on peduncles (1.5–)2–7(–11) cm long; seeds ± spherical to shortly ellipsoid, smooth, dark brown.

UGANDA. Karamoja District: Kokumongole, 28 May 1939, *Thomas* 2856!; Masaka District: Lyantonde, 26 Oct. 1969, *Lye* 4626!; Teso District: km 25, Soroti–Moroto, 13 Oct. 1952, *Verdcourt* 824!

KENYA. Trans-Nzoia District: Kitale–Makutano, 20 May 1969, *Napper & Tweedie* 2109! & ENE slope of Mt Elgon, 24 Sep. 1962, *Lewis* 5974!; Kakamega District: Kakamega Forest near Forest Station, 12 Oct. 1981, *Gilbert & Mesfin* 6654!

TANZANIA. Ngara District: Murgwanza, Bugufi, 4 Jan. 1961, *Tanner* 5579A!; Ufipa District: 36 km from Sumbawanga on Sopa [Isopa] road, 17 June 1996, *Faden et al.* 96/270!; Mbeya District: Isyesye village, ± 1.6 km SE of Ilomba local court, 12 Mar. 1963, *Harwood* 21!

DISTR. **U** 1–4; **K** 2, 3, 5; **T** 1, 4–7; Sierre Leone, Ivory Coast, Nigeria, Cameroon, Congo–Kinshasa, Rwanda, Burundi, Sudan, Ethiopia, Zambia, Malawi

HAB. Grassland, seasonally swampy grassland & depressions, roadside banks, rocky places, shallow soil over rocks, waste places, cultivation, bushland, woodland, wooded grassland, forest clearings; 900–2350 m

Flowering plants have been seen from all months except November.

SYN. *Commelina stolzii* Mildbr. in N.B.G.B. 9: 254 (1925). Type: Tanzania, Rungwe District: Kyimbila District, *Stolz* 1094 (B, holo., K!, iso.)

C. elgonensis Bullock in K.B. 1932: 506 (1932); Ogwal, Taxonomic Study Commelina in Uganda, M.Sc. Thesis: 98, fig. 19A (1977); Malaisse in Fl. Rwanda 4: 129, fig. 47.1 (1988); Ogwal in Mitt. Inst. Allg. Bot. Hamburg 23b: 580 (1990); Faden in U.K.W.F. 2nd ed.: 305, Pl. 136 (1994). Type: Kenya, Mt Elgon, *Lugard & Lugard* 549 (K!, holo.).

NOTE. Brenan (1968) noted an apparent variation in flower colour in this species in West Africa, based on a single collection from Togo that records the flowers as "pale brownish orange". The specimen was *C. gambiae* C.B. Clarke (or *C. nigritana* Benth. var. *gambiae* (C.B. Clarke) Brenan).

Subsp. *schweinfurthii* is well collected in Kenya and Uganda but less so in Tanzania. *Turner* 50T from Uganda has extremely narrow leaves and would key out to subsp. *carsonii* whereas *Eggeling* 688, also from Uganda, has extremely broad – 3 cm wide – amplexicaul leaves, the broadest seen in any specimen of *C. schweinfurthii*, and could easily be keyed out to subsp. *cecilae*. In view of the less than perfect separation of these subspecies, I see no purpose in treating these specimens as more that just oddities in subsp. *schweinfurthii*, rather than creating the appearance of disjunct distributions for the other two subspecies.

Some specimens from W Kenya, e.g. *Knight* 4404 have an unusual habit, being low growing and much branched with short leaves. That specimen in particular has its largest leaves elliptic and its habit resembling that of *C. africana*. That specimen was pressed from cultivation but *Lewis* 5974 was wild collected on Mt Elgon at 2135 m in shallow soil over rocks and has a somewhat similar habit.

Ogwal (1977) recognized *C. schweinfurthii* and *C. elgonensis* as distinct species in Uganda, separating them on the basis of a 5-seeded capsule in the former and 4-seeded capsule in the latter. She cites only one collection of the former, *Purseglove* 1547, which I have not studied. She further indicates that the flowers are blue and that the dorsal locule is "often infertile" in this taxon. She does not describe the capsule as 3-valved. Without further study I would expect that the specimen perhaps approaches the type of *C. schweinfurthii* but probably is just an atypical plant of the more usual, blue-flowered plant in our area. It is noteworthy that *C. schweinfurthii* is omitted in Ogwal (1990).

Two collections from the Uluguru Mts, the only collections seen from **T** 6, *E.M. Bruce* 876 and *Schlieben* 3207, have glabrous leaves and exceptionally large subglabrous spathes, the only pubescence being a line of short marginal hairs along the fused edge. The old capsule on the *Schlieben* specimen is definitely 4-seeded and the seeds would appear to have been ellipsoid. The flower colour is given as blue on the *Bruce* collection – although the petal remnants are more mauve than blue – and "red-lilac" on the *Schlieben* collection. All of this adds up to a plant very close to if not belonging in subsp. *schweinfurthii* but still morphologically distinct from the other collections. *Aleljung* 247 from Loleza (**T** 7) shows some similarly to the two Uluguru collections but it has hairy leaves and spathes and its spathes are smaller.

b. subsp. **carsonii** (*C.B. Clarke*) *Faden* **comb. nov**. Type: Zambia, Tanganyika Plateau, Fwambo, 1889, *Carson* s.n. (K!, holo.)

Leaves complicate, linear to narrowly lanceolate, 3–18(–22) cm long, 0.4–0.9(–1) cm wide above the base, apex acute to acuminate, base usually rounded or narrowed into the sheath, margins ciliate, at least proximally or eciliate, scabrid at least distally, rarely not scabrid, surfaces sparsely pubescent or glabrous; spathes on peduncles (1–)2–6(–9.5) cm long, spathes 1.6–3 cm long, 1–1.5 cm high; seeds subspherical, smooth, brown.

TANZANIA. Ufipa District: Mbizi Mts, 18 June 1996, *Faden et al.* 96/287! & 2 km W of Mkowe, 21 Nov. 1994, *Goyder et al.* 3765!; Mbeya District: Mbosi Circle, Msumbi Estate, 13 Jan. 1961, *Richards* 13907!

DISTR. **T** 4, 7; Congo–Kinshasa, Zambia, Malawi?

HAB. Swampy grassland and pasture, woodland near streams and miombo forest; 1500–1800 m Flowering specimens have been seen from January through March.

SYN. *Commelina carsonii* C.B. Clarke in F.T.A. 8: 52 (1901)

NOTE. Narrow-leaved specimens from Malawi such as *Phillips* 806, which were determined as *C. carsonii*, cannot be distinguished from *C. subcucullata* in the absent of seeds, so it is questionable whether *C. schweinfurthii* subsp. *carsonii* occurs in that country. Morphologically similar specimens from Zambia that lack seeds have been accepted as subsp. *carsonii* because the type is from that country, and there are additional Zambian collections with seeds.

As interpreted here *C. schweinfurthii* subsp. *carsonii* is a gracile plant with narrow, complicate leaves with veins prominent on both surfaces, leaf bases that are either rounded or else just blend into the sheath, margins that are often scabrid in the distal half of the leaf, and spathes that are often strongly marked with purple. Specimens from southern Tanzania, e.g. *Goyder et al.* 3765, were sometimes determined as *Commelina subcucullata* C.B. Clarke, a related species also gracile and sometimes with highly colored spathes; this species is known only from Malawi, where it is appears to be widespread but not commonly collected, and on the whole its spathes are smaller and less pubescent – the pubescence sometimes concentrated toward the base of the spathe – than those of *C. schweinfurthii* subsp. *carsonii*. The seeds of *C. subcucullata* are very distinctive: they are subquadrate, dark brown and have a low raised black reticulum and very fine white hairs. In the absence of seeds it would be unwise to identify any specimen from our area as *C. subcucullata*. It is interesting to note that when Clarke published *C. subcucullata*, on the very next page after *C. carsonii*, he questioned whether *C. subcucullata* was a good species or just a variety of the previous species. On an annotation label written four years earlier on one of the syntypes of *C. subcucullata*, *Scott Elliot* 8675, Clarke described the seeds as "obscurely reticulate...the reticulation minutely hairy," whereas in Clarke (1901) he characterized the seeds of *C. subcucullata* as "globose, smooth and brown." It is unclear why he omitted the most distinctive features of this species, except perhaps that he was already committed to questioning its distinctiveness.

Milne-Redhead & Taylor 7791 (**T** 8, Songea District, ± 19 km W of Songea, near Likuyu River, 930 m, 30 Dec. 1955) represents a distinct species, *Commelina* sp. aff. *subcucullata*, that resembles both *C. schweinfurthii* subsp. *carsonii and C. subcucullata* in its gracile habit and narrow leaves. Its capsules are subquadrate, 6 × 5 mm, bilocular, bivalved and 4-seeded. Its seeds show some similarity to those of *C. subcucullata* in being dark, not completely smooth and having a reticulation. However, they are ± spherical, rugose and densely white-farinose, all quite different from the seeds of *C. subcucullata*. Moreover the reticulation they show is at the cellular level, with the walls of the individual cells of the testa somewhat raised, whereas in *C. subcucullata* the reticulation is of a higher order and less formal. Finally, the seeds of *Milne-Redhead & Taylor* 7791 are not pubescent. In view of the scarcity of collections attributable to *C. subcucullata*, it is likely that we still do not know the extent of its variation. However, I do not see how this collection can be placed in that species as we presently understand it, and for now it is best treated as a related, undescribed species. The plant was erect and growing in woodland on red loam. Its spathes are medium sized (1.7–2.5 × 0.9–1.3 cm) with the margins fused for 3–7 mm. The spathes are described as very variable in pubescence by the collectors. The upper petals are recorded as pale violet or lilac with a dull red claw, and the lower petal as reduced to a white or mauve claw. I have not been able to match this collection for certain with any other. However most specimens lack capsules and seeds, without which determinations are very difficult in this group.

Mhoro 3993 (**T** 8, Songea District, Likuyu Fussi Ujamaa Village, 5 June 1981) represents a different distinct species, *Commelina* sp. aff. *carsonii*. In this case it is matched by a series of collections from northern and western Zambia, e.g. *Richards* 3810, 5162, 12644 and 17418. It is a gracile plant, with a somewhat straggling habit and small spathes (especially small in *Mhoro* 3993: 1.3–1.5 × 0.5–0.8 cm; somewhat larger in the Zambian collections). It is characterized by having a trilocular, trivalved capsule, nominally with five seeds but they rarely all seem to develop. When the dorsal locule aborts, the capsule may appear bilocular, bivalved and 4-seeded. The capsule is unusually small and subquadrate in outline, 3.5–4(–5.5) × ± 3.5 mm. The seeds are subspherical to ovoid and (at least those of the ventral locules) are particularly short: 1.3–1.7(–1.8) × 1.6–2(–2.2) mm. Also distinctive is a greyish or greyish tan testa although this may be an outer layer that can be rubbed off. The apicule on the embryotega in many of the seeds ends not in a point, but in a small, dull orange, discoid appendage. *Mhoro* 3993 was collected in grassland with scattered trees. Most of the Zambian collections, however, were growing in dambos and other marshy places. This is a distinct species when in fruit, and we do not have a name for it. The flowers are mostly described as blue (rarely mauve) but that likely is a colour more in the lavender range, not the sky blue of many other *Commelina* species. *Milne-Redhead & Taylor* 8726 (Songea District, 8 km W of Songea, 960 m, 9 Feb. 1956) is in flower but very likely represents the same species as *Mhoro* 3993. It clearly is a straggling plant and its base is shortly rhizomatous, as seen in *Richards* 17428 from Zambia. It is much more pubescent than *Mhoro* 3993 and has larger spathes (to 1.6(–2.2) × 0.9(–1.2) cm). The upper petals are described as "purplish with reddish claws".

c. subsp. **cecilae** (*C.B. Clarke*) *Faden* **comb. nov.** Type: Zimbabwe, Matabeleland, near Gwelo, *Cecil* 139 (K!, holo.)

Leaves mostly flat, occasionally complicate, linear-oblong to ovate, 3–19 × 1–2.5 cm, apex acute to acuminate, base deeply cordate-amplexicaul, margins ciliate, at least proximally, not scabrid, surfaces densely pubescent; spathes on peduncles 1.5–4 cm long, spathes 1.5–3 cm long, 1–1.3 cm high; seeds ± spherical, smooth, dark brown.

TANZANIA. Ufipa District: Kawa River Gorge, 15 Feb. 1959, *Richards* 10886!; Chunya District: Lupa Forest Reserve, 159 km N or Mbeya, 14 Mar. 1962, *Boaler* 510!; Mbeya District: Tunduma–Sumbawanga road, Ikana, 27 km towards Sumbawanga from Ndalambo, 14 June 1996, *Faden et al.* 96/195!

DISTR. **T** 4, 7; Congo–Kinshasa, Zambia, Malawi, Zimbabwe

HAB. Woodland, at edge of swamp in woodland, and swampy short grassland by stream, also "bracken hillside"; 1200–1450(–1700) m
Flowering specimens have been seen from November, February and March.

SYN. *Commelina cecilae* C.B. Clarke in F.T.A. 8: 51 (1901)

NOTE. The three collections cited are very uniform in appearance: robust plants with leaf bases deeply cordate-amplexicaul, the whole plant entirely covered by long, greyish pubescence. They match many collections from further south. The fourth collection from our area, *Verdcourt* 3402, has much shorter hairs, shorter leaves and the collector records the flowers as white, instead of the usual 'blue'. South of our area subsp. *cecilae* is much more variable, occasionally having spathes that are nearly glabrous, e.g., *Richards* 482 from Mpulungu, Zambia. On a very different looking plant, *Richards* 17234 from western Zambia, the flowers are noted as white, so *Verdcourt* 3402 does not fall outside of the morphological range of this subspecies as a whole.

NOTE (on the species as a whole). This species has presented numerous problems in trying to sort out its taxonomy. The first is what it should be called. I am following the usage of F.W.T.A. (Brenan, 1968) and Fl. Eth. (Ensermu & Faden, 1997), but in either case one has to make the assumption that either Clarke was mistaken when he described *C. schweinfurthii* as having a trilocular, trivalved capsule or else, if he was correct, then the character is not significant. I have studied the type and I have concluded that Clarke was correct in his interpretation of the capsule. A second collection from Sudan (*Myers* 9605) shows it even more convincingly. Thus we are deciding to ignore the fruits and seeds, which also differ from those of our plants, for the sake of conformity.

Having decided to accept the name *C. schweinfurthii* for the familiar plant in western Kenya and Uganda that has been known for many years as *C. elgonensis*, which is clearly the same taxon as the plants in Ethiopia, the next matter to determine was how far south this plant occurred and also what to do with plants in southern Tanzania and further south that had been named *C. carsonii* and *C. cecilae*. Because as far as I can tell there are essentially no significant differences in the flowers, capsules and seeds of the northern plants from the southern ones, it then became a matter of whether or not only combine them into one species. This seemed the best course, but there still appeared to be a need to recognize the very narrow-leaved plants and the extremely pubescent ones with deeply cordate-amplexicaul leaves. Thus I have decided to retain *C. carsonii* and *C. cecilae* as subspecies within *C. schweinfurthii*.

No further progress is likely on the taxonomy of this species group until we know the capsules and seeds of the plants called *C. schweinfurthii* in West Africa – none have been seen – and until further collections from southern Sudan can be found and studied.

This species, as we are using it, is characterized by its usually densely pubescent foliage; the upper cincinnus producing a single male flower that has an extremely long pedicel; flowers that are in the blue-purple range (not the sky blue of many other *Commelina* species); the lower petal minute and borne in a cup formed by the fusion of the lower sepals; the staminodes with a central dark spot; the medial anther entirely dark or yellow with a dark transverse band; and the subquadrate capsule that has 4 smooth seeds which, on close inspection, are not truly spherical.

44. **Commelina kitaleensis** *Faden* **sp. nov.** a *C. schweinfurthi* C.B. Clarke foliis scabris in pagina superiore floribus fulvo-salmonaceis vel brunneo-ochraceis capsulis trilocularibus trivalvibus 5-seminalibus seminibus testa armeniaceo-fulva vel subroseo-cinerea hilo pagina seminis complani (haud elevato) apiculo embryotegae apice in appendice parva discoidea saepe terminanti differt. Type: Kenya, Trans-Nzoia District, Elgon South Road, first swamp beyond the Aerodrome, ± 13.3 km from Kitale, *R.B. Faden, A. Evans & Tweedie* 69/715 (K!, holo., EA!, iso.)

Perennial herb; roots ± tuberous, tufted at the base of the plant, shoots annual, sprawling or erect to ascending, apparently not rooting, to 40 cm, sparsely to densely branched, densely white-puberulous. Leaves distichous; sheaths to 2 cm long, densely hirsute or puberulous all over with white hairs, white-ciliate at apex, often pale with dark veins; lamina linear to narrowly lanceolate, (2–)4–14 × 0.3–1.2 cm, base ususally cordate, sometimes amplexicaul or rounded, margins sometimes revolute, scabrid, apex acuminate; both surfaces puberulous. Spathes solitary or occasionally subclustered, slightly or not falcate; peduncles 1–3.5 cm long, ± densely white hirsutellous, especially distally; spathes 1.4–2.1 × 0.8–1.4 cme, base rounded to truncate, margins fused for 4.5–9 mm, ciliate along the fused edge, apex acute to acuminate, surfaces densely pilose or puberulous; upper cincinnus 1-flowered, lower cincinnus ± 9–10-flowered. Flowers bisexual and male. Sepals pinkish white, upper sepal 3–4.5 mm long, paired sepals fused into a cup 2.5–4 mm long; paired petals apricot, buff, buff-salmon, brown or brown-ochre, limb cordate at base, not notched at the apex; staminodes 3; lateral stamens with filaments 5–6 mm long, anthers ± 1.2 mm long, medial stamen filament ± 4 mm long, anther 1.3–1.5 mm long. Capsules orange-brown to reddish brown, trilocular, trivalved, oblong to subquadrate, 4–5 × 2–3 mm, apiculate, 5-seeded, dorsal locule 1-seeded, ventral locules 2-seeded. Seeds flattened spherical to ovoid, often flat to concave on one end and rounded on the other, 1.7 × 1.8 mm (dorsal locule seed), 1.2–1.6 × 1.5–1.8 mm (ventral), testa orange-tan or a dull whitish orange-brown, smooth.

KENYA. Trans-Nzoia District: E Mt Elgon, near Kitale, 7 Nov. 1954, *Irwin* 146! & Kitale, Grassland Research Station, 25 Sep. 1962, *Lewis* 5981! & Aerodrome Swamp, Kitale, Aug. 1967, *Tweedie* 3470!

DISTR. **K** 3 (vicinity of Kitale); not known elsewhere

HAB. Swamps, seasonally wet grassland and in forest edge, on sandy loam; 1800–2000 m Flowering plants have been seen from June to September.

SYN. *Commelina velutina* of Faden in U.K.W.F.: 657, illus. p. 659 (1974) & in U.K.W.F. 2nd ed.: 305, t. 136 (1994), *non* Mildbraed

C. sp. 8 of W.H. Lewis, Meiotic chromosomes African Commelinaceae, Sida 1: 283 (1964)

NOTE. I cannot recall why I selected the name *Commelina velutina* Mildbr. for this local, western Kenyan plant. Mildbraed's *C. velutina*, probably an endemic in northern Cameroon – although some plants from northern Zambia and southern Congo–Kinshasa have been so determined – has a similar capsule and seeds to *C. kitaleensis*, based on the limited material of both available for study, but it has densely clustered, very small spathes, and characteristic velutinous pubescence, unlike that of any other *Commelina*. The Cameroonian plants have blue flowers, according to the label of one collection, whereas the Zambian and Congolese collections have white or light purple flowers. In either case the flowers contrast with those of *C. kitaleensis*.

Commelina kitaleensis much more closely resembles the type of *C. schweinfurthii* C.B. Clarke from Sudan (*Schweinfurth* 2022, K) and even more so a second collection from Sudan (*Meyers* 9605) that has scabrid leaves, seeds almost identical to those of *C. kitaleensis*, and records the flower colour as "orange yellow", which was considered doubtful in a note on the label by Brenan. In contrast, Clarke (1881) recorded the petals on the type as "purple-blue", something that is notoriously difficult to determine accurately from dried flowers – except in the species that have sky blue flowers. It is interesting to note that the label of *Tweedie* 3470 mentions that the buff flowers turned purple when fixed in the liquid preservative FAA, so we know that those colours are not completely stable. Nevertheless, in all of the other collections of *C. schweinfurthii* from West Africa and Ethiopia – also see Fl. Eth. (Ensermu & Faden, 1997: 371) – in which the flower colour is indicated, it is always given as some shade of blue, so the full range of variation in this species has not been determined. No seeds or mature capsules have been seen from any collection of *C. schweinfurthii* from West of Sudan. Thus accepting the western Kenyan plant as *C. schweinfurthii* would require redefining the species without having sufficient evidence to do so.

Lewis on the label of *Lewis* 5981 and in the publication Lewis (1974) records the chromosome number as $n = \pm 30$ for this species (as '*Commelina* sp. 8').

45. **Commelina grossa** *C.B. Clarke* in F.T.A. 8: 60 (1901). Type: Zambia, Kambole, SW of Lake Tanganyika, 19 Mar. 1896, *Nutt* s.n. (K!, lecto., selected here – the sheet with "Type Specimen" label; K!, iso.)

Perennial; roots fleshy, cord-like but not tuberous; shoots erect, sparsely branched, (8–)10–35(–40) cm tall, with 1–5 nodes, glabrous or sparsely puberulous, usually only distally. Leaves mostly basal, distichous, the distalmost 1–2 leaves often bract-like and funnel-shaped, with reduced lamina; sheaths to 5 cm long, those of the cauline leaves reduced, usually split to the base (those of the cauline leaves sometimes shortly fused), glabrous or sparsely ciliate at apex, often flushed with purple; lamina linear, 20–40(–50) cm long, in cauline leaves 8–30 × 0.4–1(–2.5) cm, base narrowed then abruptly broadened into the sheath, or gradually broadened and broadest at the base, margins scabrid distally, occasionally ciliate near base, apex attenuate; surfaces usually glabrous, occasionally sparsely puberulous abaxially and near the margins adaxially. Spathes solitary or loosely aggregated; peduncles (1.7–)3.5–7.5 cm long, glabrous; spathes 2.7–4.7 × 1.5–2.6 cm, base truncate to rounded or subcordate to hastate, margins fused for ± 6–20 mm, completely glabrous or sparsely ciliate along the fused edge, sometimes reddish purple, occasionally the acute to acuminate apex curved downward, surfaces glabrous, pale green at base with darker veins, darker green above, sometimes mainly maroon in outer half; upper cincinnus 1-flowered, the flower male; lower cincinnus (10–)13–18-flowered. Flowers bisexual and male, ± 16 mm wide; upper sepal 3–5 mm long, paired sepals fused into a cup 3–5.5 mm long, the margins tinged with pinkish purple; paired petals lavender or blue, ovate-deltate, ± 12 mm long of which the claw 4.5–6.5 mm, purple at the base; lower petal enclosed within the sepal cup, very pale mauve, linear to linear-lanceolate, ± 3 mm long; staminodes 3; lateral stamens with filaments ± 7 mm long, anther 1.5–2.5 mm long, medial stamen anther 2.5–3 mm long. Capsules grey-green, heavily mottled with brown or maroon, bilocular, bivalved, (1–)4-seeded, ± quadrate and contracted between the seeds, 6.5–7.5 × 6 mm, dorsal locule lacking, ventral locules (0–1–)2-seeded. Seeds rectangular-ellipsoid, 2.2–2.6 × 2.3–2.8 mm, testa smooth, medium to dark brown with small, scattered blackish spots, sometimes sparsely white-farinose around the hilum.

TANZANIA. Mpanda District: Uzondo Plateau, 18 Apr. 2006, *Bidgood et al.* 5587!; Chunya District: Lupa N Forest Reserve, 153 km N on Mbeya–Itigi road, 19 Apr. 1962, *Boaler* 540!; Tunduru District: just E of Songea District Boundary, 5 June 1956, *Milne-Redhead & Taylor* 10058!

DISTR. **T** 4, 7, 8; Zambia, Malawi

HAB. Woodland on sandy or seasonally waterlogged soil; 850–1550 m

Flowering April and June.

NOTE. The type collection consists of three sheets, all apparently collected at the same time by Nutt, and all annotated by C.B. Clarke as "*Commelina grossa*, sp. nov." on 5 Oct. 1900. One of them has a "Type Specimen" label affixed, and I can see no reason why this should not be selected as the lectotype.

This is a very distinctive species because of its mainly basal, long, linear leaves and its large, broad, essentially glabrous spathes. The flower colour has been called blue in most collections, but, in fact, only four collectors have ever mentioned it: two called the flowers blue, one called them purple and I called them lavender. What is clear is that wherever petals have been preserved on the herbarium specimens they have dried lilac to mauve, which, in my experience, pressed sky blue *Commelina* flowers never turn. Therefore I conclude that the petals were not the sky blue colour of many *Commelina* species, such as *C. benghalensis*, but rather a more mauvy blue.

The flower colour is significant because I think that it indicates that the relationships of *C. grossa* are with the *C. schweinfurthii* species group, which is characterized by its lavender to blue-violet (or sometimes buff to orange or white), but never sky blue flowers, the sepals fused into a cup, the lower petal very reduced, the staminodes usually with a central dark spot, the medial stamen anther usually with a dark band, the capsules ± quadrate and 4(–5)-seeded, and the seeds usually with a smooth testa.

The pedicels and sepals sometimes appears to be puberulous in herbarium specimens, at least within the spathe, but these hairs have an irregular look about them. Where they can be checked, as in *Radcliffe-Smith, Pope & Goyder* 5685, close inspection suggests that they are large glandular microhairs, which, because they collapse, are not normally considered part of the pubescence. I think that their size was enhanced by dried pectin-like material that was in the spathe, which made the hairs look larger and more conspicuous than they do in life, but also irregular. I have not seen definite macrohairs, except perhaps on the upper sepal in *Richards* 4692 and in *Faden & Faden* 74/140. In the Richards collection most of the hairs seem to be microhairs and are visible only under high magnification.

Mrs. Richards reports minute white hairs on the spathes in *Richards* 4692. Even under high magnification I can confirm only very few that are not microhairs. I have no idea what she means by the description "edges of petals hairy" on *Richards* 9267.

46. **Commelina neurophylla** *C.B. Clarke* in F.T.A. 8: 53 (1901); Faden, Layton & Figueiredo in Syst. Geog. Pl. 79: 79 (2009). Type: Malawi, Tanganyika Plateau, July 1896, *Whyte* s.n. (K!, syn.); Nyika Plateau, July 1896, *Whyte* s.n. (K!, syn.); Kondowe to Karonga, July 1896, *Whyte* s.n. (K!, syn.); Mt Zomba, Dec. 1896, *Whyte* s.n. (K!, syn.); Shire Highlands, *Buchanan* s.n. (K!, syn.)

Tufted perennial to sparsely branched annual (20–)30–50 cm tall; roots thin, orange, mostly spreading horizontally; shoots wiry, moderately branched, ascending, sometimes procumbent at the base, not rooting at the nodes, glabrous or puberulous (sometimes only in a longitudinal line). Leaves with sheaths to 3 cm long, commonly green, occasionally purple or purple-striped, puberulous or the pubescence confined to a line along the fused edge, ciliate or eciliate at apex; lamina sessile, usually stiff, sometimes falcate, linear to lanceolate-oblong, 2–9.5 × 0.1–1(–1.4) cm, base rounded and sometimes ± amplexicaul or narrowed into the sheath, margins scabrid, usually ciliate or ciliolate at least proximally, occasionally eciliate, apex acute to acuminate; surfaces glabrous to hirsute. Spathes solitary or rarely subclustered, usually slightly falcate distally; peduncles (1–)1.5–4(–9) cm long, subglabous to puberulous; spathes 1.4–2.7 × (0.9–)1.1–1.6 cm high, base truncate to hastate, with a recurved lip at the summit of the line of fusion, margins fused 8–15 mm, ciliate along the line of fusion, rarely entirely glabrous, apex abruptly acute to acuminate, surfaces glabrous or less commonly puberulous; upper cincinnus lacking or vestigial and enclosed within the spathe, lower cincinnus (2–)10–15-flowered. Flowers bisexual and male, 15–22 mm wide; upper sepal 3–5.5 mm long, paired sepals completed fused into a cup, 3.5–5 mm long; paired petals violet to lavender (sometimes described as blue or lilac) or yellow-orange to apricot, ovate to reniform, ± 14 × 10 mm of which the violet to purple or reddish claw 4–7 mm; lower petal minute, ovate to lanceolate-ovate, white or whitish; staminodes 3; lateral stamen filaments 6–8.5 mm long, anthers 0.7–1.8 mm long, medial stamen with anther 1.2–2.6 mm long, pollen yellow; style ± 8 mm long. Capsules pale yellow, sometimes with dark spots, to medium brown, bilocular (to trilocular), bivalved, 4-seeded, slightly oblong to ± quadrate, (4.8–)6–6.5 × 4 mm, dorsal locule usually lacking, occasionally present, empty?, ventral locules 2-seeded (or fewer through abortion). Seeds ovoid to nearly circular or shortly transversely ellipsoid, 2.6–3.3 × 2.3–2.6 mm, testa muricate to echinate, finely mottled with various shades of brown and black.

TANZANIA. Ufipa District: Tatanda Mission, 23 June 1996, *Faden et al.* 96/360!; Njombe District: NW side of Kitulo [Elton] Plateau, road from Mporoto to Kitulo sheep ranch, 6 May 1975, *Hepper & Field* 5319!; Songea District: Matengo Hills, Lupembe Hill, 20 May 1956, *Milne-Redhead & Taylor* 10262!

DISTR. **T** 4, 7, 8; Congo–Kinshasa, Burundi, Zambia, Malawi, Zimbabwe

HAB. Moist depressions, ditches and banks, swampy pastures, swamps, grassland, and *Brachystegia* woodland; sandy soil; (1000–)1600–2300 m

Flowering February to June.

SYN. *C. bianoensis* De Wild., Contr. Fl. Katanga, Suppl. 3: 62 (1930); Faden, Layton & Figueiredo in Syst. Geog. Pl. 79: 79 (2009). Type: Congo–Kinshasa, Katanga, Plateau de Biano, Tshisinka, *Homblé* 1272 (BR!, holo., photo K!)

NOTE. Plants from wet places may have substantial bases, but it is not certain that they are perennial because the roots are thin and wiry and the shoot bases are not noticeably thickened, although the bases of some of the larger plants hardly seem as if they could have been developed in a single season. Plants from drier habitats, such as *Brachystegia* woodland, e.g. *Faden et al.* 96/196, are much smaller and are distinctly annual.

This species is unusual in having two distinct flower colours, violet to lavender and yellow-orange to apricot. Each plant has only one colour, and normally populations have purely one colour or the other. In Tanzania only yellow to orange flowers have been recorded for six collections and only violet to lavender flowers from four others. Two Tanzanian populations have been reported with both flower colours (*Milne-Redhead & Taylor* 10262 and *Faden et al.* 96/335). There is no straightforward way to formally describe one of these colour forms as a form or variety of the other, because the flower colour is impossible to determine in any of the syntypes.

The spathes of *C. neurophylla* are commonly glabrous, except for the marginal hairs along the fused edge. However, when the whole plant is pubescent, then the spathes may be pubescent as well. The hairs on such spathes are ± uniform in length, in comparison with the two hair lengths on the spathes of *C. triangulispatha*. Hook-hairs are lacking in both species.

Plants normally produce many flowers per spathe. The small number of flowers indicated parenthetically in the description comes from *Faden et al.* 96/136, which consisted entirely of small annual plants growing in *Brachystegia* woodland, a much drier habitat than is typical for this species in Tanzania. Although they had undergone a full life cycle and produced seeds, it is possible that the plants were stunted. In its annual habit, subclustered spathes, some reflexed against the peduncle, on relatively short peduncles (for *C. neurophylla*), and few flowers per spathe, *Faden et al.* 96/136 approaches *C. triangulispatha*. However, in its maximum leaf, internode, and peduncle lengths it falls outside the range for that species, and its largest spathe would match the largest dimensions for *C. triangulispatha*. In its glabrous spathes, the old spathes not showing prominently raised veins, and the relatively large, dark spots on the fruits, this specimen much better agrees with *C. neurophylla* than with *C. triangulispitha*. Overall, this specimen is best treated as an unusual form or ecotype of *C. neurophylla*.

Male flowers are commonly produced in *C. neurophylla*. A majority of flowers preserved well enough in the Kew Herbarium to determine gender were male. Judging by the floral parts that could be measured, the flowers seem to vary greatly in size. It is possible that the later flowers in the spathe are smaller than the earlier produced ones.

The syntypes from Mt Zomba and Kondowe to Karonga are mounted on the same sheet. The separate labels for each of these collections has had the original species name crudely removed. Brenan has written on the sheet, with an arrow directed towards where the name used to be, "what clot scratched out the name!". Since the handwriting on both of these labels appears to be that of Clarke, as is the separate determination label that applies to both of these collections, it is reasonable to conclude that Clarke realized that he had originally misidentified the specimens, and rather than cross out the original determination, he simply scratched it off. To support the above, the syntypes of *Whyte* s.n. from the Tanganyika Plateau and *Buchanan* s.n. are also mounted on one sheet. Both labels have the name scraped away, as above. The label for the *Buchanan* collection is separate from the printed label, and Clarke (undated) has written, "seems halfway between *latifolia* and *ecklониana* which are said to be too near" [to each other?]. Clarke's dated determination label that clearly applies to both specimens reads, "*Commelina neurophylla*, sp. nov. C.B. Clarke ms, 5 Oct 1900" clearly was written after the note on the Buchanan specimen. Upon checking the packets on the syntypes it is clear that Clarke did not have mature seeds available. Had he had them he might noted how distinct they are.

Bidgood et al. 5494 has unusually narrow leaves, some less than 2 mm wide. Except for this character it appears to be typical for the species.

47. **Commelina triangulispatha** *Mildbr.* in N.B.G.B. 84: 255 (1925); Ogwal, Taxon. study Commelina in Uganda: 103, fig. 20 (1977) & in Mitt. Inst. Allg. Bot. Hamburg 23b: 580 (1990); Faden in U.K.W.F. 2nd ed.: 305 (1994). Type: Tanzania, Rungwe District: Kyimbila, *Stolz* 725 (B, holo.; K!, MO!, iso.)

Annual (3–)10–30(–60) cm tall; roots thin, fibrous; shoots erect or ascending to decumbent, rarely rooting at the lower nodes, hirsute below the node or the pubescence confined to a longitudinal line continuous with the fused edge of the sheath above, glabrous proximally. Leaves with sheaths to 1.3 cm long, occasionally split to the base, purple-veined or entirely green, hirsute or the pubescence confined to a longitudinal line along the fused edge, ciliate at apex; lamina sessile, sometimes falcate, linear-lanceolate to oblong-elliptic, (1.5–)2–6(–7.5) × 0.2–1(–1.6) cm, base sometimes rounded, occasionally narrowed into the sheath, margins scabrid, usually ciliate at least proximally, apex acute to acuminate; surfaces sparsely to densely pubescent. Spathes clustered, funnel-shaped; peduncles 3–11 mm long, puberulous; spathes sometimes curved downward against the peduncle, especially in fruit, often slightly falcate, (0.5–)0.8–1.7 × 0.4–1.3 cm, base ± truncate, with a recurved lip at the summit of the line of fusion, margins fused for 3.5–13 mm, usually ciliate along the fused edge, rarely completely glabrous, apex acuminate to acute, often abruptly so, surface pilose-hirsute, rarely nearly glabrous; upper cincinnus abortive, lower cincinnus 2–5(–8?)-flowered. Flowers ± 10 mm wide; upper sepal ovate, 2–2.5 × ± 1.5 mm, paired sepals completely fused into a cup, individually ovate-elliptic, 2–2.5 × 1.5 mm; paired petals orange-yellow, buff or pink, ovate to ovate-reniform, 4.3–5 × ± 4 mm of which the claw ± 2 mm long; lower petal greenish white, lanceolate, ± 3 × 1 mm; staminodes 3; lateral stamens with filaments ± 3 mm long, anthers ± 0.5 mm long; medial stamen filament ± 2.5 mm long, anther 1–1.3 mm long. Ovary ± 1 mm long; style 2–2.5 mm long; stigma subcapitate. Capsules light brown with inconspicuous darker flecks, bilocular, bivalved, ± quadrate, (3.8–)4.2–5 × 3.8–4.3 mm, contracted between the seeds, not beaked, valves persistent, dorsal locule lacking, ventral locules 2-seeded. Seeds ± circular to slightly ellipsoid, (1.8–)2.1–2.3 × (1.9–)2.1–2.4 mm, testa ± muricate, dark brown, not farinose.

UGANDA. Mbale District: Elgon, Sipi, Bugishu, 31 Aug. 1932, *Thomas* 414!
KENYA. West Suk District: N end of Cherangani, Aug., 1971, *Tweedie* 4083! & Cherangani Mountains, 3.5 km on Kaibwibich [Kaibichbich] road from junction with Makutano–Chepararia road, 15 June 1971, *Faden & Evans* 71/487!
TANZANIA. Ufipa District: Hill NW of Tatanda Mission, 23 June 1996, *Faden et al.* 96/363! & Mbizi Mts, 18 June 1996, *Faden et al.* 96/274!; Njombe District: NW side of Kitulo [Elton] Plateau, road from Mporoto to Kitulo sheep ranch, 6 May 1975, *Hepper & Field* 5318!
DISTR. **U** 3; **K** 2; **T** 4, 7; Congo–Kinshasa
HAB. Moist roadside banks, pastures, damp depressions dominated by *Loudetia arundinacea*, scrub, *Faurea-Protea gaguedi* wooded grassland, and *Brachystegia* woodland; (1350–)1650–2300 m
Flowering May, June and August.

NOTE. In the Kenyan collection *Faden & Evans* 71/487 the antherodes had a central dark spot, while in two Tanzanian collections, *Hepper & Field* 5318 and *Faden et al.* 96/274, they were completely yellow. Thus the character appears to be variable in this species.

This species bears a number of striking similarities to *C. neurophylla*, particularly in the form of its spathes and in its capsules and seeds. It differs from that species by the smaller size of the plants, consistently and clearly annual habit, smaller, regularly clustered, more shortly pedunculate spathes that are often curved downward against the peduncles and almost always pubescent with hairs of two different sizes, the old spathes with prominently raised veins, and much smaller flowers that are never violet to lavender. The two species, however, may well be closely related.

48. **Commelina aurantiiflora** *Faden & Raynsford* in Novon 4: 230, fig. 3 (1994). Type: Zambia, Ndola District, Just N or Garneton (Itimbi), S of the Kafue River, ± 8 km NW of Kitwe, *R.B. & A.J. Faden, Handlos & Fanshawe* 74/184 (US! K!, holo., MO!, iso.)

Annual; roots thin, fibrous; shoots prostrate to erect, 2–20 cm long, rooting only at the base, unbranched to densely branched, with a longitudinal line of short hairs continuous with the line of fusion of the distal leaf sheath. Leaves spirally arranged on primary shoot, distichous on lateral shoots; sheaths 0.3–0.9 cm long, puberulous along the line of fusion, often sparsely to densely hirsute elsewhere, ciliate at apex; lamina sessile, oblong-elliptic to lanceolate-elliptic, 2–6 × 0.3–1.4 cm, base rounded to cuneate, often amplexicaul, margin ciliate at least proximally, rarely eciliate, scabrid at the apex, apex acute, mucronulate; adaxial surface with a line of hairs along the midvein at least proximally, rarely glabrous, abaxial surface glabrous or sparsely hirsute. Spathes solitary, terminal, becoming leaf-opposed; peduncle 0.7–2(–2.7) cm long, with a line of hairs continuous with the fused edge of the spathe; spathes 1–1.8 × 0.5–0.9 cm, base truncate to cordate, margins fused for 3.5–6.5 mm, puberulous along the line of fusion, apex acute to acuminate; upper cincinnus lacking, lower cincinnus 1–2(–3)-flowered. Flowers bisexual, 9–10.5 mm wide; upper sepal ovate-elliptic, 2.5–3.5 mm long, paired sepals fused into a cup, 2.6–4.3 mm long, sometimes with petaloid appendages; paired petals orange or buff-orange (also recorded as yellow or white), reniform to ovate-reniform, 5.8–6.5 × 4.4–5.3 mm of which the purple claw ± 3 mm long, lower petal white, linear to lanceolate, 1.2–1.7 mm long; staminodes 3; lateral stamens with filaments 4–5 mm

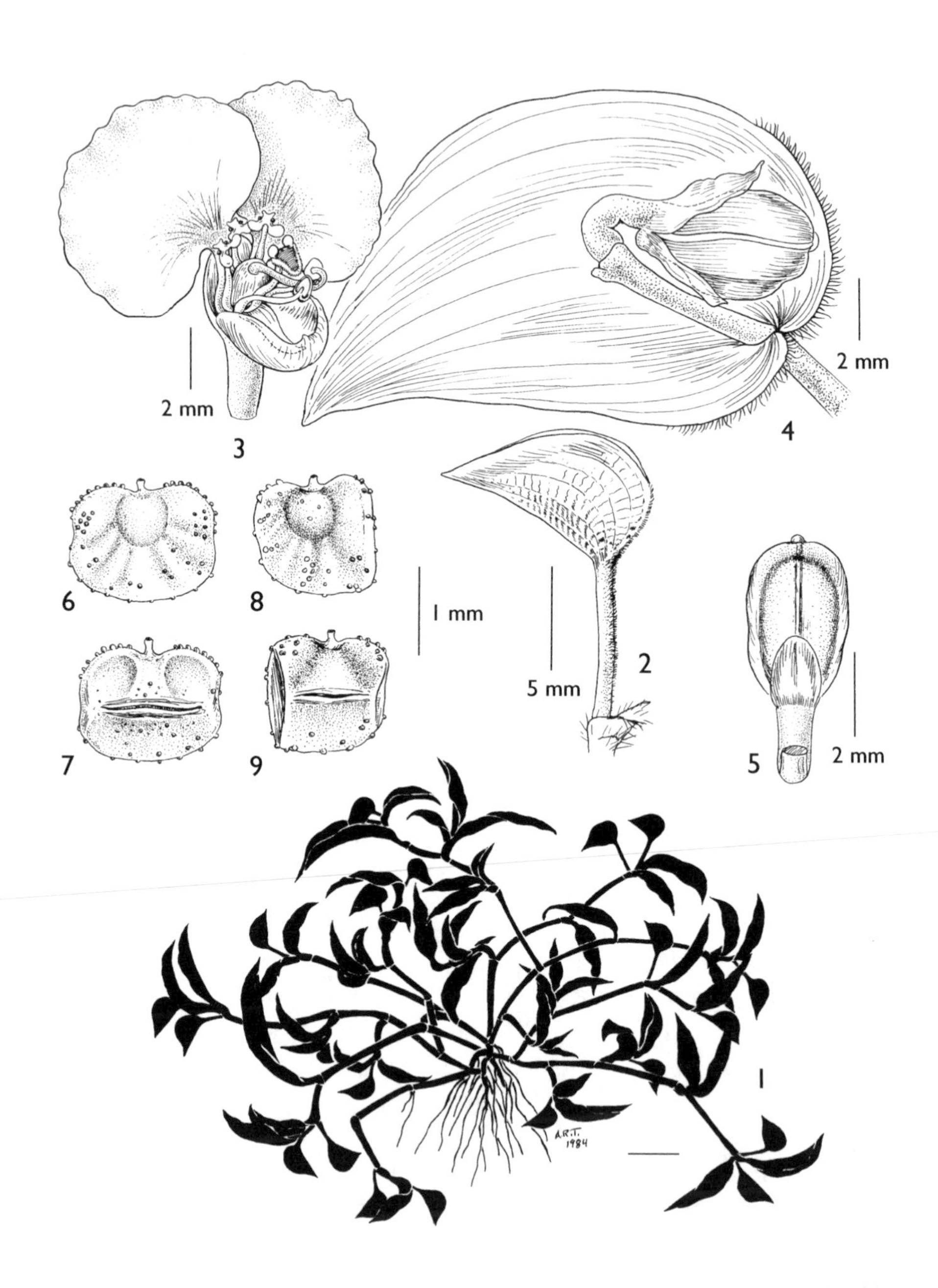

FIG. 21. *COMMELINA AURANTIIFLORA* — **1**, habit; **2**, spathe; **3**, perfect flower, front/side view; **4**, opened spathe; **5**, capsule, dorsal view; **6** & **7**, dorsal locule seed, dorsal & ventral view; **8** & **9**, ventral locule seed, dorsal & ventral view. Drawn by A. R. Tangerini. Reproduced with permission from Novon 4: 231 fig. 3 (1994).

long, anthers 0.5–0.6 mm long, pollen yellow; medial stamen filament ± 4 mm long, anther 1.2–1.3 mm long, pollen yellow. Ovary ± 1.5 mm long; style 3.5–4 mm long. Capsules pale brown to grey-green, trilocular, trivalved, 5-seeded, oblong-ellipsoid, 4–5 × 2–2.5 mm, dorsal locule (0–)1-seeded, ventral locules (1–)2-seeded. Seeds strongly dorsiventrally compressed, mostly trapezoidal, 1.5–2 × 1.5–1.8 mm, testa dark grey-brown with darked brown spots and streaks, smooth with very small scattered warts. Fig. 21, p. 232.

TANZANIA. Iringa District: Ruaha National Park, Mangangwe Ranger Post, 8 Mar. 1972, *Bjørnstad* AB1430a!

DISTR. **T** 7; Zambia, Malawi

HAB. Woodland and wooded grassland, sometimes growing in sand or in very shallow soil over laterite; 1330 m in FTEA (up to 1550 m in Zambia)

NOTE. Two additional specimens were cited as isotypes in the protolog, but they were pressed on a different date from the holotype, so they represent a different collection and are not types.

The maximum dimension given for the peduncle length is from a cultivated plant. In fact, a peduncle up to 4 cm long was measured, but because the specimen was not pressed, this figure has been omitted from the description.

Commelina aurantiiflora is known from only seven collections: one from Tanzania, five from Zambia and one from Malawi. No doubt it has been overlooked by collectors because of its diminutive size.

Commelina aurantiiflora belongs to a well-defined group of savanna and woodland annuals that is widespread in Africa. It includes *Commelina aspera, C. nigritana* and *C. saxosa*, all of which occur in our area. This species group is characterized by its annual habit, small, apricot, orange or buff-orange flowers, fused spathe margins, fused lower sepals, a very reduced medial petal, and the basic chromosome number x = 14. All these species differ from *C. aurantiiflora* by their falcate spathes, usually narrow leaves, and three one-seeded capsule locules. Five-seeded capsules are present in this species group otherwise only in *C. gambiae* (*C. nigritana* var. *gambiae*) but that species is confined to West Africa, its report from Uganda by (Kayemba-Ogwal 1994) being in error (see 50. *C. saxosa*).

49. **Commelina nigritana** *Benth.* in Fl. Nigrit. 541 (1849); Brenan in F.W.T.A., 2nd ed.: 3: 50 (1968); Ogwal in Mitt. Allg. Bot. Hamburg, 23b: 581 (1990); Kayemba-Ogwal in Proc. XIIIth Plenary Meeting AETFAT, Malawi, 1: 419 (1994). Syntypes: Nigeria, Quorra R. at Attah, *Vogel* 85 (K!, lecto., selected here) & Quorra, *Vogel* s. n. (K!, syn.)

Single-stemmed or tufted annual; roots thin, fibrous; shoots erect to ascending or occasionally decumbent, to ± 40 cm, unbranched to densely branched, usually with a longitudinal line of hairs continuous with the fused edge of the distal sheath, sometimes more densely pubescent just below the nodes. Leaves sessile; sheaths commonly split longitudinally, to 1.2 cm long, pubescent or puberulous in a longitudinal line along the fused edge or over the whole surface, ciliate or eciliate at apex; lamina linear, 2–14.5 × 0.2–0.8 cm wide, base sometimes narrowed into the sheath, usually not narrowed, margins scabrid their whole length or only distally, sometimes ciliate basally, apex acuminate; both surfaces usually pubescent, rarely both surfaces glabrous. Spathes solitary along the shoots or loosely to densely clustered at the ends of the shoots, sometimes strongly bent downwards, slightly falcate or not falcate; peduncles (2–)4–10 mm long, with a line of pubescence; spathes 1.5–1.7 × 0.3–0.8 cm, base truncate to hastate, margins fused for 3–7 mm, completely glabrous or puberulous only on the fused portion, apex acuminate, surface puberulous or hirsute-puberulous; upper cincinnus lacking; lower cincinnus (1–)2–3-flowered. Flowers bisexual and male; upper sepals ovate, 2–2.2 × 1 mm, to 2.7 mm long in fruit; paired sepals completely fused into a cup 2–2.5 mm long, each ovate, 2–2.5 × 1.2–1.5; paired petals buff or yellow-orange (orange, very pale orange, deep apricot yellow, creamy yellow), broadly ovate, 4.5–6 × 3.3–4.5 mm of which the reddish purple claw ± 2–2.5 mm; medial petal hyaline white, filiform to oblong-lanceolate, 0.7–1 mm long; staminodes 3; lateral stamens with filaments 4.5–5 mm long, anthers 0.7–0.8 mm long, medial stamen filament 3.5–4 mm long, anther

1.1–1.4 mm long. Ovary ± 1.3 mm long; style ± 4 mm long; stigma capitate. Capsules light brown or pale yellow with dark brown flecks, trilocular, trivalved, broadly ellipsoid, 3-seeded, 3–4.5 × 3–4 mm, apiculate. Seeds ellipsoid, strongly dorsiventrally compressed, 2.3–3.7 × 1.8–2.3 mm, testa medium to dark brown, orange-brown or tan-grey to pinkish grey with some underlying grey spots, or mottled different shades of brown, with a central boss on the dorsal surface delimited by 2 black pits, the whole testa uniformly echinulate, muricate or completely smooth, warted or ribbed.

subsp. **nigritana**

Spathes scattered, not clustered, pubescence variable; sheaths with a line of hairs along the fused edge; seeds with testa variable.

UGANDA. West Nile District: Valley 1.6 km N of Maracha rest camp, 3 Aug. 1953, *Chancellor* 107A!; Bunyoro District: Kabalega [Murchison Falls] National Park, 4.8 km N of road to Butiaba on Masindi–Paraa Lodge road, 3.2 km S of Paraa Ferry, 15 Sep. 1969, *Faden* 69/1056!; Masaka District: 4–5 km N of Lake Nabugabo, 25 Sep. 1969, *Lye* 4341!

TANZANIA. Kigoma District: N of Kigoma, near Mukaraganga Village, 28 Apr. 1994, *Bidgood & Vollesen* 3225!; Mpanda District: Lake Katavi, 12 Feb. 1962, *Richards* 16085!; Ufipa District: km 28 on Isopa–Sumbawanga road, 20 June 1996, *Faden et al.* 96/313!

DISTR. **U** 1, 2, 4; **T** 4, 7; Senegal, Gambia, Guinea, Sierre Leone, Burkina Faso, Ivory Coast, Ghana, Togo, Nigeria, Cameroon, Congo–Kinshasa, Burundi, Zambia, Malawi, Zimbabwe

HAB. Woodland, wooded grassland, grassland, seasonal swamp, moist bank of roadside ditch, usually in sandy or gravelly soil; 700–1800 m

Flowering plants have been seen from February to April, August and September.

SYN. *Commelina umbellata* of C.B. Clarke in DC., Monogr. Phan. 3: 179 (1881) & in F.T.A. 8: 55 (1901); Morton in J.L.S. 55: 529 (1956) & in J.L.S. 60: 190 (1967), *non* Thonn.

subsp. **aggregata** *Faden* **subsp. nov**. a subsp. *nigritana* spathis dense aggregatis apice surculorum differt. Type: Tanzania, Sumbawanga District, Matai–Nkowe road km 14, *Faden, Phillips, Muasya & Macha* 96/334 (US!, holo., EA!, K!, NHT, iso.)

Spathes densely clustered terminally on the shoots, always with numerous, long, uniseriate hairs on the surface; leaf sheaths usually hairy all over, occasionally just in a line; seeds muricate or echinulate.

TANZANIA. Chunya District: Igila Hill, Kempembawe, 22 Mar. 1965, *Richards* 19816!; Iringa District: Ruaha National Park, Magangwe, 18 May 1968, *Renvoize & Abdallah* 2242!; Ufipa District: 36 km from Sumbawanga on Sopa [Isopa] road, 17 June 1996, *Faden et al.* 96/272!

DISTR. **T** 4, 7; Congo–Kinshasa, Burundi, Zambia

HAB. *Brachystegia* woodland, moist roadside ditches and banks, and soil pockets on a granitic rock; 1500–1900 m

Flowering specimens have been seen from March and June.

NOTE (on the species as a whole). This species is characterized by its seeds, which, although very variable in testa pattern and color, have two very distinctive features: two obvious, transverse dorsal pits and a very fine pubescence that is only visible under high magnification. Within our area the testa pattern in subsp. *nigritana* shows some variation that appears to be geographic. For example, the collections with a smooth or nearly smooth testa are from Tanzania. Until recently, the only collections that have (or appear to have, in the case of immature seeds) a ribbed testa were from Uganda, but *Bidgood et al.* 5959 from Tanzania also shows that pattern, so perhaps any apparent geographic variation within our area could be the result of under-collection. The possible significance of different testa patterns throughout the extensive range of this species was beyond the scope of the present endeavor. Subsp. *aggregata* consistently has muricate or echinulate seeds.

The seeds typically have a length to width ratio of about 3:2, with one exception. In *Bidgood & Vollesen* 3225 the seeds are only slightly longer than wide, and with their dark brown, muricate testa they look very distinctive. However, there are no other unusual features of this collection and no other collections that show such seeds, so they must be considered just another variant.

The pubescence of the spathes and foliage also varies considerably in subsp. *nigritana* within our area. The spathes consistently have very short, 2-celled prickle-hairs on them – a character by which they may be separated from the West African *C. gambiae* C.B. Clarke throughout their area of overlap – which are most conspicuous near the midrib. Rarely this is their only pubescence. More typically, uniseriate hairs are also present, and these may be long or short and sparse or abundant. When more than one plant has been collected, they all appear to have the same pubescence type on the spathes. Again, within our area, there are some geographic patterns, such as none of the Ugandan specimens has abundant, long, uniseriate hairs on the spathes, whereas half the Tanzanian specimens of subsp. *nigritana* show that pattern. But the collections are too few and a wider study is needed in order to determine the significance, if any. All collections of subsp. *aggregata* have numerous long, uniseriate hairs on the spathes.

The leaves are almost always pubescent, at least on the adaxial surface. This is especially true in subsp. *aggregata*, which is much more pubescent than nearly all collections of subsp. *nigritana* in our area. However, in a number of collections of subsp. *nigritana* the abaxial surface of some or most of the leaves may be glabrous, and the hairs where present may be sparse and short, and therefore inconspicuous. I could find no hairs at all on either surface of the leaves on only one collection, *Faden et al.* 96/313. This is also the only collection from our area that completely lacked uniseriate hairs on the spathes. Two collections of subsp. *aggregata*, *Richards* 19816 and *Bidgood et al.* 5589, appear to have the abaxial leaf surface of most leaves glabrous. In *Bidgood et al.* 5589 most leaves also have the adaxial surface glabrous and the leaf sheaths with only a line of pubescence, so it is not any more pubescent than specimens of subsp. *nigritana*.

The plants with clustered spathes have been given subspecific status, as subsp. *aggregata*, for want of a better choice. Although in general appearance they much more closely resemble *C. aspera* than *C. nigritana* subsp. *nigritana*, their seeds definitely tie them in with the latter. It is only arrangement of the spathes and the general difference in pubescence that provide the characters for separating the two subspecies.

This species is seriously under-collected, which makes interpreting its variation difficult. In the current state of our knowledge it can be distinguished from *C. saxosa* only on the basis of its capsules and seeds. Kayemba-Ogwal (1994) reported *C. nigritana* var. *gambiae* (C.B. Clarke) Brenan from Uganda but the taxon is *C. saxosa* (see discussion under that species and also Rejected Species).

In spirit material of two populations of *C. nigritana* subsp. *nigritana*, one from Uganda and one from Zambia, it was observed that the first flower in the spathe is apparently always bisexual. If that flower sets fruit the next flower will be male. Few details of the flowers of subsp. *aggregata* are known, so it is unclear whether there are any floral differences between the two subspecies.

50. **Commelina saxosa** *De Wild.*, Pl. Bequaert. 5: 206 (1931). Type: Congo–Kinshasa: Mboga, *Bequaert* 4812 (BR!, holo., photo K!)

Annual (15–)20–40 cm tall; roots thin, fibrous; shoots erect to ascending, sometimes rooting at the lower nodes, glabrous or with a line of fine pubescence continuous with that of the sheath above. Leaves with sheaths mostly 0.5–1 cm long, commonly split to the base, often purple-veined, with a line of fine pubescence along the fused edge, ciliate at apex; lamina linear to linear-lanceolate, 2–12 × 0.2–0.4(–0.6) cm, base cuneate to rounded, margins scabrid distally, apex acuminate; surfaces usually glabrous. Spathes solitary, slightly to strongly falcate, occasionally not falcate; peduncles 0.3–1.2 cm long, puberulous in a line or all around; spathes (0.6–)0.7–1.2(–1.4) × (0.3–)0.5–0.7 cm, base hastate to truncate, margins fused for (3–)4–6 mm, ciliolate along the fused edge, otherwise glabrous, apex acuminate or abruptly acuminate, surfaces green with contrasting green veins, scabrid to patently pilose; upper cincinnus lacking, lower 2–4-flowered. Flowers probably bisexual, upper sepal ovate, 2–2.5 mm long, paired sepals fused into a cup ± 2.5 mm long; upper 2 petals probably apricot to orange (recorded by collectors as yellow, orange, purple, bluish and blue), lower petal minute, ± 1 mm long; staminodes 3, antherodes cruciform. Capsules pale yellow or light brown, often spotted with darker brown, trilocular, trivalved, ± obovoid, 3-seeded, 2.5–3.5 × 3.5–4 mm, all locules 1-seeded. Seeds moderately dorsiventrally compressed, ± dumbbell-shaped, 1.5–2 × 1.5–2.3 mm, testa usually pinkish grey, rarely yellowish tan, often with scattered brown flecks smooth, not farinose, rarely pubescent.

UGANDA. Lango District: Agwata, Jan. 1984, *Ogwal* 135!; Mengo District: Kipayo, Oct. 1914, *Dummer* 1065! & Mitala Maria, 31 Oct. 1975, *Ogwal* 67!

TANZANIA. Kigoma District: Kigoma, Kimirizi, 7 Apr. 1994, *Bidgood & Vollesen* 3046!; Ufipa District: 36 km from Sumbawanga on Sopa [Isopa] road, 20 June 1996, *Faden et al.* 96/312!; Lindi District: Lake Lutamba, 5 Apr. 1935, *Schlieben* 6235!

DISTR. **U** 1, 4; **T** 4, 6, 7?,8; Zambia, Burundi, Congo–Kinshasa

HAB. Woodland, disturbed and degraded woodland, bushland, stream valleys, moist roadside banks, grassland, weed in abandoned cultivation; usually in well-drained soil; 250–1700 m
Flowering specimens have been seen from January, March, April, June, and October.

SYN. *Commelina nigritana* Benth. var. *gambiae* of Kayemba-Ogwal in Seyani & Chikuni, Proc. XIIIth Plenary Meeting AETFAT, Malawi 1: 419 (1994), *non* (C.B. Clarke) Brenan

NOTE. In the absence of capsules and seeds, this species cannot reliably be distinguished from *C. nigritana* subsp. *nigritana*. However, certain characters have been found in some specimens of *C. nigritana* that have not yet been observed in *C. saxosa* and thus provisionally will serve to distinguish this species from *C. saxosa*: long hairs on the abaxial leaf surface and on the sheath surface; hairs on the internodes (in addition to the fine longitudinal line of pubescence); and hairs on the spathes >1.5 mm long. In view of the paucity of specimens of *C. saxosa*, however, it is possible that even these characters will not hold up in when more specimens of this species are collected. It is possible that floral characters might be found to help distinguish the two species.

A note from Brenan on the Kew sheet of *Richards* 698 from Zambia, alludes to the difficulty in distinguishing this species from *C. nigritana*. It reads, in part: "Compared with the type... on loan from Brussels... The very important round seeds... agree well." A second Brenan note written on the same date, however, shows that he had become a bit more skeptical: "*C. saxosa* seems not to be clearly distinguishable from *C. umbellata* Thonn. [= *C. nigritana*] except by the round seeds, and the two may prove to be not specifically distinct." Even eminent botanists do not like to rely on characters that may be lacking.

Kayemba-Ogwal (loc. cit.) recognized two varieties of *Commelina nigritana* in Uganda, namely var. *nigritana* and var. *gambiae*. She distinguished var. *gambiae* by its shorter spathes, shorter capsules, and subglobose (versus ellipsoid) seeds that were smooth to shallowly reticulate. She made no mention of the number of seeds per capsule, which would normally be three in var. *nigritana* and five in var. *gambiae*, as these varieties are recognized in F.W.T.A., ed. 2. I have examined one the specimens cited by Ogwal (*Lye et al.* 5135) and a duplicate of the other collection cited (*Ogwal* 67). They are clearly *C. saxosa*, as are *Dummer* 1065 and *Ogwal* 135, both from Uganda, which Kayemba-Ogwal did not cite. The seeds in the four Ugandan collections are smooth, which characterizes *C. saxosa*, but not *C. nigritana* var. *gambiae* – here treated as *C. gambiae* – which has a reticulate testa. It appears that Kayemba-Ogwal's mention of 'reticulate seeds' for the Ugandan plants came from the literature rather than the specimens.

Commelina saxosa can be distinguished from *C. nigritana* by the shape of its capsule. The seeds of *C. saxosa* are approximately equal in length and width, whereas those of *C. nigritana* are have a length to width ratio of 3:2. The result of this difference in seed proportions is that the capsules of *C. saxosa* are obovoid and proportionally broader than the broadly ellipsoid capsules of *C. nigritana*. The capsules of *C. saxosa* also tend to be shorter than those of *C. nigritana* (2.5–3.5 vs. 3–4.5 mm long).

Bjørnstad 1431B, from Ruaha National Park (**T** 7), might be *C. saxosa*, but its capsules and seeds are too immature to be certain. *Milne-Redhead & Taylor* 8740, which was originally determined as *C. saxosa*, lacks fruits and therefore cannot be identified to species.

51. **Commelina aspera** *Benth.* in Fl. Nigrit. 542 (1849); C.B. Clarke in DC., Monogr. Phan. 3: 180 (1881); Rendle in Cat. Afr. Pl. Welw.: 78 (1899); C.B. Clarke in F.T.A. 8: 56 (1901); Morton in J.L.S. 55: 518, fig. 14 (1956) & in J.L.S. 60: 176 (1967); Brenan in F.W.T.A., 2nd ed.: 3: 50 (1968); Schreiber et al. in F.S.W.A. 157: 7 (1969); Obermeyer & Faden in F.S.A. 4(2): 33 (1985). Syntypes: Ghana, Accra, *Don* s.n. (BM!, syn.); Nigeria, confluence of the Niger and Benue rivers, *Vogel* s.n. (K!, lecto., selected here)

Sparsely branched annual 5–30 cm tall; roots thin, fibrous; shoots erect to ascending or decumbent, occasionally rooting at the nodes, pilose or puberulous, sometimes reduced to a line of short hairs descending from the fused edge of the distal sheath. Leaves often mainly clustered at the summits of the shoots; sheaths to 1.5 cm long, often splitting longitudinally, pilose or puberulous or only ciliate along the fused edge; lamina sessile, linear-oblong to oblong-elliptic, 3–13 × 0.3–1.4(–1.9) cm, base cuneate or rounded, margins scabrid distally, ciliate proximally, apex acute to acuminate; both surfaces pilose or glabrous, or the adaxial surface pilose and the abaxial glabrous. Spathes clustered terminally on the main and lateral shoots, funnel-shaped; subsessile, pedicels up to 5 mm long, glabrous or puberulous; spathe strongly falcate to not falcate and/or bent downwards, 0.8–2 × 0.5–1.9 cm, base rounded to truncate, margins fused for 5–19 mm, ciliate or ciliolate along the fused edge, often also apparently ciliate distally, sometimes completely glabrous, sometimes scabrid at the apex, apex acuminate to attenuate; upper cincinnus lacking or vestigial; lower cincinnus 2–3-flowered. Flowers bisexual or very rarely male, ± 10 mm wide; upper sepal strongly convexo-concave, ovate, 1.5–2.8 × 1.5–2 mm, paired sepals completely fused into a cup or with an apical notch, ovate-elliptic, 1.8–2.5 mm long; paired petals orange or pale apricot to white, very asymmetric or symmetric, 4.5–4.8 × 3.5–4.5 mm of which the mauve claw 2–2.8 mm; medial petal appressed against the sepal cup, linear-lanceolate, 1.2–1.5 × 0.25 mm; staminodes 3; lateral stamens with filaments 4–4.5 mm long, anther 0.5–0.9 mm long; medial stamen filament 3.5 mm long, anther 1.2–1.7 mm long. Ovary 1–1.2 mm long; style J-shaped, 3.5–4 mm long; stigma capitate or capitate-deltate. Capsules trilocular, trivalved (or reduced by abortion), obovoid, 2.5–5 × 3.1–5 mm, pale yellow. Seeds circular to ellipsoid, convexo-planar to biconvex, 2–3.3 × 1.7–3.1 mm, testa light to dark brown, with darker flecks, not farinose, ± smooth.

var. **aspera**

Spathes usually pubescent (if glabrous then <1 cm long), 0.8–1.3(–1.9) × 0.5–1 cm; leaves variable in pubescence but often pilose on both surfaces; seeds ellipsoid to nearly circular, 2–2.9 × 1.7–1.9(–2.3) mm, testa light to medium brown with sparse, raised, welt-like ridges and patches or welt-like ridges and patches lacking; hilum in a ventral groove less than half the width of the seed.

UGANDA. Bunyoro District: Ukidi Forest in Bunyoro [Unyoro], Nov. 1862, *Grant* s.n.!; Teso District: Serere, June, 1926, *Maitland* s.n.!

TANZANIA. Ufipa District: Matai–Nkowe road km 14, 22 June 1996, *Faden et al.* 96/337!; Iringa District: Magangwe, 10 Apr. 1970, *Greenway & Kanuri* 14316A!; Songea District: Nangurukuru Hill, ± 25 km E of Songea, 8 Apr. 1956, *Milne-Redhead & Taylor* 9489 K sheets 1+2!

DISTR. **U** 2, 3; **T** 4, 7, 8; Ghana, Togo, Benin, Nigeria, Cameroon, Chad, Congo–Kinshasa, Burundi, Angola, Zambia, Malawi, Zimbabwe, Namibia, Botswana

HAB. Woodland, grassland, roadside banks and shallow soil in rock crevices; 850–1700 m
Flowering specimens have been seen from March, April, June and November.

SYN. *Cyanotis hirsuta* Baker, Trans. Linn. Soc. 29: 162 (1872), *non* Fisch., C.A.Meyer & Avé-Lall. (1841)

Commelina aspera Benth. var. *firma* C.B. Clarke in DC., Monogr. Phan. 3: 180 (1881). Syntypes: Uganda, Bunyoro District: Ukidi Forest, Nov. 1862, *Grant* s.n. (K!, lecto., selected here); Angola, Pungo Andongo, between Calundo and Mangue, *Welwitsch* 6644 (BM!, syn.) & around Pedras de Guinga and Mangue, *Welwitsch* 6645 (BM!, syn.)

C. firma (C.B. Clarke) Rendle in Cat. Afr. Pl. Welw. 78 (1899); C.B. Clarke in F.T.A. 8: 56 (1901)

var. **opulens** (*C.B. Clarke*) *Faden* **stat. & comb. nov.** Syntypes: Malawi, Khondowe [Kondowe] to Karonga, 2000–6000 ft, July 1896, *Whyte* s.n. (K!, lecto., selected here); Angola, Huilla District: Humpata, *Welwitsch* 6592 (BM!, syn.)

Spathes glabrous, 1.2–2 cm long, 1–1.9 cm high; leaves usually glabrous, occasionally sparsely pilose adaxially; seeds nearly circular, 2.4–3.3 × 2.3–3.1 mm, testa usually medium to dark brown with numerous, raised, welt-like ridges and patches; hilum in a ventral groove more than half the length of the seed.

TANZANIA. Iringa District: Ranger's Post, Magangwe, 14 Apr. 1970, *Greenway & Kanuri* 14351!; Lindi District: Rondo [Muera] Plateau, 17 Apr. 1935, *Schlieben* 6292!; Songea District: Nangurukuru Hill, 25 km E of Songea, 8 Apr. 1956, *Milne-Redhead & Taylor* 9489 K sheets 3+4!

DISTR. **T** 7, 8; Guinea Bissau, Sierra Leone, Ivory Coast, Ghana, Togo, Benin, Cameroon, Central African Republic, Angola, Zambia, Malawi, Zimbabwe

HAB. Woodland, open forest, and shallow soil in crevices between rocks; 500–1300 m
Flowering specimens have been seen from April.

SYN. *Commelina opulens* C.B. Clarke in F.T.A. 8: 58 (1901)

NOTE (on the species as a whole). All three names that are associated with *C. aspera*, namely *C. aspera*, *C. aspera* var. *firma* and *C. opulens*, are each based on specimens of more than one taxon, hence the importance of lectotypification. The *Don* collection that is a syntype of *C. aspera* is *Commelina erecta* L. sensu lato. Therefore I have selected the other syntype, *Vogel* s.n., as the lectotype of this species. Clarke (1881) cited three collections when he described *C. aspera* var. *firma*, separating it from the typical variety by its slightly longer leaves and seeds. Following Rendle (1899) in raising var. *firma* to species rank, Clarke (1901) noted that *C. firma* was stouter than *C. aspera* and had longer capsules and seeds. Clarke evidently relied more heavily on the Grant collection, which I have chosen as lectotype, rather than the two *Welwitsch* collections, as evidenced by his description of the plant as hairy, which only applies to the *Grant* specimen. I find the *Grant* collection from Uganda not to be significantly different from *C. aspera* var. *aspera*. Further evidence of Clarke's reliance on the *Grant* collection for his concept of *C. aspera* var. *firma* (and *C. firma*) is that after his description of *C. opulens*, Clarke (1901) cited in synonymy the *C. aspera* account in Clarke (1881) followed by "partly", but he failed to cite the *Welwitsch* collections from that account, to which he was clearly referring, as *C. opulens*. Instead he cited *Welwitsch* 6592, which is *C. aspera* var. *aspera*. Thus *C. opulens* has to be lectotypified by the other syntype, *Whyte* s.n., in order for it not to fall into synonymy with *C. aspera* var. *aspera*. Whether to recognize *C. opulens* at some level remains an open question but if, as in F.W.T.A. (Brenan, 1968), all variants in size, pubescence and seed characters are to be accepted as typical *C. aspera*, then *C. opulens* would not be recognized at any level. However, if *C. opulens* is accepted as a variety of *C. aspera* and defined as we have done here, then var. *opulens* proves to be widespread but unrecognized in West Tropical Africa and in much of the rest of its range. Some country records for *C. aspera* in F.W.T.A., e.g. those for Guinea Bissau and Sierra Leone, now are assignable to *C. aspera* var. *opulens* instead of typical *C. aspera*. *Commelina aspera* seems to be uncommon in our area. In a 1996 collecting trip to southern Tanzania we found only two plants on a single roadside bank. This is somewhat unexpected because there are more collections of this species from Zambia than any other country. *Commelina aspera* was omitted by Ogwal (1990) in her account of the Uganda species of *Commelina*, so we may presume that it is quite rare there.

The four sheets of *Milne-Redhead & Taylor* 9489 at Kew appear at a quick glance to be merely glabrous and hairy extremes of the same taxon, with flowers from the glabrous spathes not very different from those of the pubescent spathes. A more detailed study of the herbarium specimens shows differences in maximum spathe length, capsule length, seed size, testa color, presence and size of welt-like ridges on the testa, and in the width of the ventral groove on the seed. Thus it appears that this is a mixed collection of two closely related taxa, *C. aspera* var. *aspera* (sheets 1 & 2) and *C. aspera* var. *opulens* (sheets 3 & 4), not a collection of a variable population. Sheet 4 is the only one that lacks mature seeds, but it closely resembles sheet 3, and it is almost certainly also var. *opulens*. The raised ridge around the hilum is very characteristic of this species. Most likely it functions as an elaiosome to attract ants for seed dispersal.

Faden et al. 97/11, a collection of *C. aspera* var. *aspera* from Zimbabwe, has some floral details not mentioned by collectors in our area: "flowers 8–11 mm wide... petals orange...; antherodes yellow with a central purple spot; lateral anthers entirely yellow; medial anther with a broad violet patch; pollen all yellow." How general this colour pattern is within the species remains to be demonstrated, but it should be looked for by collectors in our area. Although the flower colour can be variable, the record of blue flowers, occasionally reported by collectors, is probably erroneous.

Bidgood et al. 5595 from Mpanda District (**T** 4) consists of very tiny, almost stemless plants. They appear very different from typical *C. aspera* plants which have elongate stems. However, in Zambia, where this species is most abundant and most variable, such dwarf plants are more common as are plants with broad leaves, both dwarf and of more normal stature. The broad-leaved plants have sometimes been named *C. aspera* var. *firma*, but that is not distinct from *C.*

aspera var. *aspera*, as we have lectotypified the taxa. Insofar as they have been checked their seeds agree with those of *C. aspera* var. *aspera*, but further study is required to determine whether additional taxa might need to be recognized. *Commelina robynsii* De Wild., from Congo–Kinshasa, is also part of the *C. aspera* complex. It is recognizeable by its tiny glabrous spathes but it needs further study before its status can be fully evaluated.

Rejected species

Commelina macrospatha Mildbr.

This species was reported from Uganda and Kenya by Morton (in J.L.S. 60: 188 (1967)) and Brenan in F.W.T.A., 2nd ed., 47 (1968). Among the three collections found in a folder for Uganda labeled *C. macrospatha* at Kew in 1998, *Chandler & Hancock* 102 and *Johnston* 113 were narrow-spathe forms of *C. diffusa* whereas *Wilson* 554 was *C. africana* var. *villosior*, despite the colour on the label having been recorded as 'mauve'blue'. No specimens from Kenya have been found that were named *C. macrospatha*, and it is unknown what was the basis for that record. It is implausible that this West African species occurs in our flora.

Commelina nigritana Benth. var. *gambiae* (C.B. Clarke) Brenan [= *C. gambiae* C.B. Clarke]

This West African plant was reported for Uganda by Kayemba-Ogwal (Proc. XIIIth Plenary Meeting AETFAT, Malawi 419 (1994)). Both specimens cited, and two others, have been examined and are *C. saxosa* De Wild. The reticulate testa pattern reported does not occur in *C. saxosa* and was not seen in the specimens that I examined (see discussion under 50. *C. saxosa*).

INDEX TO COMMELINACEAE

New names validated in this part

Aneilema ephemerum *Faden* **sp. nov.**,
Aneilema hockii *De Wild.* subsp. **longiaxis** *Faden* **subsp. nov.**,
Aneilema hockii *De Wild.* var. **rhizomatosum** *Faden* **var. nov. Aneilema termitarium** *Faden* **sp. nov.**,
Aneilema minutiflorum *Faden* **sp. nov.**,
Aneilema minutiflorum *Faden* **sp. nov.** subsp. **zanzibarica** *Faden* **subsp. nov.**,
Coleotrype udzungwaensis *Faden & Layton* **sp. nov.**,
Commelina aspera *Benth.* var. **opulens** (*C.B. Clarke*) *Faden* **stat. & comb. nov.**,
Commelina bracteosa *Hassk.* subsp. **bracteosa** var. **lagosensis** (*C.B. Clarke*) *Faden* **comb. et stat nov.**,
Commelina bracteosa *Hassk.* subsp. **rhizomifera** *Faden* **subsp. nov.**,
Commelina chayaënsis *Faden* **sp. nov.**,
Commelina eckloniana *Kunth* subsp. **claessensii** (*De Wild.*) *Faden* **comb. nov.**,
Commelina eckloniana *Kunth* subsp. **critica** (*De Wild.*) *Faden* **comb. nov.**,
Commelina eckloniana *Kunth* subsp. **echinosperma** (*K. Schum.*) *Faden* **comb. nov.**,
Commelina eckloniana *Kunth* subsp. **nairobiensis** (*Faden*) *Faden* **comb. nov.**,
Commelina eckloniana *Kunth* subsp. **thikaënsis** *Faden* **subsp. nov.**,
Commelina kitaleënsis *Faden* **sp. nov.**,
Commelina latifolia *A. Rich.* var. **undulatifolia** *Faden* **var. nov.**,
Commelina nigritana *Benth.* subsp. **aggregata** *Faden* **subsp. nov.**,
Commelina pallidispatha *Faden* **sp. nov.**,
Commelina polhillii *Faden & Alford* subsp. **kucharii** *Faden* **subsp. nov.**,
Commelina schweinfurthii *C.B. Clarke* subsp. **carsonii** (*C.B. Clarke*) *Faden* **comb. nov.**,
Commelina schweinfurthii *C.B. Clarke* subsp. **cecilae** (*C.B. Clarke*) *Faden* **comb. nov.**,
Commelina sulcatisperma *Faden* **sp. nov.**,
Cyanotis paludosa *Brenan* subsp. **bulbifera** Faden **subsp. nov.**,
Cyanotis speciosa (*L.f.*) *Hassk.* subsp. **bulbosa** *Faden* **subsp. nov.**,
Polyspatha oligospatha *Faden* **sp. nov.**,

LIST OF ABBREVIATIONS

A.V.P. = O. Hedberg, Afroalpine Vascular Plants; **B.J.B.B.** = Bulletin du Jardin Botanique de l'Etat, Bruxelles; Bulletin du Jardin Botanique Nationale de Belgique; **B.S.B.B.** = Bulletin de la Société Royale de Botanique de Belgique; **C.F.A.** = Conspectus Florae Angolensis; **E.J.** = A. Engler, Botanische Jahrbücher für Systematik, Pflanzengeschichte und Pflanzengeographie; **E.M.** = A. Engler, Monographieen Afrikanischer Pflanzen-Familien und Gattungen; **E.P.** = A. Engler, Das Pflanzenreich; **E.P.A.** = G. Cufodontis, Enumeratio Plantarum Aethiopiae Spermatophyta; in B.J.B.B. 23, Suppl. (1953) et seq.; **E. & P. Pf.** = A. Engler & K. Prantl, Die Natürlichen Pflanzenfamilien; **F.A.C.** = Flore d'Afrique Centrale (*formerly* F.C.B.); **F.C.B.** = Flore du Congo Belge et du Ruanda-Urundi; Flore du Congo, du Rwanda et du Burundi; **F.E.E.** = Flora of Ethiopia & Eritrea; **F.D.-O.A.** = A. Peter, Flora von Deutsch-Ostafrika; **F.F.N.R.** = F. White, Forest Flora of Northern Rhodesia; **F.P.N.A.** = W. Robyns, Flore des Spermatophytes du Parc National Albert; **F.P.S.** = F.W. Andrews, Flowering Plants of the Anglo-Egyptian Sudan *or* Flowering Plants of the Sudan; **F.P.U.** = E. Lind & A. Tallantire, Some Common Flowering Plants of Uganda; **F.R.** = F. Fedde, Repertorium Speciorum Novarum Regni Vegetabilis; **F.S.A.** = Flora of Southern Africa; **F.T.A.** = Flora of Tropical Africa; **F.W.T.A.** = Flora of West Tropical Africa; **F.Z.** = Flora Zambesiaca; **G.F.P.** = J. Hutchinson, The Genera of Flowering Plants; **G.P.** = G. Bentham & J.D. Hooker, Genera Plantarum; **G.T.** = D.M. Napper, Grasses of Tanganyika; **I.G.U.** = K.W. Harker & D.M. Napper, An Illustrated Guide to the Grasses of Uganda; **I.T.U.** = W.J. Eggeling, Indigenous Trees of the Uganda Protectorate; **J.B.** = Journal of Botany; **J.L.S.** = Journal of the Linnean Society of London, Botany; **K.B.** = Kew Bulletin, *or* Bulletin of Miscellaneous Information, Kew; **K.T.S.** = I. Dale & P.J. Greenway, Kenya Trees and Shrubs; **K.T.S.L.** = H.J. Beentje, Kenya Trees, Shrubs and Lianas; **L.T.A.** = E.G. Baker, Leguminosae of Tropical Africa; **N.B.G.B.** = Notizblatt des Botanischen Gartens und Museums zu Berlin-Dahlem; **P.O.A.** = A. Engler, Die Pflanzenwelt Ost-Afrikas und der Nachbargebiete; **R.K.G.** = A.V. Bogdan, A Revised List of Kenya Grasses; **T.S.K.** = E. Battiscombe, Trees and Shrubs of Kenya Colony; **T.T.C.L.** = J.P.M. Brenan, Check-lists of the Forest Trees and Shrubs of the British Empire no. 5, part II, Tanganyika Territory; **U.K.W.F.** = A.D.Q. Agnew (or for ed. 2, A.D.Q. Agnew & S. Agnew), Upland Kenya Wild Flowers; **U.O.P.Z.** = R.O. Williams, Useful and Ornamental Plants in Zanzibar and Pemba; **V.E.** = A. Engler & O. Drude, Die Vegetation der Erde, IX, Pflanzenwelt Afrikas; **W.F.K.** = A.J. Jex-Blake, Some Wild Flowers of Kenya; **Z.A.E.** = Wissenschaftliche Ergebnisse der Deutschen Zentral-Afrika-Expedition 1907–1908, 2 (Botanik).

FAMILIES OF VASCULAR PLANTS REPRESENTED IN THE FLORA OF TROPICAL EAST AFRICA

The family system used in the Flora has diverged in some respects from that now in use at Kew and the herbaria in East Africa. The accepted family name of a synonym or alternative is indicated by the word "see". Included family names are referred to the one used in the Flora by "in" if in accordance with the current system, and "as" if not. Where two families are included in one fascicle the subsidiary family is referred to the main family by "with".

PUBLISHED PARTS

Foreword and preface
*Glossary
Index of Collecting Localities

Acanthaceae
  Part 1
  **Part 2
*Actiniopteridaceae
*Adiantaceae
Aizoaceae
Alangiaceae
Alismataceae
*Alliaceae
*Aloaceae
*Amaranthaceae
*Amaryllidaceae
*Anacardiaceae
*Ancistrocladaceae
Anisophyllaceae — as Rhizophoraceae
Annonaceae
*Anthericaceae
Apiaceae — see Umbelliferae
Apocynaceae
  *Part 1
  **Part 2
*Aponogetonaceae
Aquifoliaceae
*Araceae
Araliaceae
Arecaceae — see Palmae
*Aristolochiaceae
*Asclepiadaceae — see Apocynaceae
Asparagaceae
*Asphodelaceae
Aspleniaceae
Asteraceae — see Compositae
Avicenniaceae — as Verbenaceae
*Azollaceae

*Balanitaceae
*Balanophoraceae
*Balsaminaceae
Basellaceae
Begoniaceae
Berberidaceae
Bignoniaceae
Bischofiaceae — in Euphorbiaceae
Bixaceae
Blechnaceae
*Bombacaceae
*Boraginaceae
Brassicaceae — see Cruciferae
Brexiaceae
Buddlejaceae — as Loganiaceae
*Burmanniaceae
*Burseraceae
Butomaceae
Buxaceae

Cabombaceae
Cactaceae
Caesalpiniaceae — in Leguminosae
*Callitrichaceae
Campanulaceae
Canellaceae
Cannabaceae
Cannaceae — with Musaceae
Capparaceae
Caprifoliaceae
Caricaceae
Caryophyllaceae
*Casuarinaceae
Cecropiaceae — with Moraceae
*Celastraceae
*Ceratophyllaceae
Chenopodiaceae
Chrysobalanaceae — as Rosaceae

Clusiaceae — see Guttiferae
Cobaeaceae — with Bignoniaceae
Cochlospermaceae
Colchicaceae
Combretaceae
Commelinaceae
Compositae
*Part 1
*Part 2
Part 3
Connaraceae
Convolvulaceae
Cornaceae
Costaceae — as Zingiberaceae
*Crassulaceae
*Cruciferae
Cucurbitaceae
Cupressaceae
Cyanastraceae — in Tecophilaeaceae
Cyatheaceae
Cycadaceae
*Cyclocheilaceae
*Cymodoceaceae
Cyperaceae
Cyphiaceae — as Lobeliaceae

*Davalliaceae
*Dennstaedtiaceae
*Dichapetalaceae
Dilleniaceae
Dioscoreaceae
Dipsacaceae
*Dipterocarpaceae
Dracaenaceae
Dryopteridaceae
Droseraceae

*Ebenaceae
Elatinaceae
Equisetaceae
Ericaceae
*Eriocaulaceae
*Eriospermaceae
*Erythroxylaceae
Escalloniaceae
Euphorbiaceae
*Part 1
*Part 2

Fabaceae — see Leguminosae
Flacourtiaceae
Flagellariaceae
Fumariaceae

*Gentianaceae
Geraniaceae
Gesneriaceae
Gisekiaceae — as Aizoaceae
*Gleicheniaceae
Goodeniaceae
Gramineae
Part 1
Part 2
*Part 3
Grammitidaceae
Gunneraceae — as Haloragaceae
Guttiferae

Haloragaceae
Hamamelidaceae
*Hernandiaceae
Hippocrateaceae — in Celastraceae
Hugoniaceae — in Linaceae
*Hyacinthaceae
*Hydnoraceae
*Hydrocharitaceae
*Hydrophyllaceae
*Hydrostachyaceae
Hymenocardiaceae — with Euphorbiaceae
Hymenophyllaceae

Hypericaceae — see also Guttiferae
Hypoxidaceae

Icacinaceae
Illecebraceae — as Caryophyllaceae
*Iridaceae
Irvingiaceae — as Ixonanthaceae
Isoetaceae
*Ixonanthaceae

Juncaceae
Juncaginaceae

Labiatae
Lamiaceae — see Labiatae
*Lauraceae
Lecythidaceae
Leeaceae — with Vitaceae
Leguminosae
Part 1, Mimosoideae
Part 2, Caesalpinioideae
Part 3 } Papilionoideae
Part 4 }
Lemnaceae
Lentibulariaceae
Liliaceae (*s.s.*)
Limnocharitaceae — as Butomaceae
Linaceae
*Lobeliaceae
Loganiaceae
*Lomariopsidaceae
*Loranthaceae
Lycopodiaceae
*Lythraceae

Malpighiaceae
Malvaceae
Marantaceae
*Marattiaceae
*Marsileaceae
Melastomataceae
*Meliaceae
Melianthaceae
Menispermaceae
*Menyanthaceae
Mimosaceae — in Leguminosae
Molluginaceae — as Aizoaceae
Monimiaceae
Montiniaceae
*Moraceae
*Moringaceae
Muntingiaceae — with Tiliaceae
*Musaceae
*Myricaceae
*Myristicaceae
*Myrothamnaceae
*Myrsinaceae
*Myrtaceae

*Najadaceae
Nectaropetalaceae — in Erythroxylaceae
*Nesogenaceae
*Nyctaginaceae
*Nymphaeaceae

Ochnaceae
Octoknemaceae — in Olacaceae
Olacaceae
Oleaceae
*Oleandraceae
Oliniaceae
Onagraceae
*Ophioglossaceae
Opiliaceae
Orchidaceae
Part 1, Orchideae
*Part 2, Neottieae, Epidendreae
*Part 3, Epidendreae, Vandeae
Orobanchaceae
*Osmundaceae
Oxalidaceae

*Palmae
Pandaceae — with Euphorbiaceae
*Pandanaceae
Papaveraceae
Papilionaceae — in Leguminosae
*Parkeriaceae
Passifloraceae
Pedaliaceae
Periplocaceae — see Apocynaceae (Part 2)
Phytolaccaceae
*Piperaceae
Pittosporaceae
Plantaginaceae
Plumbaginaceae
Poaceae — see Gramineae
Podocarpaceae
Podostemaceae
Polemoniaceae — see Cobaeaceae
Polygalaceae
Polygonaceae
*Polypodiaceae
Pontederiaceae
*Portulacaceae
Potamogetonaceae
Primulaceae
*Proteaceae
*Psilotaceae
*Ptaeroxylaceae
*Pteridaceae

*Rafflesiaceae
Ranunculaceae
Resedaceae
Restionaceae
Rhamnaceae
Rhizophoraceae
Rosaceae
Rubiaceae
Part 1
*Part 2
*Part 3
*Ruppiaceae
*Rutaceae

*Salicaceae
Salvadoraceae
*Salviniaceae
Santalaceae
*Sapindaceae
Sapotaceae
*Schizaeaceae
Scrophulariaceae
Scytopetalaceae
Selaginellaceae
Selaginaceae — in Scrophulariaceae
*Simaroubaceae
*Smilacaceae
**Solanaceae
Sonneratiaceae
Sphenocleaceae
Strychnaceae — in Loganiaceae
*Surianaceae
Sterculiaceae

Taccaceae
Tamaricaceae
Tecophilaeaceae
Ternstroemiaceae — in Theaceae
Tetragoniaceae — in Aizoaceae
Theaceae
Thelypteridaceae
Thismiaceae — in Burmanniaceae
Thymelaeaceae
*Tiliaceae
Trapaceae
Tribulaceae — in Zygophyllaceae
*Triuridaceae
Turneraceae
Typhaceae

Uapacaceae — in Euphorbiaceae
Ulmaceae
*Umbelliferae
*Urticaceae

Vacciniaceae — in Ericaceae
Valerianaceae
Velloziaceae
*Verbenaceae
*Violaceae
*Viscaceae
*Vitaceae
*Vittariaceae

*Woodsiaceae

*Xyridaceae

*Zannichelliaceae
*Zingiberaceae
*Zosteraceae
*Zygophyllaceae

Editorial adviser, National Museums of Kenya: Quentin Luke
Editorial adviser, Makerere University: J. Kalema
Adviser on Linnaean types: C. Jarvis

Parts of this Flora, unless otherwise indicated, are obtainable from:
Royal Botanic Gardens, Kew, Richmond, Surrey TW9 3AB, England. www.kew.org or www.kewbooks.com

*** Only available through CRC Press at:**
UK and Rest of World (except North and South America):
CRS Press/ITPS,
Cheriton House, North Way, Andover, Hants SP10 5BE.
e: uk.tandf@thomsonpublishingservices. co.uk

North and South America:
CRC Press,
2000NW Corporate Blvd, Boco Raton, FL 33431-9868, USA.
e: orders@crcpress.com

**** Forthcoming titles in production**
For availability and expected publication dates please check on our website, www.kew.books.com

Information on current prices can be found at www.kewbooks.com or www.tandf.co.uk/books/

First published in 2012 by
Royal Botanic Gardens, Kew
Richmond, Surrey, TW9 3AB, UK
www.kew.org

ISBN 978 1 84246 436 6

British Library Cataloguing in Publication Data
A catalogue record for this book is available from the British Library

Design and typesetting by Margaret Newman,
Kew Publishing, Royal Botanic Gardens, Kew.

Printed in the the USA by The University of Chicago Press

Kew's mission is to inspire and deliver science-based plant conservation worldwide, enhancing the quality of life.

Kew receives half of its running costs from Government through the Department for Environment, Food and Rural Affairs (Defra). All other funding needed to support Kew's vital work comes from members, foundations, donors and commercial activities including book sales.